The Medicinal Plant Industry

Editor
R. O. B. Wijesekera, Ph.D., D.Sc.
Senior Partner
INTECNOS Associates, International
Industrial Consultants
Rajagiriya, Sri Lanka

CRC Press
Boca Raton Ann Arbor Boston London

Library of Congress Cataloging-in-Publication Data

The Medicinal Plant Industry / editor, R.O.B Wijesekera
p. cm.
Includes bibliographies and index.
ISBN 0-8493-6669-0
1. Medicinal plants. 2. Materia medica, Vegetable. 3. Botanical drug industry.
I. Wijesekera, R.O.B., 1928— .
SB293.M43 1991
615'.32—dc20 91-25373

This book represents information obtained from authentic and highly regarded sources. Reprinted material is quoted with permission, and sources are indicated. A wide variety of references are listed. Every reasonable effort has been made to give reliable data and information, but the author and the publisher cannot assume responsibility for the validity of all materials or for the consequences of their use.

Direct all inquires to CRC Press, Inc., 2000 Corporate Blvd., N.W., Boca Raton, Florida, 33431.

International Standard Book Number 0-8493-6669-0

Library of Congress Card Number 91-25373
Printed in the United States

PREFACE

The medicinal plant industry is so wide that it can be considered to range from the collection, propagation, harvesting, and sale of crude vegetal drugs, to the production formulation, packaging and dispensing of sophisticated phyto-pharmaceuticals and herbal preparations.

The global turnover of the industry is impossible to assess with precision although there is evidence that it would exceed billions of dollars. The range of sophistication is also widest, and so also is the nature of personnel involved in its pursuit. One sees in the markets in the developing world, fresh herbs, dried herbs, and partially processed herbs, for sale. Likewise in the sophisticated pharmacies and supermarkets of the industrialised world, an array of herbal preparations impressively formulated and nowadays aggressively marketed, greet the customer. Taking part in the preparation of the range of products that characterise the industry globally, are farmers, laborers, and salespeople together wit traditional physicians, on the one hand and on the other, scientists of many disciplines such as pharmacists, botanists, chemists, technologists, etc., combined with the production and marketing machinery of the modern age. The traditional remedies of the developing world are crucial to their people. At the same time these can unfurl many remedies that are new to modern medicine. The scientific study of these medicines and the systematic uplifting of the industry to preserve the ancient and to serve the modern, is now a global challenge.

The present volume was undertaken at the request of several colleagues — those involved in science and technology, those interested in the value of traditional therapies as well as those involved in national development and technical assistance delivery. Their purpose and those of the authors was to draw attention in a single volume to the multiple dimensions of the industry and the opportunities for collaborative endeavour it affords, to serve humanity's common goal. The volume does not attempt to comprehensively cover all aspects of the industry. What it modestly purports to achieve is to sensitize the readers to the multiple dimensions of the industry and their interdependence. Several important aspects of the industry have been dealt with by authors who have had specialized experience with the subjects they have chosen. The experiences of China and India have been elaborated because of the special scientific progress made in these two great lands that is relevant to the case, and is of importance to others.

The aspect of marketing has been left out as it is too large a topic to deal with in a single volume and needs extensive study if it is to represent accurately the industry in its many forms and stages of development. The ITC/GATT Geneva is currently engaged in such a market study, albeit in a more circumscribed scope and context.

Overall, it was the view of authors, editor, and publisher that there was the need for a volume of this type which brings to focus the many facets of the medicinal plant industry, the potentialities and prospects. It is in a way an industry that has sustained itself in one form or the other over a long period of time in different geographical regions.

It is hoped that the book will be useful to many, scientists, technologists, as well as managers, decision-makers, and entrepreneurs. The authors have given gladly of their time and collaboration for which the editor is most grateful.

R. O. B. Wijesekera

THE EDITOR

R. O. B. Wijesekera B.Sc. Hons. Chemistry, was born in 1928 in Colombo, Sri Lanka. He studied at the University of Ceylon (Colombo) during the Vice-Chancellorship of Sir Ivor Jennings, later Vice-Chancellor of Cambridge University, U.K. He served as a Research Officer at the Medical Research Institute Colombo (1952—1966) pioneering research on medicinal plants and Ayurvedic pharmaceuticals. He also researched at the University of Sheffield, U.K., under Prof. R. D. Haworth, F. R.S., on the Chemistry of Tannins (Ph.D., 1958).

In 1966, he joined the CISIR Colombo and undertook the task of building a division of natural products and interacting considerably with industry. He received international recognition for his successful accomplishment of this task by the award in 1976 of the Guinness Medal, by the Guinness Awards for Scientific Achievement (London). During 1968—1969, Dr. Wijesekera widened his horizons by spending a year as a research fellow at the Department of Food Science and Technology, University of California (Davis), which assisted his research with a Fulbright Travel Award.

He was his country's national leader at a U.S.-Sri Lanka Workshop on Natural Products in 1975 alongside Carl Djerassi of the U.S. and the following year was President of the UNESCO-sponsored Asian Symposium on Medicinal Plants and Spices in Colombo. In Sri Lanka, he served on voluntary organisations of science, and on the governing boards of state industrial corporations. He also acted as Secretary-General of the National Science Council of Sri Lanka in 1976 and was a visiting university teacher at several campuses.

His international career commenced as a consultant to UNIDO-ESCAP on the essential oil industry in 1976. During 1977—1978, he was a specialist-consultant on natural products and science policy matters to the Government of Guyana. In 1978, he was selected by the World Health Organisation as Task Force Manager of their International Research Programme on Plants for Fertility Regulation. He moved over to UNIDO in 1980 as a Special Technical Adviser to the Chemical Industries Branch and has been responsible for developing and managing UNIDOs Technical Assistance Programmes on the Industrial Utilisation of Medicinal and Aromatic Plants.

Dr. Wijesekera is a Fellow of the Royal Society of Chemistry, a Chartered Chemist of the U.K., and a Fellow of the National Academy of Science (Sri Lanka). He has also ben awarded a D.Sc. Honoris Causa by the University of Peradeniya, Sri Lanka.

He and his group have published over 60 scientific papers and several patents covering a wide range of topics in chemistry and technology of medicinal plants, essential oils and spices, and in equipment design.

IN MEMORIAM

DR. HIROSHI HIKINO

While this volume was being prepared, Mrs. Y. Hikino informed me of the sad death of her husband on 8 December 1988.

Dr. Hiroshi Hikino was one of the most distinguished scientists in Japan. His research on the validation (or interpretation) of ancient oriental medicines on the basis of modern scientific concepts within chemistry and pharmacology had already reached monumental proportions by the time of his death at a comparatively young age.

He once discussed with me his ideas on the limitations of chemistry and pharmacological concepts relating to the therapeutic effects reported in oriental medicines and was planning to edit a volume himself dedicated to the validation of oriental medicines.

The untimely demise of this talented and dedicated servant of science leaves a void which will indeed be difficult to fill.

May he attain Nibbana.

TABLE OF CONTENTS

Chapter 1

PLANT-DERIVED MEDICINES AND THEIR ROLE IN GLOBAL HEALTH

R. O. B. Wijesekera

I. GLOBAL PHARMACEUTICAL REQUIREMENTS AND TRADITIONAL THERAPIES

The world's population during 1984 was estimated to be around 4.79 billion[1] and 75% lived in the developing countries. Projections indicate that the world's population will increase to 7.6 billion by the year 2020 with a greater percentage, as demographers see it, occupying the poorer countries of the developing world. Therefore, it is a stark picture that the demographers paint for us for the turn of the century, that of a diminishing percentage of the world's population living in the industrialized countries, while the majority in the developing world struggling to increase standards of living and to improve health care delivery in the face of increasing poverty, scarce resources, growing populations, and obviously needing more and more pharmaceuticals that are becoming more costly.

It has recently been recorded[3] that the world consumption of pharmaceuticals during 1983 was on the order of $88.2 billion (Figure 1). In 1977, when the corresponding global consumption figure was only $49.5 billion,[2] Africa, with <10% of the world's population, accounted for 2 to 3% of the world's consumption, with a per capita consumption average of $3. The Middle-East Asian countries did somewhat better, with 5% of the world's population accounting for 2% of the world's consumption with a per capita average of $5. The entire developing world with around 74% of the world's population accounted for <20% of the total world's consumption, with an average per capita consumption of $3, while the developed world, comprising about 25%, enjoyed the benefit of >80% of the $49.5 billion world consumption of drugs, with a per capita average of $34. A few selected countries of Western Europe and the U.S. showed per capita consumption averages of $50 to 55. This picture has remained virtually unchanged during the period up to 1983[2-5] when the industrialized nations again accounted for around 80% of the global consumption figure of $88.2 billion (Figure 1 and Table 1). If the developing countries had the same capability to acquire a similar distribution of drugs, and for the purpose of argument, we extended to them the same average per capita consumption as in the case of the industrialized countries, their total requirement in drugs would approximate to over 100 billion or more than double the world's total consumption of 1977.

In terms of the figures for 1983 (where the total world consumption valued according to 1983 prices was estimated at U.S. $88.2*) the developing world's estimated requirements, calculated on the basis of the average per capita

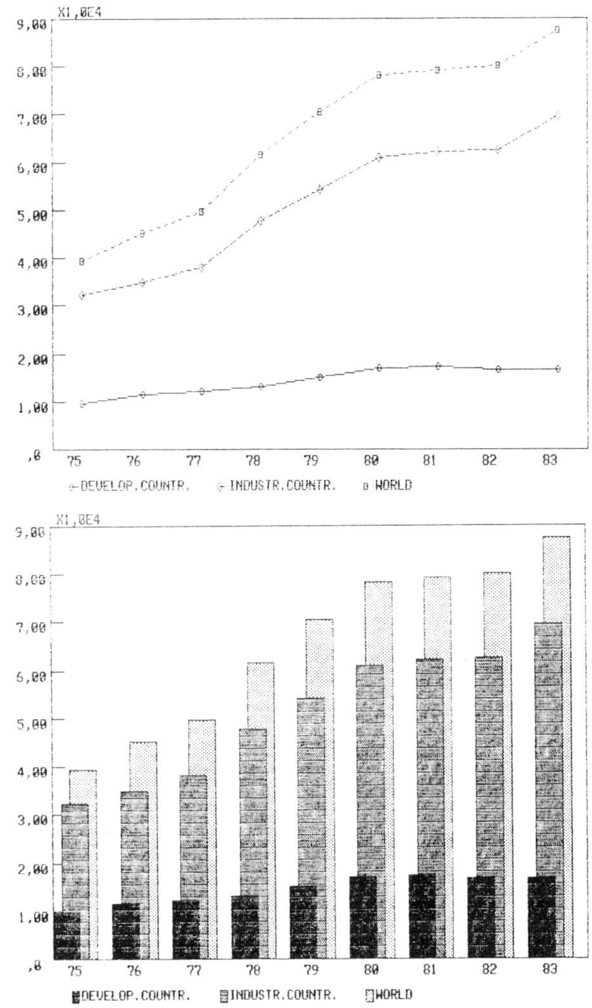

Figure 1.

requirement of the industrialized world, should then be in the range of a staggering $200 to 250 billion. What of the real situation in the developing world? The 1983 consumption of pharmaceuticals was only $17.73 billion.

Dr. F. Johnson Romuald,[6] President of the World Health Assembly (1977), has put it this way: "You only fully understand the vital importance of drugs for health care when you see the long queues of the sick in front of a little dispensary out in the bush which has nothing on its shelves not a single tablet or an antimalarial or antibiotic. Yet this is the situation in the entire regions of the world."

The sick in the long queues in front of the little dispensaries, and many more in the developing world, have often only their traditional herbal remedies for their ailments, to achieve even the modest life expectancies that are realised. Various estimates are made that indicate 70 to 80% of the populations of the developing world depend partly or entirely on the herbal

* U.S. $88,200,000,000 or 1 billion = one thousand million.

TABLE 1
World Pharmaceutical Consumption 1975-1983 by Geographic Regions (Million $U.S.)

Region	1975	1976	1977	1978	1979	1980	1981	1982	1983
North America	8,000	10,300	11,100	12,900	14,400	16,200	18,000	20,200	22,600
Europe North	10,600	11,400	12,200	16,000	18,600	21,300	19,500	19,000	20,000
Europe South	2,200	2,400	2,300	2,700	3,100	3,300	2,900	2,700	2,500
Europe East	5,200	6,200	7,200	8,600	9,300	9,800	9,300	8,700	9,100
Japan	3,600	4,000	4,900	7,500	8,100	9,600	11,500	12,000	15,000
Other industrialized	750	740	760	830	990	1,150	1,250	1,210	1,300
Total industrialized countries	30,350	35,040	38,460	48,530	54,490	61,350	62,450	63,810	70,500
Latin America	3,200	3,700	3,800	4,600	5,700	6,400	6,100	5,500	4,800
North Africa	350	390	440	560	630	780	820	980	1,230
Tropical Africa	570	650	720	790	880	980	1,130	1,110	1,170
West Asia	900	920	1,100	1,200	1,330	1,630	1,920	1,860	2,140
South Asia	600	700	800	1,000	1,100	1,290	1,300	1,410	1,500
East Asia	700	760	1,030	1,190	1,330	1,400	1,550	1,730	1,920
Southeast Asia	250	260	290	330	370	470	560	630	570
Other Asia	2,400	2,700	2,900	3,600	4,300	4,600	4,400	4,300	4,400
Total developing countries	8,970	10,080	11,090	13,270	15,640	17,550	17,780	17,520	17,730
World	39,320	45,120	49,540	61,800	70,130	78,900	80,230	81,330	88,230

Source: UNIDO, 1985 (in press) and Reference 3.

remedies and it can be safely assumed that this is at least so for more than half of the world's population.

Several reasons attribute to the continued popularity of traditional herbal remedies, particularly in the rural regions of the developing world. Anand and Nithyanand[7] cite the following: a ready availability of herbal remedies compared to modern drugs in rural and less accessible regions of developing countries, and a shortage of practitioners of modern medicine in such areas, but they reckon that the sociocultural reasons are of greater significance. Adherence to the traditional, accepted systems are certainly very strong particularly in the old cultures of Asia, the Arabic world, and Latin America (Figures 5, 6, 9, and 10).

There are scientific reasons as well. Over the millennia, man has evolved a corpus of knowledge in all geographical regions regarding the therapeutic utility of plants when applied to a wide range of diseases. Admittedly, this has been through a "trial and error" methodology, but the long and sustained use of particular species by many cultures should not be lightly passed over. Farnsworth and other authors have documented over the years[8-15] (Table 2) the fact that plants have given the modern world a considerable array of drugs, and that in the industrialized countries 25% of the prescription drugs contain active principles that are still extracted from higher plants, a situation that has persisted for at least the last 25 years. The evident conclusion is that the alleged "trial and error" methodology has provided useful drugs for man. There are also reasons why the search for new plant-derived drugs will continue to benefit mankind. First, only a fraction of plant species used as therapeutic agents all over the world have been hitherto adequately explored. The research efforts so far have been mostly directed toward the discovery of new chemical entities with novel structural features, and on the assumption that the plant constituent itself does not always possess the most acceptable pharmacological-toxicological profile, the new structures have served as leads to manipulative syntheses resulting in useful modern therapeutic agents. However, the rational design of drugs must be ideally based on a detailed insight into the biochemical mechanisms that lie behind the pathogenesis of disease. Such insights are not often available or are incomplete, and even these synthetic approaches bear a degree of trial and error methodology.

Computerized studies[15] of ethnomedical information in specialized areas have confirmed the view that plants are a reservoir of potentially useful chemical compounds to serve as drugs or useful models for drug design by synthetic means.

What has failed to receive recognition in the case of research and investigations on ethnomedical therapeutic agents are the data that have already been gathered by the "trial and error" methodology over thousands of years, such as:

1. The ethnomedical recommendations as regards gathering and preparation of plant material for dosage
2. The role of accompanying plants frequently used in multicomponent prescriptions
3. The recommended dosage regimens and contraindications of certain food items

Accordingly, the validity in the context of modern pharmacological data as a measure of clinical potential and utility has to be examined in a different light when applied to the assessment of traditional therapies. Since so large a population of the developing world is served by plant-derived pharmaceuticals based on concepts of health therapy which have been developed by civilizations over considerable numbers of years, it is worthwhile to examine (albeit very briefly) some ex-

TABLE 2A
Some Phytochemical Agents Currently Employed in Modern Medical Formulations[a]

Compounds	Plant species
Acetyldigoxin	*Digitales lanata*
Adoniside	*Adonis vernalis*
Aescin	*Aesculus hippocastamum*
Ajmaline	*Rauwolfia vomitoria, R. serpentina*
Ajmalicine	*Catharanthus roseus, Rauwolfia* spp.
Alloin (Barbaloin)	*Aloe barbadensis*, others
Anisodamine	*Anisodus tanguticus*
Anisodine	*Scopolia tanguticus*
Arecoline	*Areca catechu*
Belladonna (extract of)	*Atropa belladonna*
Berberine	*Berberis* spp., *Coscinium fenestratum*
Boldine	*Peumus boldus*
Bromelain	*Annanas comosus*
Brucine	*Brucea* spp.
Caffeine	*Camelia sinensis, Theobroma cacao*
Camphor	*Cinnamomum camphora*
Capsaicine; Capsicum oleoresin	*Capsicum spp.*
Cephaeline	*Cephaelus* spp.
Cocaine	*Erythroxylon coca*
Codeine	*Papaver* spp.
Colchicine	*Colchicum autumnale/Gloriosa superba*
Convallatoxin	*Convallaria majalis*
Curcumin	*Curcuma longa*
Deserpidine	*Rauwolfia* spp.
Digitoxin	*Digitalis lanata*
Digoxin, Digitoxigenin	*Digitalis lanata*
L-Dopa	*Mucuna pruriens*
Emetine	*Cephaclis ipecacuanha, C. acuminata*
Ephedrine, Pseudoephedrine	*Ephdra* spp.
Erysimin (=Helveticoside), Erysimoside	*Erysimum canescens*
Eugenol	*Eugenia caryophyllata Cinnamomum* spp.
Forskolin	*Coleus forskohlii*
Glycyrrhizin, glycyrrhizinic acid & glycyrrhetinic acid	*Glycyrrhiza glabra*
Guggul steroids	*Commiphora mukul*
Hyoscyamine	*Datura* spp.; *Hyoscyamus muticus*
Hyoscine	*Duboisia* spp.
Hesperidin	*Citrus* spp.
Hydrastine	*Hydrastis canadensis*
Kawain, dihydrokawain	*Piper methysticum*
Khellin	*Ammi visnaga*
Lanatosides	*Digitalis lanata*
Leurocristine (Vincristine)	*Catharanthus roseus*
Linoleic acid (GLA)	*Nonnea macrosperma, Scrophularia marilandica*
Alpha lobeline	*Lobelia* spp.
Menthol	*Mentha* spp.
Morphine	*Papaver* spp.
Papain	*Carica papaya*
Palmatine	*Berberis* spp.
Physostigmine	*Physostigma venenosum*
Pilocarpine	*Pilocarpus jaborandi*
Procillaridin A	*Urginea* spp., *U. maritima, U. indica*
Protoveratrines A & B	*Veratrum album*
Quinghausa = Artimisinin	*Artemesia annua*
Quinine	*Cinchona* spp.
Quinidine	
Rescinnamine	*Rauwolfia serpentina*
Reserpine & deserpidine	*R. canescens, R. vomitoria*, etc.
Ruscogenins	*Ruscus aculeatus*
Rutin	*Cassia* spp., *Fagopyrum* spp.
Santonin	*Artimisia cina, A. maritima*
Scillarins A & B	*Urginea* spp.
Scopolamine	*Scopolia tangutica, Datura metel*
Sennosides A & B	*Cassia augustifolia, C. acutifolia*
Silybin, silydianin, silychristin	*Silybium marianum*
Sparteine	*Spartium scoparium, Cytisus scoparius*
Strychnine	*Strychnos nux-vomica*
Tetrahydrocannabinol (THC)	*Cannabis* spp.
Theobromine	*Theobroma cacao*
Theophylline	
Tubocurarine	*Strychnos toxifera*
Vincamine	*Vinca minor*
Vincaleucoblastine (vinblastine)	*Catharanthus roseus*
Vincristine (leurocristine)	
Valepotriates	*Valeriana officinalis, V. wallichi*
Xanthotoxin	*Ammi majus*
Yohimbine	*Pausinystalia yohimbe*

[a]Mentioned in one or more internationally recognized pharamcopoeia/phytotherapeutic texts. (For a recently published comprehensive list see Reference 14.)

TABLE 2B
Some Examples of Plants Used for Production of Total or Purified Standardized Extracts for Modern Therapeutic Usage

Plant	Standard extract
Aloe spp.	Aloe for dry extract containing 20% hydroxyanthracenes calculated as aloin
Atropa belladonna	Belladonna dry aqueous alcoholic extract containing 1% alkaloids calculated as hyoscyamine
Cassia angustifolia	Senna purified extract containing 45% sennoside B
Capsicum annum	Capsicum oleoresin containing 8—10% capsaicin
Centella asiatica	Centella purified extract, containing 70% triterpenic acid
Coriandrum sativum	Total extract/oleoresin
Cephaelis ipecacuanha	Ipecac soft extract containing 6% alkaloids calculated as emetine
Cynara scolymus	Extract containing 35% caffeylquinic acids, determined as chlorogenic acid
Digitalia spp.	Digitalis total extract
Glycyrrhiza glabra	Liquorice extract, total or purified
Harpogophytum procumbens	Dry extract containing 1.5% of harpagoside
Hamamelis virginiana	Dry extract containing 15% tannins
Hyocyamus niger	Henbane dry extract containing 1% alkaloids determined as hyoscyamine
Passiflora incarnata	Passionflower soft aqueous alcoholic extract containing 2.6% flavanoids calculated as isovitexine
Panax ginseng	Ginseng dry extract containing 10% saponins calculated as ginsenoside Rg 1
Puemus boldus	Boldo soft alcoholic extract containing 0.04% boldine
Polygala spp., *P. senega*	Polygala soft alcoholic extract containing 3—3.5% total saponins calculated as tenuifolin
Podophyllum peltatum, P. emodi	Podophyllum extract

TABLE 2B (CONTINUED)
Some Examples of Plants Used for Production of Total or Purified Standardized Extracts for Modern Therapeutic Usage

Plant	Standard extract	Plant	Standard extract
Prunus africana	Purified extract	*Ribes nigrum*	Dry extract containing 25% anthocyanosides determined as anthocyanidins
Rhamnus frangula	Dry alcoholic extract containing 20% hydroxy-anthracene diglycosides calculated as glucofrangulin	*Serebia serrulata*	Soft purified extract of fruits
Rhamnus purshiana	Dry alcoholic purified extract containing 25% hydroxy-anthracene glycosides determined as cascaroside A	*Silybium marianum*	Purified extract containing silybin + silydianin + silychristin, i.e., 65% flavanoids determined as silybin
Rheum officinale and *R. palmatum*	Rhubarb: dry alcoholic extract containing 8% anthraquinones calculated as aloe-emodin	*Smilax* spp.	Sarsaparilla extract containing saponis
Ruscus aculeatus	Dry aqueous alcoholic extract containing 10% saponins calculated as ruscogenin	*Vaccinium myrtillus*	Dry extract containing 25% anthrocyanidins
		Valeriana officinalis & *V. wallichi*	Valerian dry extract containing 1.3 and 0.75% valtrate
		Zingiber officinalis	Total extract/oleoresin

amples of these concepts before addressing the question of approaches to the current problems of drugs for the majority of the world's population.

II. CONCEPTS AND THEORIES OF ESTABLISHED TRADITIONAL SYSTEMS THAT UTILIZE PLANTS AS THERAPEUTIC AGENTS

The long-established systems of traditional medicine have evolved by systematically recorded human experience over several millenia. Although not strictly based on concepts of modern science, they nevertheless are founded on a corpus of organized knowledge manifested in written documents where even the "scientific method" of hypothesis, experimentation, and confirmation is discernibly revealed. Such organized systems must be differentiated[6] from the relatively empirical folk medicines which too have indisputably useful experience to give to the world.

Among the recognized organized systems of traditional medicine,[16-25] we may briefly consider as examples the concepts of two of them, viz., the Ayurvedic system, originating in the Indian subcontinent and now prevalent in southern Asia in modified forms, and the Chinese system which is dominant throughout China and in the neighboring regions. Similar organized systems which prevail in other regions have doubtless, similar theoretical bases; special mention may be made of the concepts and beliefs that constitute the Unani-Tibb system,[19,26] a derivative of the medical systems of the great Arabic and Greek civilizations, as well as those derived from the Aztec, Inca, and Mayan civilizations, though the latter have no written texts.

A. THE AYURVEDIC SYSTEM OF MEDICINE

The Ayurvedic system of medicine[16-18] is now widely practiced in southern Asia, particularly in India, where it originated, and to varying degrees in Burma, Bangladesh, Nepal, Pakistan, Sri Lanka, and Thailand.

"Ayurveda" literally means the "science of life". The main purpose or objective of Ayurveda, or *prayojana*, is considered to be the restoration and maintenance of the metabolic equilibrium. Health, or *Swasthya*, in Ayurveda is defined as the dual state of a well-balanced metabolism, or *Dhatusamya*, and a state of well-being of mind, body, and senses, or *Prasanna atma indriya mana*.

Disease is considered to be the result of contact with *Dukka*, which connotes physical discomfort, pain, mental anguish, and an array of feelings unpleasant to the human mind and body which (for short) may be called "suffering", but reaches well beyond that term. Disease is also recognized as being fourfold, namely: *Agantuka* (Adventitious or casual); *Shariraka* (Physical); *Manasika* (Mental); and *Svabhavika* (Natural).

The *Agantuka* or adventitions diseases are the result of external injury. The *Shariraka* diseases are internal ailments caused by nutritional or metabolic imbalances, as for example growths, inflammatory conditions and infections. The *Manasika* diseases cover in addition to those considered as mental diseases in modern medicine, such states of imbalance as caused by anger, pride, greed, hate, uncurbed desires, etc. The *Svabhavika* category covers such natural phenomena as birth, aging, hunger, thirst, sleep, etc.

In general terms therapeutic approaches to disease are dependent on the category of disease; broadly, the adventitious diseases call for surgical methods, the physical diseases are treated medically, the mental category is attended to psychologically, and the natural category spiritually. The Ayurvedic concept of the treatment of disease, *Upashaya*, is a three-pronged approach with judicious use of combinations of *Aushada* (therapeutic agents), *Anna* (diets), and *Vihara* (spiritual practices).

According to the *Charaka Samhita*,[16] one of the important texts of the Ayurvedic system, a concept, to be accepted as valid, has to be supported by three *Pramanas* or degrees of proof, viz.:

1. *Pratyaksha*: Direct observation by means of the senses.
2. *Anumana*: Inference by logical deduction or induction, by analogy, or by the method of reductio ad absurdum with respect to a contrary belief.
3. *Aptopadesha*: Proof based on the word of an accepted authority.

If a concept is unequivocally upheld by the three degrees of proof after repeated observation, or *Bhuyo darshana*, it is accepted as valid. Concepts of Ayurveda consider the human body and mind (and indeed perhaps even the soul) as a

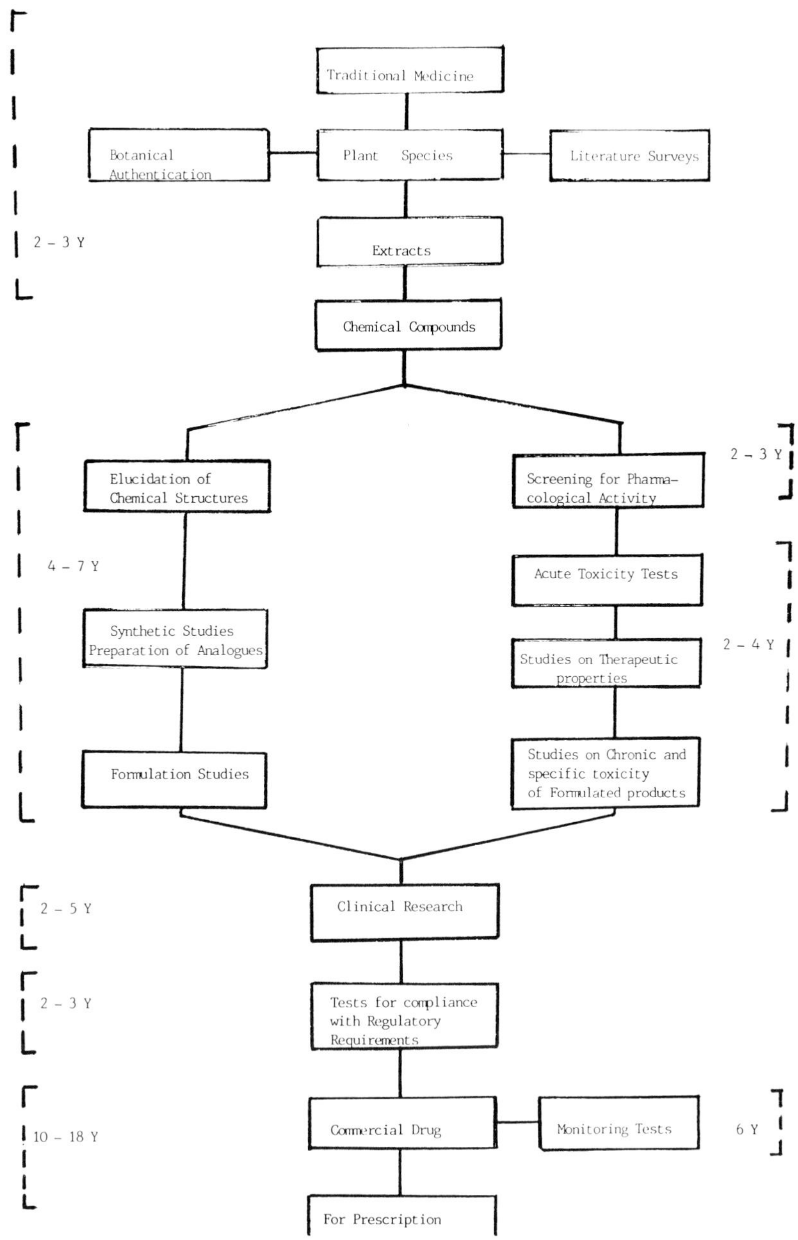

Figure 2.

composite, inseparable whole. The physiological symptoms and the pathogenesis of disease is explained in terms of the theory of the "Tridoshas" — the three basic constituent elements of the human system. These elements named *Vayu* (or *Vata*), *Pitta*, and *Kapha* (or *Shlesma*) cannot easily be rendered into modern medical terminology. Very crudely interpreted, *Vata* is responsible for the utilization of energy by cells and organs for the growth processes. *Vata* also controls the movements of *pitta* and *kapha*, and therefore all the functions and activities of the human body. *Pitta* is approximately a representation of body energy and so is considered to be the agent of metabolic changes and is the governing force behind the entire body chemistry. *Kapha* has everything to do with the cell structure of the body, and therefore its physical well-being and effectiveness.

In Ayurveda, health and disease in humans and indeed in all living matter included within the animal (*Jangama*) and plant (*Sthavara*) kingdoms are based on the normality and the abnormality of the functioning of *vata*, *pitta*, and *kapha*. Amelioration of disease conditions, therefore, is dependent on the reconversion of any abnormality of the *doshas* back to the state of normality.

Besides the *doshas*, which are analogous to the system of "humors" in the Greek system of ancient medicine, the human body is said to consist of seven *dhatus*: body fluids (*rasa*), blood (*raktha*), muscle tissue (*mamsa*), adipose tissue (*medha*), bone tissue (*asthi*), nerve tissue (*majji*) and generative tissue, sperm and ova (*shekra*), as well as several *malas* representing body waste such as stools, urine, sweat, hair, nails, mucus, etc.

The Ayurvedic definition of health is a state of the body in which there is a dynamic balance or equilibrium between the *tridoshas*, the *dhatus*, and the *malas*. Any imbalance results therefore in disease and therapy consists of measures to restore the balance.

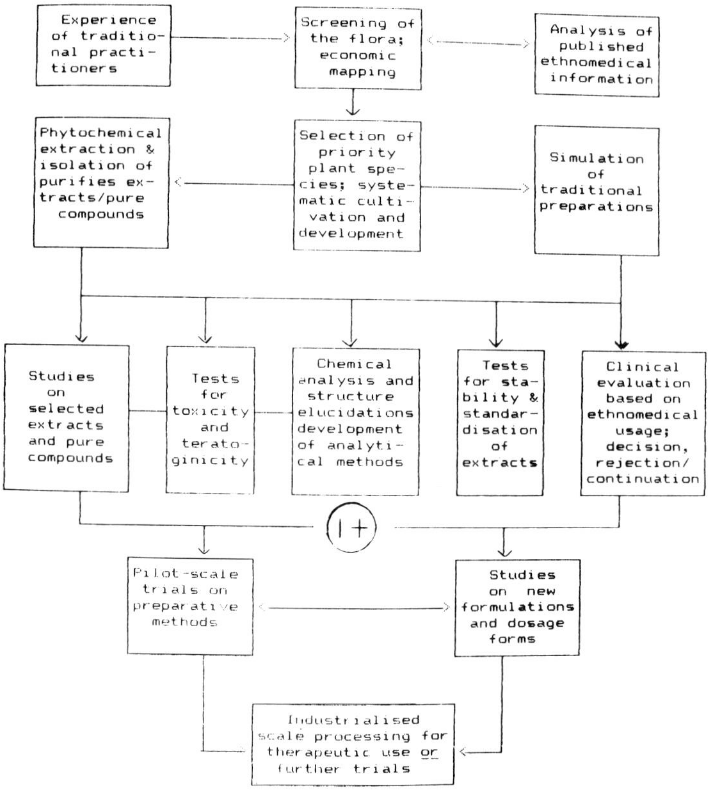

Figure 3.

In the diagnosis of diseaseAyurveda prescribes two types of diagnostic approaches, or *Pariksha*, viz.: examination of the patient, *Rogi pariksha*, and examination of the disease, *Roga pariksha*. The former approximates to the physical examination of the patient and the patient's urine and feces, and the latter determines the pathological condition of the patient.

Treatment consists in avoiding the factors that are responsible for the disease and by administration of medicine, diet, and activity regimens, restoring the balanced state of the body — a holistic rather than a symptomatic approach. Treatment of disease can be by two distinct therapeutic categories. *Shamana* therapy is directed toward the normalization of vitiated *doshas* and restoration of the body balance. The administration of restoratives, digestives, carminatives, and certain physical therapies are categorized as *Shamana*. On the other hand, *Shodana* therapy literally connotes a cleansing type of treatment and emeses, purgation, enemas, and release of blood, etc., come under this category. Surgical treatment has also been known in the classic Ayurveda and has been a recognized method of treatment where deemed most appropriate.

The regulation of diet is an inseparable part of Ayurvedic treatment and foods are classified as being "*HEATY*" or "*COLD*" and therefore indicated or contraindicated according to whether a particular disease is associated with an imbalance caused by a "*COLD*-factor" or a "*HEATY*-factor" respectively.

This type of concept appears as a characteristic feature of the Chinese system of medicine as described later. It is an interesting and noteworthy concept that should be taken into consideration in the assessment of therapies that originate within these systems of medicine.

B. CHINESE TRADITIONAL MEDICINE

Ancient Chinese concepts bear a close formal resemblance to those generated independently by other civilizations.[20-24] The first systematic account of Chinese medical theories appeared in *Huang-di Nei Jing* or *Inner Classic of the Yellow Sovereign*. In classical Chinese concepts the body is seen as composed of a number of interrelated functional entities which have been recently described by Porkert[24] as "Orbisiconography", a dynamic interplay of several functional systems, consisting of the "orbs" of the five viscera which correspond to the functions of the body organs such as the heart, liver, spleen, kidney, and lung. These are designated the *Yin* orbs, and the orbs of the six "bowels" are called the *Yang* orbs. The *Yang* orbs correspond to the functions of the gall bladder, stomach, large and small intestines, and urinary bladder. The *Yin* orbs are considered to be associated with the storing of life constituents, and the *Yang* orbs with the digestion and transport of body nutrients. The orbs are functional entities and the corresponding organs are merely the substrate or carrier. Each orb is not defined by any anatomical role but rather by its specific function such as in processing, storing, and distributing vital energy and thereby contributing to the substance of life.

According to the basic concepts of Chinese medicine[21,22] the well-being of the human body depends on maintaining homeostasis. Bodily functions are designated according to *Yin* and *Yang*, and in the normal state they work in harmony and the body acts as an organic unit, adapted to the environment. This implies that there is coordination between the various parts of the human body which then can adapt to the variations of the natural environment, and there exists a dynamic balance between functional activities and body chemistry. Health occurs when this balance is upset, i.e., when a breakdown occurs in the equilibrium within the body as well as in the balance between body and environment. Disease is the consequence of a confrontation between the harmful effects of a pathogen and the intrinsic resistance of the body and results when such a confrontation succeeds in breaking down the states of balance and equilibrium.

The pathogenic factors in Chinese medicine are classified into categories such as: atmospheric factors — wind, cold, damp, dryness fire, etc.; emotional factors — joy, anger, melancholy, anxiety, grief, fear, etc.; others — strain, stress, trauma, irregular meals, improper diet, etc.

The body resistance encompasses all the energy, essential substances, and physiological activities needed to maintain life processes. These in the terminology of Chinese traditional medicine are *Yin, Yang, Ai* (energy), and *Xue* (blood). Since *Yin* is associated with the functions of body organs *Xue* is included within *Yin,* and similarly *Ai,* which is the energy responsible for the body's metabolic processes is included with *Yang*.

The classification of syndromes in the Chinese traditional systems can be carried out in many ways, but the most popular is based on the nature of the confrontation between the pathogenic factors and the body resistance. All syndromes belong to three main categories: (1) syndromes which are due to inadequate body resistance, diminished functional activities, or deficiency in essential substances (deficiency syndromes); (2) syndromes showing overabundance or excess of pathogenic factors (excess syndromes); and (3) combined syndromes of deficiency and excess.

Treatment is chiefly based on the differentiation of the syndromes. In deficiency syndromes the general principle is to enhance resistance. In excess syndromes, the therapy is aimed at dispelling the pathogenic factors. For deficiency of *Yin* he ttreatment must by directed toward replenishing *Yin* with tonics and other products, for deficiency of blood, and toning up the blood; for deficiency of *Yang* the therapy for invigoration of vital functions is employed. For excess syndromes the methods of therapy vary considerably as different methods are required to dispel different pathogenic factors.

Traditional Chinese drugs are classified according to the groups and subgroups as in Table 3,[20] in terms of classical Chinese pharmacology. The characteristics of a drug are described in terms of "four temperaments" (properties) designated as cold, hot, warm, and cool (*Han*, *Re*, *Wen*, and *Liang*), and "five tastes" — pungency (acridity), sourness, bitterness, and saltiness. Both "temperaments" and "tastes" are correlated with pharmacological effects. Cold and cool drugs belong to the *Yin* category, and differ in the degree of potency. These drugs were employed to treat "heat syndromes" such as fever with reddened eyes and thirst. Conversely, warm and hot drugs belong to the *Yang* category and were employed in treating a cold syndrome, such as those with symptoms of chills, fatigue, failing pulse, etc.

According to guidelines for the formulation of drugs that originated[25] with the celebrated Li Shizen's work, *Ben Cao Gang Mu*, Chinese traditional medicine also had well-recognized and clearly identified principles for formulation of dosage forms (Figure 4). A prescription contained four categories of ingredients (there could be one or more components within each category).

1. The *jun* or principal ingredient which was responsible for the main therapeutic action
2. The *chen* or second ingredient which enhances the action of the principle ingredient
3. The *zuo* or auxiliary ingredient which relieves secondary side effects or tempers the potency of the principle ingredient
4. The *shi* which directs the therapeutic effect to the desired site, or acts as an auxiliary ingredient

The principles governing the formulation also took account of the possibilities of drug interaction, and the possible modes of drug interaction were identified. For example, synergistic effects were recognized and employed in calculated fashion. Drugs were deliberately employed to counteract the undesirable effects of another ingredient or the toxic effects of a

TABLE 3
The Pharmacological Classification of Drugs[20]

Drugs for dispelling pathogens from the exterior of the body, i.e., diaphoretics (*jie-biao yao*):
- With pungent taste and warm temperament (*xin-wen jie-biao yao*)
- With pungent taste and cool temperament (*xin liang jie biao yao*)

Antitussives (anticoughing), anti-asthmatics, and drugs for resolving phlegm:
- Warm drugs for resolving cold phlegm (*wen hua-tang yao*)
- Cool drugs for resolving heat phlegm (*qin hua re-tan yao*)
- Antitussives (*zhi-ke yao*)
- Anti-asthmatics (*pin-chua yao*)

Drugs for clearing internal heat, i.e., antipyretics (*qin-re yao*):
- Drugs for clearing up and purging intense internal heat (*qin-re xie huo yao*)
- Drugs for clearing up heat and cooling the blood (*qin-re xie liang-xie yao*)
- Drugs for clearing up heat and dampness (*qin-re zao-shi huo yao*)
- Drugs for clearing up heat and pathogens (*qin-re jie-du yao*)

Drugs for dispelling internal wind dampness, i.e., antirheumatics (*qie-fong-shi-zao*)

Warm drugs for dispelling internal cold (*wen-han yao*)

Aromatic drugs for resolving dampness (*fang-xiang hua shi yao*)

Diuretics and hydrogogues (*li-niao shu-shui yao*):
- Dierutics and mild hydrogogues (*li-niao sen-shi yao*)
- Hydrogogues or drastic purgatives (*zhu-shui yao*)

Drugs for regulating the flow of vital energy (*li-qi-yao*)

Drugs for regulating blood condition and circulation (*li-xie-yao*):
- Hemostatics (*zhi-xie yao*)
- Drugs invigorating blood circulation and resolving blood clots (*huo-xie hua-yu yao*)

Aromatic drugs for stimulating resuscitating (*fang-xiang kai-qiao yao*)

Drugs stabilizing the spirit (*an-shen yao*), essentially, sedatives and tranquilizers

Drugs for subduing hyperactivity of the hepatic orb and endogenous wing (*pin-gan xi-fong yao*)

Drugs reinforcing vital functions, i.e., tonics and aphrodisiacs (*bu-yang yao*):
- Drugs for replenishing vital energy (*bu-qin yao*)
- Drugs for nourishing the blood (*bu-xie yao*)
- Drugs for replenishing vital essence (*bu-yin yao*)
- Drugs for reinforcing the vital function (*bu-yang yao*)

Drugs arresting discharges (*gu-se yao*), including astringents and hemostatics

Digestive and evacuants (*xiao-dao yao*)

Purgatives and laxatives (*xie-xia yao*)

Anthelmintics (*qu-chong yao*)

Drugs for external use (*wai-yong yao*)

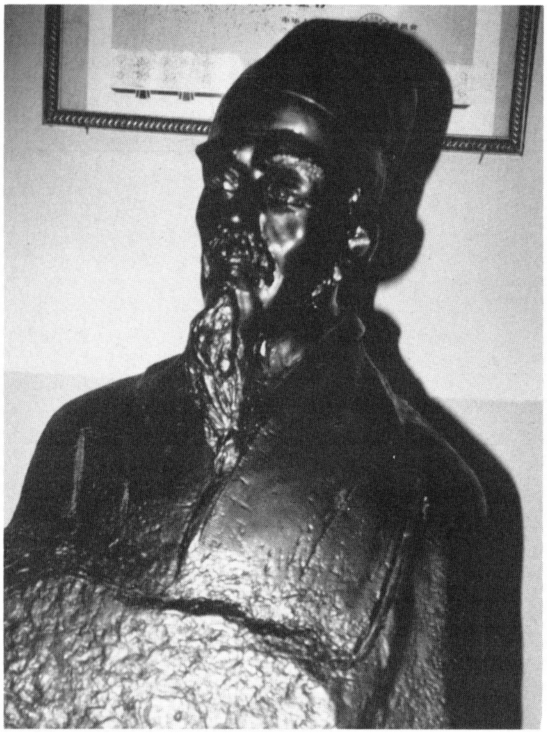

Figure 4. Sculptured bust of Li Shizen, a great physician, naturalist, and author of *Ben Cao Gang Mu*, the comprehensive pharmacopoeia of Chinese medicine, containing a total of 1894 drug items and 11,096 related formulations, and published in 1596. *Ben Cao Gang Mu* still serves as a valuable reference for teaching, practicing, and as a guide for drug research. The sculpture lies in one of the Chinese State Pharmaceutical Administrations Traditional Medicine Processing Factories in Hangzhou. (Photo by Wijesekera, 1982.)

poison. It was also recognized the some drugs which were individually safe could interact to produce a toxic effect.

C. THE OTHER TRADITIONAL THEORY-BASED SYSTEMS

These brief discussions on the theoretical bases of the Ayurvedic and the Chinese systems of medicine are sufficient to give the reader an idea of their depth and their systematic approach to therapy. Other systems of equal content and stature are also known: (1) the Unnani-Tibb system,[19,26,28] on which considerable documentation is still available in the original Turkish script at Istanbul, its evolutionary center,[27] (2) the medical theories of the Babylonian-Assyrian periods and the Egyptian civilization to the later theories of Hippocrates (460—377 B.C.) and Galen (130—200 A.D.);[29] (3) the medicinal theories of the indigeneous Amerindian peoples of the Latin American region;[30] and (4) the *siddha* system of medicine practiced by the people of Dravidian origin in Southern India.[28]

There are several common features among all these systems that call for attention in assessing their therapies for modern usage. As mentioned earlier they are based on the hypothesis or theoretical considerations that have been logically derived from systematically accumulated observations representing the experience of mankind over considerable periods of time. There are many concepts that have been

Figure 5. Mali: a stall for the sale of herbal drugs in the outskirts of the capital, Bamako. (Photo by Wijesekera.)

arrived at independently by cultures that had no communication one with another. The concept that disease is caused by the change in the "balanced state" of a body is not in conflict with the position of modern medical science, but how may we interpret in modern terms the identity of the balancing forces, and could there be a modern concept that could replace that of *Yin* and *Yang*, and Hot and Cold? Should not the theories, diagnoses, and therapies be interpreted and understood in the attempt of modern science to examine traditional medical systems and their therapies?

Recently, a view was expressed that "contrary to what is often advocated by believers in herbal medicine and traditional medical systems, natural products and isolates are rarely optimal for therapeutic purposes". What "therapeutic purposes", one may ask? Purposes as devised for modern drug therapy unmindful of the context of the traditional method? A lack of appreciation, certainly, of the traditional milieu is evident in this view, as indeed in the common approach toward obtaining new natural products with therapeutic activity. Again, in a recent context it was observed:

> "We have kept the tenor of this symposium on a high level of molecular biology and have helped to avoid the misleading and perilous stigma of the daily newspapers expressed in a possible dreaded headline: Academy Embraces and Endorses Herbal Medicine. Several questions beg themselves: What is the stigma in endorsing herbal medicine where it is scientifically established as effective? Is the high level of molecular biology the sole realistic approach or is it just one preferred approach? Is this attitude the reason why modern science has missed much? What are the other approaches possible? These are some of the questions that will have to be considered if the real value of traditional systems is to be harnessed, for the benefit of modern man."

III. THE REALIGNMENT OF STRATEGIES FOR RESEARCH AND DEVELOPMENT

A. STRATEGIES AND PRIORITIES

Several publications[8-14] pointed out that drugs derived from plants continue to be a valuable source of therapeutic agents in

Figure 6. Guatamala City: herbal drugs in packets labeled in an orderly fashion are sold on the main street sidewalks. (Photo by Wijesekera, 1983.)

Figure 7. Large-scale harvesting of *Pimpinella anisum* in an experimental farm station, Fundeleka, Romania. Children are employed during school vacations and taught the elements of botanical identification as well. (Photo by Wijesekera, 1984).

the armory of modern medicine itself. Some examples of both pure substances of plant origin and plant extracts or isolates that are currently in therapeutic use in Europe, the U.S., and elsewhere are indicated in Table 2A and 2B. In general, natural products that have come into modern therapy are the result of an approach[35-37] that has been adopted during the past 50 years and essentially has had the goal of acquiring new chemical compounds with biological activity. This approach consists of the sequence represented in Figure 2. Drug development in this way takes 8 to 18 years, and the costs of such a development program has been variously estimated[31-40] as being on the order of $15 to 50 million per drug, sometimes even more. This range of values is based on the experience of the development of pharmaceutical products in industrialized countries such as the U.S., the U.K., and Japan.

Even at the lowest levels, developing countries, i.e., more than the world's population, can afford neither the money nor the time for drugs based on this existing strategy and pattern of development. Obviously then, the situation needs to be regarded with all the seriousness it deserves, and a contingency strategy needs to be evolved to help meet what is a dire situation. The position was well put recently[40]:

> "About 50 m pounds has to be spent to bring an innovative product to the market. For this to be worthwhile, the market must be huge or rich. Not surprisingly there are a number of rare diseases for which a specific drug would not be commercially viable. These are the so-called orphan drugs. A similar situation obtains in Third World Medicine. Third World Medical services are too poor to pay for a new drug even if it is discovered. So any way of using existing pharmaceuticals is a bonus."

Given the situation that between 50 to 80% of the developing world depends on traditional therapies for their health care, then the only answer is an effort toward improving these therapies as much as possible with the aid of modern science and technology. In doing so some of the factors that make these traditional remedies popular and acceptable should not be ignored. Furthermore, a new approach based on the urgent needs of a large though poor sector of the global population must indicate the elements of this new strategy. The strategy for development of plant-based pharmaceutical products would undoubtedly vary from situation to situation, but there are some common elements. One of the most important requirements is that plants for industrial scale processing must be systematically cultivated after proper authentication of the correct species used. Gathering the plant material from the wild flora is unacceptable for a number of very important reasons dealt with elsewhere in this volume. Obviously, each country must decide on the priority species to be cultivated and the choice of species must depend on factors such as:

- The climatic conditions
- The most abundantly used species
- The species utilized for curing diseases which have no effective therapy in modern medicine
- The species utilized for curing diseases for which modern therapy is disproportionately costly
- Species that have export potential

Cultivation of the plant species makes it possible to produce raw material that is both reliable and consistent with respect to its chemical constituents.

As noted before, the method of drug development from plant sources that has hitherto been popularly adopted is based on a sequence of operations leading mainly toward the isolation of pure natural products. Though it has served to identify novel chemical structures with biological activity, this methodology nevertheless has severe limitations. Generally speaking, organic chemists will tend to find such compounds which are easily isolated and purified into crystalline form for comparatively rapid structure determination. It is likely that compounds which by virtue of their structural characteristics have presented more formidable problems of isolation have in the past been missed and these may have been the very ones that possessed the biological activity attributed to the plant. Isolation methods monitored by test for biological activity have only recently been adopted, but these can only be adopted if the nature of the bioactivity is known, and the test is suitable for

Figure 8. Large-scale cultivation of *Coriandrum sativum* in an experimental station, Fundeleka, Romania. (Photo by Wijesekera, 1984.)

Figure 10. Classified and packed herbal medicines on sale on a sidewalk in Eskirshie, Anatolia, Turkey. (Photo by Wijesekera, 1986.)

Figure 9. A herbal dispensary in Anatolia, Turkey. (Photo by Dr. Bayhan Cubukcu.)

Figure 11. An ancient herbal Ayurvedic text in Nepal. (Photo by Wijesekera, 1981.)

repetitive application. In the standard strategy of drug development which is adopted by the large transnational pharmaceutical organizations the costliest operations in time and money are those related to tests for toxicology, clinical research, and regulatory requirements.[31-35] This process would generally be prohibitive for any developing country, even if the expertise were available; hence, new approaches must be considered for developing countries. The strategy that must be chosen should initiate some visible impact in minimum time without expenses that will be out of reach of the country's exchequer. Such an approach in practice would involve the stages represented in Figure 3. After assessment of the ethnomedical information, the plant species used in a preparation should be authenticated and then placed on a crop cultivation basis. Preparations based on traditional medicine should be simulated using modern methods of processing and subjected to clinical trials. Following indications of therapeutic potential, they should be formulated into modern dosage forms, and compared by clinical trials with the simulated preparations. Simultaneously, methods of analysis and quality control should be developed, and after efficacy has been established, isolation procedures for chemical constituents could be initiated, while successful preparations are prepared and formulated in bulk for distribution to health programs. This is now the considered approach recommended to Third World countries.[41]

The chemical studies will have dual benefits: besides bringing up active chemical compounds with perhaps new structural features, they would enhance the development of quality control methods based on new analytical techniques such as are available today, and these methods may be extended for the quality control of extracts of multicomponent preparations as well. Such studies would also enable the development of standardized, internationally recognized techniques for the quality assessment of plant-derived products. Some advances toward this have already been made.[36]

B. NEEDS AND CONCERNS OF A MEDICINAL PLANT INDUSTRY

The development of a medicinal plant industry becomes for many countries of the Third World both a health and an economic necessity. The benefits of such an industry can be manifold in the context of the Third World and may be outlined as follows:

1. Crop diversification and increase of agro-production
2. More productive use of available land resources particularly in the case of lands which are only marginally suit-

Figure 12. Weighing out herbal drugs prior to dispensing a prescription at a Chinese herbal pharmacy in Beijing. (Photo by Wijesekera, 1982.)

able for other types of economic crops such as food grains, fruits, tea, rubber, coconut, cocoa, and spice crops
3. Stimulation of rural-based industrial efforts which will serve as a means of offering greater stability to farmers
4. Generating employment in area where there is unemployment or underemployment; the opportunity that such an industry offers for self-employment of educated youth in developing countries is also significant
5. Creating new export opportunities and providing existing industries with alternative locally produced raw materials, that would lead to import substitution; examples could be products such as oils and fats, essential oils, pigments, gums, tannins, and other phytochemicals which may result as by-products of a medicinal plant industry and be gainfully utilized in the cosmetics, paints, and leather industries, in place of imported materials

Besides these benefits the industry would serve as a provider of:

1. Plant-derived medicines at cheaper rates than imported alternatives
2. Therapeutic preparations specific to maladies that are localized or prevalent within a particular region
3. Plant-derived biocides and pest repellants which are relatively inexpensive, and are biodegradable, and would cause little long-term ecological damage, and hence would serve as alternatives to hazardous synthetic pesticides

The industry would also increase the local, regional, or national self-sufficiency with respect to therapeutic and related pharmaceutical products.

A

B

Figure 13. (A) Ancient equipment used in the preparation of ayurvedic drugs in a traditional-style factory in Kathmandu, Nepal. (Photo by Wijesekera, 1981.) A large pan and spoon used to concentrate aqueous extracts. (B) A mortar and pestle.

In developing a medicinal plant industry in Third World countries, the regions of the world that most require such an industry, certain constraints will have to be overcome. Some of these are of a general nature and have already been mentioned, such as the need for systematic programs for cropwise cultivation of medicinal plant species (Figures 7 and 8). An industry cannot draw on spontaneously growing plant species for reasons of species depletion as well as the fact that the raw materials for processing will vary with regard to quality. Cultivation on a cropwise basis involves a long research and development effort consisting of the following elements:

- Selection of plant species based on ethnomedical usage, modern assessments, as well as chemical and biological characteristics of the species
- Crop development, including genetic improvement of species, agronomic research, harvesting techniques, and postharvest methods
- Farm and estate management including methods of transport and storage, as well as marketing and distribution to central authorities

- Technology, which would include basic processing and quality assessment of raw materials as well as the first-stage products, be they dry raw materials or simple products such as essential oils, extracts, or formulated drug delivery forms

A second set of constraints exists which is particularly acute though the constraints can vary considerably with the different situations prevailing in the Third World. The major constraint for almost every developing country is the shortage of funds. Entrepreneurship is based on the visibility of the profit prospect and in much of the Third World entrepreneurship, even if it exists, is limited and will only try less risk alternatives. State entrepreneurship is sometimes the only possibility, but here too, foreign exchange is scarce and this is a barrier to the acquisition of equipment, expertise, and training. Research and development efforts therefore cannot be productively initiated and sustained. Often the lack of local capital hinders all efforts directed toward the necessary research and development. The lack of expertise in developing countries is a serious barrier to technology transfer in all areas such as agronomy, processing, and quality control. In the attempts to initiate such technology transfer one frequently encounters the lack of infrastructural necessities such as ongoing relevant research programs at local institutions, universities, polytechnics, or technical schools graduating quality personnel to staff a developing industry, and managerial expertise at all levels. Finally, one of the most important considerations is a political vision to develop a medicinal plant industry in a given situation. The presence of such a political manifest must be translated into state policy responses favorable to the industry, protective acts for products and operations, and above all the availability of investment incentives, without which the industry, even if initiated satisfactorily, will soon cease to exist.

The ideal conditions that would warrant investment in an industry based on the production of plant-derived medicines follow:

- The existence of a population where a significant proportion uses traditional herbal therapies and where such therapies are socially accepted and believed in.
- The existence of well-organized traditional systems of medicine based on authentic documents, medical theories, and groups of systematically educated and trained practitioners whose practice is regulated by codes and systems of ethics.
- The lack of a competing synthetic drug industry or the situation being such that there is little possibility of one being established due to lack of petrochemicals or a coal-based chemical industry.
- The existence of systems of modern medical education and practitioner training that includes phytotherapy and traditional medical theories.
- The existence of a research infrastructure that can provide for the scientific functions of the industry (Table 4), combining good university education and research, particularly in the disciplines that are relevant to an industry utilizing medicinal plants.

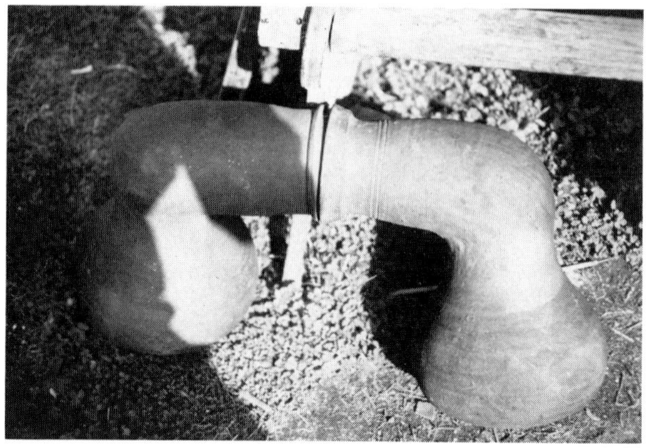

A

B

Figure 14A and 14B. Traditional equipment and technology employed by a traditional-style factory in Kathmandu, Nepal, for processing Ayurvedic preparations. Various types of traditional equipment for conducting extractions and for distillation. (Photos by Wijesekera, 1981.)

- The availability of medicinal plant species in abundance both with a rich and varied spontaneous flora and in cropwise cultivation.
- The availability of local expertise in crop management, skilled labor, artisans, and technicians.
- A milieu for favorable policies, legislative enactment, and incentives for industrial development utilizing agro-based raw materials and products that must be standardized and are quality sensitive.

C. IMPORTANCE OF RESEARCH IN THE TRANSFER OF TECHNOLOGY

Research is vital to the successful transfer of technology in any field of activity, especially in the medicinal plant industry. The countries of the Third World are generally what have been described as "pre-science cultures" and scientific research, though now practiced in almost all of them, is still somewhat of an alien concept. Initiation of research activity has its own set of problems which are at best difficult to surmount.[42-44]

The medicinal plant industry utilizes plant materials with such a diversity of physical properties, e.g., delicate flowers, soft leaves, succulent fruits, fibrous materials and grasses, hard woods, brittle barks, oily seeds and nuts, and so on. The

TABLE 4
Scientific Functions of the Medicinal Plant Industry

Research	Development	Production and quality control
Natural product chemical — investigation — structures Elucidation	Scale-up of extraction methods and then	Process finalization
Synthetic studies Development of analytical pharmacological and metabolic studies; toxicology	Analytical methods for monitoring Pilot plant studies on process development	Preparation scale separations, extractions, etc. Pilot scale and semi-commercial plant designing
Clinical studies, teratology	Scale-up of plant propagation with analytical monitoring: methods of harvesting and postharvest preparation and storage	Techno-economic studies
Botany, agronomy and plant propagation Climatic and varietal differences of chemical constituents	Genetic improvement of plant species Formulation studies: *In vitro* testing; Drug stability/exipient interaction studies; Formulation stability studies; New formulations	Analytical control methodology for: Raw materials; Isolated fraction or total extracts; Pure compounds; Formulated products

Figure 15. A covered mortar and pestle and weighing scales.

processing of such material imposes a wide range of requirements due to both physical diversity and chemical variety, and Figures 13 to 20 depict some of the traditional equipment and technology used in the processing of Ayurvedic drugs in a drug factory in Nepal. Particularly in China and India, and in other countries as well, relatively modernized factories produce similar preparations as seen in Figures 21, 22, and 23.

There is also a wide range of agronomic requirements and thus the research necessities become accordingly complex and multidimensional. A concerted research program is required, involving the disciplines of botany, plant genetics, agronomy, entomology, phytochemistry, pharmacology, chemical engineering, pharmacy, and clinical and marketing studies oriented toward the common goal of product development. The research therefore compliments the scientific functions of the industry (Table 4).

For the transfer of technology to be accomplished and sustained, particularly in the context of developing countries, research expertise has to be built up over several years — perhaps decades — involving a number of existing institutions and perhaps with the collaboration of one or more funding agencies. The competence of the national institutions has to be slowly developed to enable them to undertake indigenous goal-oriented research. The institutions may be universities, research institutes, or private sector research organizations.

Several problems are understandably encountered in developing indigenous research competence[42-46] and not all of these problems are peculiar to the medicinal plant industry. Some, however, are characteristically so. Much effort in planning is needed and even formal interinstitutional links must be forged to ensure the coordination of research activities to enable goal-directed priorities to be maintained. For example, agronomic

Figure 16. A type of vessel in which herbs are "cooked" under pressure. (Photos by Wijesekera, 1981.)

Figure 17. The manual preparation of pills at a traditional Ayurvedic factory in Kathmandu. Note modern equipment at left which is used sometimes. (Photo by Wijesekera, 1981.)

Figure 18. Packaging done manually at a traditional Ayurvedic factory. (Photo by Wijesekera, 1981.)

research expertise as generally available in developing countries is more often that not related to the major crops such as rice, coffee, tea, rubber, and cocoa and located in departments or institutions that are responsible for the development of these crops; botanical expertise is usually located in established botanical gardens, universities, and forestry institutions; chemical expertise will be found in university chemistry departments, industrial, or technological research institutions, and pharmacological and pharmaceutical expertise is found within medical and pharmaceutical research organizations and universities.

Thus, the nucleus of personnel, funds, and equipment which may be utilized for an industry-oriented research exercise on medicinal plants would be divided between several institutions. Given the different accepted priorities of each individual institution, this makes coordination of research activities a goal common to all, but of lesser priority to each, a most difficult proposition. Yet, coordination of activities within an existing infrastructure is the most economical way open to developing countries, given the constraints of time and funds, and must be so regarded. Furthermore, as often seems the case, the public sector must shoulder the burden of laying the foundations of basic research and provide the leadership for the development of research to initiate and sustain an endogenous crop-oriented medicinal plant industry. Most often, the private sector tends to wait until sufficient basic and applied research has been done to demonstrate profit potential before they invest in further research. Once this is done and profit potential can be estimated, the "visibility" of product-development possibility would entice the private sector to become interested in being fully vested in product research and development. New products (which are what interests the private sector) have to be conceived based on laboratory research activity. They must be formulated; tested; manufactured, first in pilot scale amounts on a trial basis, then in bulk; clinically evaluated; launched; and marketed. This activity can be the responsibility of the private sector. Governments of developing countries could, if their political policies permit it, commit such activity to the private sector on prearranged terms and even stimulate private sector involvement by legislative and policy incentives, tax benefits, low interest loans, market safeguards against imported competitor products, and the like. Bridging the vast gap between basic research, in many instances just phytochemical or pharmacological screening, and product development is the problem of developing Himalayan countries within the medicinal plant sector, with the possible exception of the People's Republic of China and perhaps India.

There are many constraints that prevent research from being profitably commercialized in developing countries. These constraints fall into two categories: constraints in crop production and constraints in process technology. Constraints in crop production involve those discussed earlier and factors such as land availability, transportation, irrigation, fertilizer input, labor requirements, etc. The constraints in process technology also involve those described in connection with the transfer of technology. Even if crop development and research in the basic sciences relevant to the industry may be

Figure 19. Herbal and mineral materials are mixed and incubated in sealed clay vessels buried in sand. Furnace temperatures are sometimes used by application of heat over the clay sand pit. (Photo by Wijesekera, 1981.)

Figure 20. Grinding the dried herbal material with inorganic minerals. (Photo by Wijesekera, 1981.)

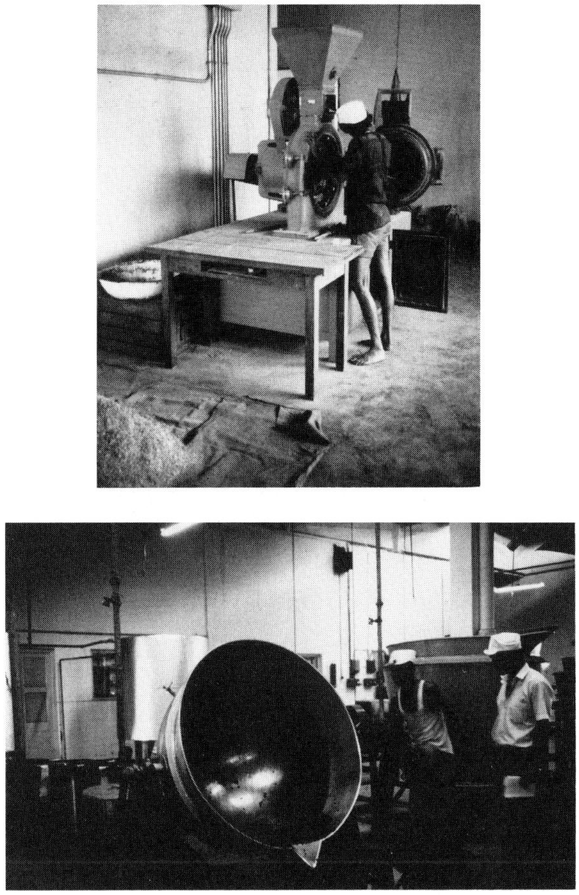

Figure 21. Inside an ayurvedic drug factory (Nawinna, Sri Lanka) where the equipment has been modernized. (Photos by Wijesekera, 1985.)

achieved, the lack of expertise and facilities for development of process technology remains a major constraint to the development of a medicinal plant industry in many developing countries. Product processing methods at the laboratory level must be tested on a pilot scale for process suitability as well as to ensure some measure of economic feasibility. Acquisition of such pilot plants is generally the major obstacle for research and development institutions in developing countries. Versatile pilot plants that can be used for solvent extraction, percolation, steam distillation, and concentration of extract with recovery of solvent are generally produced in developing countries, for capacity ranges of 50 to 250 L and a cost of between $30,000 to $150,000. It is here that international funding agencies can fill a major need. Some developing countries are now able to construct such prototype pilot plants themselves, which can be equally effective for product quality, though perhaps not product economy. Sometimes processing technologies that appear feasible at the pilot level may encounter difficulties on scaling up to commercial production. Problems of magnitude, such as handling large quantities of plant material, waste material disposal, solvent recovery, product drying, etc., can become disproportionately large when moving from pilot scale to commercial levels. It is also necessary at the commercial level to match processing with the availability of plant material to ensure economic operation.

Marketing must also be emphasized if the industry is to sustain itself in a Third World context. Investors always seek existing or potential market availability before investing in research and development ventures, and in the medicinal plant industry it is not often easy to possess such availability. Development of a market depends on the production of a quality product on a viable commercial scale. Research and development is, however, a necessary prerequisite to producing a quality product which will stimulate market interest. Accordingly, the need exists for adequate and therefore costly investment in research and development in Third World countries if they are to reap a substantial part of the benefits themselves.

This "which comes first" situation exists for all new products but is more complicated in the medicinal plant industry because:

- It is an agro-based industry

Figure 23. Ginseng and other plant-derived medicines on sale at a counter in Europe (Copenhagen). (Photo by Wijesekera, 1986.)

Figure 22. Part of the processing equipment in a State Herbal Pharmaceutical Factory in Hanghou, People's Republic of China. (Photos by Wijesekera, 1982.)

- The marketing and crop production activities need to be coordinated
- Products must be tested for toxicological and teratological manifestations
- Products face severe competition from synthetic alternatives

Even if markets can be acquired by developing countries for products based on medicinal plants such as speciality items, several formidable regulatory barriers will have to be overcome.[44]

The legal and ethical requirements for products include ensuring that products are safe when taken within the dosage stipulations by large sections of people. Such insurance involves quality assessment of an order of magnitude far more demanding than in the case of products not internally consumed by man. In order to satisfy regulatory needs, developing country investment levels can never be sufficient. Even in the instance of a developing county acquiring a situation where marketing of a product is a distinct prospect, it faces yet another barrier in the form of uncertainty of patent protection, due both to the complexities of obtaining such protection and the costs.

The transfer of technology to Third World countries must in reality recognize marketing and regulatory aspects as well.

In summary, therefore, a new strategy (Figure 3) must be adopted toward developing the medicinal plant industry. This strategy must address itself to the contigency situation existing in Third World countries and should have the following approach:

1. The generation of factory produced traditional medicines with attendant standardization and quality control as well as acceptable manufacturing practices
2. The formulation of new products (preparations) based on traditional remedies, including modern dosage forms, and at economic prices in the Third World context
3. Natural product drug development

The strategy is aimed at benefitting not only the developing countries, but those of the industrial world, particularly where adequate therapy for particular diseases is lacking. The strategy should be designed to draw support from all interested, on humanitarian grounds, concern for the health of the underpriviliged, the need to develop appropriate industries in the poorer parts of the world, and the global benefits that can accrue by sharing a research and development enterprise.

IV. CONCLUSION

Given the present scenario, it is worthwhile to examine future prospects. Several things become immediately evident. It is well established that plants both terrestial and marine could continue to provide mankind with valuable therapeutic agents for a long time to come. The use of plant species in cultures where such therapies were developed should be scientifically examined as well as the theories on which they are based. There are numerous scientific instances when the dosage regimen or preparative methods prescribed in old texts have been proved to be rational and valid. They may be stimulated using modern technology, but it would be unwise to ignore the rationale. Medicinal plant species grow abundantly in many developing countries and modern processing technology may not often generate the cheaper and reliable drug preparations needed so desperately for their health care needs. A contingency need exists for adopting such processing methods which would produce reliable drugs economically and no less effective, particularly for Third World needs. However, there is always the unfortunate possibility that a valuable lead that may come from the indigenous, traditional therapy of a developing country, would, while generating an effective and profitable modern drug, fail to benefit the country of origin. New crop introductions from developing countries, though relatively costly, are carried out in many developed countries. Where successfully accomplished, it follows that the developing country concerned may have totally lost an existing or potential cash crop.

Therefore, the improvement of the medicinal plant industry in developing countries is best achieved by way of a partnership with westernized countries, whereas much of the processing as is possible will be carried out in the developing country itself.

There is undoubtedly a wave of enthusiasm within industrialized nations for herbal remedies, though at the same time, understandably suspicion and skepticism co-exist. One senses disproportionate emphasis in many semi-scientific as well as popular publications on the "toxicity" in plants, and not nearly as much on the toxicity of synthetic drugs or other synthetic products in everyday use.

A partnership between industrialized and developing nations could realize the best in traditional medicines for the world, as well as to meet the contigency of the woeful shortage of therapeutic agents for the majority of the worlds' sick who inhabit developing countries. A global effort will surely be needed in developing the medicinal plant industry for all mankind's future needs. The profit motive should not solely determine strategies concerning the health of the worlds' population.

REFERENCES

1. G. I. Lythcott (1985). U.N. World Health Day 35th Keynote Address: Childrens' Health Tomorrows' Wealth.
2. A. M. Sallam. Perspectives on the Distribution and Availability of Pharmaceuticals in Africa and the Middle East in Conference Proceedings: Pharmaceuticals for Developing Countries, Division of International Health, Institute of Medicine, National Academy of Sciences, Washington D.C.
3. **Anon.** (1986). UNIDO Document: IO/R.219. Pharmaceutical Industry Development in Developing Countries and the Concept of Total Quality Control. United Nations Industrial Development Organisation, Vienna, 1986.
4. A. Tcheknavorian-Asenbauer (1985). Public Pharmaceutical Industry in the Service of Health: *Ann. 1st Super Sanita*, 21(4), pp 565-572.
5. J. Pogany (1986). Private communication from UNIDO Data Base.
6. F. Johnson (1978). Presidential Address to the World Health Assembly. W.H.O., Geneva.
7. N. Anand and S. Nithyanand (1984). Integrated Approach to Development of New Drugs from Plants and Indigenous Remedies, in Natural Products and Drug Development: Alfred Benzon Symposium 20. Krogsgaard-Larsen, Christensen, and Kofod, Eds., Munksgaard, Copenhagen, pp 78-92.
8. M. F. Balandrin, J. A. Klocke, E. S. Wurtele, and W. H. Bollinger (1985). Natural plant chemicals: sources of industrial and medicinal materials, *Science*, 228, 1154-1160.
9. N. R. Farnsworth (1984). Medicinal Plants and Drug Development: Alfred Benzon Symposium 20. Krogsgaard-Larsen, Christensen, and Kofod, Eds., Munksgaard, Copenhagen, pp 17-29.
10. B. J. Spalding (1985). Modern drugs from fold remedies, *Chem. Week*, Feb. 27, pp 52-53.
11. N. R. Farnsworth (1977). The current importance of plants as a source of drugs, in *Crop Resources*, D. S. Siegler, Ed., Academic Press, New York, pp 61-73.
12. N. R. Farnsworth and R. W. Morris (1976). Higher plants — the spelling giant of drug development, *A. J. Pharm.*, 147, pp 46-52.
13. N. R. Farnsworth and A. S. Bingel (1977). Problems and prospects of discovering new drugs from higher plants by pharmacological screening, in New Natural Products and Plant Drugs with Pharmacological Biological or Therapeutic Activity, H. Wagner and P. Wolff, Eds., Springer-Verlag, New York, pp 1-22.
14. N. R. Farnsworth, O. Akerele, A. S. Bingel, D. D. Soejarto, and Z. Guo (1985). *Bull. WHO*, 63(6), 965-981.
15. N. R. Farnsworth, A. S. Bingel, D. D. Soejarto, R. O. B. Wijesekera, and J. Perea Sasian (1980). Prospects for higher plants as a source of useful fertility regulating agents for human use, in Symposium on Recent Advances in Fertility Regulation Beijing. Chang Chai Fer and D. Griffin, Eds., World Health Organization, Geneva, pp 331-364.
16. Caraka-Samhita: Vimanasthana in 6 volumes, P. M. Mchta Ed. (English Version), Shree Gulabkunverba Ayurvedic Society, Jammnagar, 1949.
17. P. N. V. Kurup (1983). Ayurveda in *Traditional Medicine and Health Care Coverage*. R. H. Bannermann, J. Burton and Ch'en Wen-Chieh, Eds., WHO, Geneva.
18. S. Shiv Sharma (1971). Ayurvedic medicine, *Progress Drug Res.*, 15, 12.
19. Mohamed Said (1983). The Unani system of health and medicare, in *Traditional Medicine and Health Care Coverage*, Bannerman, Burton, and Ch'en, Eds., WHO, Geneva, p. 61.
20. Shen Jiaxiang (1985). An introduction to Chinese Materia Medica, in *Proceedings of the Workshop on the Pharmaceutical Industry (Combined Modern — Traditional Pharmacy) for Promoting Technical Cooperation among the Developing Countries*, UNIDO/IO.614, pp 5-28.
21. Xie Zhufan (1985). Basic Concepts of Chinese Traditional Medicine, *ibid.*, UNIDO/IO.614, pp 29-36.
22. Wang Pei (1983). Traditional Chinese medicine, in *Traditional Medicine and Health Care Coverage*, Bannerman, Burton, and Ch'en, Eds., WHO, Geneva, p. 68.
23. Huang Liang (1984). Drug research based upon the leads of Chinese traditional medicine, in *Natural Products and Drug Development*, Alfred Benzon Symposium 20. Krogsgaard-Larsen, Christensen, and Kofod, Eds., Munksgaard, Copenhagen.
24. M. Porkert (1983). *The Theoretical Foundations of Chinese Medicine: Systems of Correspondence*. MIT Press, Cambridge, MA, as quoted in ref. 18.
25. Hu Changhong (1983). Traditional Chinese dosage forms and their modernization, ref. 20, UNIDO IO-614, pp 59-71.
26. M. Said (1983). Unani-Tibb in Pakistan, *World Health*, 6, 13.
27. B. Cubukcu (1983). Personal communication.
28. R. Kumaraswamy (1983). Medicinal plants and indigenous medical systems in health care, in *Report of the Seminar on Medicinal Plants Phytochemicals and Bulk Drugs*, Basic Chemicals Pharmaceuticals and Cosmetics Export Promotion Council, Bombay, pp 62-66.
29. C. C. Mettler (1946). *History of Medicine*, Blakiston.

30. C. Goldwater (1983). Traditional medicine in Latin America, in *Traditional Medicine and Health Care Coverage*, Bannerman, Burton, and Ch'en, Eds., WHO, Geneva, pp 37-49.
31. A. Selby (1985). Analytical chemistry: the hub of pharmaceutical development, *Chem. Br.*, p. 606.
32. R. W. Hanson (1981). The Pharmaceutical Development Process: Estimates of Development Costs, and Terms and Effects of Proposed Regulatory Changes in *Issues in Pharmaceutical Economics*, Reprint Ser. No. C.81. Centre for Research in Government Policy and Business, University of Rochester, Rochester, NY, p. 151-181.
33. U. Reinhardt (1983). in *Drug Therapy and Its Price*, K. Von Grebmer, Ed., Medicopoca International, Inc., 1983.
34. N. Rogerson (1986). Licensing — A solution to product continuity, *Manufact. Chem.*, 57(7), 41-43.
35. C. Djerrassi (1979). *The Politics of Contraception*, W. W. Norton, New York, pp. 68-69.
36. R. O. B. Wijesekera and A. Tcheknavorian-Asenbauer (1982). UNIDO, IO. 505 Vienna Industrial Utilization of Medicinal and Aromatic Plants — A Review.
37. Sukh Dev (1983). Natural Products in Medicine — Present Status and Future Prospects, in Report on the Seminar on Medicinal Plants, Phytochemicals and Bulk Drugs, BCPC Export Promotion Control, New Delhi, pp 10-28.
38. Association of the British Pharaceutical Industry, Annual Report for 1976-1977, London.
39. Industrial Bank of Japan: Quarterly Survey April 1985 No. 62. Japanese Finance and Industry, pp 6-7.
40. Anon. (1985). Tropical drugs, in *Chem. Br.*, 21, 234.
41. R. O. B. Wijesekera (1983). Industrial utilization of the phytopharmaca, in Report of the Seminar on Medicinal Plants Phytochemicals and Bulk Drugs, Basic Chemicals Pharmaceuticals and Cosmetics Export Promotion Council, New Delhi, pp 103-114.
42. R. O. B. Wijesekera (1978). Building national scientific and technological research capability in the context of under-development, in *Integration of Science and Technology with Development*, B. Thomas and M. S. Wionceck, Eds., Pergamon Press, Oxford, pp 17-30.
43. R. O. B. Wijesekera (1976). Science in a small developing nation, *Sci. World*, No. 1, pp 6-9.
44. R. O. B. Wijesekera and M. Wijesekera (1977). *Sci. World*, No. 2, pp 21-23.
45. D. C. Jayasuriya (1985). *Regulation of Pharmaceuticals in Developing Countries — Legal Issues and Approaches*, WHO, Geneva.

Chapter 2

METHODOLOGY OF ECONOMIC MAPPING OF MEDICINAL PLANTS IN THE SPONTANEOUS FLORA

Ovidiu Bojor

I. INTRODUCTION

At present in the developing countries within Asia, Africa, Latin America, and the Pacific Ocean islands, only about 20% of the population benefit from modern drugs. These people use traditional, natural remedies for curing different types of diseases, and even by the year 2000 a radical change cannot be visualized. There is no doubt that herbal drugs are and will still be one of the main resources of raw materials for obtaining medicines during the next millenium, in addition to the substances obtained by biosynthesis "*in vitro*", chemical synthesis, and mineral sources.

Numerous developing countries of the world are rich in medicinal and essential oil plants (aromatic plants), but this richness, especially the stock of the spontaneous flora, is limited. The merchants of medicinal plants wishing to obtain large profits do not protect nature and maximally exploit the existing stocks in different developing countries. The medicinal and essential oil plants — trees, shrubs, herbs — belong to an ecosystem, and if its biological balance is impaired by irrational exploitation, this action will have disastrous repercussions on the other factors of the ecosystem.

A. THE IMPORTANCE OF ECONOMIC MAPPING TO FOUND A SMALL-SCALE INDUSTRY IN DEVELOPING COUNTRIES

It is not possible to establish a small- or medium-scale industry based on wild flora without assessing the quality and quantity of the raw materials available for industrialization, and without taking into account the protection of the environment. Neither is it possible to program the exploitation of the wild medicinal and essential oil plants (aromatic plants) without knowledge of the useful constituents present in the species of medicinal herbs individually. Establishing the exploitable amounts from a locality, from a geographical area, implies necessarily allowing the species to perpetuate itself. To schedule the exploitation and to establish an annual plan of wild flora evaluation, there must be close cooperation between those responsible for the production and the research and development aspects. Sowing the exploited field with the same species as those already collected from an area is compulsory for the needs of the future as well as for protecting the wild medicinal plants.

From an ecological point of view, special attention must be paid to the soil from which some species, with a role in fixing the soil and preserving it from erosion, have been collected. One should not neglect the role of some medicinal shrubs and trees in favoring the multiplication of birds or insects or animals useful in natural pest control.

Mapping the spontaneous medicinal flora within a geographical area may also generate important information which will bring about decisions regarding locating drying equipment, the postharvest conditioning and primary processing of plants, or in locating some small industrial units which can extract all or only some active substances. This information is very important, especially for the zones where conventional transportation is difficult or the distances to the industrial centers are too great. For some geographical zones such small extraction units close to the sources of the raw materials are very important from both an economic and social point of view. Processing fresh or dried raw materials — in corresponding conditions — without storing too long is highly efficient and helps to cut down transport charges of raw material. From a social point of view, besides providing labor opportunities in areas remote from the agricultural and/or industrial centers, economic mapping allows people to become skilled workers in a new branch of activity. Local energy sources (water, micro-, hydroelectric power, solar energy, wind energy, local fuel) can become profitable as well.

Scientifically, economic mapping of spontaneous flora helps researchers achieve a better phytosociological and phytogeographical understanding of the respective zones and a better delimitation of areas suited to growing specific species. Simultaneously, it is possible to have a better genetic understanding of conserving plants having a high active substance content which are able to provide multiplication material with a view toward improving it by means of chemotaxonomic selection. It is also possible to obtain information on the edaphic factors of the studied area through those species that give details on soil moisture, pH, and nitrogen content.

B. DEFINITION AND OBJECT

Economic mapping of the spontaneous medicinal flora is defined as identification, notation, and transposition of conventional signs onto topographical maps of the quantities of raw material of one or several medicinal species recommended to be harvested in a certain area. A vital aspect to economic mapping is that the dictates and requirements of environmental protection are studiously observed. Within this action is the evaluation of the therapeutic nature of the chosen medicinal plants, particularly with respect to the content of active substances. This is achieved either by phytochemical determinations (analyses) carried out in the field or by drawing (collecting) samples for laboratory tests to be conducted.

The object of economic mapping in medicinal plants (herbs) is to identify quantitatively and qualitatively the natural resources of a certain area with a view to rationally rendering its economic potential profitable by organizing collecting, drying, and conditioning centers and by establishing pharmaceutical industrial units for the extraction of action substances.

In order to preserve the medicinal and essential oil plants and to ensure their natural multiplication at the same geographical location, the rhizomes, bulbs, roots, and aerial parts of the plant are partially (30%) harvested at approximately 5-year intervals. The flowers, fruit, and seeds are harvested yearly, at 50%, without endangering the perpetuation of the species.

Another aspect of economic mapping is to locate the small-scale processing units at which collecting and drying take place.

It is also possible to obtain information on the edaphic factors of the study area, several species giving details on soil moisture and pH, calcium, potassium, nitrogen, and trace element contents. All factors are of importance for cultivating medicinal plants. For example, *Acorus calamus*, *Valeriana officinalis*, *Menyanthes trifoliata*, and other medicinal plants from Umbelliferae, Ranunculaceae, or other families are able to grow in swampy land; other species, e.g. *Matricaria chamomilla*, can grow in slightly salted soils; and other, e.g., *Lavandula vera*, *Thymus* sp., *Rosa* sp. *Hippophae rhamnoides*, and *Hyssopus officinalis*, can reclaim natural dry slopes.

As mentioned earlier, medicinal and aromatic plants represent the principal means of therapy in traditional medical systems. By comparison with cosmopolitan medicine, the traditional form is very complex and varies from one country to another, from one medical system to another, from one geographical area to another, and depends on the economic and historical development of the human communities in the various regions of the world. To take into account all the data concerning the use of medicinal plants in local traditional medicine, the botanical survey and economic mapping of the territory is very important.

C. ECOLOGY AND UTILIZATION OF PLANT RAW MATERIAL

As mentioned previously, intensive exploitation of medicinal and aromatic wild plants jeopardizes not only some important species, but can also have severe implications on large territories. Some medicinal species are known to have been uncontrollably exploited resulting in gradual decrease in number or even in extinction. In such cases it has become necessary to protect them legally by declaring these species "endangered".

In the same way, by ruthlessly exploiting some shrubs, as for instance *Pinus montana*, as a source of essential oil, large areas in the alpine zone, or high tablelands, have been denuded. This resulted in water source evaporation and severe erosion. Presumably in that case, recommendations to collect alternatively by strips in order to allow the shrubs to grow continually were not heeded when the raw material was collected. The cultivation of medicinal and aromatic plants can be programmed and controlled, whereas the irrational exploitation of wild flora has doomed them to extinction.

To reiterate, one of the main aims of economic mapping is to maintain a favorable balance between exploitation and protection of plants and the environment.

II. WORK METHODS

A. FIELD STAGE

1. Geographical Survey of the Territory and Delimitation

Within the framework of the general information on the area to be studied, and after determining geographical coordination, relief is discussed taking into account the existing data in the specialized literature. The orographical data are completed with particular consideration for work in areas having a richer wild flora.

The surface of the territory to be mapped should cover territorial-administrative units (regions, counties, villages, etc.), hydrographical basins, mountainous massifs, islands, etc., which, owing to their relief or geographical location, permit a strict delimitation in square kilometers.

The geographical survey should contain a broad description of the forms of relief; altitude; rivers and lakes; the nature and structure of the soil; the location in terms of cardinal points (N, S, E, W); and delimitation of the cultivated zones.

The richest spontaneous flora is normally found in hydrographical basins, forests, brushwoods, plantations, glades, natural lakeshores, uncultivated steep coasts, pasture land, and mountain pasture. The shores of artificial lakes, plantations, or orchards may also represent great sources of medicinal flora. Small quantities of spontaneous medicinal plants are also found among fruit trees, food crops, and ornamental or industrial species (banana tree, coffee shrub, etc.).

Polluted areas, road borders (except for some arborescene shrubs), natural reserves (plants protected by law), natural parks, and public gardens should *not* be listed. In the survey, some stations of meteorological points or stations from which the climatological data regarding the investigated area may be collected, should also be indicated. Some species do grow in depressions having microclimates different from the main region.

To these general data, meteorological and pedoclimatic information, such as rainy seasons, drought periods, etc., should be added.

2. Literature and Documentation

Before evaluating medicinal flora, interested parties must study the previous literature on the flora of the concerned geographical area. Although many publications do not always cover the necessary elements for evaluating the quantities of medicinal and aromatic plants, they supply even less information on their active principle content. However, these publications may contain general information on the propagation of these species.They usually mention the name of the species, often indicating quite accurately the phenotypes and ecosystems, and provide useful information for listing the existing flora. Any work containing information about the flora of the investigated territory is interesting not only scientifically, but also because it may contain detail on the species.

When herbals with samples of the listed species are available, they should be consulted as they too provide valuable information. Besides locating the species, the herbals can also indicate the regional vernacular names, the vegetative stage of a plant, the precise time and conditions it may be harvested, etc.

Another source of information is the work dealing with the medicinal and aromatic plants of a certain area, data collected from the experience of an organization and/or herbs companies, and interviews with the inhabitants of different zones of the region or country. On the basis of such information, explorations are carried out to detect the areas where medicinal plants grow, but take into account only the areas that might give important yields.

The data obtained from publications and herbals are useful for compiling an index of medicinal and aromatic plants. This index should contain new information uncovered during listing.

3. Research Programming in the Field: Duration and Tracks

To judiciously use the time alloted to the field assignment, sensible programming is required. In temperate, continental, and submediterranean zones, the listing work should take into account the relief and the season.

In the plain areas of the temperate zones, work begins in the second half of April with the forests where the vegetation cover is rich until May.

The listing in hilly and subalpine zones will occupy the next 2 months, and in the mountainous zone, up to 1500 to 2000 m altitude, during July. In August and September (sometimes even in October, depending on the meteorological conditions), the zone between 2000 to 4000 m should be listed.

In the equatorial, tropical, and subtropical zones, where definitive wet and dry seasons exist, the listing will be programmed depending on the season. In the monsoon zones, herbaceous vegetation develops rapidly during the first weeks of rain, and however difficult, listing cannot be entirely interrupted as some species have only a very short growth period. This vegetative stage is particularly important from a phytochemical standpoint, i.e., the content of active substances. For the tropical and subtropical areas affected by monsoons, two field trips for economic mapping are necessary: one before and one after the monsoon season. Only in this way can conclusive results be drawn with respect to the amount of active substances in the plant species.

Listing the flora over a zone of 8000 to 10,000 km^2, with different relief forms and difficult access roads, in which only 50% of the surface is cultivated, requires 5 to 6 trips of 20 d each and 3 teams of 2 to 3 people each. The teams will trek on a rotasystem developed by the head of the research team. Programming is essential so as to avoid unlisting some areas or twice listing the same area.

For an effective data gathering program, it is essential to have access to a suitable vehicle. When approach roads or railways are available, the distances can be covered by a four-wheel drive vehicle, fitted with a laboratory compartment. When there are no striking differences in the altitude of the relief forms, listing takes a rather short time, while it takes a longer time when the approach roads are poor or nonexistent.

Before beginning the field work, the leader of the research team will organize an orientation trip. This means that the team has already studied the bibliographic documentation of the geography and flora of the investigated area and all maps to find out the approaches, communal roads, forest tracks, etc. In connection with this, the team leader must plan for expenditures for fuel and supplies for land cruisers and if necessary, an aeroplane charter (or helicopter) to reach difficult areas.

4. Team Composition: Scientists, Technicians, Auxiliary Staff

Teams usually include three or four graduate personnel, technical assistants, laborers, and drivers. Two main aspects should be considered concerning the team members: their physical fitness and professional background. The listing team should include research workers thoroughly conversant with the medicinal and aromatic plants both botanically and phytochemically (pharmacists/phytochemists/biologists/technicians). A team of 8 to 11 people should have 3 to 4 university graduate specialists and 4 to 5 technicians. Of the five technicians at least four should be drivers experienced with rough terrain and one should be an engine mechanic. At least two technicians are necessary for collecting and drying samples necessary for laboratory tests. They will dry and keep in good order samples for the Dry Herbarium and will preserve some of the samples in alcohol or formalin for anatomo-morphological study.

Finally, the team leader is responsible for planning the field work, including the formalities required by the local authorities, camping facilities, and the welfare of the team.

5. Means of Conveyance, Outfit, Camping, and Supply Bases

It cannot be overemphasized that in order to carry out the listing of the wild flora in the best possible manner, the team should have their own means of transportation.

Taking into account the difficulties of some tracks, the author recommends four-wheel drive (land cruisers) vehicles. Good results may be obtained by using either four-wheel drive microbuses outfitted with a laboratory compartment or suitable cars with one-axle caravans. Convenient conveyance will greatly enhance the mapping work, and increase the efficiency at the site. Besides reducing the duration of the journey, a satisfactory means of conveyance will permit the work in the field to be accomplished in one trip, as a central location for all technical materials and equipment within the mapping zone will considerably facilitate the effort of the investigating team. Ideally, a mobile unit should include four vehicles, the first carrying three members of the team and a driver, and each of the other three carrying a member of the team and a driver.

Before going into the field, the mechanical condition of the vehicle must be checked. Make certain the land cruisers carry essential spares in case problems occur during the trip, the majority of the mapping zones being remote from any service station. The driver must be able to cope with all kinds of difficult roads and have a thorough knowledge of the mechanical and electrical needs of his vehicle.

One land cruiser is for the team leader and two members of his team, including the driver. Two land cruisers should carry minilaboratories where, by means of suitable reagents, and perhaps thin layer chromatography (TLC), rapid chemical determinations can be performed in the field. The last vehicle, fitted with a larger laboratory, should have a refrigerator, heating equipment and carry the literature, the samples to be analyzed, and the food stocks.

For mountainous and subalpine zones, mountaineering boots are necessary. Light and comfortable wool or cotton clothes are preferable. Synthetics should be avoided. All members of the team must have a pair of rubber boots, e.g., Wellingtons, for protection from the morning dew or for use in marshy zones, wet river meadows, peat bogs, and snake-infested regions. They must also have raincoats, torches, cold food, and water for 1 d in their foodbags or haversacks. All teams must carry a first aid kit. Other precautionary measures concerning the handling of potentially toxic herbs must be observed as well.

Our own experience showed that maximum efficiency can be obtained by round trip mapping. To this end, at least two tents each accommodating two people, or three tents, in case the car cannot accommodate one to two sleeping places, are necessary. In such instances, the base camp will be the parking place of the team vehicles, which must be within members' reach. Thus, the work may be resumed on the site first thing each morning without covering distances to and from the base.

When the mapping teams are without the necessary equipment for camping, a town, village, motel, or chalet will be chosen as a base. In that case, the work will progress "in star" fashion, i.e., the teams will cover the mapping points within a radius of about 50 km from the base and return to base each time along another track. The starting point will be moved to another locality at the end of the work.

It should be mentioned that plant listing requires considerable trekking on foot; the vehicles serve only to carry the equipment and food as well as the members on long distance, from one listing zone to another. In the field, some analyses should be carried out in the in-van laboratories.

In both round trip mapping or starting from a base, adequate stocks of food, water, and fuel need to be secured. The food must be of good quality, resistant to deterioration, and sufficiently nourishing to meet the exacting requirements of the team.

With regard to the means of conveyance, one must take into account that a four-wheel drive vehicle can cover 500 to 600 km without being refueled. When the refueling points are far off, refueling will be done by the driver within the interval of 2 to 3 h necessary for a team to identify the flora of a certain locality. An auxilliary canister for petrol should be also provided for each vehicle as well as several water containers.

6. Equipment

In order to efficiently accomplish mapping work, the following items are necessary:

- Topographical or distinct maps on the scale of 1:100,000 or 1:50,000 (for difficult alpine zones, 1:25,000 or 1:10,000 scale maps are preferable)
- Guide books for identification of plants
- Compass, altimeter, field glasses, magnifying glass, and SLR cameras with close-up attachments and an adequate supply of film
- Stakes and string for laying off frames
- Notebook and pencils (black lead pencils are preferred, as lead is rain resistant)
- A pick or a garden trowel, pruning shears, a knife
- An inflatable boat when the mapping zone is a delta, lake, or river with a difficult approach
- A kit with the reagent required to identify alkaloids, glycosides, tannins, anthraquinone derivatives, etc.
- A first-aid kit, and supplies of prescribed medicaments
- A pressing machine for preparation of vouchers, paper, paper bags, string

When round trip mapping is the methods of choice, tents, traveling bags, inflatable mattresses, alcohol, or paraffin lamps and gas cookers with a spare gas cylinder should also be included.

7. Local Contacts

After satisfying all legal formalities within the respective zone and announcing to the local authorities the aim of the mission, general information should be gathered concerning prohibited areas, natural reservations, road conditions, accommodation, and provisioning facilities. Any information on the zones rich in medicinal and aromatic plants and their use by the traditional medicine contribute to a better listing. Some of this information can be obtained from the teaching staff of the communal schools or from the population within the mapping zone. In any case, all data obtained in this way, while informative, will have to be checked in the field.

8. Methodology for Quantitative Evaluation of Medicinal and Aromatic Plants — Examples

There are two main options: (1) listing a limited number of plants intended as candidate-plants for industrial utilization and (2) listing all the spontaneous medicinal flora.

The concept of medicinal value being relative, one must come to an agreement with the beneficiary or sponsors of the work by drawing up a minimum list of medicinal species. The list will include not only the classical species or those used at the national level, but the herbs that have international market prospects. This list may be supplemented by including the plants used in the local, popular medicine, some species under study, and even species which were not considered medicinal at the time but which are present in great abundance in the area under surveillance. At any rate, this list should be adhered to by all groups or mapping teams, so that the data gathered will concern a limited number of species. Every species in the minimal list should be known by each member of the working team in the field at all their vegetative stages. Should an obscure point concerning the species arise, this must be resolved by a discussion with all the members of the team. The species not determined or those apt to be confused will be pressed into vouchers in the herbarium and examined at a later time. The field worker will record the indeterminate species with the respective specification and a question mark (?). The second important element that has to be recorded is the plant part to be utilized.

It should be re-emphasized that evaluation should be carried out on dry raw material whose exploitation could be rendered profitable in the respective area. The economic mapping of the spontaneous medicinal flora does not mean, therefore, that all species and their existing amount from a region should be inventoried; only the amounts of raw material of medicinal herbs proposed to be rendered profitable must be taken into account. Generally speaking, with obvious variations from species to species, the subterranean parts represent 30 to 40%, aerial parts of the evergreen species 30 to 40%, and the flowers, leaves, fruits, and seeds represent another 30 to 40% of the whole plant.

In order to render the evaluation as objective and uniform as possible, each member of the research team should be cognizant of two important parameters concerning plant raw materials used in the medicinal plant industry. The term of the "vegetal raw material" must not be mistaken for "vegetal product", the latter being the result of a processed raw material. The term "raw material" or "raw drug" means plants and/or a part of a dry plant that must meet certain quality specifications concerning the external appearance, maximum percentage of impurities and foreign matter, moisture or other physicochemical properties, as well as the minimum amount of the active substances or volatile oil. For industry, the crude (not dried) raw material may also be delivered which should also comply with the quality specifications. Accordingly, the team members must have knowledge of:

1. What amounts of raw material can be obtained from a specific plant, e.g., from *Atropa belladonna*, during the vegetative stage of flowering, 4 to 6 g of dry leaves and 20 to 25 g of dry root can be obtained. With this knowledge one can estimate the amounts obtainable per single hectare. Thus, if on a surface of 1 ha 660 plants have been identified (15 m^2 for one plant), one may obtain between 3 to 4 g *A. belladonna* dry leaves and 14 to 26 kg dry root from the same species. Suppose the brushwood where the belladonna has been identified is 100 ha and the density nearly the same. From this surface one could harvest 300 to 400 kg leaves and 1400 to 2600 kg dry root. The recommended amount to be rendered profitable in this case (taking into consideration the protection of the species) is 466 to 866 kg root (about 30% from the whole) and 150 to 200 kg leaves as dry raw material (about 50% from the whole) from the 100 ha surface.
2. The next important parameter that the team must be aware of is how much a plant part loses after being dried. There are studies and special books from which such information may be predetermined.

After the medicinal species to be mapped have been detected and the two above-mentioned parameters studied, the most difficult operation to follow is to identify from a quantitative standpoint the raw material proposed to be rendered profitable within a region which represents a definite surface unit.

There are at present several systems from quantitative and qualitative evaluations of phytoconosis. These end plant lists are drawn up on the site with reference to the stationary factors (in French, this kind of evaluation is called "relevé"). The work generates an ensemble of ecological and phytosociological observations concerning a certain locality.

The species abundance is evaluated on the basis of several degree scales, such as, for instance, the Braun-Blanquet scale:

1. Very scarce individuals
2. Scarce individuals
3. Few individuals
4. Numerous individuals
5. Very numerous individuals

Emberger and co-workers devised a scale of 10 degrees of abundance in the relevé:

1. 1 individual
2. 2 to 4 individuals
3. 5 to 9 individuals
4. 10 to 19 individuals
5. 20 to 49 individuals
6. 50 to 99 individuals
7. 100 to 199 individuals
8. 250 to 499 individuals
9. 500 to 999 individuals
10. 1000 individuals and over

The number of the individuals on the studied surface being known, one can evaluate the density with the following formula:

$$d = \frac{n}{s}$$

where d = density, n = number of individuals, and s = surface.

Special frames of 50 or 100 m each side, divided into squares of 1 dm^2, help to establish accurately the abundance or the preponderance of the species. At the beginning of our own mapping work we adopted the delimitation of 1 m^2 surface via 4 sharp stakes of 1 m height, tied with string at their upper ends. The stakes driven into the ground form a frame delimiting a visible surface of 1 m^2 for the low vegetation as well as the high plants (see Figure 1).

With at least two frames of 1 m^2 each, the team may carry out relevé, following and recording the medicinal species within each frame. The relevé number varies as a function of the frequency and depends on the constancy and geometrical shape of the surfaces of the mapped area. For example, let us suppose that between a river flowing downward without any loops and a highway there is a water meadow of 10 m/0.5 km. Let us also suppose that six herbaceous species of medicinal

Figure 1. The delimitation of a typical relevé frame of 1 m², surface by fixing of stakes of 1 m height tied with strings at the upper end.

plants grow in that water meadow, which on the whole distance, repeat themselves with an almost similar frequency: *Petasites hybridus*, *Uritica dioica*, *Eupatorium canmabinium*, and *Equisetum arvense*, and others such as *Symphytum officinale* and *Heracleum sphondylium* are sporadic. In such a case, ten relevés are sufficient by placing the frames at a distance of 50 m from one another in a zigzag line to obtain an estimated value of medicinal flora in our imaginary water meadow. The ten relevés are as follows:

No. of species	Relevé no.	1	2	3	4	5	6	7	8	9	10
....											
....											
....											

The mean amounts of dry raw material for each individual from six species are 1 = 20 g; 2 = 3 g; 3 = 8 g; 4 = 4 g; 5 = 5 g; 6 = 3 g (fruit).

The absolute value of the dry raw material for the species from the 10 m² proposed to be rendered profitable, according to the plant protection index, would be 1 = 125.4 g; 2 = 324 g; 3 = 210 g; 4 = 412 g; 5 = 13.6 g; 6 = 27 g.

The relative value proposed to be rendered profitable for 1/2 ha mapped in the respective water meadow would be

- *Potesitos hybridus* (Rhizomes) = 62.7 kg (30% of the individuals)
- *Urtica dioica* = 162.0 kg (whole harvest)
- *Eupatorium cannabinum* (Radix) = 105.3 kg (30% of the individuals)
- *Equisetum arvense* (whole plant) = 206.0 kg (50% of the individuals)
- *Symphytum officinale* (Radix) = 13.5 kg (whole amount of fruits from all individuals)

As we have already shown, the number of relevés and the 1 m² frames used for our evaluations differ following the frequency and consistence of the species and the geometrical shape of the study area. Let us suppose we are in a circular shaped water meadow of 1 ha surface, where there is only one medicinal species, *Colchicum autumnale*, from which the seeds are gathered. In a typical case, about 80% of the plants have capsules which will reach maturity. To map this surface and obtain results as conclusive as possible, place the relevé frames as indicated by Figure 2. From 20 relevés diagonally placed on the studied surface and on plots (14, 15, 16, 17, 18, 19, and 20 frames), an estimate of the number of individuals existing in the water meadow can be obtained. Following the relevé, a mean density of 5.5 plants per square meter was obtained. Only 80% of the plants bear fruit, therefore, the plants that can be evaluated are 4.4 m².

To harvest the *C. autumnale* seeds the plant protection coefficient is not necessary because it multiplies by bulb. For the 1 ha water meadow it means that there are about 44,000 plants providing about 66 kg of *Semen colchichi* calculated as 1.5 g seed per capsule. Some examples of this calculation are given in Table 1.

Figure 2. Indicates the manner of placement of relevé frames within a selected area. The relevé frames are placed diagonally on the studied surface and on plots (14 to 20 frames).

TABLE 1
Some Typical Examples of Quantitative Estimation of Medicinal and Essential Oil Plants to Illustrate the Methodology Adopted

1. Locally	Jumla-Uthu
Plant	*Thymus serpyllum* L.
Parts used	Whole serial parts
Total length covered by the plant	6 km
Average breadth covered by plant	400 m
Total area covered by plant	2,400,000 m^2
Average density of the plant	1 plant/m^2
Total number of plants in the area	2,400,000 plants
50% of the plants being left for regeneration of the species, the total number of plants that can be harvested	1,200,000 plants
Average fresh weight of one plant	2.5 g
Moisture content in the fresh material	60%
Average dry weight of one plant	1 g
Total amount of dry aerial part suggested to harvest from the area	1,200 kg
2. Locality	Kagebada-Harigaon
Plant	*Abies spectabilis* (D. Don) Spach
Parts used	Essential oil
Average amount of fresh leaf-tops and young twigs/tree	250 kg
Average essential oil content in the fresh materials	0.5% w/v
Average amount of essential oil obtained from one tree	1.25 L
Total length covered by the trees	5 km
Average breadth covered by the trees	2 km
Total area covered by the trees	10 km^2
Average density of the tree	1 tree/1000 m^2
Total number of trees in the area	10,000 trees
Total amount of leaf-tops and young twigs available in the area	2,500 tons
Total amount of essential oils available in the area	12,500 L
By harvesting only 20% of the existing leaf-tops and young twigs, the total amount of essential oil that can be obtained from one tree	0.25 L
Total amount of essential oil that can be exploited under controlled condition from the area	2,500 L
3. Locality	Deochaur-Lamri
Plant	*Juglans regia* L.
Part used	Leaves
Total length covered by the trees	1 km
Average breadth covered by the trees	200 m
Total area covered by the trees	200,000 m^2
Average density of the tree	1 tree/500 m^2
Total number of trees in the area	400 trees
Average amount of dry leaves that can be harvested safely from one tree	10 kg
Total amount of dry leaves that can be harvested from the area of every year	4 tons
4. Locality	Jumla-Uthu
Plant	*Berberis asiatica* Roxb. ex DC
Parts used	Root-bark, stem-bark, and leaves
Total length covered by the plants	6 km
Average breadth covered by the plants	60 m

TABLE 1 (CONTINUED)
Some Typical Examples of Quantitative Estimation of Medicinal and Essential Oil Plants to Illustrate the Methodology Adopted

Total area covered by the plants	360,000 m^2
Average density of the plant	1 plant/600 m^2
Total number of plants in the area	600 plants
Average amount of dry root-bark present in one plant	100 g
Average amount of dry stem-bark present in one plant	200 g
Average amount of dry leaves present in one plant	200 g

Note: Only 50% of the existing raw materials is harvested from each plant, the dry raw materials that can be harvested from each plant is 50 g dry root-bark, 100 g dry stem-bark, and 100 g dry leaves. Thus, the total amount of raw materials that can be harvested annually from the area is 30 kg of dry root-bark, 60 kg of dry stem-bark, and 60 kg of dry leaves.

When other economical medicinal species are present in the mapped water meadow, the species will be recorded individually within every frame and the estimation will be carried out by observing the environmental protection regulations in the area.

After several thousands of relevés carried out by each team in different zones, and after several years of work in the field, research workers are in a good position to evaluate the amounts of raw material recommended to be rendered profitable for every species in the respective region. They can do this without resorting to relevé frames, the exception being the cases where the results of the evaluation give rise to doubt.

For a quantitative assessment it is important to obtain results as accurately as possible. To achieve this, one has to cover a long distance on foot. The longer the distance crossed on foot, the more frequent are the records, and therefore the estimates on the species' abundance will be close to the true amount.

In mapping the spontaneous medicinal flora, one must take frequent note of the aim of that mapping. It is carried out with the object of estimating the natural resources that can be rendered profitable within a certain geographical area, district, or country. The data obtained will be useful to the department that will draw up annual plans aiming at rendering the wild medicinal flora profitable. One should not forget that after acquiring the data, a harvesting team should be sent to verify the accuracy of the team's evaluations.

The preservation of the spontaneous flora and the possibility of recovery in an area which has been exploited depends to a great extent on how a reasonable exploitation is organized. The natural recovery of the economically exploited areas is possible through the plants left on the site as a multiplication material and by means of sowing or planting seedlings in the exploited areas.

Species protected by the law such as *Gentiana lutea*, *Arctostaphylos-uva-ursi*, *Angelica archangelica*, and others (which undoubtedly vary from one country to another) should not be mapped, but their presence, frequently, phenological data, or other observations concerning them would be recorded. From the species protected by law one could not but take some multiplication material with a view to try to cultivate some of them.

Other categories of medicinal herbs that are not protected by law but are very scarce in some regions and which could disappear should they be intensively exploited ought also to be protected. It is also unlawful to map or harvest any species from natural reservations or national parks. These reservations should be accurately delimited in economic mapping.

9. Data Recording in the Field

After carrying out the relevés and estimations with regard to the frequency and number of species on a delimited area, all data are recorded in duplicate by two members of the team. Should one record be lost, the other would enable the team to have the data, because it is quite impossible to reconstruct them from memory. Species recognized in the area are recorded and on the same line, +, ++, or +++ are noted, as the species are found repeating themselves during field observation. After having traversed the respective region, the team will meet and discuss estimates proposed as usable within the area. The ciphers within brackets following the name of the species are employed to represent these estimates of raw material. When these amounts do not represent values higher than 10 to 20 kg (other than, for example, *Flores verbasci*, *Semen colchici*, *Fructus carvi*, *Flores rhoeados*, and others from which only small amounts can be harvested), their presence has been noted by (+), (++), or (+++), as agreed.

The following are other examples of data recording:

District, commune, locality...
Althaea officinalis ++ (500) Radix
Lisimachia numularia (150): Herba
Calystegia sepium +++ (500); Herba cum radicis
Salix sp. +++ (10,000) Cortex
Equisetum arvense (100): Herba
Arctium lappa (50); Radix
Lythrun salicaria + (100): Herba
Mentha longifolia (150) Herba
Artemisia absinthium (100) Herba
Symphytum officinale + (100) Radix
Chrysanthemum vulgare (100) Herba
Potentilla reptans (++) Herba
Rheum emodi (100) Rhizoma
Taraxacum officinale + (100) Radix

Valeriana walichii (100) Radix
Glycyrrhiza glabra (100) Radix, etc.

The areas where the occurrence of medicinal herbs has been observed should be delimited as accurately as possible with such information as the name of the commune or village, the locations where the species have been identified, the name of the rivulet, hill, mountain, glade, or forest, as well as the number of kilometers to the highway or railway. Any specification or indication concerning the mapping will facilitate the harvesting of the medicinal herbs.

Besides the places and the amounts proposed to be rendered profitable, all of which are mandatory data for economic mapping of the spontaneous medicinal flora, the workers may also record other observations in their notebooks:

- Great number of species not considered to be medicinal at mapping
- Phenological data: the vegetative stage, the abundance or absence of inflorescences, the maturity stage of fruit, and the presence of a cryptogenic virus disease or pest; when only plants during their first year of life are seen — as it often happens in *Atropa belladonna* from fresh brushwoods — one must note "growing in progress", etc.

Another series of data refers to the local popular names of plants, which must be checked with at least three to four different people from the region. From an ethnobotanical point of view, these data must be recorded because otherwise, their names may be lost permanently. Any other indications on the uses of plants must be recorded whatever their medicinal virtues, as well as the mode of preparation, the dosage regimen, the mode of administration, and the empirical therapeutic indications of plants specific to a region. Information should be acquired concerning the disease(s) reputed to be cured by the respective plant(s), taking into account that the disease(s) may have a popular name. It is not surprising in such cases to discover that a heart disease is interpreted as a "stomach" ailment, etc. An example of recording such information is given in Table 2.

10. Collecting, Drawing, and Preserving Samples for Laboratory Testing

As we have already shown in the introductory part of this work, the second important objective of economic mapping is the qualitative evaluation of the medicinal herbs. In order to obtain the most reliable data on their quality, the samples should be sent and tested after having been collected at the optimum moment. As it is not possible to find all species at their "optimum moment", it is important to record the vegetative stage of the plants at the time of collection.

For the majority of species the samples will be collected from a great number of individuals of the same species, the sample representing the mean value of a given population. Whether attention is focused on one or several species, the samples will be collected from only one individual which must be marked and accurately located on-site. If there are only a few individuals of a therapeutical valuable species in the respective zone, the samples must be collected very carefully from these individuals for cultivation purposes, in case the analytical results are very important.

Below is an indicative list of species and the amounts necessary for laboratory tests in terms of dry material:

Flores Chamonillae	50 g
Herba Origani	50 g
Herba Thymi sp.	50 g
Folium Menthae (from different well-determined, separate species)	50 g
Flores Millefolii	50 g
Folium Hyosciami	40 g
Folium Daturae stramonium	30 g
Folium Belladonnae	30 g
Cortex Frangulae	20 g
Rhizoma Rhei	20 g
Rhizoma Hydrastidis	30 g
Fructus Rosae sp.	10 g
Radix Liquiritiae	50 g

Very special attention must be paid to rapid drying of the samples or, in other cases, their preservation in a fresh condition. Most samples are tested after they have been dried. For the subterranean parts, breaking up or cutting of the plant material is advisable. In order to obtain more rapid drying even the aerial parts may be broken up when the whole plant is collected. Under no circumstances should the samples be collected in plastic bags as in them plant material begins to deteriorate in a few hours. Greaseproof paper bags are the best for this purpose. The samples should be spread on sheets of paper at the base station or in the mobile laboratory as soon as possible after harvesting. Sample collecting should not be done on rainy days. Special conditions may be required in certain cases, for example, in order to assess the ascorbic acid content, fruits should be kept in the refrigerator located in the mobile unit. Every time the base station is changed, one must spread the samples to dry.

To the same subject of sample collecting also belong the preparation of herbarium voucher samples of some uncommon or less known species as well as the preservation of some parts of the herbs in formalin or in alcohol for didactic purposes. Particular attention should be given to avoid the deterioration of the plants by moisture.

11. Laboratory Test On-Site

When the mobile unit is equipped with adequate apparatus, a series of tests may be carried out on-site by medium-skilled technicians trained in this field.

Carrying out tests on the active principles directly on the site may result in important data on the conservation limit as well as on the dynamic variability of the active principles, depending on different factors.

Identification of some large groups of active substances may be performed by means of general reagents that do not need a special apparatus. Identification of alkaloids, anthracene derivatives, tannins, polyphenols, etc. on-site is optimal,

TABLE 2
Field Card (Model) for Recording Information on Traditional Medicines

Code no.	Scientific name of plant and family	Vernacular names	Plant part used	What disease used for	Empirical use	Name of persons who supplied the info.	Name of the researcher who gathered the info.	Precise place where it grows	Other remarks
1	2	3	4	5	6	7	8	9	10

Directions for completing the card

1. Each species should have a code number which is also mentioned in the herbal sketch and on the sample to be studied.
2. The scientific name should be given as complete as possible; where synonyms exist, they should be also mentioned. When a species cannot be identified in the field, only code number should be written as the botanical identification is done later in the laboratory. Mention should be made whether the species belong to the spontaneous or cultivated flora.
3. Names of the plant in the local dialect.
4. Root, stem, leaves, etc., and vegetable period when the plant is collected.
5. All available information about the disease and of the affected body organ, taking into account the vernacular name.
6. Infusion, decoction, external use, etc., and directions for use.
7. Name, surname, and exact address.
8. Very accurate description of the zone where the species grows, adding if possible a sketch of the locality.

especially when medicinal plants are studied. For this, I recommend the use of the practical manual "Methodology for Analysis of Vegetable Drugs", by Cuilei, published by UNIDO-Vienna and the Joint UNIDO Romania Centre, Bucharest.

In a special case where active substances are identified in the plant previously considered devoid of medicinal properties, the species should be authenticated according to the established botanical criteria; it should be pressed into herbarium vouchers, modified by mentioning the place where it was found, and sufficient samples for further laboratory tests should be collected.

B. LABORATORY STAGE

1. Sample Preparation in Order to Test Their Active Principles Content

The second main objective of economic mapping is to establish the active substances content of the mapped plants from a distinct area. The collected samples should be dried by natural or artificial means. Should some obscure points arise concerning the identity of a plant, the worker must ensure that the prepared voucher specimens have all the parts necessary to enable its authentication. In most cases the raw material collected from common species do not give rise to difficulties in identification. There are still some exceptions such as *Herba millofolii*, *Herba serpilli*, *Fructus cynosbati*, etc. for which some special treatises must be consulted. Since the dynamics of active substance accumulation in plants depends on the vegetative stage of each plant, this information must also be recorded. The table containing the samples collected in the field should have the following information: name of species, locality, collection date, vegetative stage at collection, moisture, active principles content, and date of the test. The qualitative evaluation of the active principles should be carried out for the various species following the methods recommended in national or international pharmacopoeas, national or international standards, or the routine method of the laboratory where it is performed.

Besides classic and common medicinal species, others may be tested if they are phylogenetically related to established species or following other intuitive criteria. It has happened indeed that species considered not to be medicinal proved to have more active principles than the classic medicinal herbs after having been tested. A preliminary screening of the cytostatic properties of some extracts should also be carried out.

C. SUMMARY

After the mapping work in the field is concluded, or even when it is ongoing, all data obtained must by recorded in final form. We recommend the use of register cards containing the collection locality and the amounts which could be rendered profitable for the systematic user of each species (see Table 3). The card comprises:

1. Scientific name of the medicinal species (i.e., *Semecarpus anacardium*).
2. Vernacular name of the plant (Marketing-Tree).
3. Part of the plant that has to be evaluated (*Radix*, *Folium*, etc.)
4. Precise statement of the locality where the plant has been evaluated to enable a rapid identification
5. Quantity of dried raw material, in kilos or tons, recommended for each species to be turned to account.

To exemplify, the following represents a typical register card:

Usneaceae

Usnea barbata (L), Hoff. (ind.*U.Florida* [L.] *U. longissima* ach.). Oasa Mica to Surianu (going up to "Schit") (1500m); Oasa Mica, on the peak (500), Oasa Mica to Slimoiu (500). Valea Dobrei, at Joagare toward Valea Hosului (5000); from Marginea Din man to Gungurezul-Derbata Mare through Gropata Lunga to the Frumoasei plateau (5000); Valea Frumoasei (10,000); between Oasa Mica and Oasa Mare (1000); from Vlaicu's Sheepfols do Slimoiu (500); the

TABLE 3
Examples of Recording Data Concerning the Medicinal Plants in A Field in Romania

POLYPODIACEAE

Dryopteris filix-mas (L.) Schott — Rhizoma = 1125 kg

Porcu Valley (50). Hârâbaru Valley (25). Susitei Vally (200). Suseni Valley (200). Sohodol Valley (50). Bilta-Arcanu forest track and branchings (150). Bistrita Valley, Bistricioara Valley (350). Pocruia Valley (100).

Pteridium aquilinum (L.) Kuhn — Radix = 22,050 kg

Jiu Valley right bank (250). Porcu Valley (100). Virjoaia Valley, Viezuroiu Valley, Hârâbaru Valley (1550). Schela-Vilcan track as far as Zânoaga Peak (10,000). Susita Valley (500). Suseni Valley, Igirosu-Fintina Plaiului forest track (500). Bilta-Arcanu forest track (5000). Bistricioara Valley (50). Rotocol Hill, Piscuri Peak, Sura Plaoului Hill (1100). Pocruia Valley (700). Motru Valley, left bank (300).

BETULACEAE

Corylus avellana (L.) — Folium = 28,150 kg.

Jiu Valley right bank, Bratcu Valley (1600). Pocruia Valley (759). Susita Valley (2500). Bilta-Arcanu forest track and branchings (750). Bistrita Valley, Bistricioara Valley (4700). Piscuri Valley, Gura Plaiului Hill (500), Tismana Valley (8000), Pocruia Valley (1500). Alun Valley, Motru Valley, left bank (4150).

Betula verrucosa Ehrh. — Folium = 19,000 kg.

Jiu Valley, right bank, Bratcu Valley (2000), Pocruia Valley as far as Zânoaga Peak (1500). Susita Valley (13,750) Sohodol Valley (1000).

Prigoanei springs (20,000); from the Runcu seed-bed to the Riul mare Valley (2000); Thal = 55.500 kg.

For species, e.g., *Crataegus monogyna Jac*., or *Prunus spionsa* L. from which only flowers and fruits are collected, or others, e.g., *Althaea officinalis* L., or *Bombax malabaricum* DC, collected for their roots, leaves, or flowers, mention must be made of the part of the plant used for purposes of the evaluation.

When some species (*Papaver rhoeas* L., *Centaurea cyanus* L., and *Delphinium consolida* L.) are frequently met within the whole mapped district or zone, an overall estimation should indicate their presence among the cereal crops of the respective locality. Taking into account the rotation of crops, it would be impossible to accurately give the names of the localities. Other species grow in small amounts all over the district. In such a case, one must mention "sporadic species all over the investigated territory". Any other details or observations the research team may consider necessary should be recorded and presented in the final work on the mapped zone.

A table containing the test results and all the above-mentioned data must be accompanied by a brief commentary at the end of the work.

In all economic mapping, the work should also generate a phytogeographical-economic map with the main localities where medicinal plants are growing and the respective amounts proposed to be rendered usable for industrial processing. The map should be reproduced directly from the worker's notebook. Figure 3 shows a section from a map where the species are recorded in squares of 5 × 5 km each and in Figure 4 in circles. The whole amount of raw material proposed to be valued may be reckoned as a function of the diameter of the circle. Within the circle there is a proportional representation of the amounts of medicinal species proposed for a definite locality. The species are displayed as symbols which represent abbreviations of the respective genus and species. In any case, the abbreviations will be put down in an explanatory legend or table. The size of the circles should be in some proportion to the mapped area. Thus, a circle 4 cm in diameter on a map on the scale of 1:100,000 represents an area of about 12.5 km^2.

Maps referring to a valley or a mountain mass, where the species and amounts have different conventional signs (ciphers, letters, etc.) may be useful as well. Whatever map is employed, the important thing is that it should provide data concerning the number and amounts of species to be utilized.

All data referring to economic mapping of the spontaneous medicinal flora within a distinct geographical zone or area will present an overview of a whole territory or country.

After having concluded the economic mapping of a country or of a large and precisely delimited geographical region, a map of the country or of the geographical zone should be drawn from each species, marking the frequency (frequent, rare, or intermediary) of a species by means of symbols. In order to draw such maps, we recommend the UTM Projection (Universal Transverse Mercator), also known under the name Gauss-Kueger. This system has many advantages, for instance, the possibility of comparing these maps to the continent or global maps.

III. PRACTICAL AND THEORETICAL ADVANTAGES OF ECONOMIC MAPPING

As we have already seen in the first part of this study, the main objective of economic mapping is to identify quantitatively and qualitatively the natural resources in medicinal plants of a certain area for assessing their economic potential in terms of profitability.

For purposes of rational use of medicinal plants in the flora, we must first understand the requirements of environmental protection. It may be achieved by:

1. Programming the exploitation of medicinal species from a locality or district so that it allows normal vegetative recovery. This is not possible without knowing all the valuable raw material for all medicinal herbs. Recommend that one not return to the same locality for 4 or 5 years to gather species collected for their subterranean parts, bark, or even the aerial parts (except the flowers, fruits, and seeds).
2. Establishing the exploitable amounts from a locality and leaving enough individual plants to allow the species to perpetuate; scheduling the exploitation and establishing the annual plan of spontaneous flora evaluation and rational utilization of a given species requires close cooperation between the production and research departments.

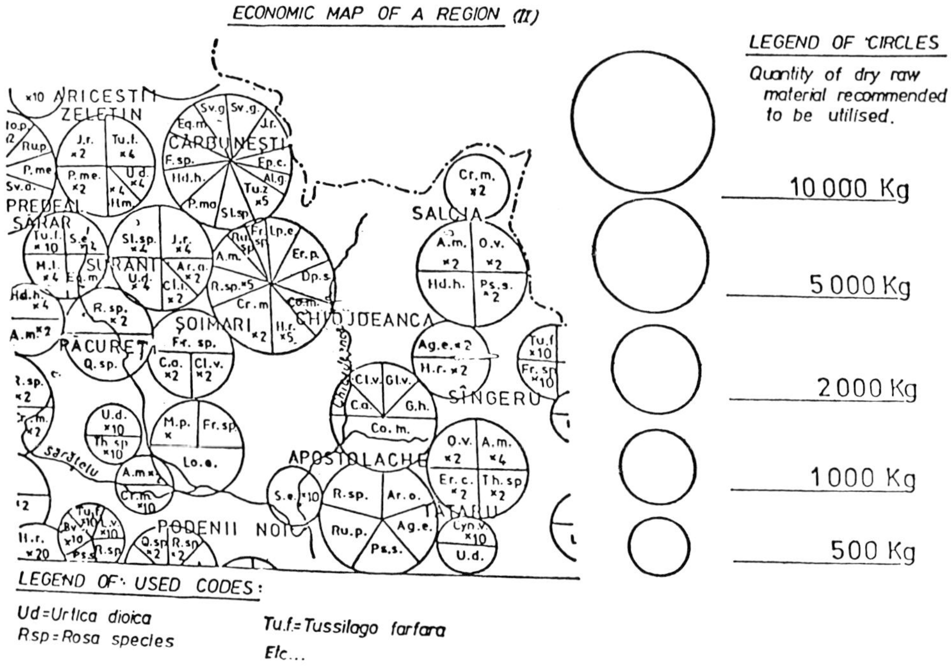

Figure 3.

3. Sowing the exploited field with the same species as those already collected from a locality is mandatory and affords another means of conserving the medicinal species.

Special attention must be paid to the soil from which some species (with a role in fixing the soil and in preserving it from erosion) have been collected. One should not neglect the role of some medicinal shrubs in aiding the multiplication of birds or some animals useful in pest control.

Rational exploitation of the spontaneous medicinal flora is possible only by observing the environmental protection laws, because for many countries the flora represents important natural wealth besides the beauty it confers to the scenery.

Mapping the spontaneous medicinal flora within a geographical zone may also provide important information allowing the location of the drying equipment, the conditioning and primary processing facilities, or some small industrial units that wholly or selectively extract some active principles. This information is very important, especially for the regions in which the approaches are unfit for walking or when the industrial centers are too distant. For some geographical zones, such small extraction units not far from the sources of raw material are very important socially and economically. Decentralized processing of fresh or dried raw material, without too long a storing stage, is highly economical and helps to reduce transport charges. Socially, besides providing labor opportunities for people in zones removed from the industrial or agricultural centers, it allows people to become skilled workers in a new branch of production.

The data used for the economic listing are also useful indicators for the future fields where medicinal plants may be cultivated, taking into account the pedoclimatic conditions required by some plants. Owing to the analytical information obtained on the spontaneous flora, valuable material for plant propagation can be procured. Local energy sources (water, hydroelectric power, local fuel, etc.) may be rendered profitable as well.

From a scientific standpoint, economic mapping of the spontaneous medicinal flora leads to a better phytogeographical understanding of the respective zone and

- To a better delimitation of zones suited for some species.
- To a better genetic understanding and conserving of some taxons or local medicinal plants which have a high active principle content intended to provide multiplication material for improving it by means of phytochemical selection; the data obtained by the laboratory tests enable the researchers to characterize some individuals (different from "chemataxons") within the same species. All these data lead to the establishment of a genetic bank of very important varieties of medicinal and aromatic plants.
- To obtain information on the edaphic factors of the studies localities and zones, several species giving details on soil moisture and pH, nitrogen content, etc.

Economic mapping of the spontaneous medicinal flora within a geographical area implies investment in specialist equipment, time, and passion for work. The mapping, besides providing new and interesting theoretical results, may help to organize better evaluation and protection of natural resources based on scientific criteria. Finally, I want to emphasize that listing and mapping of the spontaneous medicinal flora are especially aimed at protecting these species, at protecting nature. This goal can be reached only through the data supplied by these activities since the true meaning of accounting for the flora of a country means protecting it by rational exploitation of the natural wealth.

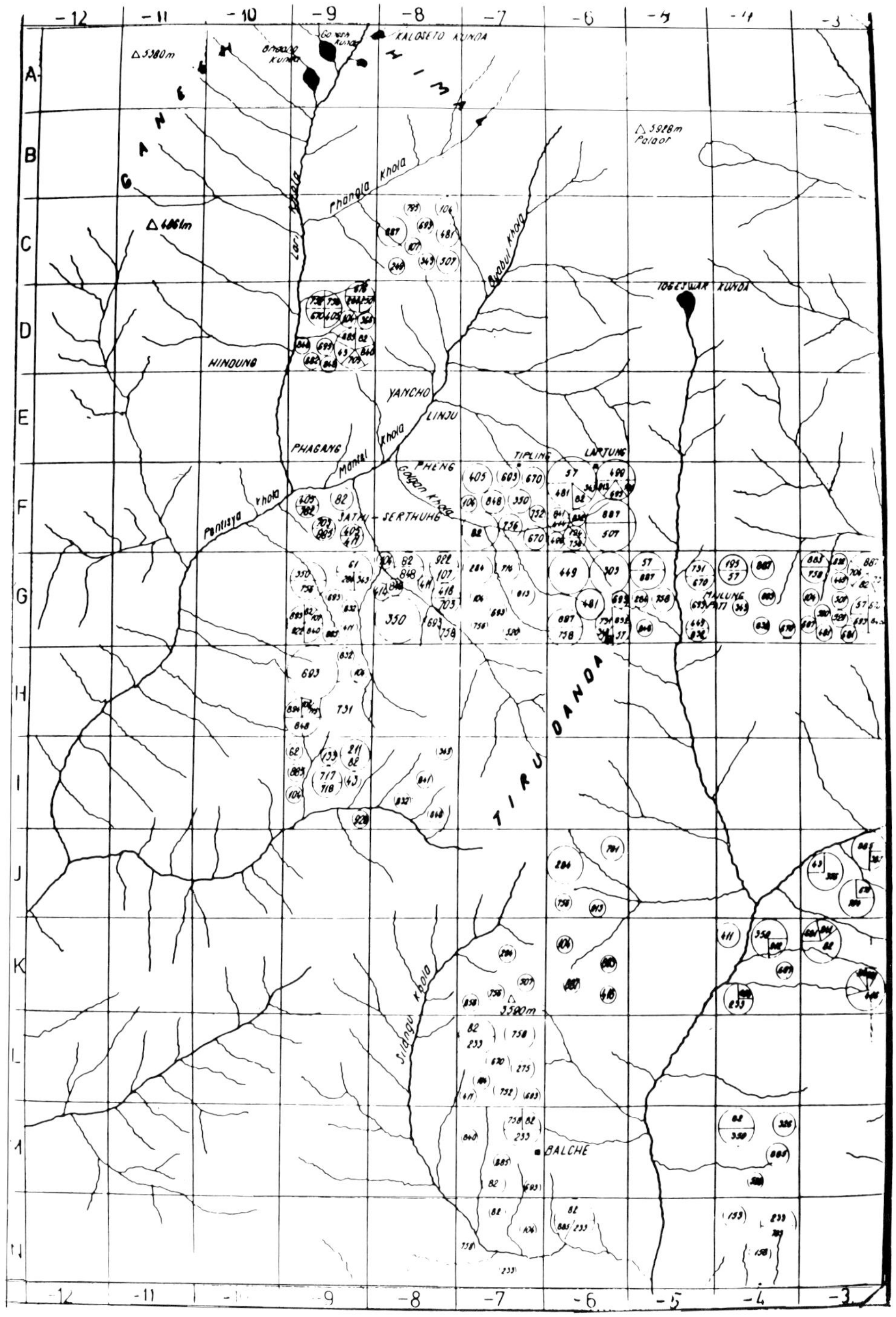
KALOSETO KUNDA
HINDUNG
YANCHO
LINJU
PHAGANG
Mantal Khola
JATH - SERTHUNG
TIPLING
TOGESWAR KUNDA
TIRU DANDA
Siangu Khola
BALCHE
Phangla Khola
Pantisya Khola

Figure 4.

Chapter 3

BIOLOGICAL PRECONDITIONS FOR CULTIVATION AND PROCESSING OF MEDICINAL PLANTS

P. Tetenyi

I. INTRODUCTION

The characteristics of biological phenomena are closely related to the chemical and physical characteristics of macromolecules occurring in living organisms. Their dynamic equilibrium may be attributed to the alterations that continuously occur in the living organisms; to the specified enzymes that control these alterations; and to the different levels of maturity, ratio, and activity of the enzymes, depending on well-defined conditions. Enzymes are not the only source of inherited characteristics, because the sum total of chemical characteristics ("chemism") is actually the full system of biochemical processes, characteristic of the particular structure. This includes DNA and RNA, which are the bearers and processors of genetic information, proteins (enzyme stock), and the universal steps of metabolism (photosynthesis, respiration) and the special steps of metabolism (biosynthesis, accumulation, decomposition of alkaloids, flavonoids, terpenes, etc.).

Owing to the large number of processes in chemism and to the variability in the interrelationships between these processes, the study of any system of chemical characteristics is still performed through the intermediate and end products. Investigation of amino acid sequences[1] can be regarded as a first step toward understanding chemism. This can also be said about the study of isozymes. Products of metabolic processes, however, can be directly observed (color, taste, odor) if these products accumulate in plants, and can be revealed indirectly by the methods of chemical analysis. Differences between these products are attributed to the dissimilarities of metabolism and chemism occurring in plants, and for this reason the term "chemosyndrome" may be accepted to indicate this phenomenon.[2] It is evident from these phenomena that chemical characteristics are as important as other features, which are more easily observable. Their inheritance and variability are governed by the same laws.

The differentiation of chemism is the general feature of living organisms as a variability determined by heredity. Chemical characteristics and their differences are especially important in the case of medicinal plants (gathered or cultivated for their active substances) because these serve as the real biological base for their propagation.

The efficiency of medicinal plant cultivation fundamentally depends on the productive ability of plant material from which the desired products are processed. Medicinal plants available for cultivation always represent only a small part of the diversity that exists worldwide. Thus, the research work being carried out with different collections of medicinal and aromatic plants are of special importance and great significance in nature conservation. This kind of work is far different from the mixed cloister garth in the Middle Ages, as its main task is to preserve the world treasure of medicinal plants and their germplasm for posterity. Among several ongoing projects, I especially cite the gene bank of 220 species in the Philippine Islands, the special germplasm bank for medicinal and aromatic plants in India, and because of its peculiar interest, Paulrott's study in Germany dealing with the medicinal plants grown in the past. It is well known that such collections as these ensure the definition of the infraspecific differences of medicinal and aromatic plants, and furthermore provide an appropriate starting point for systematic breeding work.

II. PLANT BREEDING AND PROPAGATION FOR INDUSTRIAL USE

The aim of plant breeding is generally the improvement of producing capacity, related to one or several desired constituents; in other words, a profitable yield. Sometimes intervention has the goal of improving the reliability of the yield, as for example, in the case of breeding for a more resistant or immune plant. A decrease in the yield is also sometimes permitted, if the yield can be achieved sooner, or if it is rendered in a more suitable state for processing.

Accordingly, recent efforts for plant breeding do not simply aim at greater yields, but the extent to which the active principles are technically and biologically exploitable is taken into consideration. Thus, in industrial practice, plants are judged according to the content of material able to be processed present in them. In the case of medicinal plants, the therapeutical value, in addition to the yield, is the important material. This value, however, can be determined only by biological methods, and the classical methods of chemical analyses are not used extensively. Recent developments in instrumental techniques of phytochemical analyses allows for the chemical evaluation of raw material, and the use of costly biological means only where such are deemed necessary.

The biochemical differences or the dissimilar taxawith respect to chemism, are the result of spontaneous processes in nature. The required target, i.e., a new medicinal plant variety, can be achieved only by the consideration and proper application of this law. The new plant may remain the same morphologically, but its chemical properties should be improved, and for this purpose breeders should monitor the inheritance and fixation of biochemical features.

The determination of the active principle content is a very difficult task, however, because living plants and their metabolism are to be considered and data based on a single growth stage of the plant is not often valid. Therefore, transitory differences resulting from the daily cycle of the active principles and from the dissimilarities of ontogeny should be excluded. The environmental effect should be balanced, devoid of any extraneous influences that would lead to false conclusions.

In the evaluation of medicinal plant varieties, the primary parameter is the quality of the active material, and in this respect, factors such as resistance, uniform ripening, freedom from seed spreading and from thorns and the possibility of mechanized harvesting are secondary parameters. The same alkaloid or essential oil yield can be reached either by a large amount of plant material containing a small amount of active principles or by a smaller plant yield which contains a high percentage of the required ingredients. The industrial viewpoint is such that the achievement of the latter is required, because the specific parameters for processing are enhanced in case of a concentration of active principles. The total content of active principles is frequently not so important as that of its individual constituents. The ergot containing ergocristine has a lower amount of total alkaloids and of auxiliary alkaloids than the ergot containing ergotamine. Despite this fact, the rye areas infected with ergot containing ergocristine are continuously increasing in Hungary because the alkaloids of this ergot serve special pharmacological purposes. This is also the situation in the case of lavender: the French varieties, having less foliage yield and essential oil content, are cultivated in preference to the high-yielding varieties, because their greater ester content, expressed as linalyl-acetate, better suits the requirements of industry.

The active principles content, the main pharmaceutical constituent, and the practically important chemical properties should thus be considered most of the time when medicinal plants are bred. The chemical differences, however, can only be observed by sensitive serial analyses.

The efforts of the medicinal plant breeder are aimed at increasing the final yield of the special active principles and enhancing the metabolic functions that result in their accumulation. Morphological differences, which may aid in the breeding of other cultivated plants, can be insignificant or misleading in the case of medicinal plants. Poppy breeders, for instance, wanted to establish an ovate capsule with a large number of valves in order to establish a poppy variety of high morphine content, but did not achieve the desired result. The breeders of *Digitalis* assumed a positive correlation between a high glycoside content and thick wide leaves, but did not obtain the expected results. Morphological features are sometimes determinants, especially in the case of raw materials for direct consumption such as spice. The large-fruited coriander is more popular than the high essential oil-bearing varieties, which have small fruits. Among the large-fruited coriander varieties, however, those with higher essential oil content are preferred. Morphological features are occasionally of significant help to the breeder; among essential oil-bearing plants, such as lavender, the selection by the number and size of follicles has led to considerable success. Where the morphological features are not correlated with biochemical characteristics, these should be neglected; some lavender follicles have been observed to be empty or only half full.

The morphological characteristics of a new variety, having a more valuable active principle, can be ascertained in relation to the initial plant. Thus, the breeder cannot be relieved of the obligation to describe these differences. In the examples cited here, the variation in the active principle was initially studied quantitatively. In the angelica rhizome, in belladonna leaf, and in rhubarb root, differences have been observed. In angelica, the changing parameter was the level of essential oil content, while the alkaloid content was different in belladonna. It was the anthraquinone level in the rhubarb that determined a difference in quality; it was further observed that in the meristem of the bud only rheumemodin accumulates, while in the root, emodin, chrysophanol, and aloeemodin accumulate in addition to rheumemodin. In the leaf of the Kangaroo apple, quantitative differences were observed (Figure 1); the tip and upper part of the leaf blade contained more alkaloid glycoside than the middle, underside, and veins of the leaf blade. The variations of anthraquinone glycoside in the internode of the peduncle and leaves of rhubarb was called "morphogenetical variability" by Schratz,[3] and he considered these variations a general occurrence.

One of the difficulties in breeding medicinal plants arises when the active principles are not exactly known, and thus, the aim of the breeding cannot be defined adequately (for instance, in the case of chamomile or catnip). Another situation may also occur where the content of the active principle and the biological effect itself are not proportional; the root of rhubarb also contains anthraquinones, but their amount is not correlated with the effect.

Although the cultivation of medicinal plants is several thousand years old, certain species have not been systematically bred at all. For this reason most of the species are not classified into varieties according to their active principles; only populations exist. In the last 25 years, research has led to the establishment of almost 50 new varieties with higher content of active constituents, but the statement of Schratz in 1960 is still justified — the breeding of medicinal plants is still far behind the breeding of other plants.

Despite the significant development of analytical techniques, analyses of reproducible reliability were generated only recently. These are generally rather expensive, and require sophisticated equipment. Thus, the selective breeding of medicinal and aromatic plants is relatively costly. To avoid this inconvenience, most breeding stations have developed and adopted the use of tissue cultures, a technique given consideration later in the chapter. There are also high expectations for the improvement of radioimmunoassay (RIA) methods as they afford better prospects for the determination of constituents in a greater number of individuals and for repetitive assays of the active substance either in *Digitalis* or in other similar plant species.

The variability of medicinal plants is more extensive than that of other species. Populations found in spontaneous growing locations are not at all balanced in terms of chemical characteristics and active substances. At the first level of breeding, the mixture of chosen, infraspecific groups is already different from that of the spontaneous populations of the species. They are bred in this way to some extent, although not purposefully. The commercial propagating material taken is

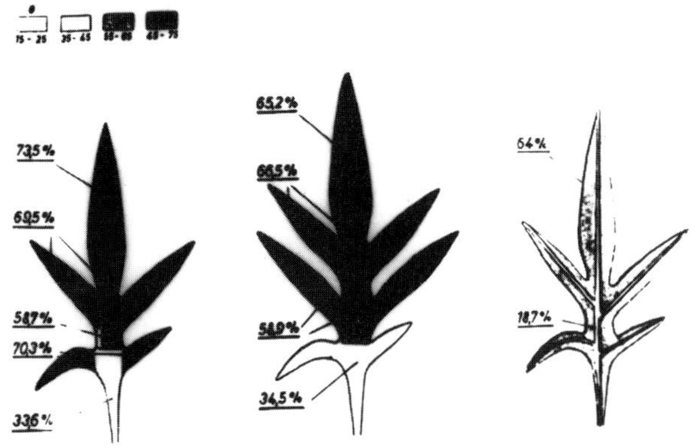

Figure 1. Quantitative differences in the solasodine content of leaves at *Solanum laciniatum* (100% is equivalent of 3% glycoalkaloid in the drug).

selected randomly, and it is almost impossible to guarantee the inheritance of compositional properties. For example, the peppermint retained for 4 years in the same plot contained varying amounts of essential oil between 1.45 and 3.19%, depending on the weather.

The next level of breeding is represented by landraces. These are also the mixtures of infraspecific groups, but the regional feature appears in chemical characteristics. In this case, the variability is somewhat restricted and the inheritance of chemical characteristics can be expected. A number of poppy varieties of this level are cultivated in Hungary, including the *Soproni* landrace, which is rich in codeine.

If the extent of variability is restricted by multiple self-pollinations, then an autogamous variety can be bred. The essential oil content of the French lavender variety bred in this way in Hungary is reproducible. The variability is minimal in the case of clones, as in English lavender. The essential oil content of peppermint clones changed to a lesser extent (2.06 to 2.67%) during 4 years of study than that of the mixture of infraspecific groups.

The breeding of medicinal plants should always begin from populations. According to population genetics, the goal is not the panmictic propagation of every individual to as great an extent as possible, rather the ability of propagation shall be controlled so that one of the genotypes is favored, and the selection pressure is exerted in this way. This positive selection is successful, mainly for the regional varieties. The selection of individuals is successful only if it is made according to chemical data. During the selection of an individual, care should be taken to prevent migration by the action of foreign pollen. The selection of an individual can be made more efficient by inserting drift. For this purpose, the population is divided into parts, thus increasing the number of homozygotes, and the material will be more uniform. The breeding period is rather long, as relatively small amounts of individuals can be analyzed in each case. Generally the compromise of proximate analysis can be chosen, in which groups of 20 are formed according to the active substance content, and the procedure can also be hastened by preselection.

In the case of medicinal plants, which have relatively recently been introduced into cultivation, it is generally regarded that the maintenance of a large mixture of individuals is necessary. The three levels of the selective, family-centered breeding system are based on the maintenance of genetic differences, while performing a restricted negative selection. The three levels ensure the continuity of the breeding process. The involvement of new material, on the other hand, retains the fitness of the plant. In this case the groups of 10 are the most appropriate for the selection of medicinal plants.

In breeding using the method of selection, it is very important to retain fitness, especially of the medicinal plants. The uniformity can be improved by inbreeding, but the deterioration of cross-pollinated plants increases while vitality decreases. Strict inbreeding is, for instance, harmful for *Digitalis;* the second generation plants are more sensitive to diseases, and will not produce seed if inbreeding continues. The deterioration caused by inbreeding as well as the heterosis effect of subsequent crossing on the total glycoside content have been proven many times.

Clone selection is regarded by many as a procedure that predetermines the uniformity of descendant plants. It was shown, however, that active substances of *Lavandula* and *Geranium* clones (the ratio of essential oil and of linalylacetate) are dissimilar for each individual plant. Clone selection also is not an absolute method. French researchers have grafted clone individuals on a *Citrus* clone base; in spite of this, the active principles content of the individuals and the body of the original mother plants were dissimilar, and the organs are different in stages or in heredity. This is confirmed by chimeric and bud mutation phenomena, which are induced by environmental effect, mostly after cross-breeding, and appear as vegetative splitting. This fact is worthy of attention during medicinal plant breeding.

Up to this time more efforts have been made to "loosen" the heredity than to retain the vitality in the breeding procedures. A promising method was remote cross-breeding. In this process the different content of active principles present in the parent plants might appear simultaneously in the descendants. Also, the chemical characteristics of one parent may suppress those of the other. A later generation of descendants can differ in characteristics of active substances, as was observed in the case of the hybrids of poppy with *Papaver pseudo-orientale.* The advantage of remote cross-breeding may be heterosis (as with English Lavender), but sterility is a disadvantage in propagation. This can, however, be overcome by vegetative propagation and polyploidization. A new method of producing ploid varieties is the valence cross-breeding. Materials having different degrees of ploidity are cross-bred, which results in plants having advantageous amounts of desired active substances. The number of polyploid plant varieties is small, but the Hungarian polyploid chamomile has a larger inflorescence and has advantageous characteristics in terms of active constituents.

A popular method of plant improvement is irradiation. Its direct effect is accompanied by harmful biological consequences, but the larger number of variations facilitates the breeding. Irradiation may also cause a heterosis effect. The treatment is successful only if the "loosening" of heredity is followed by continuous individual selection. In this way, a new variety can be produced, as, for example, the breeding of new Hungarian varieties of ergot, nightshade, and poppy.

Breeding by hybridization seems to be a very long-term method. Sobti[4] in India has produced hybrids of *Ocimum canum* x *O. basilicum* with a camphor content of 47.7% in the essential oil. A difference has been reported in the linalool and methylchavicol content from that in the parents, but the synthetic allohexaploid is far from being stable. Mechler and Haun[5] reported on trisomic mutants of *Datura stramonium* with an alkaloid content 155 to 227% higher than that of the standard line. Kashikar et al.[6] found in mutants of *Capsicum annum* in the third to fourth generations following ethylmethanesulfonate (EMS) treatment, the alkaloid content increased by more than 20 to 60% in comparison with the initial material.

Clone selection has resulted in the greatest number of new varieties. Hussain[7] has selected new clones of *Dioscorea* with 6% and *Costus* with 3% diosgenin content. Topalov[8] has produced mint clones with a 110 kg/ha yield (0.6 to 0.8% volatile oil content), with 46% carvone in the cultivar Mehta, and 90% linalool in the cultivar Lina. New *Nepeta transcaucasica* clones in Bulgaria and new clones representing six chemotypes of *Thymus vulgaris* have been bred in France.[9] Unfortunately, in the latter case, the clone with thymol, linalool, and geraniol content must be cultivated together because the commercial interests demand the mixed rather than the homogeneous compounds.

Some other medicinal plant breeding results also deserve mention: Astadjov's[10] new cultivar, Snejinka, of *Leocojum aestivum* as compared with the spontaneous population can ensure a twofold amount of galantamine (0.4 to 0.6%) and thereby a 13 to 21% increase in drug production in Bulgaria. Indian plant breeders[7] reported a new *Hyoscyamus muticus* cultivar with 4% and an *Atropa* cultivar with 0.8% alkaloid content.

To classify the newly bred plant material and to ensure its survey always requires meticulous work, but the results facilitate further progress for others. Wagner[11] (Yugoslavia) achieved, through several years of study, the chemotaxonomical system of hops cultivars, based on the α-acids, the proportion of α- and β-acids, and cohumulone content.

When the breeder stabilizes the chemical and morphological characteristics of the new cultivar, he must request official registration. States monitor different kinds of institutions where comparative trials of a new medicinal plant variety can take place over several years. If testing is positive, state agricultural offices accept the cultivar and issue the registration certificate. Breeders are also obliged to maintain the varieties by continuous selection after cultivar approval. This work is also controlled and supervised by the testing institutes, and only the identical quality propagating materials are allowed to be cultivated. If germination capacity, purity, health, identity, vitality, or uniformity of seed material cannot be proven, then the authorities exclude the lot; if it deviates repeatedly, the variety is also excluded. Therefore, it is of primary interest for breeders and institutions to maintain their varieties. The maintenance of seed and other propagative material for the production of medicinal plant varieties is a delicate procedure, best realized in specially equipped farms, where the personnel are able to meet the responsibility for future cultivation practice.

III. TISSUE CULTURES

The use of cell and tissue cultures in plant breeding and genetics is a relatively new technique whose scientific basis appears to be in a preliminary stage; however, use of the technique cannot be delayed for complete clarification. This section does not deal with its use for biosynthesis or production of active principles, but its adaptation for breeding medicinal plants and aromatic plants.

This problem can be approached by theoretical as well as by practical means. One may first pose the question: why use cell and tissue cultures? The answer: to obtain a higher degree of homogeneity in plant material or conversely to raise heterogeneity heredity, and to initiate mutagenesis.

Achieving more homogenous plant material has long been practical. We saw in Section II how different levels could be reached in this respect from landraces, until clones are obtained that force the continuous diminution of natural variation. Cell cultures provide the most potent possibility at hand, because they can produce an unnatural state of chromosome material — the haploid stage. Normal plants have a double chromosomal helix and this double status is, of course, mostly heterogeneous. This natural structure can be reduced to two homogenous unical helixes with two different heredities. Microsporogenesis, the production of pollen and pollentube with internal haploid spores, is widely known to have occurred daily in billions of cases since the evolution of seed plants. The use of these natural phenomena for plant breeding arose only from cell cultures. This method assures the healthy development of haploid plants until flowering. Still, since seed progenies from haploid plants are to be derived, chemical agents for diploidization must be employed.

The tissue culture method has proven successful; many plant species produced haploid plants. Among medicinal plants, about 20 results were published, most in the Solanaceous species. A success story is that of Abou-Mandour et al.[12] in their work with *Catharanthus roseus* (Apocynaceae). They regenerated plants from haploid and diploid calluses, propagated them; most interestingly, the plants of the progeny contained a higher content of active principles than the parent from which they were derived.

Another aim in breeding is to have more genetically heterogenous plant material, because later selection is facilitated. Positive or negative selection, en masse or individually, is good, but time intensive and laborious, until the creation of new breed varieties with better qualities. To hasten and increase the probability of this type of breeding, two new methods in the field of tissue and cell cultures have been developed: (1) protoplast fusion and (2) genetic engineering. We can differentiate between these as follows. In the first case, we use whole protoplasts, enforcing their fusion; in the second, only a portion of a foreign protoplast/nucleus, chloroplasts, mitochondria, nucleic acids, plasmids, etc., is inserted or exchanged in the experimented cell. Consequently, we have two different expressions: "somatical hybridization" and "gene-operation". Both techniques are very adaptable for broadening the genetic material, which results in a higher variability. This leads to better conditions in selection work, so that breeding will be positively influenced, made more effective and quicker.

Bajaj et al.[13] produced protoplasts from the mesophylla of leaves and from the pollensacs of *Atropa belladonna* and *Nicotiana tabacum.* These protoplasts were fused and afterward developed into plants. The authors could determine that these plants were different in chromosomal behavior, being haploids, triploids, and also normal diploids. In morphological and chemical characteristics the authors observed greater variability in the plants, showing that they had obtained hybrids between two genera.

The third theoretical purpose of tissue cultures can be the initiation of mutagenesis. Mutagenetical methods have been known for decades, and have been used in medicinal plant breeding. Tissue cultures are more sensitive when affected by irradiation or chemical agents. Their utilization seems therefore more promising than the methods of dealing with entire plants. Indeed, research has confirmed this and breeders use this combination of methods successfully. Induced mutants were most effectively and frequently found in the case of haploid cell and protoplast cultures because the recessive ones could be detected easily and quickly.

The use of this method for medicinal plants is well documented, but the most interesting was published by Yoshikawa and Furuya.[14] They produced callus culture from germinating plants hypocotils of *Papaver somniferum* (Papaveraceae). They could regenerate whole plants, and cultivate them in soil after treatment with a special combination of hormones. Plants were flowering and gave fertile seeds. Interesting physiological phenomena were detected in capsules, containing no morphine alkaloids as in the mother plant, but instead the more industrially valuable codeine.

Cell and tissue cultures can be used with high efficiency to realise practical purposes in breeding. Embryocultures, vegetative propagation, assuring freedom from virus and/or mycological infections, as well as cryopreservation, are the different fields where these new methods have been introduced.

Embryocultures can help to combat incompatibility or autoincompatibility and elimination of valuable hybrids when successful hybridization was followed by physiological disturbances. As this technique avoids the rest periods of seeds, the number of progenies can be multiplied. In the field of medicinal plants, Furmanova et al.[15] arrived at a very good regeneration by this method in the case of *Carum carvi* (Apiaceae).

There are many examples of multiplying positive factors of propagation in cases where tissue cultures were used. This technique is most appropriate to cloning the bred individuals and to have as homogenous progeny as is practically necessary. For example, in the case of *Urginea indica* (Liliaceae), Jha and Sen[16] produced, from one bulb, 400 little bulbs by tissue culture during 120 days and 90% of these produced normal plants when developed in soil. Tissue culturing can be effective against infections caused by viruses or by fungi. With the help of growing point meristems breeders can avoid these inconveniences and assure the freedom of their plant material. Shah and Dalal[17] produced cultures from buds of *Glycyrrhiza glabra* (Papilionaceae) and cloned these quickly. They developed plants that were resistant against pathological infections.

There is a very interesting utilization of the technique of tissue culture, viz., that of cryopreservation. This method will find use in gene banks or for any other preservation purposes for specially bred material or strains. For example Dietrich et al.[18] preserved tissue cultures of *Digitalis lanata* (Scrophulariaceae) in liquid nitrogen for 3 years. The authors tested the material afterward and found that it could regenerate and develop into whole plants. Neither viability nor cardenolid production were affected, and plants realised normal life cycles.

To summarize the theories and practical uses of cell and tissue culture methods, we can judge these to be a new development in breeding work. They offer vast perspectives and quicker results in the field of medicinal and aromatic plant breeding.

IV. ECOLOGY FOR CULTIVATION

Ecology in this respect means the rationalization of medicinal plant cultivation. For the sake of commercial viability, medicinal plant production must be located in a region where the yields are usually high and the geoclimatic conditions over several years are deemed conducive and favourable.

How and in what manner may these ecologically optimal conditions be established? Two ways are open for medicinal plant cultivation. First is the empirical, because fields where they were cultivated with satisfying results over a certain period of time (at different places, depending on cultivars) might be favourably considered. Observations of the spontaneous flora, the regional conditions, and microclimate, taking into consideration each taxa within the flora, are very important.

A complex study of the region passes well beyond this empirical stage or level. To successfully introduce a species of medicinal plant for practical large-scale production, the most appropriate region must be decided upon by studying the nature of the soil, conditions of light, and other essential factors. For example, Dános and co-workers[19] studied the poppy cultivar, SB, in Hungary. It was established that capsule yield (as the most important from the standpoint of the pharmaceutical industry) is high in the eastern parts of the country. For purposes of morphine production, the southeastern parts of the country, along the mid-region of the River Tisza, as well as in certain regions of the Great Hungarian Plain were most favourable. On the other hand, the morphine production of the plain in the northwestern Hungary and the northeastern country districts is unfavourable, and is even worse in the Transdanubian districts. With respect to cultivating possibilities and locating the plantation, special attention to daylight *exposure* is most advisable. There is much literature dealing with the problem of reactions to increased height above sea level by different plant species and this must be taken into consideration for medicinal plants. The material dealt with by the conference in Wageningen[20] (1957) gave many examples as to plants containing essential oils, alkaloids, and glycosides.

In the case of *Thymus serrulatus* it was observed by Rovesti[21] that the linalool content of the volatile oil might attain even 20%, depending on the appropriate humidity of the soil; on the other hand, in case of dry conditions, it may not total more than 3 to 5%.

The introduction of spontaneous medicinal plant species into cultivation can bring considerable change. Successful experiences in this field from recent years are the results obtained on *Anthemis ruthenica* and *Salvia* sp. in Servia, the introduction of *Origanum vulgare* in Israel, and finally that of *Solanum khasianum* in the Philippines.

Experimental ecological studies are needed to establish the optimal conditions for medicinal plant propagation. Neubert[22] found that the optimal habitat of *Arnica montana* is at a height of 1800 m above sea level, as the cultivation of this species produced the highest quantity of total flavonoid in such a territory. Chatterjee[23] established the optimum conditions of interaction between the soil, rainfall, and height above sea level in the case of *Dioscorea* and *Cinchona*. Hussain[7] described the introduction of *Duboisia* and the adaptation of peppermint to the subtropical climate in India. The species *Heterotheca imloides*, native to the high mountains of Mexico, has now been successfully acclimatized to mid-Europe.

There is also a correlation between the water supply of the soil and the xerophylous effect, thus relating strong reactions to the plant to changes of climate. A discussion on the effects of water stress on plants with essential oil content was initiated recently by Box,[24] who opined that the optimal yield in the case of marjoram and pelargonium is produced only by the effect of the above-mentioned stress, but it is considered necessary to prove this hypothesis by further experimental studies.

With regard to the optimum ecological conditions, the nature of the radiant energy source is most important. Since reactions to *temperature* are well known, it may be sufficient to note that optimal cultivations of warmth-demanding plants could evolve only in regions where these demands could be met. Therefore, some kinds of plants are suited only to specific temperature and climatic needs (for instance, coffee, tea, and plants supplying subtropical essential oils, etc.) and these can be cultivated only in territories and places where the relevant temperature conditions prevail. There are vast countries (e.g., the U.S. and the U.S.S.R.) that are able to select cultivating conditions in case of numerous floral "dominions", i.e., to choose the productive habitat according to the optima of the plant in questions. Thus, Mentha plantations are to be found in the southeastern regions of the U.S., or production of *Solanum laciniatum* takes place in the Kazakh Soviet Socialist Republic; in both of these cases it is obvious that the temperature optimum for the plant in question is satisfied in these regions. In the same manner, the conditions in southern France were unfavorable for poppy cultivation, so production has been introduced in the northern regions with much cooler summers.

The influence of light is much less known compared to temperature. True enough, there is a long-lived theory concerning alkaloid formation, according to which this must be related to the intensity of light (and this theory declares also that UV rays stimulate alkaloid formation); there is, however, very scanty experimental evidence to support it.

Máthé's[25] ecological data as to yield in alkaloids of *Vinca minor* should be noted, i.e., allowance should be made concerning enrichment in active principles in the case of plant material brought out of the forest and planted for tillage in ploughlands.

Another well-known example is the alkaloid level of poppy in Europe which, owing to capsule ripening, increases as the cultivation advances in the eastern direction, where sunshine is of higher frequency and intensity.

The evident influence of light, however, could not be first proven other than by phytotronic experiments. Cosson et al.[26] proved that *Datura innoxia* does not form scopolamine (besides hyoscyamine) except when the light intensity surpasses 100,000 lux.

Our phytochamber experiments in Hungary made it evident that in the case of *Solanum laciniatum*, the amount of steroid-glycoalkaloids diminished to a fourth part when light intensity was low. However, when illumination comes nearer to the blue, alkaloid production increases. This observation is in agreement with results by Kuznietsova and Khazanov;[27] according to them, *Solanum laciniatum*, *Stephania glabra*, and *Rauvolfia canescens* are stimulated to form active principles by radiations in the shorter wave range.

In our studies it was observed that the qualitative composition of the aglycone content in *Solanum dulcamara* changed, depending upon the wavelength range; while with the plants raised in composite (white) and violet light, 35 to 40% of the aglycone was saturated. The saturated part diminished to around 20% in those raised under green or blue light.

Our latest results in experimental ecology proved the correlation between light-temperature effects and alkaloid production on poppy grown in phytochambers (Figure 2). These conclusions were also proved by long-term field experiments, comparing the circumstances in autumn and spring sowing with spring, autumn-spring, and autumn type poppy cultivars.

V. MECHANIZATION AS APPLICABLE TO SELECTED PLANTS

Medicinal plant cultivation has now come into a new stage of development in many parts of the world. Earlier, in the overwhelming majority of cases, these plants were partly collected or produced in household plots; during the last few years it became necessary in more and more countries to produce medicinal plants and spices in up-to-date large farms. This development became necessary because of the changed economic conditions, competitive challenges, as well as in order to achieve greater commercial viability of the industry.

It was conceivable earlier to carry out large-scale production and harvesting by making use of the "extensive model"; later, it was proved that intensification gave better prospects. There were some developing countries which for justifiable reasons wanted to intensify labor, i.e., using larger numbers of

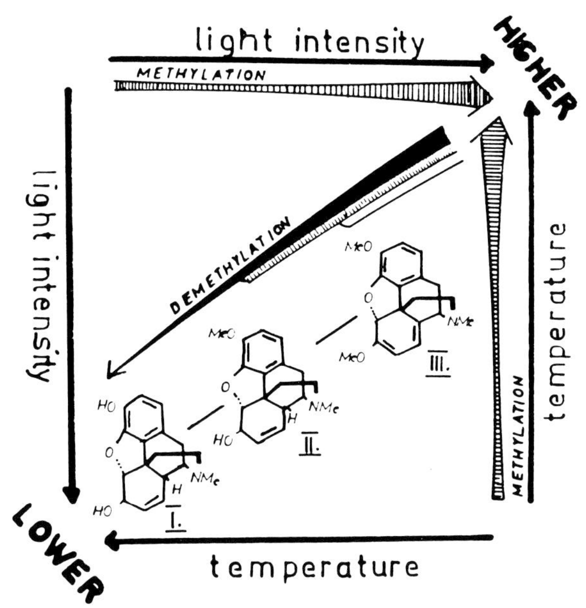

Figure 2. Influence of ecological factors on alkaloid biosynthesis in *Papaver somniferum*.

people for manual labor; on the other hand, this procedure in certain circumstances has limited utility.

To reduce the employment of superfluous manual labor, some materials and fixed assets are absolutely necessary; so too is the use of mechanization and chemical fertilizing and protecting agents. Chemical fertilizers as well as the so-called "microfertilizers" have already been an object of study, pursued over several decades. On the other hand, herbicides, regulators, and plant-protecting materials were introduced only during more recent years. As to reactions of these chemicals with the active principles of medicinal plants, knowledge is still scarce; i.e., it is possible that their interactions raise difficulties against their utility as therapeutic agents. On the other hand, the mechanization of medicinal plant cultivation does not adversely affect the therapeutic functions, but the output may be less. In order to eliminate this possible loss in output, it would be necessary to bring about suitable biological conditions, taking into consideration consequences of mechanization.

In modern large-scale farming one always expects large risks; reduction of these hazards as far as possible should be the goal.

The cultivation of plants by mechanical means should be initiated by laying a foundation upon ecological as well as biological optima. To form plantations of medicinal plants (in order to grow them in large-scale farming) for factory-like production, they need an ecological optimum to comply with their demands for light (concerning light intensity and the most suitable spectrum); in other words, a mechanized production requires a higher number of individuals. Satisfactory yields cannot be attained except by increasing the number of individuals within each of the rows. (A mechanized system of tillage can only be brought about by rows, and distances between each stock or stem within the row must be reduced.) On the other hand, this solution leads to worse light conditions, which has a directly negative bearing upon yield. This is the reason that the closely "packed" stems (crowding each other) become unfavorable, and mechanization cannot bring satisfying results.

Another biological prerequisite of mechanization for medicinal plant production is the suitable variety to be cultivated. There are certainly losses caused by mechanization; to offset these by compensation, there remains only the possibility of making use of a plant with higher productivity. Whether some cultivars are able to withstand all demands as to mechanization, there are some questions to be answered first:

1. Is it suitable for cultivation?
2. Will harvesting of the crop by easy enough?
3. Can it be processed (distillation, etc.) without difficulties?

All these questions are equally important concerning suitability and must draw positive answers.

A part of mechanized cultivation of medicinal plants is more or less easy to accomplish with the machinery generally used in agricultural large-scale farming. Tractors and machine tools are able to carry out preparation of the soil, making ready seed-beds, effecting plant protection, applying fertilizers, etc.; to some extent they may also carry out sowing and planting.

Sowing presents certain problems, i.e., up-to-date, large-scale farms require a precision, with manpower-management using the utmost minimum in labor. In order to comply with such demands, the general agricultural machines for plant cultivation must be adapted for the specialized purpose required by the nature of the crop. For medicinal herbs propagated by stolons, machines must be expediently transformed. These machines for crop care, harvesting, and chemical plant protection for rowcrop cultivation must be available so that the tasks can be carried out easily and efficaciously.

There are several kinds of machines for plant protection (they can also be employed for other special purposes), including land machines, aircraft, and helicopters. All may comply with the requirements of the large-scale farming establishments to produce medicinal plants. Equipment for application of both bio- and inorganic fertilizer is available that can be used without any changes or adaptation.

After harvesting, most medicinal plant-raw materials carry a humidity content of 60 to 80%; but because of this, they cannot be stored without drying. According to a general rule neither seeds nor drugs must contain more than 10 to 14% humidity. During the operation of artificial drying the plant material must not be damaged or suffer losses that would prevent it from conforming to pharmacopoeal or other accepted standards. The majority of medicinal plants require relatively mild conditions for drying (within the temperature range of 40° to 60°C at most). Recently, Hungarian large-scale farms have employed several types of driers, i.e., working with hot air, conveyor belts, pushing trays, etc., which may also be used for other purposes in the same way.

The primary processing of medicinal plants requires solvent or water extraction of active agents and/or distillation. In the case of aromatic plants, for instance, modern equipment for distillation ensures preservation of the quality of the essential oil, the extraction of which is the entire purpose of cultivation and production. Distillation of the volatile essential oil can be accomplished by batchwise distillation or by a continuous one. In the most advanced distillation methods developed in the U.S., harvesting and distilling in containers is carried out.

No variety of plant yet exists with suitable biological characteristics complying with all the requirements as enumerated above, for mechanization, and so compromises must be made. In addition to development of mechanization technologies producing cultivars with higher productivity in the biological sense would be the aim of the plant breeder. This means the content in active principles should be higher, or the composition more favorable, or the organ which furnishes a higher yield of active constituent(s) should be produced in a higher proportion. In case the variety carries in itself the propensity for greater productivity, the grower will be better able to realise the best conditions by employing suitable cultivating technologies, soil strengthening, plant protection, etc.

Harvesting the medicinal plant differs primarily from any other mechanized procedure. To harvest the yield, especially in particular cases, special methods, equipment, or machinery may be needed.

The following may be cited as examples for the attainment of biological optima, varieties, and conditions for producing plants suitable for mechanized cultivation.

First, the cultivation of ergot is a complex task. The rye plant as well as its contagious fungus must be selected in a manner so as to have (as far as possible) a most advantageous combination from the standpoint of maximum yield in active ingredient. Therefore, in addition to the propensity for high yield in the rye variety, its susceptibility must be taken into consideration. In the same manner, an important factor is the slow ripening of the rye cultivar. It is well known that mechanized harvesting must be carried out several times; this can be attained if rye does not suddenly ripen but rather during a longer time.

On the other hand, ergot must have a suitable virulence. If this fails or the strain degenerates, high yields cannot be expected. Two kinds of rye having the same susceptibility give different yields because of the different ergot strains that have been injected; they cause on the ear or the plant individual different numbers of sclerotia. A characteristic of the strain is also the capacity of after-development. The big ergot sclerotia is collected first; it is necessary that the sclerotia (smaller ones that remained in the rye plant) develop further on the living rye, making possible a later harvest.

Complying in a complex manner with these demands, a biological optimum may be ensured for mechanization. It is by far not accidental that the polyploidic rye varieties with big seeds (and slower development) proved good. They have the advantage that they do not grow in length as the diploidic rye varieties. Thus, they are more easily accessible for the mechanized infections as well as for the propitious moment of harvest.

When wheat varieties are to be improved, the grower strives to produce a kind of wheat with a minimum amount of straw and a yield as high as possible in ears (or, respectively, in seeds). In the breeding rye the researcher strives to have rye stems half as high as normal, as a special variety for ergot production and adapted to mechanized harvest.

As a second example, the harvest of poppy capsules may be mentioned. An earlier requirement (being fully independent of mechanization) is that the poppy capsules should be perfectly closed. This is increasingly true for mechanized harvest; the machine promotes breakup of the capsules as well as the straw part. For harvesting poppy capsules in large-scale farming, this is still not the single biological prerequisite, but the individuals of the poppy cultivar in question should produce a few capsules only. Otherwise, ripening cannot be simultaneous, and the same height of the capsule above-ground necessary for mechanical harvesting cannot be achieved. Therefore, as Böhm[28] pointed out, plant improvement must aim for individuals with two to three capsules as a maximum. This effect can be realized by a compact poppy stand, and with this solution side inflorescence do not grow. Biologically, the breeding of a special poppy variety with few capsules answers the above requirements. However, the "real yield" must not decrease, the lower number of capsules must be compensated by bigger poppy-heads, and in them the poppyseed weight should be higher. Researchers of DDR have already described some such varieties; in all likelihood we could produce suitable poppy cultivars for mechanized large-scale farming at the Institute of Hungary.

As a third example, a biological approach to mechanical cultivation of *Chrysanthemum cinerariaefolium* in large-scale farming should be mentioned. The plant breeder has to fulfill a double request in the case of this plant. First, the plant habitus must be homogenous. For example, there are individuals to be found with fully prostrate inflorescences; on the other hand, there are also more forcible individuals growing upward. Selecting the latter as the base material may be biologically more suitable; the upstanding inflorescences are easier to harvest. There is also another requirement, i.e., most of all inflorescences must stay in the same level and bloom simultaneously. A failure of this basic characteristic in a variety will result in the harvest of individual inflorescences in different stages of ripeness and having different amounts of active principles, and this will cause lowering in the final content of pyrethrum.

Another possibility of increasing commercial viability (although not of mechanization, but a better exploitation of existing machinery available) is to be found in the breeding of this plant for pyrethrum. A variety should be bred with a regular second blossoming, taking as a starting point individuals occurring up to this time only sporadically. This second flowering should give an additional yield as needed.

The harvest problem and its solution by breeding adapted for mechanization of some other perennial plants of the Labiatae genera (e.g., lavender, hyssop, and thyme) is more or less the same as that of *Pyrethrum.*

After careful consideration of the examples above, it appears that the characteristics of plant cultivars must in certain milieus conform to progressive mechanization. A prerequisite of modern production of medicinal plants in large-scale farming is to produce very adapted varieties, in order to ensure economic viability and higher yields in situations where manual labor continually decreases.

The knowledge of applied biology, the newest achievements of biological basic research (irrespective of the circumstance whether the object of research is botanical, ecological, or biological production, in relation to molecular biology or chemotaxonomy), contribute in the same way to the result that active principles in medicinal plants should be available for humanity in qualities and quantities as necessary.

REFERENCES

1. Boulter, D. Meatyard, B. T., Seawen, M. D., and Ramshaw, J. A. M., Cytochrome C from *Rhodymenia palmata* and *Porphyra umbilicalis* and the amido acid sequences of their N-terminal regions, *Phytochemistry,* 14, 1493, 1975.

1a. Ramshaw, J. A. M. and Boulter, D., The amino acid sequence of cytochrome C from Niger-seed *Guizotia abyssinica, Phytochemistry,* 14, 1945, 1975.

2. Culberson, C. F. and Culberson, W., Chemosyndromic variation in plant taxonomy, *Lichens Quarl. J. Am. Soc.,* 1, 331, 1976.

3. Schratz, E., Arzneipflanzen, in *Handbuch der Pflanzenzüchtung,* Vol. 5, Kappert, H. and Rudolf, W., Eds., P. Parey, Berlin, 1960, 383.

4. Sobti, S. N., Chemical and genetic investigations in essential oils of some *Ocimum* species. Their hybrids and synthesized allopolyploids, *J. Nat. Prod.,* 41, 50, 1978.

5. Mechler, E. and Haun, N., Alkaloid content in trisomic mutants of *Datura stramonium, Planta Med.,* 42, 102, 1981.

6. Kashikar, S. G., Vywahre, M. K., and Umalkar, G. V., New pepper mutants with high capsaicin content, *Planta Med.,* 39, 291, 1980.

7. Hussain, A., Current Status of Research on Cultivation of Medicinal Plants in India, presented at 4th Asian Symposium on Medicinal Plants and Spices, Bangkok, 1980, p. 66.

8. Topalov, V., Bulgarian Clones of *M. arvensis* L. with High Essential Oil Contents, presented at 1st Int. Symp. Aromatic Plants, Halkidiki, 1981.

9. Lamy, J., Passet, J., Vernet, P., Gouyon, Ph., and Valdeyron, G., La variabilitä chimique chez le thym (*Thymus vulgaris* L.) base scientifiques et application agronomique, *Sci. Ann. Aristotelian Univ. Thessaloniki,* 22, 96, 1983.

10. Astadjov, N., New Variety of Summer Snowflake "Snejinka", presented at 1st Int. Symp. Aromatic Plants, Halkidiki, 1981.

11. Wagner, T., Effect of cultivar on the essential oil composition in the hop *(Humulus lupulus). Sci. Ann. Aristotelian Univ. Thessaloniki,* 22, 52, 1983.

12. Abou-Mandour, A. A., Fischer, S., and Czygan, F. G., Regeneration von Intakten Pflanzen aus Diploiden und Haploiden von *Catharanthus roseus, Z. Pflanzenphysiol.,* 91, 83, 1979.

13. Bajaj, Y.P.S., Gosch, G., Ottma, M. Weber, A., and Grübler, A., Production of polyploid and aneuploid plants from anthers and mesophyll protoplasts of *Atropa belladonna* and *Nicotiana tabacum, Indian J. Exp. Biol.,* 16, 947, 1978.

14. Yoshikawa, T. and Furuya, T., Regeneration and *in vitro* flowering of plants derived from callus cultures of opium poppy (*Papaver somniferum), Experimentia,* 39, 1031, 1983.

15. Furmanova, M. and Oledzka, H., Sowinska D., Regeneration of plants by embryogenesis with callus cultures of *Carum carvi, J. Plant Physiol.,* 115, 209, 1984.

16. Jha, S. and Sen, S., Mass propagation of Indian squill, *Urginea indica* Kunth *in vitro, Acta Horticult.,* 1985.

17. Shah, R. R. and Dalal, K. C., *In vitro* multiplication of *Glycyrrhiza, Curr.Sci.,* 49, 69, 1980.

18. Dietrich, B., Neumann, D., Popov, A. S., Butenko, R. G. and Luckner, M., Cryopreservation of *Digitalis lanata* Cell Cultures, presented at Int. Symp. Plant. Tissue Cultures, Olomouc, 1984. 42.

19. Dános, B. Alberovsky, E., and Sárkány, S., Der Sammen und Kapselertrag ferner die Morphingehalt des Gratenmohns, *Bot. Kozlemenyek,* 49, 201, 1962.

20. Hegnauer, R., Arbeitstagung über Arznei- und Nutzpflanzenkultur in Wageningen, September 9-11, 1957, *Pharm.Weekbl.,* 92, 761, 1957.

21. Rovesti, P., Recherches sur les incidences écologiques en rapport a la composition des huiles essentielles, *Pharm. Weekbl.,* 92, 792, 1957.

22. Neubert, U., Ökologische und ontogenetische Einflüsse auf Blótenertrag und Flavonoidgehalt von Arnikaarten, *HGK Mitteilungen,* 33, 1981.

23. Chatterjee, S. K., Cultivation of Medicinal Plants in India. Special Reference to Eastern Himalaya Region, presented at 4th Asian Symp. Medicinal Plants and Spices, Bangkok, 1980, 100.

24. Box, E. O., Productivity of aromatic plants: climatic models, in *Aromatic Plants: Basic and Applied Aspects,* Margaris, N., Koedam, A., and Vokou, D. Eds., Martinus Nijhoff, The Hague, 1981, 87.

25. Máthé, I., Über die Sprorskoloinen von für Arzneiindustrielle Zwecke angebaute *Vinca minor* L., *Agrartud.Kozl.,* 26, 27, 1967.

26. Cosson, L., Chouard, P., and Paris, R., Influence de l'éclairement sur les variations ontogéniques des alcaloides de *Datura tatula, Lloydia,* 29, 19, 1966.

27. Kuznietszova, G. K. and Khazanov, V. S., Spectral effectiveness of photosynthesis of some alkaloid-bearing plants during adaptation to light of different quality, *Fiziol.Raszt.,* 20, 554, 1975.

28. Böhm, H., Ergenbnisse und möglichkeiten der Arbeit an einem Arzneimohn, *Planta Med.,* 19, 93, 1970.

Chapter 4

AGROTECHNOLOGY OF MEDICINAL PLANTS

Rajendra Gupta

I. INTRODUCTION

Medicinal plants can be considered to include all plant materials such as foliage, root, flower, fruit, and seed which may be used as such or in the form of their extracts and chemical compounds isolated from them to produce drugs for human and veterinary medicine. These plants are closely related to those that produce stimulants, condiments, spices, essential oils, and such other higher forms of plant life that produce specific influences on cell metabolism. Medicinal plants were among the first from the vegetable kingdom to be used by man. Though most countries of the world have their national formularies, the oriental countries were known to be particularly rich in diversity of plants found in their flora and the knowledge on their utilization in medicine. The World Health Organization (WHO) has complied a list of over 20,000 common medicinal plants used in different parts of the globe and many of these are known for their properties in several countries for use against different maladies. A survey of major world markets for drug raw materials suggest that over 100 botanicals enter into regular trade and these have a consistently large demand in the organized sector of the plant-based pharmaceutical industry. However, the list of major medicinal plants in the international trade has always remained fluid, since some of these lose their relative importance to others, whereas as new ones are discovered, their contents are identified and efficacy established. The plant-based recipes are still very popular and largely used in most developing countries of the Old World. As much as 95% of the drugs manufactured in the African continent is derived from plants.[17] In the industrialized countries as well, the plant-based drugs are becoming important sources of medicaments. In these countries compound galenical preparations are now increasingly used in domestic medicines. In the U.S., for example, 25% of all prescriptions dispensed from community pharmacies from 1959 to 1980 contained plant extracts or active principles prepared from higher plants. Further, in the year 1980, U.S. consumers paid more than U.S. $8000 million for prescriptions containing plant-derived active principles.[6] There are several estimates available on world trade in medicinal plants and their derivatives. According to one, the trade in crude drugs and herbs in the major drug markets of Europe is estimated at U.S. $551 million (1980).

The six major developed countries, U.S., U.K., France, Germany, Switzerland, and Japan, are reported to import medicinal herbs, vegetable saps, extracts, and enzymes and their derivatives for a total amount of U.S. $700 million (1976) of the total of U.S. $1147.4 million imported by all Organization of Economic and Commercial Development (OECD) countries annually. The demand estimates may go up when nonpharmaceutical use of certain medicinal plants is taken into consideration such as liquorice in confectionery, cinchona bark in soft drinks, caffeine in beverages, oil of peppermint and menthol in food flavours, β-ionones in perfumery and for manufacture of vitamin A, etc. Obviously the growing interest and increasingly larger use of medicinal plants and their products offer good prospects for bulk production of these raw materials.

II. NEED FOR CULTIVATION OF MEDICINAL PLANTS

A large variety of medicinal plants continue to be gathered from their wild growth in the country of their origin, both for home consumption and export. The tropical forests, in particular, often possessed a luxuriant distribution of medicinal plant species, and this has served as a natural repository of raw materials for the pharmaceutical industry. The last few decades have witnessed several rapid changes in the land-use pattern in developing countries. It is estimated that about 11 million ha of forests are annually eliminated in the tropical regions of the world for various purposes, rendering large-scale loss of plant life. The second concern is the large-scale and often ruthless exploitation of plants for use in the organized sector of the pharmaceutical industry to meet the sudden surges in demand of selected herbs and herbal products. In this manner several valuable species have come under the virtual threat of extinction, at least in easily accessible forest ranges that served the genuine health-care needs of the local population.[10] Some of the widely known sources of drugs such as *Rauvolfia serpentina* and *Dioscorea deltoidea* in India, *Ephedra sinecia* in China, *Artemissia maritima* in Pakistan, *Hyocyamus muticus* in Egypt, *Cassia acutifolia* in Ethiopia, and *Catharanthus roseus* in the West Indies have become endangered species in their original habitats.[6] The loss of plant life does not simply mean a perpetual loss of raw material source for any medicinal plant, but it means a greater loss of valuable genes of diverse traits such as for yield and quality characteristics which could have been harnessed for future benefits. The tropical forests have remained the seat of variation and evolution in the plants. As yet, there are few studies made in the population structure of medicinal plants found in their wild growth in different centres of origin and diversity to determine the potential for transfer and utilization of these traits of quality from the ancestral genepools and landraces into the cultivated varieties. The pharmaceutical organizations have depended too long on the cheap raw materials from natural resources. There is a growing realisation that this dependence should end to relieve pressure on collection of herbal materials and allow the plants to evolve and further synthesise new and useful variations in their natural habitats. This wealth of genetic diversity in medicinal plants is our common heritage and should be protected. The conservation of economic plants is thus interrelated to protection of the natural ecosystem. This protection of natural plant strains could be ensured only when industrial needs of these raw materials is met through other means of production at reasonable prices.

It is true that the last 5 decades saw epoch-making discoveries in pharmaceutical chemistry. Organic chemists have been able to synthesise in production units countless chemical versions of natural products in bulk at relatively low prices. At one time, chemical laboratories rivalled nature itself in obtaining a wide range of derivatives from a single plant substance. However, users in the industrialised countries soon were alerted to the potential of the adverse effects of many of these synthetics and antibiotic drugs. Thus, simple forms of herbal remedies such as laxatives, decoctions, beverages made of leaf mixture, restoratives, stimulating or invigorating plant extracts, hot fomentations and oily massages, sedatives, and health foods have rapidly gained acceptance as mild remedies for common ailments; their demand is growing in bulk. Side by side, there is now refinement in instrumental techniques and analytical methods which have come to the rescue of manufacturers with regard to the need for standarisation of complex natural products for drug formulations. It is now possible both to identify and quantify the ingredients in accurate doses for several plant products within safe limits. Work in the field is progressing at a rapid pace, and soon we may have complex organic formulations to cure relatively more serious and dreadful diseases. This will harbinger a new era in medicare when more and more medicinal plant species may need commercial cultivation in larger bulk in different parts of the world.

It may be stated in all fairness that our knowledge of genetics and physiological make-up of most medicinal species is poor, and we know still less about the biosynthetic pathways leading to the formation of active constituents for which these crops are valued.[12] Commercial cultivation of medicinal plants demands strong and continuing attention to these diverse fields. However, the contemporary literature on agrotechnology of several medicinal crops carried out in several countries has amply demonstrated that useful active principles present in a species could be produced and improved upon under cultivation in an alien habitat. Knowledge of breeding behaviour and genetic architecture of a species has helped to evolve new cultivars of these species with reduced crop duration in the field, increased fertilizer responsiveness, and in which one can find eliminated any harmful constituent from the overall composition together with improved synthesis of the desired secondary metabolite. Although agricultural research in medicinal plants has not advanced as much as food crops, we have evolved high yielding cultivars in several medicinal plants and developed good management practices to produce these raw materials at reasonable input cost. All this calls for further support through interdisciplinary research in several fields of agricultural research on medicinal plants to enable bulk production at reasonably low prices.

III. CHOICE OF SPECIES

Cultivation of medicinal plants is currently limited to a few commodities, regulated by bulk demand generated in international trade, manufacturing drug units, and national pharmacopoeias. It is easy under cultivation to maintain the purity profile and the quality parameters of a product prescribed by the regulatory laws of the land. The cultivations bring to bear a salutory effect on price and supply schedule for trade to regulate the market. By and large, developing countries such as China, India, Thailand, South Korea, Brazil, Mexico, Egypt, Indonesia, Nepal, the Philippines, and Kenya grow several medicinal plants in the farm sector. Similarly, most Eastern European countries such as Bulgaria, Hungary, U.S.S.R., and Romania grow many species, mainly to meet the requirements of their domestic pharmaceutical industry. The countries in Western Europe, the Mediterranean region, and the U.S. own cultivations of a few species only, but these countries are large importers of these raw materials annually. The commodities offered for sale in crude drug markets makes a large and impressive list of botanicals from all parts of the world, though many of these are not known under cultivation anywhere. A list of major drug plants under cultivation is therefore given in alphabetical order below to provide an idea of the range of plant products in demand at different trade centres. Although literature on agrotechnology of several of these listed species is not available in detail, it is desirable to compile and update such information from diverse nationalities and regional languages where these plants are grown for the benefit of user world community. The drug (plants) from cultivation are Anise, Artemisia, Aconites, Aloes, *Ammi majus & A. visnaga,* Belladonna, Basil, Buchu, Cinchona, Celery, Casara bark, Colchicum, Chamomilla, Costus, Dioscorea, Datura, Dill, Duboisia, Ergot, Ephedra, Foxglove, Ginseng, Gentians, Henbane, Hydrastis, Ipecac, Liquorice, Lemon grass, Mints, Male fern, Opium poppy, Periwinkle, Pyrethrum, Psyllium, Papain, Polygala, Podophyllum, Rauvolfia, *Solanum* spp., Sweet Flag, Smilax, Senna, Strophanthus, Squill, Thyme, Vinca, Valerian, and Withania.

Some new medicinal plants are appearing on the horizon promising a large demand and therefore pointing to the need for cultivation. Prominent among these are *Artemisia annua, Papaver bracteactum, Glaucium flavum, Eleutherococus senticosus,* and *Coleus forskolii,* etc.

IV. FACTORS AFFECTING CULTIVATION

The synthesis and accumulation of secondary metabolites used in drugs and pharmaceuticals are governed by the laws of genetics. Tetenyi[18] has described the occurrence of chemical races in several medicinal plant taxa and emphasised their utility in upgrading quality and yield of the active principles from the farm sector. The selection of cultivars of a crop, therefore, is of utmost importance to ensure a particular minimum quality and content of a given chemical in the produce. In crop plants where a species essentially resorts to cross breeding with allies such as in genus *Cinhona, Dioscorea,* and *Mentha,* a plantation is raised through vegetative propagation to maintain uniformity of crop strain and its yield. In a few crops, breeders have succeeded in evolving new hybrid lines by exploiting heterosis, e.g., in basil. The hybrid seeds are therefore produced every year for distribution to growers. In most cross-bred species, plant breeders have worked out special procedures to carry out artificial pollination and rogueing of off-types during seed production in isolated fields so that growers are able to produce high yields and the desired quality of produce through cultivation. In essence, climate, irrigation, and nutrition cannot change the content and composition of a crop but only modify it within certain limits. By and large, the distribution of a species in its natural habitat can provide guidelines on suitable soil and climatic conditions for its domestication under cultivation. Franz[7] discussed the genetic and environmental factors that influence quality of produce of certain medicinal and aromatic plants under cultivation. The environmental factors include soil (type, texture, pH, fertility and drainage), altitude, seasonal variation in temperature, atmospheric humidity, daylength hours, rainfall pattern, associate plant species, shade, dew/frost, and protection from high winds. Economic botany laboratories in several countries document data on ecological distribution of these species to analyse and determine ideal growing conditions of different medicinal species.[5]

A plantation industry based on medicinal plants would need utilization of basic information on several such environmental parameters affecting their growth and development. For example, liquorice, chicory, and hops produce flowers profusely and set seeds only around 40° latitude. Similarly, most *Mentha* species demand a minimum of 14 h daylength at harvest to produce a high menthol content in the oil; the menthol is converted into menthone under cloudy weather, which spoils the quality of the oil for commercial purposes. Foxglove and Gentians produce a lower content in the foothills, perhaps because the synthesis of active principles does not keep pace with faster plant growth at lower altitudes. Within these limitations, genotypes have been evolved and cultural practices modified such that it is possible to grow these species in less than ideal conditions profitably. The opium poppy is a classic example; it is a biennial crop in its temperate homeland, but through selection in tropical conditions the crop varieties are tailored so as to produce a high content of opiates in the latex within 140 to 160 days under conditions prevailing in India. Similarly, biennials such as celery, dill, and peppermint are grown as annuals in tropical climates and most of these remain in the field 5 to 7 months only. A selection of henbane *(Hyocyamus niger)* is identified in India which produces a high biological yield in only 100 to 110 days. It is thus possible to fit this crop profitably in the cropping system of a region where competition from cash crops during the growing period is not severe. The pharmaceutical industry in general continues to be reluctant to invest in agricultural research on medicinal plants owing to the fluctuation in demand of some botanicals, longer gestation period for development, and large demand for farmland which it finds difficult to acquire due to land laws in developing countries. Stringent regulatory laws are being enacted in user countries demanding studies on better drying of produce and storage, shelf life under warehouse conditions, and factors such as the status of herbicidal residues and the effects of heavy metals on produce.

In the following treatment, details are recorded of good agrotechnological practices (GAP)* on ten important medicinal plants whose volume in trade is large and thus are grown on commercial scale in several countries. These are the steroidal yam dioscorea, cinchona, ginseng, belladonna, foxglove, opium poppy, psyllium, solanum, periwinkle, and mint. These crops represent annuals, biennials, and perennials adapted to different agro-ecological conditions. It covers primitive cultivars that have not changed much from their original progenetors in the wild state in nature, such as psyllium and ginseng to the highly developed and evolved varieties of opium poppy and mint crops. Information on all those parameters that are known to influence growth, yield, and quality under cultivation are, within the limitation of space, included.

V. STEROIDAL YAM (*DIOSCOREA FLORIBUNDA* MART. AND GAL.)

A. GENERAL

Among the many sapogenin-containing plants, the tubers of *Dioscorea* sp. have been found to be a most promising commercial source. The world pharmaceutical industry needs over 1000 tonnes of diosgenin and its analogues for the production of corticosteroids, pregnenes, androstenes, 19 norsteroids, and other similar chemicals. Of the 600 *Dioscorea* species found in the world, only 15 possess the sapogenins in their tubers. These are predominantly tropical in distribution, with a few habitating temperate regions. These yams are herbaceous, diocious vines. China, Mexico, Guatemala, India, and a few other countries are major producers of this raw material. The wild-growing populations of these steroidal yams have depleted considerably, and accordingly, a few *Dioscorea* species have been domesticated as a successful plantation crop.

B. THE PLANT

Dioscorea floribunda is a Central American (Mexican) species belonging to family Dioscoreaceae and is considered best for cultivation because it develops a diosgenin content of around 3 to 4% in tubers within a 2-year period. *D. composita* Helms. is another Central American species with similar requirements of climate and soil conditions, and is also cultivated in some countries. These species need highly fertile irrigated lands with high cost of input on trellis work to support the growing vines. *D. deltoidea* Watts. is another promising species distributed in northwestern Himalayan regions (India). It produces hard, compact tubers containing a maximum diosgenin content of 5 to 7%, but this species is very slow growing and has a duration of 5 to 6 years under cultivation. The *D. floribunda* produces compact soft tubers at a shallow depth; two cultivars have been developed in India: one 'FC (c)-1' is a composite variety, hardy, and produces vigourous growing vines yielding 5 tonnes of fresh tubers per hectare in 2 years. It has 3 to 3.5% diosgenin on a dry weight basis.[4] The other is 'Arka upkar' which also produces robust vines, bearing broad, dark green leaves, uniform growth habit of tubers

* The unit quintals/hectare (q/ha) is employed in the metric context = 100 kg/ha.

which spread on all sides. The tubers are broad, thick, deeply lobed, and yield 5 to 6 tonnes tubers with a 3.5 to 4% diosgenin within a 2-year period.

C. SOIL AND CLIMATE

Dioscorea sp. prefer rich, well-drained, medium-light soils in mild tropical to sub-tropical climatic conditions, but avoids heavy clay and highly acidic soils. If sub-soil is hard, the tubers grow horizontally.

D. LAND PREPARATION AND PLANTING

Three to four mould-board ploughings are needed to prepare a uniform fine tilth. 18 to 25 tonnes of FYM or compost is mixed into the soil at land preparation. Fields of convenient size are prepared for irrigation. A new plantation can be raised from seed, leaf node, and tuber pieces. Seed set is profuse in a plantation where male and female vines are allowed to intertwine.[1] Flowering commences in September and onward and ripe seed pods may be collected in January under conditions within peninsular India The seedlings are raised in plastic bags under shade in nurseries where 2 seeds per bag are placed at 0.6 to 1.0 cm depth in February and kept moist by using a fine spray for watering. Germination commences in 2 to 3 weeks and plants are ready for field planting with the commencement of the rainy season. The seed-raised crop does not yield uniform progeny and usually takes a 3-year duration for harvest. Leafnode planting could be adopted for faster multiplication of selected clones in glass houses. Pieces of tuber are the planting material of choice. It is divided into three parts, viz., crown, which is a swollen hypocotyle (preformed bud); it takes 30 days on planting for sprouting, whereas median and tip parts take 90 to 100 days. In practice, the harvested fresh tubers are stored as a layer between moist sand in well-aerated fields. The planting is done in the spring season such that the crown pieces are removed and planted in late April, whereas median and tips are planted early in February so as to obtain a uniform crop. The planting time can be adjusted according to local conditions. The crop is planted at 60 × 45 cm in rows. The size of tuber for planting is generally within 50 to 70 g in weight, placed 5 to 7.5 cm deep in furrows; small-sized tuber pieces produce poor growth and lower yield. The tuber pieces should be treated in a solution 3000 ppm. (3 g in 1 L water) of benamyl (bentate fungicide) and the cut ends dusted with 0.3% benamyl talcum to protect it from soilborne fungus which causes rotting. The growing vines need stakes for support. After the sprouting is complete, the soil is furrowed.

E. FERTILIZER, IRRIGATION, AND INTERCULTURE

This is a high nutrient feeding crop and needs 300 kg N, 150 kg, P_2O_5, and 150 kg K_2O_5/ha. The P and K are introduced basically before planting while N-fertilizer is applied in four split doses during the months of June and August both years. The newly planted crop needs light irrigation at an interval of 7 to 10 days, whereafter it needs two to three irrigations in a dry season, depending upon atmospheric humidity. The plantation requires frequent weeding in the first 3 months when the vines are weak. Once plants climb over their supports, weeds are smothered due to shading. It is noted that cowpea, horsegram, clusterbean, and french bean could be intercropped with *Dioscorea*. The growing vines need strong support and a system (developed in India) of stone pillars tied with galvanised wire on four ends of the field, fixed at 10 m distance in rows, and the plants are pegged to the ground by gunny twine and tied with overhead wire. Bamboo stakes are provided where this is cheaply available.

F. PESTS AND DISEASES

There are no major diseases reported on the crop. Aphids and Red Spider mites sometimes attack and cause damage. The aphids attack in the early stages of growth and feed on young leaves and kill the vine tips. These are checked by spraying with Roger (0.05%) or Kelthane (0.5%).

G. HARVESTING AND YIELD

It is found that maintaining a 2-year crop cycle is economical. The tubers are dug out with the help of pickaxes in the dry season when growth is minimum. The tubers lie at 20 to 30 cm deep in soil and mechanical digging by a heavy moulded board plough is possible under proper soil moisture. The fresh yield varies from 5 to 6 tonnes of tubers containing a minimum of 3.0% (dry weight) of diosgenin. The fresh tubers are cut into convenient pieces and dried in the sun where they lose 70 to 80% to their weight due to moisture.

VI. CINCHONA (*CINCHONA LEDGERIANA* LINN.)

A. GENERAL

The bark of a number of *Cinchona* species (Rubiaceae) contain quinine, quinidine, and related alkaloids, known for their use as antimalarials and other uses, and therefore command a worldwide market. The quinine salts are natural bitters and also find increasingly larger use in the manufacture of soft drinks, whereas quinidine sulfate has found utility in the treatment of cardiac diseases such as auricular fibrillation and ventricular tachycardia. The commercial supply of *Cinchona* bark and alkaloids is obtained from exploiting trees in Andean Mountain countries. Indonesia, India, and Zaire possess large plantations of *Cinchona* trees that yield commercial quantities of raw material to meet the growing world requirement.

B. THE PLANT

There are 65 species of genus *Cinchona,* distributed from 800 to 2800 m elevation in the Andean Mountains of Peru, Bolivia, Colombia, and Ecuador in South America. These range from evergreen shrubs to small trees, but mainly the bark of *C. ledgeriana, C.officinalis, C. calisaya,* and *C. succirubra* together with their numerous hybrids are marketed due to their high alkaloid content. The *C. ledgeriana* Moens. is a small tree grown between 1000 to 1900 m in the hills of west Bengal and Tamil Nadu in India. The estimated total alkaloid content in a plantation of trees between 10 to 12 years old, averages 7.21, 6.01, and 4.0% in the root bark, trunk bark, and branches stem bark, respectively, of which 5.4 and 1.98%, respectively, is quinine. A hybrid *(legeriana × succirubra)* is found to produce more robust tree growth even under less favourable growing conditions and is preferred despite its lower yield. *C. officinalis* Linn. is a weak, struggling tree reaching 6 m height, preferring higher elevations (2000 to 2600 m) and possesses a quinine content range of 1.75 to 4.16, 2.56 to 4.42, and 1.44 to 2.35% in the bark of root, trunk, and branches. A few hybrids called 'robusta' between *officinalis* × *succirubra* and between *officinalis* × *ledgeriana* have found better adaptation to a wider range of elevations than their parents for raising plantations. *C. succirubra* Pavon. reaches 18 to 20 m height and is more adapted to middle ranges (1200 to 2000 m) for withstanding high humidity and drought conditions. It is rich in total alkaloids. It has a high cinchonidine content but is low in quinine. *C. calisaya* Wood prefers a lower altitude (400 to 1000 m), higher humidity and is grown largely in Indonesia.

Besides several hybrids known in cultivation, a number of high-yielding clones have been selected in Indonesia that produce between 4 and 7.5% quinine content. These high quinine yielding trees are raised through vegetative propagation to maintain purity of their population.

C. SOIL AND CLIMATE

The cinchona trees prefer deep, well-drained, rich loamy to clay-like, acidic soils (pH 4.5 to 6) on sheltered slopes. These species thrive well in subtemperate to temperate humid climate, high rainfall (150 to 400 cm), and warm weather with minor variation in maximum and minimum temperatures during growing period. It avoids waterlogged conditions and frost is known to kill the growing plants.

D. LAND PREPARATION AND PLANTING

The land for plantation should be well prepared 1 year ahead of the actual planting. The soil is exposed to the summer sun and all weed growth is removed. Usual precautions such as gully plantation, bunding, etc., are carried out to protect soil erosion at land preparation. Shade trees such as Silver oak *(Grevillea robusta)* and Pangra *(Erythrina suberosa)* are planted at wide distances over this land to provide partial shade to the growing trees. Pits of 30 × 30 × 45 cm are dug out at contour from 1.25 × 1.25 m to 2.0 × 2.0 m distance, depending upon slopes and allowed to weather for 1 month. These pits are filled with a mixture of leaf mould, compost, and soil in which 45 g per pit of rock phosphate is mixed.

Seed and vegetative propagation methods such as rooted cutting, mould layering, patch and slit budding, air layering etc. are used for raising a plantation. Seeds are small and light (350 to 400/g) and lose viability soon (5 to 6 weeks). About 50 g seed is sown from February to April by broadcast in well-prepared, highly manured seed beds of 2 × 1 m dimension, depending upon weather conditions. The seed beds are kept moist and are provided with shade. The seeds begin germination in 20 to 25 days and the seedlings grow 3 to 4 pairs of leaves in 4 to 6 months when these are transplanted during the month of June into 30 × 12 cm polythene bags. The soil in the bags are filled with leaf mould and sand in a 5:1 ratio. These seedlings are kept moist in shade for hardening. It takes 4 to 5 months to grow to 25 to 40 cm high when these are transferred to pits in the field at the commencement of summer rains. The Silver oak trees provide partial shade to the young growing trees.

E. PESTS AND DISEASES

Tea fly damage is reported in the Indian subcontinent for which the spraying of systemic fungicides such as dimenthoate or dimethylphosphoric ester (0.2%) is recommended as a control. Damping-off occurs in seed beds occasionally and any copper fungicide is used. Root rot *(Fomis lamaysis),* canker, and dieback are also reported from *Cinchona* plantations. A high incidence of these diseases is favoured by waterlogged conditions. Improvement in drainage conditions and spraying of bordeaux mixture helps to prevent the damage.

F. HARVESTING AND YIELD

The plantations are thinned at 6 years of age when 50% of trees are uprooted and debarked. This is repeated at 8 years of age and finally the remaining trees are uprooted at the end of 12 years. Wherever the land is used for a new plantation, the trees are cut at 5 cm from ground level, taking precautions to avoid splitting the stump. These stumps regenerate into several sprouts during the next rainy season. Usually two sprouts are maintained to grow so as to have another cycle of tree growth to remain in the field for the next 8 to 10 years. The bark is removed by ringing the stem at a height of 60 cm from the ground and is dried in the open. The drying reduces weight by 60 to 70%.

The root bark is taken out similarly. An average yield of 5 tonnes of dry bark per hectare is obtained from a plantation. The bark contains around 30 chemically related alkaloids, of which quinine is the most important. The concentration is maximum in the collar part (30 to 45 cm length) and the root bark; it declines in trunk bark and is low in branches and twigs.

VII. GINSENG (*PANAX GINSENG* NEES.)

A. GENERAL

Ginseng root (Araliaceae) is basically an adaptogen that reportedly improves body functions. In Chinese literature it is considered a panacea for all ailments. It is used as a demulcent, tonic, and stimulant, and is considered to reduce depression and prolong life expectancy. The roots contain the saponins (ginsengenin, stigmasterol, phytosterol, sitosterol, glucosides, and deucosterin) besides starch, resin, volatile oil, panaquilon, mucilage, and other substances. The root as such has a world-wide demand for use in health foods.

B. THE PLANT

The Korean species *Panax ginseng* var. *coriaiensis* Nees. is considered to be the true (high quality) ginseng of commerce. Several allied species are reported from China *(P. notoginseng, P. pseudo-ginseng),* Japan *(P. japonicum),* far eastern parts of the U.S.S.R., Canada, and the U.S. *(P. quinquifolium).* It is a slow-growing, glabrous, perennial herb reaching from 30 to 60 cm in height at 5 years of age. It possesses thick, corpulent roots. The plant bears three to five palmately compound, stalked leaves, each with five leaflets. The leaflets are simple, ovate to oblong in shape, acuminate, 15 × 8 cm in dimension, with a serrated margin. In the third year during June and July the plant bears 6 to 20 white to greenish white flowers in an umbel. It forms oval berries, 1 cm in diameter, turning red when ripe. The berry contains two to three grayish-white, wrinkled seeds about the size of a small pea. Fresh seed does not yield to germination for 12 to 18 months. It should not be dried but kept for stratification in cool moist soil in between layers of sand and covered with sawdust for a few months. This process of stratification allows the seed embryo to develop.

C. SOIL AND CLIMATE

The ginseng is found over rich, cool, woody shaded temperate forests. It prefers well-drained, deep acidic (pH 5.5 to 6.5), light sandy-loam soils, rich in organic matter and potash. Although shade loving, the planting site should be faced northeast to allow exposure to the warm sun. The annual variation for maximum temperature of localities chosen to raise plantations should be around 20° to 25°C in summer and 9° to 14°C in winter. It grows well in localities receiving 80 to 110 cm rain and these locales are covered with snow in winter.

There are no distinct known practices for different varieties as the literature is very scanty in regard to the agrotechnology of ginseng cultivation. The countries growing ginseng maintain strict secrecy on cultivation practices.

D. LAND PREPARATION AND PLANTING

Usually, forest land or land kept fallow for 1 year is chosen for its cultivation. The land is given 6 to 8 cross ploughings and harrowings and exposed to sun in dry weather. FYM or compost or green grass at 50 to 80 tonnes/ha is mixed in the soil at land preparation. The seed beds are heavily manured similarly and given a slope facing the northeast sun. The top layer of the seed bed is made of an equal quantity of sand and humus or forest litter. The stratified seed is sown during October to November in seed beds in ridged rows as a distance of 10 to 12 cm. Germination is very slow and takes 2 to 4 months, when about 30% of the seeds germinate. The nursery beds are kept moist and provided with shade from mid-April onward. Seed treated with GA_3 improves germination. Usually 1-year-old healthy rooted seedlings are transplanted. The seedlings are dug out carefully and planted at a permanent site at 10 cm distance in rows over ridges of 60 cm width and 15 to 20 cm height. The transplanting can be done any time between October to November and March to April, but autumn planting is preferred. These plants are thinned after 1 year in rows to maintain a distance of 15 to 20 cm. The plants shed their foliage annually and the dead leaves are removed from the fields. The growing plants are provided shade by raising trees or with a lathe system with reed roofs, such that 75% light is cut during summer, but it allows free circulation of air. In winter, these beds should be covered with 4 to 6 cm layers of dry leaves to protect the underground roots from freezing temperature and frost injury. This leaf cover is removed during the spring season.

E. FERTILIZER, IRRIGATION, AND INTERCULTURE

The crop is not treated with any inorganic fertilizers. The seed bed is provided with dung and bone meal from May to July and the plantation is given compost/green grass three times between May to July each year. Weeding is done such as to keep the plantation completely weed-free. Irrigation is provided twice a month in the furrows.

F. PESTS AND DISEASES

Mice and other rodents damage the growing plants and feed on its roots. A large number of fungi are reported on ginseng plants such as *Alternaria panax, Collectotrichum dematium, Phoma panacicola, Phylloststicta panax, Rumularia destructans, Verticillium, Albo-atrum, Rhizactoria solani,* etc. Clean cultivation is recommended to maintain healthy plants for obtaining higher root yield.

Several nematodes have also been identified from ginseng eg. *Aphelenchoides panax, Heterodera marionl, Meloidodync* sp., *Caconema radicicola* which cause root knot and reduce yield.

G. HARVESTING AND YIELD

The plants are dug out after they remain in the field 4 to 5 years after planting. The roots are dug out carefully without damaging their forks. The digging is carried out in the dry season from July to October. The mature roots are 10 cm long and 2.5 cm in thickness. An average of 2 to 2.5 tonnes/ha of dry root is obtained. The dug out roots should not be scrubbed but spread over wide netting at 40° to 44°C in well-ventilated rooms for a few days and then at 35°C for the next several days. It thus takes 10 to 15 days to dry the roots completely. The produce should maintain its appearance in order to command high market prices. It should be protected from mould or overheating which spoils colour, appearance, and texture.

VIII. BELLADONNA *(ATROPA BELLADONNA* LINN.)

A. GENERAL

The leaves and roots of *Atropa belladonna* (Solanaceae) are the age-old remedy of neuralgic and muscular pains. It possesses tropane alkaloids of which the hyoscyamine content is high, whereas atropine and hyoscine are present in smaller quantities. The root is essentially used for external application. The hyoscyamine alkaloid in particular is given as a pre-anesthetic agent to check secretion in throat and respiratory passages and also to relieve spasm and soothe muscles. Its cultivation is carried in several countries of Europe and South-east Asia.

B. THE PLANT

Atropa belladonna Linn. is a small, perennial branched herb which grows up to 1.5 m high all over temperate Europe. Leaves are alternate, shortly stalked, abovate-lanceolate, 20 to 32 × 10 to 12 cm in dimension. It bears bell-shaped, purple flowers in two to four cymes in axil of leaves during June to September. The fruit is a globose, many-seeded berry, which on ripening in autumn changes to a purplish-black colour. The root is long, thick, and tapering, and goes 12 to 20 cm deep in the soil. An allied species, *A. acuminata* Royle ex. Lindley, occurs in coniferous forests of the western Himalayas; it has a similar alkaloid content and is recognised as its local substitute.

The belladonna crop remains for 3 to 4 years in the field, yielding two to four foliage crops annually. A number of cultivars of *A. belladonna* are grown in several European countries and are known for producing high foliage yields and alkaloid content.

C. SOIL AND CLIMATE

The belladonna plant grows well over well-drained, deep silt-loam to clay-loam, acidic soils, rich in humus. Though tolerant to partial shade, its growth is profuse under clear, sunny, warm weather conditions where rainfall is high (160 cm) and evenly distributed throughout the growing period. Waterlogged and frost conditions damage the growing plants considerably.

D. LAND PREPARATION AND PLANTING

The soil should be well worked up to a depth of 30 cm by repeated cross-ploughing and harrowing until all clods are broken and perenniating roots are removed. The land should be treated with 20 tonnes of FYM with 25 kg of 5% aldrin or 10% BHC at the final land preparation, and planked. The fields should be laid out into a convenient size and given a slight gradient for drainage.

A plantation can be raised conveniently through seed. The seeds are small, brown to black in colour, and weigh 700 pg. Both direct sowing in field and nursery raising of seedlings and their replanting in fields are practiced. About 4 kg of seed is needed to raise enough plants in a nursery for planting 1 ha of land. The seed coat possesses water-soluble substances that inhibit germination and should be removed by washing the seed in running water for some hours. Seed treatment with ethyl alcohol for 3 h, followed by washing in running water improves germination. The seed requirement for a direct sown crop is around 20 kg/ha. Seed may be treated with Captan or Agrasan at 4 g/kg and sown in lines in well-prepared nurseries either in mid-autumn or in the early spring season. Germination commences from 10 to 21 days, depending upon soil temperature and is up to 40 to 50% in the next 30 days. The seedlings grow to two to three leaves in a 3-month period when it could be planted in the field at 60 × 45 or 60 × 60 cm spacings at the commencement of summer rains (June to July). Ridge planting is recommended for high rainfall tracts.

E. FERTILIZER, IRRIGATION, AND INTERCULTURE

The crop is a heavy feeder on nutrients. Depending upon fertility status of the land, it is given 20, 48, and 30 kg of N,P, and K fertilizers at land preparation. Thereafter, annually 60 kg of N/ha is top-dressed in two to three splits at suitable intervals. The crop is irrigated at 10 to 15 day intervals in dry weather with two to three weedings cum hoeings in 1 year.

F. PESTS AND DISEASES

Cutworms *(Agrotis flammantra)* attack tender seedlings in nursery and young growing plants in the field. The soil treatment with Aldrin protects the growing plants. If damage persists, an application of 1:19 wettable Chlorodan through irrigation two to three times at 10-day intervals is required. Damping-off *(Pythium* spp.) is often reported. It is controlled by spraying chloropicrin solution (0.2%). Root rot *(Fusarium* spp.) is another common disease where water stagnates for some time in the field. It is difficult to eradicate completely, but improved drainage and drier field conditions minimises the intensity. The affected plants should be pulled out and burned with adhering soil to check any spread of the disease.

G. HARVESTING AND YIELD

The foliage crop has a maximum alkaloid content at the early flowering stage and reduces progressively such that it is lowest when fruits ripen in autumn. Two to four coppicing of leaf crop is taken from year 2 onward at short intervals. The plants are coppiced at 20 to 25 cm above ground in sunny weather but the autumn harvest is cut at 4 to 6 cm close to ground. The harvested material is rapidly dried by spreading over field/threshing floor in the sun for 2 to 3 days and mixed periodically so as to maintain its usual green colour. The leaf stalks are removed after the produce is completely dry. The roots are dug out after 3 or 4 years of age in the autumn season. The fleshy mature roots are cut into small pieces and dried in the sun. Drying by means of a hot blast of air at 40°C ensures the quality of the produce. The fresh crop of leaves and root lose 80 and 70%, respectively, of its weight during drying. The crop yields about 1 ton/ha of foliage annually and 3 to 4 q/ha of dry roots at the end of the season. It should contain a minimum of 0.3 and 0.4% total alkaloids in leaves and roots, respectively, calculated as hyoscyamine. The dry produce should be packed in polythene-lined sacks in cool, well-ventilated warehouses.

IX. FOXGLOVE (*DIGITALIS LANATA* EHRH.)

A. GENERAL

The leaves of several *Digitalis* species *(Scrophulariaceae)* contain cardiac glycosides such as digoxin, digitoxin, gitoxin, and lanatosides. These glycosides are used in medicine to regulate and stabilize irregular heartbeats.

B. THE PLANT

Foxglove is a biennial temperate herb, and a native of western and central Europe. Although most of the 25 species contain cardiotonic glycosides, the two species that are relatively widely cultivated as a source of these glycosides are *D. lanata* Ehrh. and *D. purpurea* Linn., which produce vegetative growth (rosette leaf) in the first year, and give out flowering shoots in the second year. *D. lanata* has long narrowly linear serrated leaves and a higher acceptability as a drug due to its wider range of safety. *D. purpurea* bear broadly lanceolate and larger leaves. The flowers are large, bell-shaped, white-purple to violet-blue in colour borne in a raceme during early autumn. The fruit is a cylindrical capsule filled with small, angular brown seeds.

C. SOIL AND CLIMATE

Foxglove plants grow well over well-drained, sandy-loam, slightly acidic fertile soils. They prefer warm temperate weather with plenty of sunshine and evenly distributed rain for most part of the growing period.

D. LAND PREPARATION AND PLANTING

The land is given three cross-ploughings and harrowings to pulverise the soil to a fine tilth and levelled. The land is then laid out into fields of convenient size for interculture and irrigation. FYM, 20 tonnes/ha with 25 kg of 10% BHC (insecticide) should be applied during final land preparation. Usually raising seedlings in a nursery is preferred, although direct sown cropping is possible. The seeds are minute, comprising 1500 to 1 g and the seed rate is 4 kg/ha for direct sowing. The seed must be soaked in water for 12 to 24 h, dried in the shade, treated with Captan (3 g/kg), and mixed with 10 times its weight of sand before directly sowing in the field at 45 cm distances in rows at 0.5 to 1.0 cm depth. About 0.5 kg seed is sufficient to cover 100 m^2 of nursery land to provide sufficient seedlings to plant 1 ha. The seed should be sown in lines at 10 cm distances in a well-prepared seed nursery during the early spring season. The seed germination takes 3 to 8 days and the seedlings grow 8 to 10 cm high in 35 to 45 days. At this stage, the seedlings are ready for planting in field from mid-April to July, depending upon location and season (temp. 20° to 30°C). A spacing of 45 × 30 cm is usually maintained and the freshly planted crop is given light irrigation; closer spacing of 30 × 30 cm have been reported to produce higher yield over soil of medium fertility.

E. FERTILIZER, IRRIGATION, AND INTERCULTURE

The crop makes a moderate demand on inorganic nutrients. A fertilizer application of 100, 50, and 25 kg N, P, and K is recommended for the crop over medium-fertile soils. The P and K fertilizers are applied basally before planting, whereas N-fertilizer is applied in four splits (following 60 days, after the first harvest in October), and the remaining two splits in February to March, and the month of May in the second year. The crop needs light irrigation at 10 to 15 days during summer months and two to three intercultures annually. The flowering shoot is removed at an early stage to induce more leafy growth. One to two sprayings of MH (500 ppm) early in the second year before initiation of flower buds is found to delay bolting and improve foliage growth.

F. PESTS AND DISEASES

The crop is relatively free from damage due to pests and diseases. Leaf blight *(Alternaria* spp.) is common. Brown spots appear over leaves which enlarge and cover the entire leaf surface and kill the foliage in a few days. Spraying of copper fungicides (0.2%) control the infection. Curly top also is reported; this is caused by a viral infection. The affected plants should be pulled out and burned.

G. HARVESTING AND YIELD

Generally one or two leaf pickings can be carried out in the first year and three to four pickings in the second year. The crop is harvested by picking three fourths of all mature green leaves on clear, sunny days. The spring harvest is found to contain a higher content of glycosides under Indian conditions. The plants bolt during the latter part of the second year and bear small capsules. A number of vigorously growing fruiting plants are allowed to remain for ripening to collect seed. The harvested leaf crop should be rapidly dried within 36 to 38 h by spreading over ground on a threshing floor and turning it frequently to prevent fermentation. The leaf glycosides are thermolabile and hygroscopic in nature and disintegrate fast due to enzymatic action in contact with moisture. Rapid drying by a hot air blower at 60° to 80°C temperature is considered safe for drying. The produce should possess a minimum of 5% moisture. The yield is 6 to 10 q/ha in the first year and about 25 q/ha in the second year. The leaves contain 0.1 to 0.3% total glycosides. Higher yields are reported in several central and eastern European countries. The dry leaves should be packed in polythene-lined bags and stored in a cool, dark place, removed from air circulation.

X. OPIUM POPPY *(PAPAVER SOMNIFERUM* LINN.)

A. GENERAL

The shell of poppy capsule contains a mixture of 23 opium alkaloids in its latex called "opiates", of which morphine, codeine, papaverine, narcotine, and thebaine are commercially important sources of pharmaceutical raw materials. These alkaloids are used as unique pain relieving drugs in medicine, and despite new sources of drugs and synthetics discovered for this purpose, the opium husk and its latex remain to be the most important source of these phytochemicals in the drug industry.

B. THE PLANT

Six of 50 species of *Papaver* genus (Papaveraceae) produce latex of economic value. However, *Papaver somniferum* Linn. remains the most important species and is cultivated for this purpose. This is a temperate, rarely branched, biennial plant with large, linear-oblong lobed, dentate or serrate leaves, and indehiscent globular capsules. It is a native of east Mediterranean Europe and west Asia from which it has extended to cultivation all over the world. The original genotype has undergone many changes in the plant habit through long selection pressure as adaptation to different climatic conditions such that both cold-weather annual and biennial cultures are available for cultivation in tropical and temperate regions. The plant grows to a height of 0.8 to 1.4 m at bloom. As a biennial, the plant remains in rosette leaf stage in the first year and produces flowering shoots in the next season. It produces large white, pink, blue, or mauve flowers in varying shades and large,elongated, cylindrical to globular terminal capsules of 2.5 cm or more in diameter on the main stem and lateral branches. All parts except the seed contain latex but the capsule contains about 70% of the total morphine content of the plant. India, Turkey, Australia, U.S.S.R., Hungary, Bulgaria, Romania, Yugoslavia, France, Spain, the Netherlands, and a few other countries raise opium poppy for production of opium alkaloids for medicine. The seed is a rich source of edible fatty oil (50%). Thus, a large number of non-narcotic cultures *(P. somniferum* var. *nigrum* Dc) are grown as oilseed crop in central and eastern Europe. These cultures are found to break down under tropical climates and produce the usual narcotic latex in varying quantity.

The crop has been a subject of intensive agricultural research. We, therefore, possess a large number of high yielding cultures of biennials such as *Kak-Duna, Indra, hatvan, Tienshan-27,* and *Manjoirkij* which are grown as summer crops and of annuals such as *Dhowla chola gothia, Telia, Ranjhatak, Kalidandi, Dhaturia,* and *Jawahar aphim-16* which are grown in India. The annuals remain in the field for 140 to 160 days. An Iranian species, *P. bracteatum* Linn., has recently come into prominience as it produces the alkaloid thebaine but no morphine; since thebaine can be converted chemically into morphine, this species has opened up a new possibility of eliminating illicit extraction and sale of narcotic products from cultivations.

C. SOIL AND CLIMATE

The opium poppy crop needs deep, clay-loam fertile, and well-drained soils (pH 6 to 7.5) rich in humus/organic matter and with adequate irrigation supply. The light sandy-loam lateritic soil can also be managed to produce higher yields with judicious use of manure, fertilizers, and irrigation. Warm, humid weather favours luxuriant vegetative growth but dry, clear, sunny days with high temperature (30° to 38°C) are needed during the reproductive phase as it favours higher synthesis of the opiates. Frost, wet weather, hailstorms, high dusty winds, particularly at capsule-bearing stage, damage the plants and reduces the morphine content considerably.

D. LAND PREPARATION AND PLANTING

The land is well pulverised by five to six cross-ploughings and harrowings to make soil of fine tilth. This is divided into small fields of convenient size to facilitate interculture and irrigation. Between 10 to 20 tonnes of FYM or compost is mixed in the soil with 25 kg of 10% BHC at final land preparation. The organic matter improves soil texture enabling it to retain moisture and facilitate healthy initial growth. The seed rate is 5 to 8 kg/ha, sown in lines (30 cm apart) or by broadcast. It is covered with a light film of soil and sown fields are given a light irrigation at a very slow rate of discharge to avoid silting. The seed must be treated with Thiram (4 g/kg) before sowing. The sowing time is generally the commencement of the summer season in central and eastern European countries, and between the last week of October to the second week of November in the subtropical plains of India when maximum and minimum soil temperature is around 20° to 8°C, respectively. The seed germination commences in 7 to 12 days and takes 3 weeks for completion depending upon soil temperature. The initial growth is very slow. The annual cultures take 40 to 45 days to come to the rosette stage under tropical conditions whereas biennial cultures remain in the vegetative phase all through the autumn in the first year. In India, a plant population of 350,000 plants per hectare is maintained through thining to obtain maximum crop yield.

E. FERTILIZER, IRRIGATION, ANDINTERCULTURE

The opium poppy crop is a high feeder on nitrogenous nutrients. Trials carried out in several countries suggest that a high nutrient supply improves morphine yield. The biennial cultures need 100 to 140 kg of N and the annuals need 90 to 100 kg N and 45 to 60 kg P/ha to promote healthy plant growth. The potash fertilizer and zinc-sulfate (12 to 20 kg/ha) are mixed in the soil with P-fertilizer basally as per requirement of the land. The plant is found to utilize more nitrogenous fertilizers during and after the rosette stage. The fertilizer is given in three splits, i.e., 30% at sowing, 50% at the rosette stage, and the rest in the next spring season. In tropical conditions such as in India, nitrogenous fertilizer is divided into three splits and given at sowing followed by 30 to 40 days and 60 to 65 days after sowing. The nitrogenous fertilizer improves both the total latex yield and its alkaloid content significantly. The first interculture is done 3 to 4 weeks after sowing and should be adjusted such that it is completed 4 to 5 days before irrigation. A second interculture is done at 55 to 60 days after sowing. The crop is thinned out to maintain plants at 10 cm in a row during both these operations. In the period following two to three weedings may be given to maintain clean, weed-free cultivation. It is noted that pre-emergent application of Dicuron/ chlorotulon at 1.5 kg/ha over heavy soils and *Asulex* (6 L/ha) over light soils controls weed growth during initial growing stages. Higher doses produce toxic effects on the crop. A good soil moisture level is maintained throughout the growing period of the opium poppy. The crop needs 10 to 12 irrigations in light soils and 7 to 8 over medium heavy soils. In particular, in the late rosette stage, the initiation of flower bud and capsule formation stages are crucial because lack of soil moisture reduces the latex yield substantially. The crop is ready in 110 to 120 days for lancing and 120 to 140 days for picking of mature capsules from which opium is obtained through solvent extraction in annual cultures. The capsule crop is harvested when fully developed and green in the second year of biennial cultures for solvent extraction of opium alkaloids.

F. LANCING

In annual cultures, the opium plant comes into the flowering stage at 90 to 95 days after sowing. The flowers remain open for 5 to 6 days and produce capsules in the next 20 to 25 days after the fall of petals. At this stage, the capsules have developed to their full size, but are green and studded with immature seeds. The latex is present in laticiferous vessels in the mesocarp region of the capsule. These capsules are ready for lancing. The lancing is a traditional labour-intensive manual operation to collect opium latex. This is practiced only in India, as most opium growing countries have resorted to picking mature capsules for extraction of opium alkaloids through solvent extraction in large processing units. In lancing the capsule wall is given a straight incision all through the length with a special knife having three to four thin-edged sharp blades fitted at a distance of 1.5 to 2 mm from each other. The cut is made during the day in clear, sunny weather at a maximum depth of 0.4 cm so as not to pierce the wall deeper than its thickness. The cut ends ooze out milky or coloured latex and deposit it over the capsule shell on exposure to air. This is scrapped the next day in the early hours of the morning. Each capsule is thus lanced four times in a period of 25 to 30 days. In India, the opium poppy crop is cultivated only under valid license issued annually for growing over a specific land, and all opium so produced is to be deposited under law at government depots, opened in the growing areas during harvest period, and is acquired at a fixed price. The lanced capsules are left in the field for the next 20 to 25 days so that they develop seed crop to full maturity. The dry capsules are then picked, spread in open yards for further drying, and threshed to obtain the seed crop. The seed is completely devoid of any narcotic principle and has good demand for culinary purpose.

G. PESTS AND DISEASES

Aphids, thrips, cutworms *(Agrostis suffuosa)*, and capsule borer *(Heliothis armigera)* attack the growing plants and cause losses in plant populations and consequently yields. Treatment of some with BHC (or Aldrin) protect it from cutworms. Spraying of 0.03% Dimecron or Nuvan is recommended to control thrips and aphids. Similarly, two to three sprayings of Endosulphan (0.25% carbaryl) at 10 to 15 day intervals after formation of capsules protect plants from capsule borer damage. A root parasite *(Orobanche papavaris)* has localised endemic occurrence in parts of central India which reduces crop yield considerably.

Diseases such as downy mildew, powdery mildew, and root rot cause higher and much serious loss of crop. Downy mildew *(Peronospora arborescens)* appear in two forms at any stage of growth. The primary infection is systemic and infected leaves develop chlorosis and curl. Seed treatment with 4 to 6 g/ha of Apron-35 control this infection considerably. The secondary infection occurs as brown or black spots over leaves of 1-month-old plants. These become larger and cover most of the five to eight lower leaves. The disease saps its vigour, stunts growth, and inhibits capsule formation and development, producing little or no latex in the shell. It has defied control, though three to four sprayings of 0.2% Ridomil at 15-day intervals from appearance of symptoms, or four sprayings of Dithane-M-45 (0.3%) were found to give, respectively, 85 and 55% control over the disease.

The infection pereniates in the soil, and crop rotation is recommended. Powdery mildew *(Erysiphe polygoni)* appears as a white mycelial growth on stems and lower sides of leaves at the flowering stage; it soon turns into a black, powdery mass, causing losses in yield. At the early stages of infection, the plants are recommended to be pulled out and burned. At later stages dusting of Sulfex (5 kg/ha) or Microsulf (5 L in 500 L water) or two to three sprayings of Dithane Z-78 prevent the spread of infection and controls the disease. Root rot *(Rhizoctonia bataticola)*, leaf blight *(Helminthosporium papaveris)*, and curly leaf virus are also reported to attack the crop and cause some damage.

H. YIELD

On an average 8 to 10 q/ha mature capsules are produced containing 0.5% morphine. Solvent extraction yield 10 to 12 kg/ha of morphine. About 7 to 10 cm length of upper capsule stalks containing total alkaloids of 0.2% and 0.08% morphine is also harvested along with the capsules. The yield of opium latex through lancing range from 40 to 60 kg latex of 70% consistency, 25% of which constitute the weight of alkaloids. The usual content of Indian opium latex (in storage) is 10 to 14% morphine, 2 to 3.5% codeine, 5.5 to 7.5% narcotine, 0.5 to 0.8% papaverine, and 1 to 2% thebaine and other related alkaloids.

XI. PERIWINKLE *(CATHARANTHUS ROSEUS* G. DON)

A. GENERAL

Periwinkle is a hardy, evergreen, branched perennial herb grown as a pot herb for its rosy and white flowers borne for most part of the year. It is a native of Jamaica (West Indies) but is now indigenous over the east African coast and Southeast Asian tropics. This is a unique species containing 100 alkaloids in all parts of the plant. The leaves contain the dimeric (VLB) alkaloids such as vincristine and vinblastine in concentrations of 0.0004 to 0.0003% which find use in cancer therapy. The root and basal stems are rich in monomeric alkaloids such as raubacine (ajmalicine) and serpentine (0.03 to 0.1%) which are employed to reduce high blood pressure. The demand for VLB alkaloids is low as a few manufacturing companies alone produce the final drug. However, the demand for total ajmalicine is estimated around 35 tonnes and large areas are annually cultivated in several tropical countries.

B. THE PLANT

Catharanthus roseus G. Don *(Apocyanaceae)* occurs in three forms, distinguished by the colour of their flowers. The pink flowered plant represent the holotype species; its white flowering form is called *albus*, and wherever the white flower has a yellow or pink ring over the corolla tube, it is called *ocillatus*. It grows up to 90 cm high, bearing simple cauline, opposite, ovate-oblong leaves. The flowers bloom continuously all through the year; these are both axillary and terminal grouped in two to three cyme clusters. The fruit is a cylindrical follicle (pod) filled with small black seeds that fall out when the pods dehise on maturity. Tetraploids developed in the species produce larger leaves and flowers, but apparently bear no qualitative differences.

C. SOIL AND CLIMATE

This is essentially a tropical species, but could be grown up to 1500 m elevation in the lower hills. It needs tropical climate and sunshine. It sheds leaves wherever temperatures fall to 5°C or below in winter and new shoots appear in early spring. It grows well on poor to medium fertile, well-drained soils; the yield is fairly high over marginal lands under good management. A well-distributed rainfall during the growing season produces good yields. It stands drought conditions well.

D. LAND PREPARATION AND PLANTING

The crop stays in field from 200 days in the tropical regions to 300 days in the semitropical regions, and is grown both as a rainfed as well as an irrigated crop. Land can be prepared as usual and fields laid out into convenient sizes for interculture and irrigation. It is advisable to add 5 to 10 t of FYM or compost together with 25 kg of 10% BHC or Aldrin. Green manuring on marginal lands is found to increase yields by over 20%. A crop is best raised from seed. The seed rate is 2.5 kg/ha. Dry seed water soaked for 24 h could be sown directly in the field after mixing with 10 times its weight of sand in rows 45 cm apart during June and July. The plants are thinned in rows 30 cm apart during August. If the nursery is raised, 0.5 kg seed would be sufficient to sow in rows at 10 cm distance within 200 m^2 of land. This will produce enough seedlings to plant 1 ha of land. Germination commences in 7 to 10 days and the percentage varies from 40 to 90% depending upon temperature and seed viability. The fresh seed gives better germination. Seedlings are ready for planting in the field at 30 to 45 days after the onset of the tropical rainy season any time between June and August. There is no special culture for commercial cultivation and farmers are advised to raise pure culture of forms.

E. FERTILIZER, IRRIGATION, AND INTERCULTURE

This crop responds significantly to application of N, P, and K fertilizers over low to medium fertility soils.[11] The P and K are given basally, whereas the N fertilizer is delivered in two equal doses mixed in moist soil in rows at 15 and 45 days after planting. It is noted that nitrogenous fertilizer improves the dry matter weight of leaves and roots, as well as their total alkaloids. Generally five light irrigations are recommended for this crop and given at an interval of 30 days in the dry season. The irrigation improves branching, induces luxuriant foliage growth, and enables the plants to utilize nutrients effectively. However, trials carried out using different soil moisture levels showed that partial drought conditions led to higher synthesis of total alkaloids and is at a maximum at 40 to 45% soil moisture deficit (SMD). Talha et al.[19] implied that under high levels of SMD, the metabolic pathways of the plant are directed toward alkaloid productivity. Detopping plants by 2 cm at 50% flowering (60 to 75 days of age) supports growth, increases root yield, and increases uptake of nitrogen.[16] The crop needs two weedings, generally at 30 and 50 days after planting, when the plants cover the land and smother further weed growth.

F. PESTS AND DISEASES

Although a number of diseases and pests are reported on this crop, these cause relatively little damage. Occasionally, termites and cutworms attack seedlings for which soil is treated with Aldrin/BHC at the final land preparation stage. Thrips and aphids are noted to attack growing tips for which one or two sprayings of 0.02% Malathion is recommended. A dieback disease *(Pythium butleri)* is reported during rainy seasons for which spraying 0.2% Dithane Z-78 at 10 to 15 days interval is found to give effective control. A mycoplasma disease called "Green Rosette" is observed to induce atrophy in all growing organs of the plant. Generally, three to four sprays of 0.5% nicotine sulfate at 15 days interval is recommended as prophylatic treatment, but once the plants are affected, these should be dug out and burned. Chlorosis is also observed in some fields in patches, but no cause could be assigned to it. Similarly, powdery mildew *(Leveillula* spp.) and leaf blight *(Alternaria tauxica)* are also reported on this crop in peninsular India.

G. HARVESTING AND YIELD

It is noted that the root growth is more during the latter part of the growing period. Studies on the alkaloid content of root vis-a-vis plant growth and age have shown that the content of raubacin and serpentine increased with age and reached a maximum at 210 days; the serpentine content increased further until 300 days of age. Wherever the demand for foliage crop exists, the plants are stripped of its leaves at 150 to 180 days of age, and a second crop is obtained when the plants are finally dug out for root crop. The leaves are separated from the twigs and dried in the shade for 3 to 5 days. Irrigation is withheld 1 month before root harvest and the plants are then pulled or dug out. A 10-cm long thick basal stem is cut and dried separately; it has commercial quantities of monomeric indole alkaloids and is therefore saleable. The roots are cut into 15 cm long pieces and dried separately. The basal stem and roots have 80 and 70% moisture, respectively, and are dried to a level of 8 to 10% moisture, when they break with a fracture. The dry leaves, stems, and roots can be stored in polythene-lined jute sacks in well-ventilated warehouses for over 2 years.

The total crop yield is 22 to 25 q/ha from irrigated crop and around 15 q/ha from rainfed crop. The root accounts for 30 to 35% of the yield, varying from 5 to 6 q/ha and basal stem between 1 to 1.5 q/ha.

XII. STEROID-BEARING SOLANUM SPECIES (*SOLANUM VIARUM* DUNAL.)

A. GENERAL

There are 1500 species in the genus *Solanum (Solanaceae)* of which 22 nontuberous species have shown varying percentages of solasodine in their berries and foliage. However, *S. viarum* Dunal. from India and *S. laciniatum* Ait. and *S. aviculare* Forst. from New Zealand have come to be recognised as commercial sources and are cultivated in several countries. The latter two species inhabit temperate climates and our experience on yield and content on the two species in the Himalayan region (1500 to 2000 m) showed that *S. laciniatum* is superior for cultivation. The solasodine is a nitrogen analogue of diosgenin and could be used similarly as a starting raw material for production of corticosteroids.

B. THE PLANT

Solanum viarum Dunal. (syn. *S. khasianum* Clarke) is a stout, many branched, woody subtropical perennial undershrub, attaining 0.75 to 1.5 m in height under cultivation. In its wild state, the plant bears long, straight spines all over its stem and leaves. The leaves are obovate and lobed, bearing spines on both surfaces. The flowers are small, white, borne in groups of one to four in a raceme. The fruit is a green berry, 2.0 to 2.4

cm in diameter when mature, and turns yellow on ripening. It contains abundant small brown seeds embedded in the mucilage. The berries contain 2 to 3% solasodine. The two other temperate species, viz., *laciniatum* and *aviculare,* are perennial, attaining 1 to 1.4 m in height, but are less branched with large lanceolate leaves and larger bluish-white *(laciniatum)* to pale-purple or deep violet *(aviculare)* flowers. The berries are large, oblong in shape, and blue when ripe. The berries possess 3 to 5% solasodine and their leaves contain 3 to 3.4% solasodine on a dry weight basis; the *S. aviculare* is late flowering and its berries contain lower glycoalkaloids.[15] The Indian species, a hardy plant, is drought tolerant and can grow on poor marginal lands, remaining for 6 to 7 months in field. The temperate species demand well-drained, fertile soils needing adequate irrigation and remain in field for 7 to 10 months under cultivation in India. These two species are grown as biennial crops in central and eastern Europe. In comparison to *Dioscorea floribunda* this crop is more hardy, demands low input, and has a shorter growing period, and therefore deserves better attention for commercial cultivation.

Several cultigens of *S. viarum* (Glaxo *var.,* BARC *var.* RRL (J), RRL (B), and *Arka sanjeevani*) have been evolved through breeding and mutation, mainly to eliminate its spinescent habit. These varieties have almost negligible to few spines on stem/leaves; the mutants have a curved spine. Tetraploids (autotetraploids, aneuploids) have been evolved, but these produce berry yield at part with diploids and have no advantage in total solasodine content.

C. SOIL AND CLIMATE

The *S. viarum* grows luxuriantly in moderate climates over sunny tropical locations with maximum and minimum temperatures, reaching around 35° and 20°C under cultivation. Higher temperatures during flowering reduces fruit set. This species needs a medium-fertile, well-drained soil, with a pH of 6 to 8.5, and could grow equally well over marginal lands under good management. Clayed, waterlogged soils and frost-receiving localities should be avoided. *S. laciniatum* needs fertile, well-drained medium-light soils (pH 5.0 to 7.5) in warm temperate regions with well-distributed rainfall of 150 to 200 cm during vegetative growth.

D. LAND PREPARATION AND PLANTING

Land for planting of seedlings is prepared as usual to make a fine tilt and 10 tonnes of FYM with 25 kg of 10% BHC or Aldrin is mixed at final land preparation. The fields are then laid out in convenient sizes for interculture and irrigation. The crop is raised from seed and planting of seedlings is preferred. Seed rate is 1.2 kg for sowing in 50 m^2 well-prepared nursery beds. The seed is required to be soaked in water for 24 to 30 h and sown during May-June in rows by 10 cm at 0.5 cm depth. It commences germination in 6 to 8 days and requires another 10 days for completion. The seedlings grow to 10 to 12 cm height in the next 4 weeks and should be planted in field with the onset of summer rains (June-July) at 75 × 75 cm which allow the bushes to produce lateral branches. The seedlings can be planted in the September-October months as well, mainly as an irrigated crop (in the tropics), but these latter-sown plants are prone to viral infection. The *S. laciniatum* seed nursery in India (2000 m elevation) is raised in mid-October (July in Europe) in the same manner and seedlings are transplanted in field at 60 × 45 cm spacing before the end of December. Spring nurseries can also be raised and the seedlings are planted in field during April (India), but this reduces the growing period to 7 months, affecting the total yield considerably. In central Europe, seedings are planted in August-September and three coppicings of foliage and fruits are obtained the following year, i.e., in March, April, and May.

E. FERTILIZER, IRRIGATION, AND INTERCULTURE

The viarum crop is treated with 80 to 100 kg/ha of N and 40 to 60 kg/ha of P over land of medium fertility. The nitrogenous fertilizer is divided in two equal splits and delivered at planting and 60 days after. Green manuring of marginal lands is recommended for obtaining higher yield. The *laciniatum* demands high nitrogenous fertilizers and trials carried out at Solan (Himachal Pradesh, India) showed increasingly beneficial effects up to 150 kg of N and 30 kg each of K_2O and P_2O_5. The N is more effective when given in several small doses periodically. Kurnosov et al.[13] reported the use of a mixture of 40 and 60 kg/ha of N and P basally at planting, supplemented with 30 kg of N/ha at 3 to 4 leaf stage and 60 kg of N/ha at 10 to 12 leaf stage, and after first coppicing of the leaves. The nitrogen application together with P under mild climate increases plant height and branching; it helps to produce early and profuse flowering and a larger number of berries in the produce.

The viarum can be raised as a rainfed crop wherever rainfall is around 100 cm and is well distributed in the summer season. Life-supporting irrigation may be given when rainfall is erratic or low. Usually two weeding cum hoeings are given to the crop at 30 and 55 to 60 days. The *lacinatum* needs two irrigations per month in the beginning and once a month thereafter until the final harvest. It needs four weeding cum hoeing operations. It is noted that GA_3 (250 and 1000 ppm), CCC (1600 ppm), and Kinetin (50 ppm) increased solasodine but reduced berry yield in trials of the *laciniatum* crop in India.

F. PESTS AND DISEASES

Wilt *(Fusarium oxysporium)* may attack the crop in the rainy season wherever drainage is poor. The plants droop and finally succumb. Two to three sprayings of 0.2% Thiram or Dithane-M-45 is recommended as control. The fields may be kept dry and crop rotation should be practiced over such lands to avoid its recurrence. Mosiac is caused by a virus and occur at the seedling stage. It produces stunting of growth and chloratic leaves. This is transmitted by aphids and one to two sprayings of 0.2% malathion protect the growing plants from aphids. Late blight *(Phytophthora* spp.) is reported to attack the laciniatum crop, causing up to 30% loss in crop strand. Spraying of copper fungicides (such as Blitox) 2 to 3 times at an interval of 7 to 10 days is recommended. This crop is also susceptible to frost injury and attack of the *Epilechana* beetle.

G. HARVESTING AND YIELD

The viarum comes into flowering between 55 to 65 days and fully formed berries are ready for first picking at 80 to 90 days after planting. The berries are picked when they just begin to change colour from green to yellow; about 60% solasodine content is in the mucilage surrounding the seed and 40% is in the pericarp. The flowering commences after 165 days in *laciniatum* and fruits are ready to 190 to 200 days after planting. Picking of green berries is done manually, and is a cost-intensive operation. The berry formation continues for 2 to 2.5 months as growth in this species is indeterminate, allowing continuous fruiting. The fresh fruits contain 80% moisture. These are cut into halves and spread in thin layers over mats on the floor and turned frequently for rapid drying. This prevents the berries from turning black, which reduces the solasodine content.

The dried raw material should be protected from rain and dew as solasodine is soluble in water. The dry berries should retain their original colour, break with cracking, and may have up to 10% moisture. They have 2 to 2.5% solasodine. These dry berries are packed in polythene-lined jute sacks and are transported to manufacturing units or storage godowns. The yield is 7 to 8 tonnes fresh berries or 1.2 tonnes dry berries. The laciniatum crop gives three to four pickings per coppicing of berries and leaves from March to May in Europe, but September to October in India. The total foliage and fruit yield vary from 2 to 2.5 tonnes dry weight, of which 60 to 65% account for berry yield.[3]

XIII. PSYLLIUM (*PLANTAGO OVATA* FORSK.)

A. GENERAL

There are some 250 species in the genus *Plantago* (Plantaginaceae) found in the world, but the seed and seed husk of *P. ovata* Forsk and *P. psyllium* Linn. alone are in demand as organic laxatives in medicine. The *P. ovata* produces a higher yield and has a large seed size and therefore is cultivated, mainly in the arid parts of northern Gujarat and southern Rajasthan states of western India. The seed is emollient, demulcent, and promotes bowel movement, as it cures inflammation of the mucous membrane of the gastrointestinal tract. The seed husk has the property of absorbing and retaining water and therefore is employed in treating constipation, chronic dysentery, and related ailments. Large quantities of seed and seed husk are imported by countries of Western Europe and the U.S. annually.

B. THE PLANT

Psyllium is a stemless herb covered with soft hairs. It attains a height of 30 to 45 cm, with large narrow linear leaves of 7.5 to 20 × 0.6 in dimension, borne in whorl. Approximately 60 days after sowing, the plant gives rise to flowering shoots that bear short, 1.2 to 4 cm × 0.5 cm, flowering spikes. These spikes bear 45 to 69 flowers; the flowers are minute, tetramerous, anemophilous, and protogynous, hence favouring outcrossing. The fruit is an ellipsoidal capsule, about 8 mm long, containing 2 to 3 mm boat-shaped, rosy-white seeds. The seedcoat is a membranous mucilagenous covering called husk and is separated on milling. The seed husk is a translucent, odourless mass and constitutes about 30% of seed by weight. The 'Gujarat Isabgol-1' and 'Gujarat Isabgol-2' are two high-yielding varieties cultivated in India.

C. SOIL AND CLIMATE

Psyllium is grown in arid to semi-arid climates in the cold season where some arrangement for irrigation is possible. The crop prefers light, well drained sandy-loam to loamy soil of pH varying from 7 to 8.5 and of poor to low fertility. It needs sunny days at ripening. Cloudy weather, dew, high winds or light rain, etc., during the reproductive phase causes heavy loss due to seed shattering.

D. LAND PREPARATION AND PLANTING

This is a short-duration (120 days) crop grown as a late winter annual in India. The land is ploughed, harrowed, and planked and divided into small flat beds of 8 × 3 m to facilitate interculture and irrigation. The seed can be sown from mid-October to the end of December, but early December sowing, when soil temperature is between 10° and 5°, is preferred; this enables the crop to escape loss of yield partly due to downy mildew disease. The seed is treated with 3 g/kg of Seresan (fungicide) and sown by broadcast at a rate of 4 kg/ha. The seed is small, weighing 600 to 660 pg, and must be mixed with 10 times its weight of sand before sowing. In actual practice the fields are accorded light irrigation and sowing is done 3 to 4 days after. The sown fields are covered with a light film of 0.5 cm soil through the use of a sweeping broom. Germination commences in 5 to 8 days and is around 90% 2 weeks after sowing.

E. FERTILIZER, IRRIGATION, AND INTERCULTURE

The crop has a shallow, fibrous root system and responds only to low doses of fertilizers. Trials on the crop over the years have shown that 25 to 30 kg each of N and P is sufficient over poor lands under Indian conditions where the soil is rich in potash. It is found that wherever cowpea or cluster bean-like legume crops are grown preceding psyllium, the latter needs no nitrogenous fertilizer. The P-fertilizer is applied together with 25 kg of 10% BHC (insecticide) at the final land preparation basally, whereas the N-fertilizer is top-dressed in moist soil after 30 days of sowing and mixed in the soil. It needs two weeding cum hoeings at 30 and 50 days. Pre-emergent application of 0.75 to 1.0 kg of Isoproturan (weedicide) maintains weed-free conditions for the initial 30 days and thus reduces cost on weeding considerably. A minimum of three light irrigations are recommended at about 30, 60, and 75 days after sowing.

F. PESTS AND DISEASES

White grub sometimes attacks the seedlings for which soil treatment with BHC/Aldrin is prescribed. Downy mildew (*Peronospora plantiginiae*) causes serious losses to this crop. It appears at the initiation of spike; the symptoms first appear over leaves as small patches, which spread throughout the leaf, destroying it. Humid weather and high dew conditions favour high disease incidence. Three sprayings of any copper fungicides such as copper oxychloride (0.2%), Dithane-M-45 (0.2%), or Bordeaux mixture after 30 days of sowing and then at an interval of 15 to 20 days is recommended mainly as a prophylactic treatment.

G. HARVESTING AND YIELD

The crop is harvested when spikes begin changing to a reddish-brown colour. At this ripening stage the lower leaves dry up and upper leaves turn yellow. The plants are cut at this stage with a sickle after the sunrise (10 a.m.) at ground level and bundled in small lots for transport to a threshing shed. After 2 to 3 days of drying in open fields, these are threshed with the help of tractors or bullocks in the morning hours to allow even separation of spikes. The seed yield is about 1 ton/ha. On milling of the dry seed in a series of specially built stone crushers, the seed-husk separates out; the seed husk is a rosy-white, light, odourless substance, and its yield is around 3 q/ha by weight; the rest of the seed is rich in protein and starch and is used as a poultry feed.

XIV. PEPPERMINT (*MENTHA PIPERITA* LINN.)

A. GENERAL

The mints are extensively cultivated both in temperate and tropical countries of the world for production of aromatic volatile oils, and contain a wide range of aroma chemicals. Among the several species grown commercially in different regions of the world, the oil of the peppermint *(Mentha piperita* Linn.) is more widely used in pharmaceutical preparations. To a lesser extent, the essential oil and menthol derived from Japanese mint *(Mentha arvensis* Dc. *var. piperescene* Holms) and the oil of spearmint *(Mentha spicata* Linn.) are used in the pharmaceutical industry. Oil of peppermint is an excellent carminative and gastric stimulant; it imparts an acceptable flavour and cooling effect to the pharmaceutical preparations given orally. It is more extensively used in confectionery, alcoholic drinks, cough drops, chewing gums, dental creams, mouthwashes, and a large number of toilet preparations.

B. THE PLANT

There are over 40 species of *Mentha* (Labiatae) distributed mainly in Mediterranean Europe, south, southeast, and far eastern Asian countries. These are essentially moisture-loving perennial herbs that cross among themselves freely, producing several intermediate forms that contain several aroma chemicals in varying percentages in their volatile oils. Of these, *M. piperita* is essentially a temperate species which grows 50 to 60 cm in height. It has a weak stem, producing few branches with opposite ovate thick leaves, covered with glandular hairs. It yields 0.2 to 0.3% oil (fresh weight) of characteristic minty odour containing 50 to 55% of menthol. There are two major subspecies known under cultivation: the black mitcham called *vulgaris* ($2n = 76$) is a hardy plant which produces a higher oil content and a higher herb yield. The other white mint called *officinalis* ($2n = 36$) is less hardy, and a low oil yielding herb, but it has oil of superior aroma. The Japanese mint ($2n = 96$) is a hardy, robust plant forming a bushy growth which attains about 80 cm height. It has a purple stem bearing thick coraceous, broadly ovate dark green leaves containing a higher oil content (0.5%) and the oil has from 64 to 89% menthol. The spearmint ($2n = 48$) is similar to peppermint in growth habit, but the oil has a characteristic odour of dill oil and is rich in carvone (59 to 69%). These species come into profuse flowering wherever the crop is exposed to 14 h or more daylight. The volatile oil occurs in numerous minute glands distributed all over the surface of the leaves; the stem contains less oil. The flowers are light pink to mauve in colour, but it seldom produces viable seeds. The plantation is raised mainly through fresh stolons (underground stem) in the spring season or suckers formed during the rainy season.

Several commercial cultivars are found in cultivation in different countries adapted to specific growing conditions. In temperate regions, the crop is maintained for 2 years; the first year's crop remains in distinct rows called row mint, but it covers the field in the second year when it is called meadow mint. In subtropical climates and the foothills of India the growth is faster such that the plant covers the entire field after rains and two to three herbage crops are taken in a period of 240 to 260 days.

C. SOIL AND CLIMATE

Peppermint can grow over a wide range of soils and climatic conditions, however, it prefers deep, moist, fertile loam to clay-loam soils varying from pH of 6.0 to 8.5, rich in humus and organic matter or muck. Growth is rapid in warm temperate weather and the plant sheds leaves, becomes dormant in winter, and pereniates through underground stolons.

D. LAND PREPARATION AND PLANTING

All peppermint species demand a well-worked fertile soil with high organic matter. After soil working, the fields are treated with between 20 to 40 tonnes/ha of FYM or compost at final land preparation. The seed rate is 250 to 400 kg/ha of fresh stolons. The fresh stolons are terete white juicy underground stems, and are treated in 0.3% solution of Agallol for 3 min and broken into 10 to 12 cm pieces before planting 1 to 2 cm deep in rows at 30 to 45 × 10 cm during February-March, depending upon soil temperature. A light irrigation should follow immediately. The stolons commence sprouting in 6 to 8 days and 90% of these stolons sprout in 10 to 15 days which cover the entire field.

E. FERTILIZER, IRRIGATION, AND INTERCULTURE

The crop responds to application of high nitrogenous fertilizers under luxuriant irrigation supply.[2] Usually 120 to 150 kg of N, 60 kg of P, and 40 kg of K/ha is recommended for use over medium fertile soils. About 20% N, together with the entire quantity of P and K fertilizers is mixed in the soil at planting. The remaining N-fertilizer is divided in two or four equal splits and top-dressed in moist soil in rows 60 days after planting after the first harvest and at suitable intervals, say, in February-

March and June in the second year. The crop needs two to three irrigations per month during the dry season and two to three interculture operations before each harvest. It is noted that preemergent application of 1.5 kg of Terbacil (Sinbar) weedicide, provides weed-free conditions in the initial growth period. In India, the peppermint crop is rotated with wheat or pea in cold weather followed by an oilseed crop (rapeseed, soybean, sunflower) and paddy in the next year to reduce weed growth in the fields.

F. PESTS AND DISEASES

This crop is damaged by a number of pests and diseases in different countries. By and large, army caterpillar *(Diacrisia obliqua)* attacks the crop and defoliates the plants in a few days. Spraying 0.2% of Folidol or Malathion is recommended for controlling the pest. Stolon rot *(Macrophomina phasioli)* is another common disease of peppermint. The infected stolons show brown lesions, turning dark, and later increasing in size, causing soft decay. It produces yellowing of leaves, stunted growth, and eventually in the death of the affected plants. Mint anthracnose disease is common in the U.S., caused by *Sphaceloma menthae*. It appears as small brown spots over leaves, extending to new shoots, causing defoliation.

Leaf spot *(Alternaria* spp.) is also reported, causing irregular brown spots that coalesce, become bigger, and destroy the leaves. Spraying of any copper fungicide (0.2%) is found to be effective.

G. HARVESTING AND YIELD

The crop should be harvested at the early flower bud stage, utilising harvester or sickle (manually). Wherever the flowering is shy, the presence of a minimum of 0.2% oil (fresh weight) estimated through sample distillation in a Clevenger's apparatus indicates its maturity for harvest. The oil content decreases rapidly after the full bloom stage. In subtropical conditions, the first harvest is made 120 days after planting and a second after 90 to 100 days of the first harvest. Similar harvests are obtained in the second year in temperate regions. The presence of weeds in the harvested material affects the odour and flavour of the oil and this should be avoided. The harvested crop is allowed to wilt in (dry) field for 2 to 4 h before transport to distilleries; it is chopped into pieces and charged into the stills immediately to avoid loss due to fermentation in heaps. The loss in stalking of the produce before distillation could be as much as 50% of its oil. The yield of fresh herb is 20 tonnes/ha/year, which in turn produces 40 to 50 kg of oil on distillation. It is found that harvesting of the crop during sunny days yields oil containing between 51 to 65% total menthol and 2.5 to 5.3% menthone, whereas shaded plants yield oil containing 47 to 53% total menthol and 11.8% of menthone; the menthone possesses a bitter and rather harsh odour, an excess of which is undesirable in high-quality peppermint oil. During cloudy weather the menthone content of oil is high.[8] The oil yield of Japanese mint is high, 100 to 150 kg/ha/year. The oil, after being completely freed from adhering moisture, is a pale yellow, aromatic, mobile liquid, slightly acidic in reaction. It is stored in airtight, galvanised (or steel) drums filled to the brim and placed in a cool, dry place.

REFERENCES

1. Bammi, R. K., Rao, G., and Randhawa, G. S. Commercial cultivation of *Dioscorea floribunda,* Report of Ford Foundation Project, 1-6, 1972.
2. Bharadwaj, S. D., Kaushal, A. N., Katoch, P. C., and Gupta, R., Level and time of nitrogen application to peppermint *(Mentha piperita* Linn.). *Indian Perfum.,* 24(1), 27-30, 1980.
3. Bharadwaj, S. D., Srivastava, L. J., and Puri, S., Studies on different yield parameters in *Solanum laciniatum* Ait. as influenced by different levels of nitrogen, in *Proc. of Improvement of Forest Bio-mass,* P. K. Khosla, Ed., Indian Society of Tree Scientists, Solan, Himachal Pradesh, India, 1982, 261-264.
4. Chaddah, K. L. and Rao, M. V., Cultivation of wild yam for steroid drug industry, *Indian Fmg.,* 28(4), 13-14, 1984.
5. Duke, J. A. Ecosystems data on medicinal plants, in *Cultivation and Utilization of Medicinal Plants,* C. K. Atal and B. M. Kapoor, Eds., Regional Research Laboratory, Jammu-Tawi, India, 1982, 13-23.
6. Farnsworth, N. R. and Soejarto, D. D., Potential consequences of plant extinction in the United States on the current and future availability of prescription drugs, *Econ. Bot.,* 39(3) 231-240, 1985.
7. Franz, Ch., Nutrient and water management for medicinal and aromatic plants, *Acta Hort.,* 132, 203-215, 1983.
8. Guenther, E., *The Essential Oils,* Vol. 5, D. van Nostrand, New York, 1949, 586-621.
9. Gupta, R., Medicinal and aromatic plants, in *Handbook of Agriculture,* Indian Council of Agricultural Research, New Delhi, 1980, 1180-1224.
10. Gupta, R., Medicinal plants component of Indian forests, *Indian Fmg.,* 25(11), 95-98, 1977.
11. Gupta, R. and Pareek, S. K., Status of fertilizer use in medicinal plants, *Fertilizer News (India),* 26(3), 8-16, 1981.
12. Gupta, R., Recent advances in cultivation of Isabgol *(Plantago ovata* Forsk.) in India, in *Cultivation and Utilisation of Medicinal Plants,* C. K. Atal and B. M. Kapoor, Eds., Regional Research Laboratory, Jammu-Tawi, India, 1982, 406-417.
13. Kurnosov, V. V., Kondraterko, P. T., and Pilkova, L. A., Nitrogen nutrition and irrigation regimes of *Solanum laciniatum* Ait., *Soil Fert. Abstr.,* 35(2), 1779, 1972.
14. Kalyansundaram, N. K., Sriram, S., Patel, B. R., Patel, R. B., Patel, D. H. Dalal, K. C., and Gupta, R., Psyllium: a monopoly of Gujarat, *Indian Fmg.,* 28(4), 35-37, 41, 1984.
15. Mann, J. D., Production of solasodine for the pharmaceutical industry, *Adv. Agron.,* 30, 207-245, 1978.
16. Mandal, S., Virk. S. S., Maheshwari, M. L., Srivastava, V. K., and Gupta, R., Effect of age and top removal on alkaloid biosynthesis of *Catharanthus roseus* roots, *Acta Hort.,* 188(June), 207-214, 1986.
17. Seigler, D. S, Ed., *Crop Resources,* Academic Press, New York ,1977.
18. Tetenyi, P., *Infra-Specific Chemical Taxa of Medicinal Plants,* Chemical Publishers, New York, 1970.
19. Talha, M., Radwan, A. S., and Negin, S., The effect of soil moisture deficit on growth and alkaloid content of *Catharanthus roseus* G. Don., *Curr. Sci.,* 44(17), 614-616, 1975.
20. Growing Ginseng, Farmer's Bull. No. 2201, USDA. Science and Education Administration, 1-8, 1978.
21. Markets for Selected Medicinal Plants and Their Derivatives, International Trade Centre. UNCTAD/GATT, Geneva, 1974.

Chapter 5

CLIMATIC AND GEOGRAPHICAL ASPECTS OF MEDICINAL PLANT CONSTITUENTS

M. Vanhaelen, J. Lejoly, M. Hanocq, and L. Molle

I. INTRODUCTION

"Secondary compounds may be defined as those natural products usually of plant origin, which do not function directly in the primary biochemical activities that support the growth, development and reproduction of the organism in which they occur"[1].

These compounds represent the most useful substances for chemical plant taxonomy. Furthermore, a number of secondary compounds or more correctly, nonprimary compounds, are biologically active constituents, as they derive from the primary metabolism. Ecophysiological factors that affect the general metabolism can also modify qualitatively and quantitatively the occurrence of these therapeutically interesting constituents. Individual factors involved in the biosynthetic pathways of biologically active constituents are essentially genetic and ontogenic.

However, ecophysiological factors (physical and environmental factors) can radically modify, within the limits set by the genetic makeup of the taxon, these specific chemical characteristics, as well as the morphological characteristics. Main extrinsic factors which are involved in determining the morphological aspects of a plant and the production of secondary compounds have been respectively discussed by Davis and Heywood[2] and Flück[3] and their explanations follow.

A. CLIMATIC AND GEOPHYSICAL FACTORS

1. Temperature is the major factor controlling the development and the metabolism of plants: temperature periodic effects, seasonal variations (the optimal temperature for the formation of one constituent should be considered as the sum of all temperature parameters recorded by the plant from the seed germination).
2. Light (intensity, radiation wavelength, photoperiodism).
3. Rainfall, water supply.
4. Altitude.
5. Wind.
6. Soil (physical, chemical, and microbiological properties, presence of pesticides).
7. Nutrition conditions, including minerals. For example, Chatterjee et al.[4] observed a significant enhancement of essential oil biogenesis in *Cymbopogum winterianus* induced by Mn, Mo, Mg, and B, while Brachet et al.[5] observed that the percentage of tropane esters in *Datura innoxia* was highly sensitive to NaCl stress. The addition of growth hormones can modify the total alkaloid content as described by Nandi and Chatterjee.[6,7] Both high (200 μg/ml) and low (25 μg/ml) concentrations of GA_3 (gibberellic acid) increased alkaloid synthesis in *D. innoxia* and *D. metel* during vegetative, reproductive, and postreproductive stages; high concentrations of GA_3 applied in the vegetative stage were found more effective as well for *Rauwolfia serpentina,* but alkaloids of *Nicotiana plumbaginifolia* were increased by high concentrations of GA_3 added in the postreproductive stage.

B. BIOTIC FACTORS

1. Fungal, bacterial, and viral infections. An example of fungal contaminations associated with rainfall conditions has been published by Hofman and Menary.[8] High rainfall and relative humidity, which permitted heavy fungal growth and a 28% decrease in the dry weight per sample of straw (capsules of *Papaver sommiferum*), led to losses of 35 and 32% in the morphine and codeine concentrations. Another example which related viral infection to the alkaloid content of *Vinca rosea* (actually *Catharanthus roseus)* has been established by Gheorghin and Ionescu-Matiu;[9] however, no anatomical or morphological modifications were observed, marked losses of vincaleukoblastine and of total alkaloids yield were recorded when *Vinca rosea (Catharanthus roseus),* under field conditions, was infected by a virus (Aster yellows group).
2. Presence of insects or deposition of eggs (e.g., galls).
3. Grazing, browsing, and trampling, fire.
4. Planting density. Yamada et al.[10] have showed that better plant growth, but with no significant effect on plant yield, was observed when *Swertia japonica* was grown at lower planting densities. On the other hand, total yields of bitter iridoids (amaroswerin and amarogentin) was the highest at a higher planting density of 30 plants per pot; the latter value was recommended by the authors as suitable for pot culture of *Swertia japonica.* This factor can obviously modify light exposure of plants.
5. Competition with other plant species. The influence of the living environment on medicinal plant production has been reviewed and studied using *Matricaria chamomilla* and *Sinapis alba* by Bernáth;[11] either dry matter or biologically active constituent production can be modified by the presence of other plant species in the cultures.

C. ONTOGENIC FACTORS AND SEEDING TIMES

Their influence has been well documented mainly in the case of medicinal plants containing essential oils and alkaloids.

D. GENETIC FACTORS

A plant population is a dynamic entity. Its geographical repartition could be different from one year to another. The separation into two remote new populations can be observed.

Even if mutations usually produce a one-gene change, metabolic variations can occur that modify the chemical characteristics of a plant. These modifications can be nonrecessive and confer to the plant, after genetic recombination, an adaptability to ecological differences.

The influence of environmental and genetic factors on the production of secondary compounds by plants in general and, in some instances only by medicinal plants has been discussed by many authors.[3,11-29]

Studies on the influence of climatic and geographical factors on the constituents produced by medicinal plants are rarely consistent with regard to plant material or the experimental conditions, and thus cannot lead to significant conclusions.

The comparison of results of different investigations shows frequently conflicting and even contradictory findings. The planning of each investigation should require the following conditions:

1. Plant material which is genetically uniform.
2. Collection of the same part of the plant, at a given moment of the day in the same conditions and period of development; the biologically active constituents are effectively organ and ontogenic dependent. As an example, translocation of alkaloids from roots to aerial parts is increased with temperature.[27] It should be kept in mind that the biogenesis of secondary metabolites is often correlated with periods of high metabolism, as during extensive growth.
3. As far as possible, identical conditions of drying and storage; indeed, many secondary metabolites are not stable (oxidation, hydrolysis, enzymatic reactions) or they are volatile, as, for example, essential oils.
4. Computing achieved with a sufficient number of samples to obtain significant results (statistical requirements). The expression of the results itself can present some difficulty. For example, a particular climatic factor can lead to the growth reduction of a plant which, when analyzed on a percentage dry weight basis, shows a high proportion of secondary metabolites while the yield per plant can be very low. In the same way, large plants can be obtained with an apparent low percentage of active constituent when expressed vs. dry weight but the yield per plant may exceed that of the control plant. Moreover, a negative correlation was observed between the size of *Mentha piperita* leaves and essential oil content, and between the size of *Coriandrum sativum* fruit and essential oil content.[29] Flück[3] recommends that the analytical results are best expressed in relation to the crude fibre or dry weight of the drug.
5. Reliable analytical methods for phytochemical analysis. More reliable results are obviously obtained when the studies are performed with phytotrons in which many ecophysiological factors are independently controllable such as the Phytotron of the National Center of Scientific Research (CNRS) of Gif-sur-Yvette (France) or such as the apparatus recently designed by Bernáth and Tétényi[30] (Climatron) to specifically study the effect of light, temperature, air, humidity, water supply, and other environmental factors on the secondary metabolites production by medicinal plants. However, some improvements of the culture conditions can be provided, e.g., irrigation, co-cultures with trees to reduce light exposure, and soil modification, but it is obviously not possible to radically modify climatic conditions. Thus, information about the influence of ecophysiological factors on the production of medicinal plants are more useful in order to select suitable habitats for the culture of known medicinal plant species. In this respect, Duke's concept[25] is of the highest interest. Computerized data related to precipitation, temperature and pH (mean, minimum, maximum) enabled this author to use precise numerical data for comparing the climatic adaptabilities of species, genera, and families and to bracket the ecosystematics of an area from climatic data sources; additionally, reported uses in traditional medicine have also been computerized in order to facilitate various screening programs of medicinal plants.

In this context, known phytogeographical zones through the world are first described.

II. MEDICINAL PLANT DISTRIBUTION THROUGHOUT THE WORLD

In order to study relationships between medicinal plants and general environmental factors, two distribution maps have been established, the first for the tropical and subtropical zones (Figure 1) and the second for the temperate zones (Figure 2).

The distribution of medicinal plants is considered in relation to their occurrence in main potential vegetation zones (climax). This mapping gives an insight into their potential zones for cultivation, even when they are mapped only as a point. The characteristics and distribution of the world vegetation zones were taken from Walter[31] and Duvigeneaud.[32] The original areas of plant species were compiled from different floras and cultivation areas from Hoppe,[33-35] Paris and Moyse,[36-38] Trease and Evans,[39] Tyler et al.,[40] and Wagner.[41]

A. TROPICAL AND SUBTROPICAL VEGETATIONAL ZONES

1. Evergreen Tropical Rain Forests (Pluviisilvae)

It is the wettest of all vegetational zones; the annual rainfall varies from 1800 to 5000 mm and more. Typically, the monthly rainfall does not fall below 100 mm; nevertheless, this vegetation type is also found in the zone with a short 1-month dry season.

Mean temperature of the coldest month is always above 18°C and is only 1° to 2°C lower than the mean temperature of the warmest month. On the contrary, the daily temperature variations are important mainly on sunny days; it is a diurnal climate.

As can be seen from Figure 1, many medicinal plants of this zone are Apocynaceae (*Rauwolfia vomitoria, Voacanga africana, Strophanthus hispidus, Holarrhena floribunda, H. antidysenterica, Tabernanthe iboga, Catharanthus roseus*), Piperaceae (*Piper nigrum, P. methysticum*), Arecaceae (*Areca catechu*), and Fabaceae (*Physostigma venenosum*).

2. Transitional Zone between 1 and 3

When the dry season lasts 1 to 3 months, semievergreen forests occur which can be replaced by anthropogenic savannas. Medicinal plants are the same as in zone 1.

3. Tropical Broadleaf-Deciduous Forest (Hiemisilvae)

The climate of this vegetational zone shows, like zones 1 and 2 cited above, a coldest month mean temperature above 18°C, but is characterized by a distinct annual temperature cycle with heavy zenithal rains in the warm, perhumid season. The dry period varies between 3 and 7 months. Leaf-sheeding in the tropical zone is an adaptation to the dry season. Many medicinal plants in this zone are Solanaceae (*Datura stramonium, D. innoxia, D. metel, Duboisia myoporoides*), Dioscoreaceae (*Dioscorea composita, D. floribunda, D. prazeri, D. deltoidea*), Zingiberaceae (*Curcuma longa*), and Cannabaceae (*Cannabis sativa var. indica*).

4. Tropical Deciduous Spiny Forests (Spinisilvae) and Thornbush Savannas

This vegetational zone possesses a climate characterized by a 8 to 10 month long dry period. Many degradations of the tropical deciduous spiny forest result in a thornbush savanna. The occurrence of microphylly and spines is the most striking adaptation to the extremely dry season. Typical medicinal plants are Mimosaceae (*Aracia senegal* and *A. seyal*) and Caesalpiniaceae (*Cassia acutifolia, C. angustifolia,* and *C. italica*).

B. TEMPERATE VEGETATIONAL ZONES (Figure 2)

1. Sclerophyllous Woody Plants (Durisilvae and Durifruticeta)

The climate of this zone is termed "Mediterranean" with a warm, dry summer and a mild, wet winter. Winter temperatures of 10°C and lower (too cool for plant growth) and frost are rare events. The most propitious periods for growth are spring, when the soil is moist and when the temperatures are rising, and after the first autumnal rains. Sclerophylly is the main adaptation to the long summer drought. This bioclimate occurs outside the Mediterranean region in California, Chile, South Africa (Capenland), and South Australia where many species with essential oils are found, mainly Lamiaceae (*Lavandula angustifolia, Rosmarinus officinalis, Thymus vulgaris, Salvia officinalis*) and Apiaceae (*Pimpinella anisum, Foeniculum vulgare, Archangelica officinalis, Coriandrum sativum*).

2. Temperate Evergreen Forests (Laurisilvae)

This forest type corresponds to a warm temperate climate without well-marked dry periods and without winter frost. The mean annual temperature generally varies between 12° and 18°C. The regions involved lie on the eastern parts of the continents between latitudes 30° and 35° in both the Northern and Southern hemispheres (Florida, southwestern China, southern Japan). This bioclimate occurs also in the tropical mountains between 1500 and 2500 m. Medicinal plants are Lauraceae (*Laurus nobilis, Cinamomum camphora*), Rubiaceae (*Cinchona ledgeriana, C. succirubra*), Theaceae (*Thea sinensis*), and Aquifoliaceae (*Ilex paraguariensis*).

3. Temperate Broadleaf-Deciduous Forests (Aestisilvae) and Other Typical Temperate Forests with a Short Period of Frost

This vegetational zone is characterized by a marked, but not too prolonged cold season (winter) and a warm, humid summer. Deciduous trees require a growing season lasting at least 4 months, with a mean daily temperature >10°C. Leaf-shedding in the temperate zone is an adaptation to the cold season.

Mixed deciduous-coniferous forests occur in temperate mountains and northward in transitional zones to taiga. The main medicinal plants of this zone are Polygonacea (*Rheum officinale, R. palmatum, R. rhaponticum*), Liliaceae (*Convallaria majalis, Veratrum album, Colchicum autumnale*), Ranunculaceae (*Aconitum napellus, Hydrastis canadensis*), and Brassicacae (*Brassica nigra, B. juncea*).

4. Temperate Steppe

The cold winter and the late drought summer limit the growing season of the steppe plants. Only about 4 months of favorable growth conditions occur in spring and in early summer. Only herbaceous plants can develop. Medicinal plants of this zone are Ephedraceae (*Ephedra sinensis*), Papaveraceae (*Papaver sommiferum*) and Cannabaceae (*Cannabis sativa var. indica*).

5. Boreal Coniferous Forest (Aciculisilive)

The true boreal zones begin when the climate becomes unfavorable for the broadleaf species, that is to say when summers are too short and winters too long; the duration of the period with a daily average temperature of >10°C drops below 120 days/year and the cold season lasts longer than 6 months (a cold month is a month with a mean daily minimum temperature <0°C). No medicinal plants are found in the coniferous forests.

III. ECOPHYSIOLOGICAL FACTORS AND PRODUCTION OF BIOLOGICALLY ACTIVE COMPOUNDS BY MEDICINAL PLANTS

A. ALKALOIDS

1. Tropane Alkaloids

Studying the synthesis of active constituents in relation to different photoperiodic and GA_3 (gibberellic-acid) treatments, Nandi and Chatterjee[6,7] have showed that long-day enhancement of alkaloid production was clear in *Nicotiana plumbaginifolia* and to some extent in *Datura innoxia*, whereas higher photoinductive cycles only were sufficient for the production

Figure 1. Medicinal plant distribution in the tropical and subtropical vegetational zones. Equatorial with diurnal climate: (1) evergreen tropical rain forests; (2) transitional zones between 1 and 3. Tropical with summer rains: (3) tropical broadleaf-deciduous forests; (4) tropical deciduous spiny forests and thornbush savannas. Subtropical arid: (5) subtropical desert vegetation.

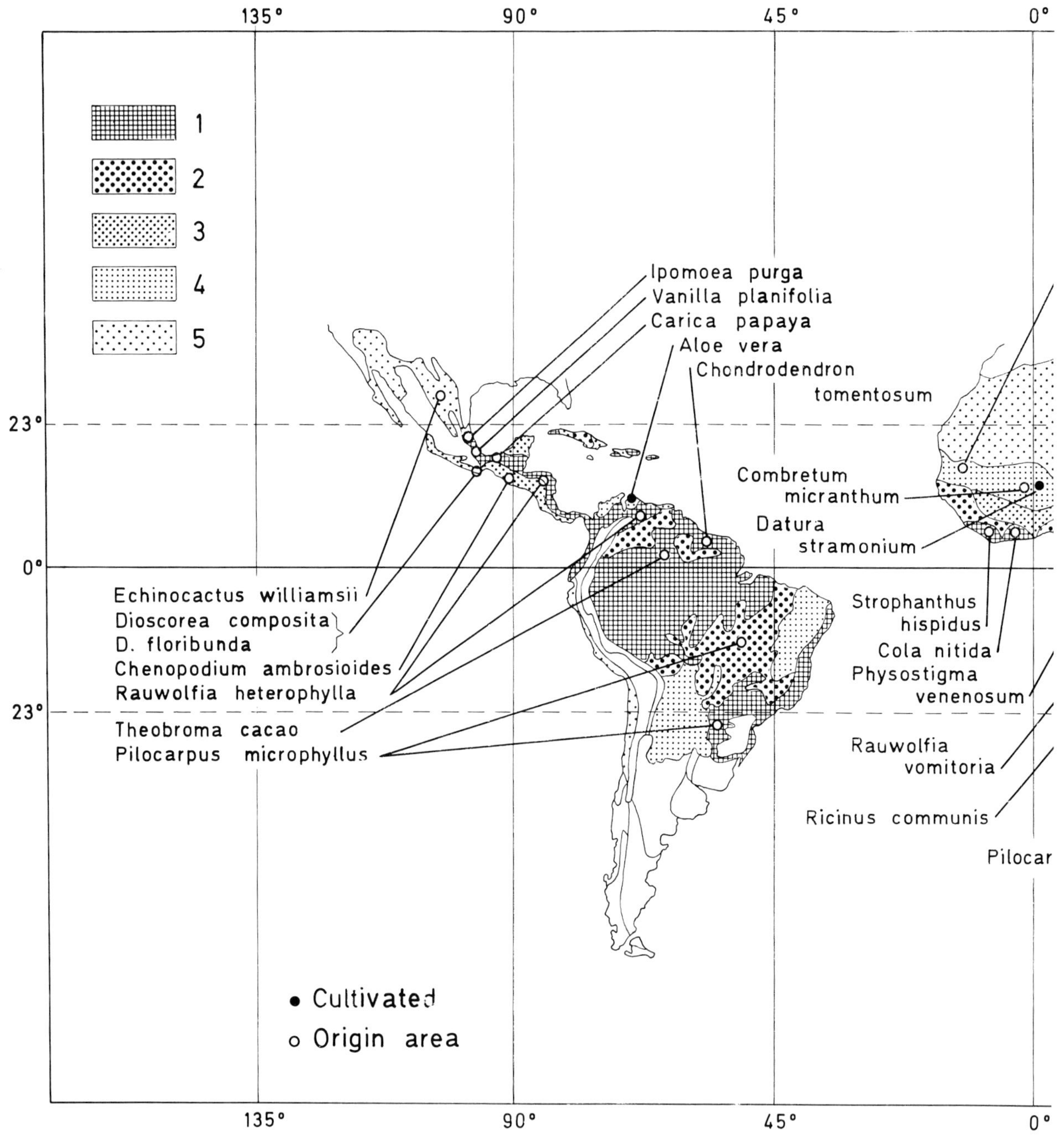

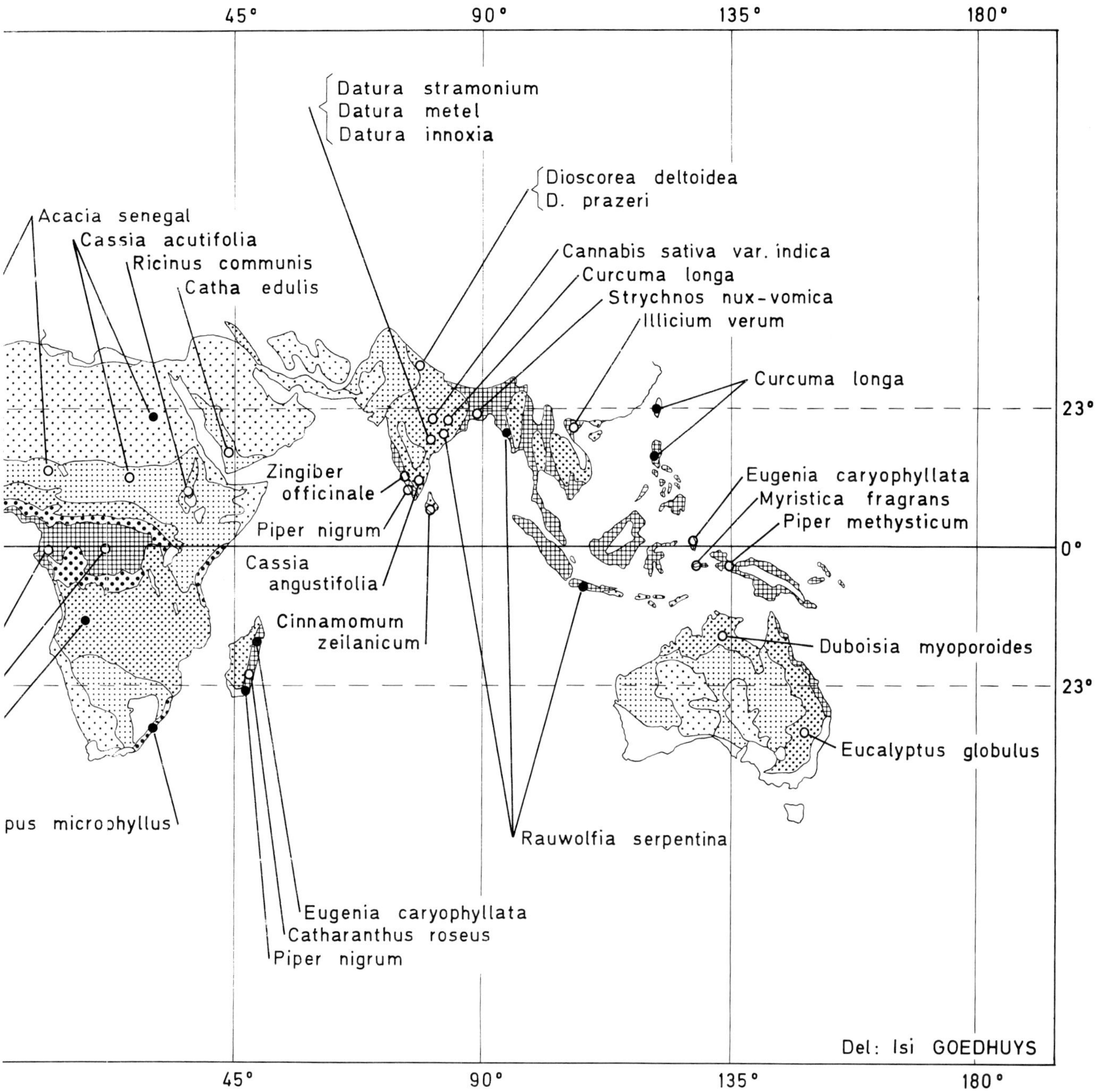
45°
90°
135°
180°
Datura stramonium
Datura metel
Datura innoxia
Dioscorea deltoidea
D. prazeri
Acacia senegal
Cassia acutifolia
Ricinus communis
Catha edulis
Cannabis sativa var. indica
Curcuma longa
Strychnos nux-vomica
Illicium verum
Curcuma longa
23°
Zingiber officinale
Piper nigrum
Eugenia caryophyllata
Myristica fragrans
Piper methysticum
0°
Cassia angustifolia
Cinnamomum zeilanicum
Duboisia myoporoides
23°
Eucalyptus globulus
pus microphyllus
Rauwolfia serpentina
Eugenia caryophyllata
Catharanthus roseus
Piper nigrum
Del: Isi GOEDHUYS
45°
90°
135°
180°

Figure 2. Medicinal plant distribution in the temperate vegetational zones. Winter rain and summer drought: (1) sclerophyllous woody plants. Warm-temperate humid climate: (2) temperate evergreen forests. Typical temperate with a short period of frost: (3) temperate broadleaf-deciduous forests; (4) mixed forest (deciduous-coniferous forest); (5) humid coniferous forest; (6) montane mixed forest. Arid-temperate with a cold winter (continental): (7) temperate desert vegetation; (8) temperate steppe. Cold temperate (boreal): (9) boreal coniferous forests (taiga); (10) transitional taiga-tundra.

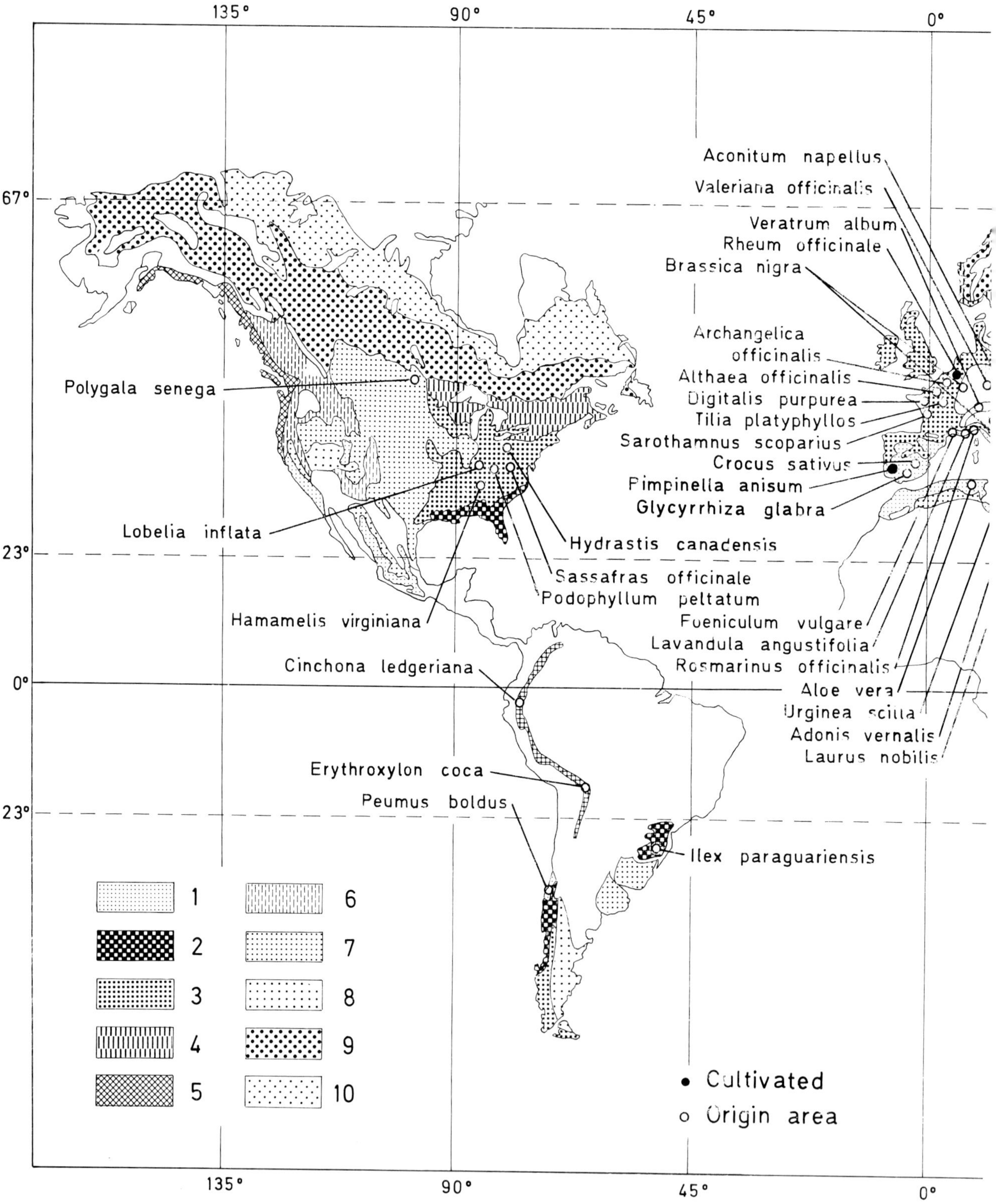

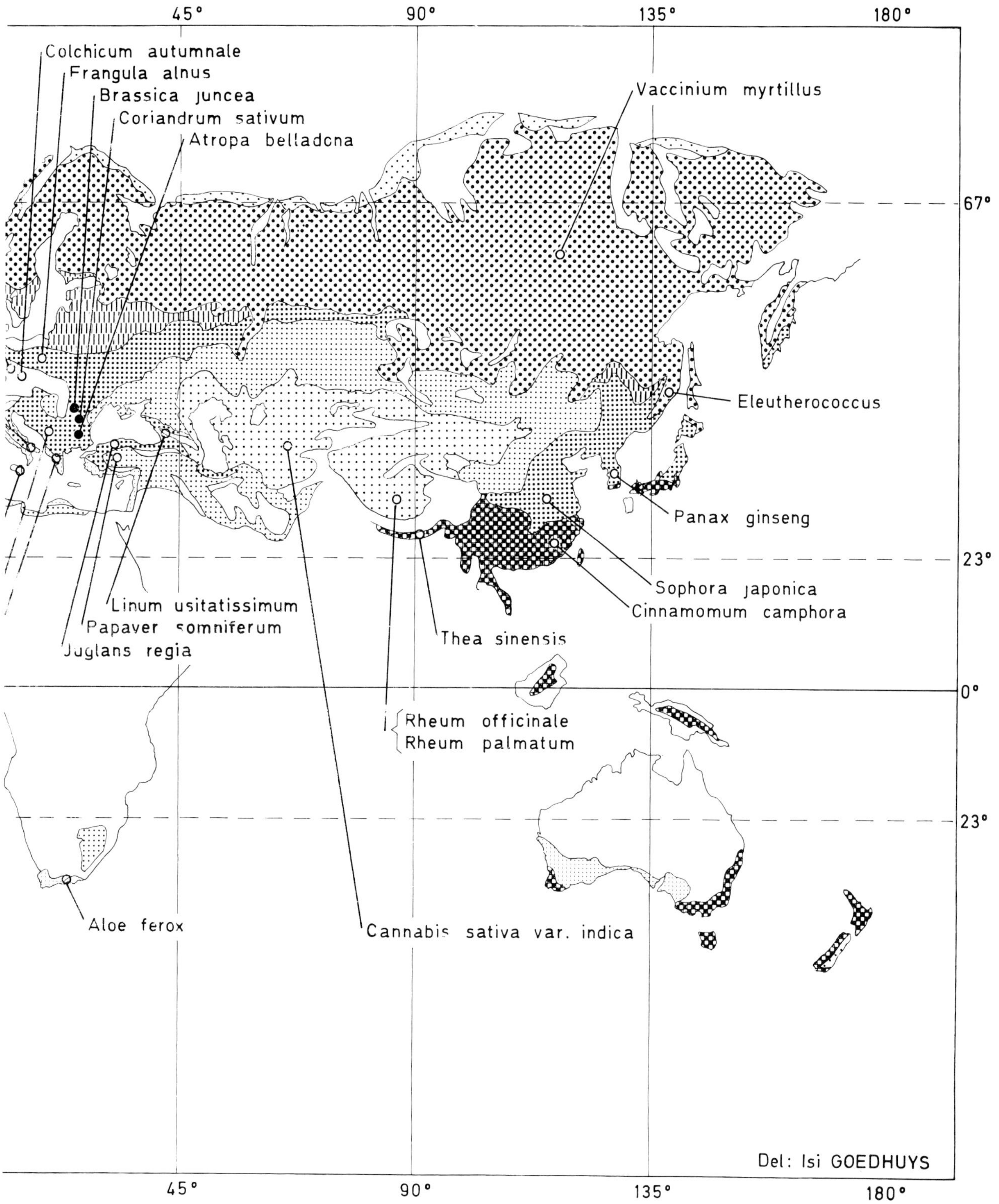
45°
90°
135°
180°
Colchicum autumnale
Frangula alnus
Brassica juncea
Coriandrum sativum
Atropa belladona
Vaccinium myrtillus
67°
Eleutherococcus
Panax ginseng
23°
Sophora japonica
Cinnamomum camphora
Linum usitatissimum
Papaver somniferum
Juglans regia
Thea sinensis
0°
Rheum officinale
Rheum palmatum
23°
Aloe ferox
Cannabis sativa var. indica
Del: Isi GOEDHUYS
45°
90°
135°
180°

of alkaloids by *Rauwolfia serpentina*. On the other hand, increase of the alkaloid production in *Datura metel* was observed with short-day photoinductive cycles. Total alkaloid content in *Datura innoxia* was highest during the reproductive stage and lowest in the postreproductive stage.

Influence of ecophysiological factors on the alkaloid content of *Datura* species has been extensively studied by Cosson et al.[42-47] using the various climatic conditions available in the phyotron of Gif-sur-Yvette. In the case of *Datura metel* and *D. tatula*, yields of scopolamine were only dependent on light which could modify the general metabolism of the plant (e.g., inhibition of scopolamine by esterase) or more specifically the epoxidation of hyoscyamine. Increase of scopolamine content in *Datura tatula* at flowering was observed only with long-day exposure; light intensity was found more important for the latter species, and in opposition to *Datura metel*, the highest yield of alkaloid was observed when flowering was delayed. Thus, the alkaloid yield in the leaves of one plant species was strongly related to both the physiological stage of the plant and to the light, as shown in Table 1. Studies by Nandi and Chatterjee[48] led to the same conclusions. The alkaloid production was increased by light (long-day treatment) when compared with decreasing effect of short-day treatment. Light has been shown to have stimulatory effects on the levels of some enzymes in plant tissues as does phenylalanine ammonia lyase (PAL).[49] Such effects on the plant metabolism could explain the above findings.

Further studies of Cosson et al.[50,51] on *Datura tatula* and *Duboisia myoporoides* showed that hyoscyamine and scopolamine can be considered markers of the ecophysiological events leading to ontogenesis; the ratio of scopolamine to hyoscyamine in *D. myoporoides* leaves decreased during ontogenesis.

Related to genetic heterogeneity, a great variability of the alkaloid content was observed in *D. myoporoides* from Australia and not from New Caledonia; furthermore, bacterial infection *(Agrobacterium tumefaciens)* led to a 2- to 10-fold increase of the alkaloids contained in infected tissues.

Altitude can also influence the alkaloid content of *Datura metel*;[52] the total alkaloids were comparatively lower at sea level than at high altitude, as shown in Table 2. An opposite influence of altitude on the alkaloid constitution of *Datura innoxia* was observed by Pelt et al.[53,54] This effect was attributed to the decrease of the vegetative cycle at higher altitude; in addition, the comparison of the total alkaloid yield in *Datura stramonium* growing at different sites but collected at the same ontogenic stage showed little difference when the climatic conditions were similar (Costa Brava, Spain and Afghanistan).

Another climatic factor that can generally increase the alkaloid content is water stress.[24] According to Gupta et al.,[55] the alkaloid content in leaves of *Datura metel* was essentially determined by the growth stage of the plant (Aslanov[56] shared this conclusion in his detailed study of the alkaloid content in different organs and at different phases of development of an other species, *Physochlaina orientalis*); however, alkaloid yield decreased with rains. Similar observations have been published for other Solanaceae such as *Datura stramonium*, *Hyoscyamus niger*, and *Duboisia myoporoides*,[57-59] the total alkaloid content in the latter species has also been compared at two sites in Japan.[60] Finally, important data concerning the qualitative and quantitative evaluation of tropane alkaloids for Indian raw material *(Atropa acuminata, A. belladona, D. innoxia, D. metel, D. stramonium,* and *Physochlaina praealta)* growing wild in various parts of India have been collected by Sarin.[61]

2. Alkaloids from *Papaver somniferum* L., *P. bracteatum* Lindl., *P. orientale* L., and *P. pseudo-orientale* L.

Genetic factors essentially influence the alkaloid content (total yield and relative concentration of each chemical group) of *Papaver somniferum*,[62] *P. orientale*, *P. pseudo-orientale*,[63] and *P. bracteatum*;[64] distinct alkaloid spectra characterized each strain. As the alkaloid differentiation occurs 18 to 26 days after germination, it seems relevant to study first the influence of climatic factors on the germination; the effects of temperature (isotherms from 6° to 35°C) and light effects (darkness with or without addition of red or far-red light) on the germination rate of *Papaver* spp. seeds have been extensively studied by Bare et al..[65] The authors were able to determine the optimal conditions of germination for *Papaver somniferum*, *P. bracteatum*, and *P. orientale*. As previously reported, significant losses of alkaloids can occur when *Papaver somniferum* is infected by fungi (Pleospora);[8] heavy fungal growth promoted by high rainfall and relative humidity could lead to confusing conclusions in relationship with the actual effects of the climatic conditions. Another source of confusion is provided by the diurnal variations of the alkaloidal content of latex samples from *Papaver somniferum* and *P. bracteatum*;[66,67] however, diurnal variations in the thebaine content found in *Papaver bracteatum* was much smaller (±20% of the daily mean value), the morphine variations found in *P. somniferum* can reach ±90%!

The stimulation effects of light and temperature on alkaloid biosynthesis in *Papaver* spp. have been established by many authors. A high percentage of morphine in capsules of *Papaver somniferum* was found 98 days after germination and under 18 h daylength and higher in capsules and roots at 10°C than at 21°C.[68]

A 14% increase of the alkaloidal content was observed when *Papaver somniferum* was cultivated under dry conditions.[69] Meanwhile, Tétényi and Vaguyfalvi[62] have found that ratio of the individual alkaloids within a chemical group was strongly dependent on the climatic conditions. Tookey et al.[70] showed that the ratio of codeine:morphine was higher in field conditions (Arizona) than in controlled environment.

When *Papaver bracteatum* was cultivated in controlled environments, the best development up to the age of 78 days was found with a 20 h daylength and light intensities of 8000 lux or above; at a photoperiod of 24 h, an inhibiting effect appeared.[71] Bernáth andTétényi,[72] also using a phytotron, confirmed that the alkaloid biosynthesis in *Papaver somniferum* increased two to three times at higher light intensity of long day treatments (2.4 to 3.2 10^4 lux) and at higher temperature (up to 26°C); at the same time, methylated alkaloid yield increased so that the presence either of morphine or codeine became char-

TABLE 1
Yield in Alkaloids Production from Young Leaves of *Datura tatula* C. at Different Stages of Development and Under Two Different Light Conditions (mg alkaloids/100 g leaves)[47]

Light period (h)	Alkaloid	Stage 3	Stage 4	Stage 5	Stage 6
	Scopolamine (S)	3	13.5	11	12
16	Hyoscyamine (H)	0.5	3	8.5	17
	Total	3.5	16.5	19.5	29
	S/H	6	4.5	1.3	0.7
	Scopolamine (S)	2	4	9	7
9	Hyoscyamine (H)	0.5	3	5	9
	Total	2.5	7	14	16
	S/H	4	1.3	1.8	0.8

Stage 3: appearance of the first flower bud (juvenile stage); stage 4: blooming of the first flower (stage of metabolic maturity); stage 5: beginning of the fruit maturation (senescence stage); stage 6: plants from 3 to 5 months

TABLE 2
Influence of Altitude on the Alkaloid Content (%) in Different Parts of *Datura metel*[52]

Site and altitude	Roots	Stems	Leaves	Flowers	Fruits	Seeds
Bombay (sea-level)	0.27	0.19	0.25	0.69	0.060	0.09
Poona (563 m)	0.52	0.29	0.32	0.86	0.079	0.10
Pachmari (716 m)	0.71	0.43	0.54	0.95	0.084	0.14
Darjeeling (2166 m)	0.89	0.46	0.58	0.99	0.097	0.19

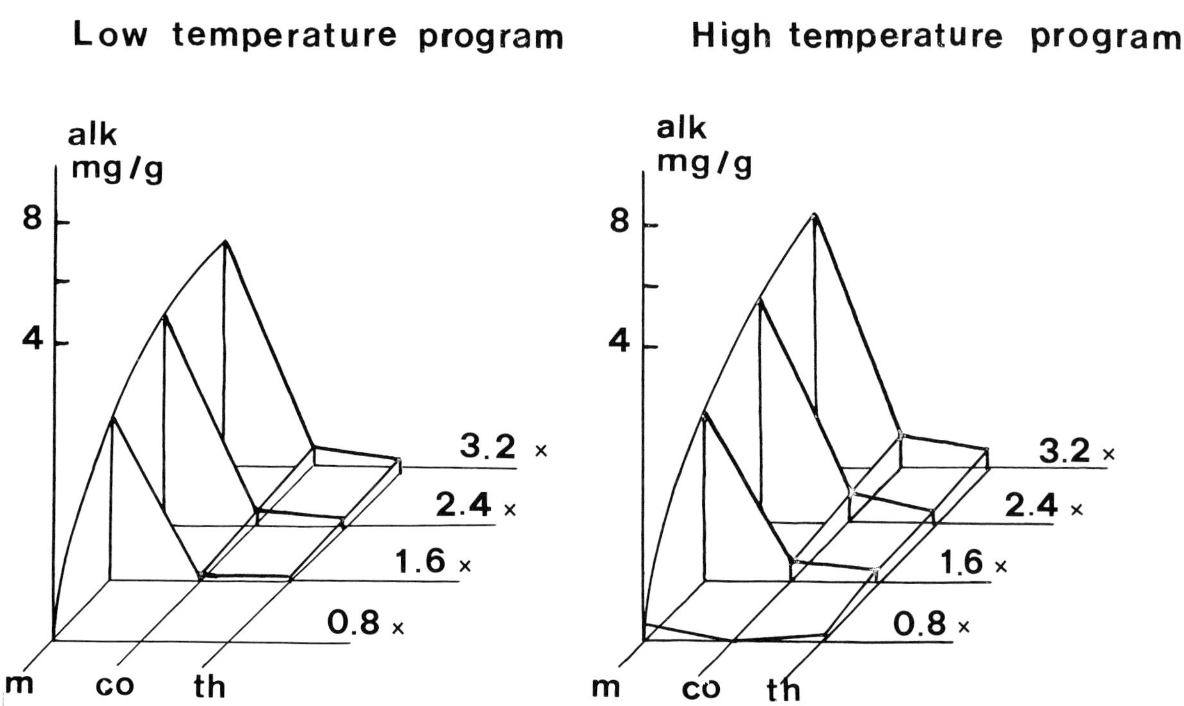

Figure 3. Morphinane alkaloids spectra of dry capsules of *Papaver somniferum* L.; interaction of light and temperature factors (cv. 'Reading'). m: morphine; co: codeine; th: thebaine.[74]

acteristic of the light conditions (Figure 3).Taking into consideration the biogenetic pathway of these alkaloids, it was postulated that a temperate decrease could promote the return of alkaloids into the general metabolism of the plant and that ecophysiological factors could regulate the stereospecific enzyme reaction of reticuline.[72-74]

Finally yields of morphine and codeine were found strongly dependent on climatic conditions of the culture year (five to ten times different yields for one year to another),[75] and Gracium and Dumitrescu[27] record that cultures of *Papaver somniferum* yield 8 kg alkaloids per hectare in northern Europe while they may generate as much as 20 kg alkaloids per hectare in subtropical regions.

3. Steroidal Alkaloids

Because of the demand for steroid raw material by the pharmaceutical industry, numerous studies on the glycoalkaloids of plant origin have been published. According to Gauhl[76] and Honváth et al.,[77] besides geographical factors, site conditions also play an important role in the distribution of morpho- and chemotaxons of *Solanum dulcamara* and may result in the development of ecotypes independent of these taxons; under this condition an adaptational nature occurs in the metabolic processes.

Both the plant weight and the alkaloid production were considerably influenced by ecophysiological factors for one *Solanum dulcamara* taxon containing tomatidenol (main aglycon), solasodine, and soladulcidine.[78] Bernáth et al.[79] carried out comparative studies on the formation of steroidal alkaloids in two *Solanum* spp. (the heliophyton *S. laciniatum* and the ubiquitous sciaphyton *S. dulcamara*) with different ecological demands. Production of dry matter and alkaloids was higher with increased light intensity; with low intensity light the alkaloid yield decreased to one quarter in *S. laciniatum* which has lower adaptability to weak light supply, while it remained unchanged in the successfully adapted species *S. dulcamara*. In the same way, the alkaloid level was more dependent on other ecological factors (temperature, water, soil, and nutrient supply) in *S. laciniatum*. The alkaloid content expressed by leaf area unit increased from 1.6 to 5.2 $\mu g/cm^2$ for *S. dulcamara* and from 2.4 to 4.6 $\mu g/cm^2$ for *S. laciniatum* when light of shorter wavelength was used. Continuing their previous studies, Bernáth et al.[80] established that irrespective of the origin, individual clones of *Solanum dulcamara* showed an increasing total steroid aglycon content when exposed to short wavelength light; however, the rate of change depended on the chemism of the individual clones. The difference was found to be of the highest rate in the solasodinic clone where a higher ratio of solasodine:soladulcidine was found (increased unsaturated character of the aglycons). These reported modifications supported the hypothesis of Máthé and Máthé,[81] according to which ecological factors are of significance in the production of chemotaxa.

Chatterjee et al.[82] observed that higher alkaloid yields are obtained with *Solanum* growing in higher altitudes and collected at the postreproduction stage; light enhanced alkaloid synthesis in these plants.

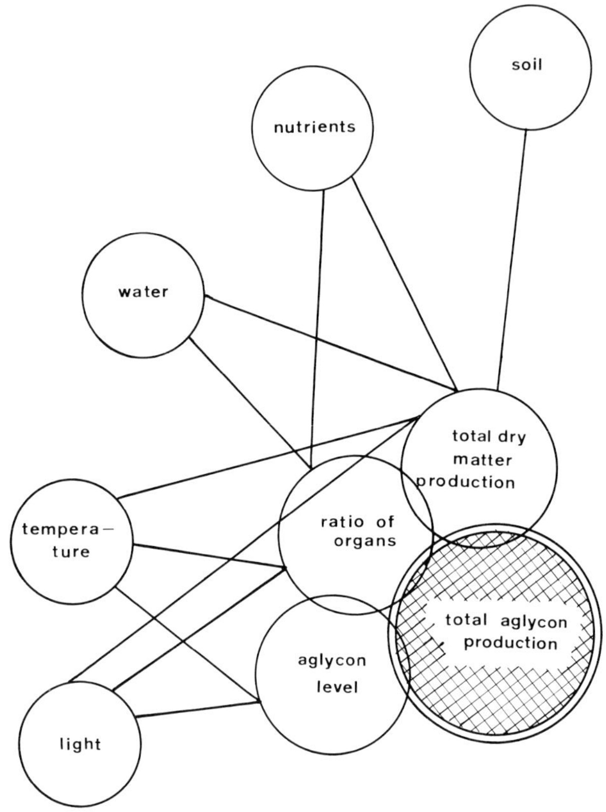

Figure 4. Factors determining alkaloid production in *Solanum dulcamara* L. and *Solanum lacianatum* Ait.[83]

Summarizing the results of their investigations, Bernáth et al.[11,83] concluded that it is more reasonable to deal with the related ecological factors-alkaloid type and with primary reactions — alkaloid production depending on the ecotype; following Figure 4, it can be stated that the spectrum composition of the light was the most unambiguous factor, since it was effective irrespective of the ecotype of the species; however, light intensity, photoperiod, and temperature also have an effect on the alkaloid levels.

Finally, Yaniv and Palevitch[24] (and references therein) emphasized the importance of drought on the production of alkaloids.

Dioscorea deltoidea and *D. prazeri* from different localities of India, grown in culture, yielded varying percentages of sapogenin content at various stages of growth. *D. deltoidea* growing at Darjeeling showed 4.8% sapogenin while the same species collected from Kashmir showed 3.8% sapogenin.[84] All the constituents determined in *Dioscorea deltoidea* (total sapogenin and phenolic compounds) were found at a higher concentration in short-day plants but their ratios were similar.[85] According to Chatterjee et al.[82], a better yield in alkaloids was obtained at low altitude with plants collected at the reproduction stage; light enhanced synthesis of plants constituents.

4. Ergot Alkaloids

Detailed data on the variability in the content and composition of alkaloids found in Canadian ergot were published by

Young et al.[86-88] The total alkaloid content as well as the concentration ratio of the individual alkaloids determined by HPLC (ergometrine, ergosine, ergotamine, ergocornine, ergocryptine, and ergocristine) were highly variable; the total alkaloid content ranged from 0.011 to 0.452% for rye ergot, from 0.013 to 0.307% for wheat ergot, from 0.042 to 0.752% for Triticale ergot, and from 0.082 to 1.04% for barley.

5. *Cinchona* Alkaloids

The dependence of altitude on the alkaloid production of *Cinchona* spp. has been observed for a long time and represent a good illustration of the influence of climatic factors on the production of biologically active constituents by medicinal plants. The best information in this field has been collected by Chatterjee,[89] who has intensively studied the cultivation of Cinchona in West Bengal (India). This review describes

1. Conditions (temperature, rainfall, altitude, soil, slope) that were required for successful growth of *C. ledgeriana*
2. Morphological variations and bark yield in *C. succirubra, C. ledgeriana, C. robusta,* and *C. hybrida*
3. Altitudinal variations of bark yield of these last four *Cinchona* spp.
4. Variations of bark yield in *C. ledgeriana* following altitude
5. Alkaloid yield data of different regions of West Bengal *Cinchona* cultures compiled from 1945 to 1975

Once more, the effect of drought was predominant,[24] as *Cinchona* produced no quinine during the rainy season.

Investigation of lead alkaloids of three *Cinchona* spp. was recently achieved using HPLC.[90] The leaves were collected from four plantations: Kenya and Zaire *(C. ledgeriana),* Thailand *(C. succirubra),* and Guatemala *(C. succirubra* × *C. ledgeriana);* samples from Kenya and Zaire showed the greatest variations in the alkaloid content.

6. *Tabernaemontana* Alkaloids

Variations in six indole alkaloids in *Tabernaemontana dichotoma* collected at different sites in Sri Lanka were determined by HPLC analyses.[91] The amounts of these alkaloids in the different extracts showed a wide range of variation (total alkaloids content and individual alkaloids) (Tables 3 and 4). The variations were primarily related to 12-methoxyvoaphylline which is the major component in most of the leaf extracts; qualitatively the differences were only small and concerned only minor peaks in the chromatograms. Thus, factors such as chemical races, environmental conditions, and different stages of development could influence the alkaloid content.

7. Miscellaneous Alkaloids

Positive effects of limited water supply on alkaloid biosynthesis have been surveyed by Yaniv and Palevitch[24] and references therein. In addition to plants already reviewed *(Atropa belladona Datura* spp., *Hyoscyamus* spp., *Solanum* spp., *Papaver somniferum),* such positive effects have been also observed for *Lobelia sessilifolia, Scopolia carniolica, Phalaris aquatica, Tabacum,* and *Lupinus* spp. The effects of altitude and light on alkaloids production have been also reported by Chatterjee et al.[82] for *Catharanthus* and *Rauwolfia* spp.

B. GLYCOSIDES

Using TLC as the analytical method, Pimenova et al.[92] have shown that underground parts of *Patrinia intermedia* (Horn.) Roem. et Schult. from different regions had different compositions of triterpene glycosides; this was attributed to the existence of several chemoforms.

Yield of glycosides, tannins, and ascorbic acid in *Vaccinium myrtillus* was found dependent on ecologic factors.[93] The highest yields were observed in plants collected in open places with medium shade and in moderately humid ground. In the course of experiments on growing *Gentiana lutea,* six different ecotypes (A, B, C, J, R, and S) have been found (Figure 5).[94] Depending on ecotype, 3-year-old roots contained 0.05 to 0.33% amarogentin, 13 to 28% total sugar, and showed distinct qualitative and quantitative differences in the composition of essential oils (Table 5).

According to early studies,[95] plants containing anthracene derivatives, especially rhubarb, were all the more rich in active constituents as they were grown at higher altitudes. However, McCarthy and co-workers[96,97] found that relative humidity, atmospheric pressure, and wind had no effect on the aloin as well as on the organic acid production in *Aloe* spp. *(A. ferox, A. marlothii, A. africana, A. speciosa* and *A. arborescens).* An additional important characteristic was observed by these authors.[98] Although *Aloe speciosa* and *A. africana* grown in drums exhibited diurnal fluctuations in acid content, they do so to a far lesser degree than their counterparts grown in a natural habitat (Figure 6). This may suggest that results of experiments on plants not growing in their natural habitats should be generally viewed with caution.

Growth of *Digitalia* spp. as well as the biogenesis of cardiac glycosides are subject to the influence of ecophysiological factors. Active constituents can be produced by leaves without flowering, as previously described by Tonn[99] for *D. purpurea* cultivated in Indonesia at 1220 m height. The influence of altitude can be variable depending on the species. A 167-m height was found more favorable to *D. lanata* than to *D. purpurea,*[100] and several observations indicated that maximum levels of glycosides were obtained in well-defined climatic conditions, e.g., temperatures that must be between 20° and 30°C. Long rainy periods can reduce the active constituent levels.[27] However, according to Silva,[101] climatic and edaphic factors do not affect markedly the concentrations and relative proportions of lanatosides A, B, and C in *Digitalis lanata,* and chemotypes can be detected.

C. ESSENTIAL OILS

Studies on the ecophysiological factors that influence the composition and the production of essential oils are particularly difficult because chemotypes are frequent, constituents are volatile (high precautions must be taken for the storage before analysis), and extremely large variations of composi-

TABLE 3
Environmental Conditions at the Collection Site and Total Content of the Tertiary Alkaloids in the Leaf Extracts of Different Leaf Samples of *Tabernaemontana dichotoma*[91]

Site of collection	Environmental conditions and stage of development	Tertiary alkaloid content (%)
Colombo	Coastal shrub; flowering	0.13
Galle	Coastal, littoral woodland, flowering	0.15
Augusta	1700 ft, waste ground; vegetative	0.34
Hantane	2000 ft, under - shrub in planted mahogany forest with natural undergrowth; vegetative	0.43
Madugoda	2500 ft, secondary montane forest; vegetative	0.20

TABLE 4
Amounts of Different Alkaloids Present in Various Extracts of *Tabernaemontana dichotoma* Leaves from Different Collection Sites[91]

	Alkaloid content mg/g fresh leaves and % of each alkaloid relative to 12-methoxyvoaphylline									
	Colombo		Galle		Augusta		Hantane		Madugoda	
Alkaloid	mg	%	mg	%	mg	%	mg	%	mg	%
12-Methoxyvoaphylline	0.20	100	0.42	100	0.28	100	0.83	100	0.31	100
19-epi-Voacristine	0.31	155	0.04	10	0.34	121	0.70	84	0.30	97
Vobasine	0.25	125	0.09	21	0.86	307	0.37	45	0.18	58
Perivine	0.03	15	0.03	7	0.28	100	0.28	34	0.07	22
19-epi-Iboxygaine	0.02	10	0.01	2	0.03	11	0.13	16	0.03	10
(-)-Apparicine	0.31	155	0.20	48	0.54	192	0.72	87	0.20	64

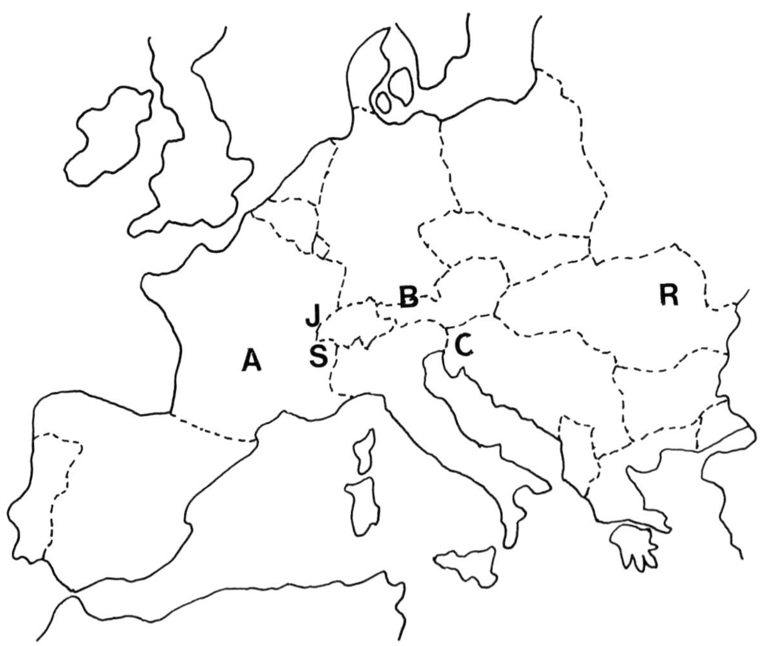

Figure 5. Geographical distribution of six ecotypes (A, B, C, J, R, and S) of *Gentiana lutea*.[94]

TABLE 5
Yield in Amarogentin, Total Sugars and Essential Oil from 3-Year-Old Roots of Six Ecotypes (A, B, C, J, R, and S) of *Gentiana lutea*[94]

Ecotype	Yield (%)		
	Amarogentin	Total sugars	Essential oil
A	0.14	27.8	0.005
B	0.15	23.5	0.007
C	0.33	19.0	0.002
J	0.05	13.3	0.005
R	0.27	12.5	—
S	0.10	15.4	0.007

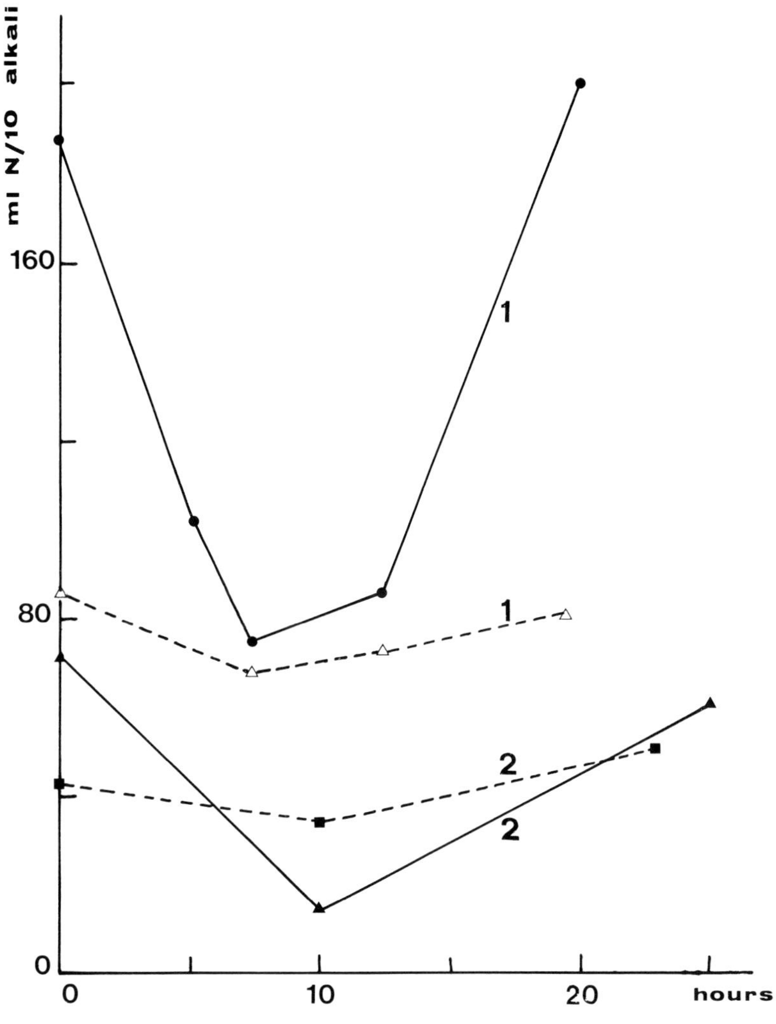

Figure 6. Diurnal fluctuations of organic acids production in *Aloe africana* Mill. (1) and *Aloe speciosa* Bak (2). (- - - -) grown in drums; (—) grown in a natural habitat.[98]

tion occur during ontogenesis; in this connection, the studies of Shah et al.[102] and Tóth[103,104] on the essential oil of *Foeniculum vulgare* subsp. *vulgare* are interesting. Furthermore, it had been reported that cultivated *Arnica montana* contained less essential oil in rhizomes (2.5%) as compared to plants growing in their natural habitats (4.7%).[105] Effects of dry conditions (decreasing rainfall or low irrigation at determined development stages) on the composition and the yield of essential oils have been summarized by Yaniv and Palevitch,[24] who report that under these conditions, the biosynthesis and the accumulation of the essential oils tend to increase (experimental data expressed as relative values) (e.g., *Artemisia absinthium, Matricaria chamomilla, Lavandula spica, Cymbopogon citratus).*

As a general rule,[27] northern regions are unfavorable and association of drought and wind led to substantial losses of essential oils; light and temperature promoted the production of essential oils so that *Salvia officinalis* cultivated in Dalmaty

(Yugoslavia) yielded 2.5% of essential oil compared with 1.4% when the plant was cultivated in northern Europe.

The photoperiodic dependence on the biogenesis of essential oils constituents has been well established in the case of *Mentha piperita*.[106] The leaves of plants grown under long-day conditions contained menthone, menthol, and traces of menthofuran. Meanwhile, the latter terpene became the major constituent when short-day conditions were used.

Franz et al.[107] reported that two strains of *Matricaria chamomilla* showing different essential oil compositions, cultivated in a phytotron, reacted upon daylength and temperature in growth and essential oil composition. Chamazulene was slightly influenced by climatic factors but the relations between bisaboloids varied (Tables 6 and 7; Figure 7).

Studying the variability of the *Juniperus communis* essential oil, Hörster[108] reported that statistical tests showed no conformity between the large variability of the essential oil constituents and either the different habitats or type of growth form. Only the biosynthesis of sabinene seemed to be favored by plants receiving more light.

D. CANNABINOIDS

According to earlier studies, cannabinoids ("resin") production from experimental cultures of *Cannabis sativa* underwent very large variations; the most important factor was the geographical origin of the seeds, climatic factors being of lesser importance.[109-111] Stephanou et al.[112] found that qualitative differences observed in the cannabinoids from *Cannabis sativa* leaves must be assigned both to genetic factors and to the plant polymorphism which must be taken into account to explain the high sensitivity of the secondary metabolism of the plant to ecophysiological factors. However, many authors have recognized the existence of several hemp chemotypes; the general opinion is now that following the climatic conditions, *Cannabis sativa* has the chemical capacity to produce either fiber or resin because the precursors for producing fibers and resin are continuously present; phytotronic studies using two climatic conditions, the first one at 32°/12°C day/night cycle and the second at 22°/12°C day/night cycle, confirmed these conclusions.[113-117] Finally, the cannabinoid profile of *Cannabis sativa* can be modified by soil characteristics.[118]

IV. CONCLUDING REMARKS

From the above data and particularly from phytotronic studies it can be safely concluded that ecophysiological factors must be taken into consideration for a better production of medicinal plants and that the management of agroecosystems may be brought to an optimal level by the exploitation of these ecophysiological factors.[11] However, no general rule can be stated; the positive influence of light and temperature on the biosynthesis of biologically active constituents is frequently observed. Other climatic factors have variable effects and depend very much on the particular plant species. The existence of chemotypes and numerous other parameters must be taken into consideration before definite conclusions on the actual relationships between ecophysiological factors and biosynthesis of active constituents in medicinal plants can be formulated. Furthermore, ecophysiological considerations could, in some cases, catalyze the development of more productive species of medicinal plants and profitable geographic (re)distribution of chemotaxa.

TABLE 6
Influence of the Daylength and Temperature on the Yield on Essential Oil Production (%) from *Matricaria chamomilla*[94]

Daylength and temperature	Strain E 29-1	Strain 38/72
8 h day	—	0.42
13 h day	0.68	0.47
18 h day	0.92	0.99
17 h day with a temperature difference between day and night	0.82	0.83
17 h day with constant temperature (24°C)	0.76	0.76

TABLE 7
Quantitative Constitution (%) of Essential Oil from *Matricaria chamomilla* Strain E 29 - 1 and 38/72[107]

Terpene	Strain E 29-1 (%)	Strain 38/72 (%)
Farnesene	5	5 — 15
Bisabolol	40 — 55	5 — 25
Bisabolol oxide	Traces	35 — 60
Bisabolonoxide	—	5
Chamazulene	20 — 30	5
En-In-Dicycloether	15 —25	15— 25

(Page 73) Figure 7. Influence on the essential oil composition of *Matricaria chamomilla* L. of daylength and temperature.[107] (1) Farnesene; (2) Bisabolol; (3) Bisaboloxide; (4) Bisabolonoxide; (5) Chamazulene; (6) En-in-Dicycloether. Daylength (a) 8 h; (b) 13 h; (c) 18 h; (d) 17 h with temperature difference between day and night; (e) 17 h with constant temperature (24°C).

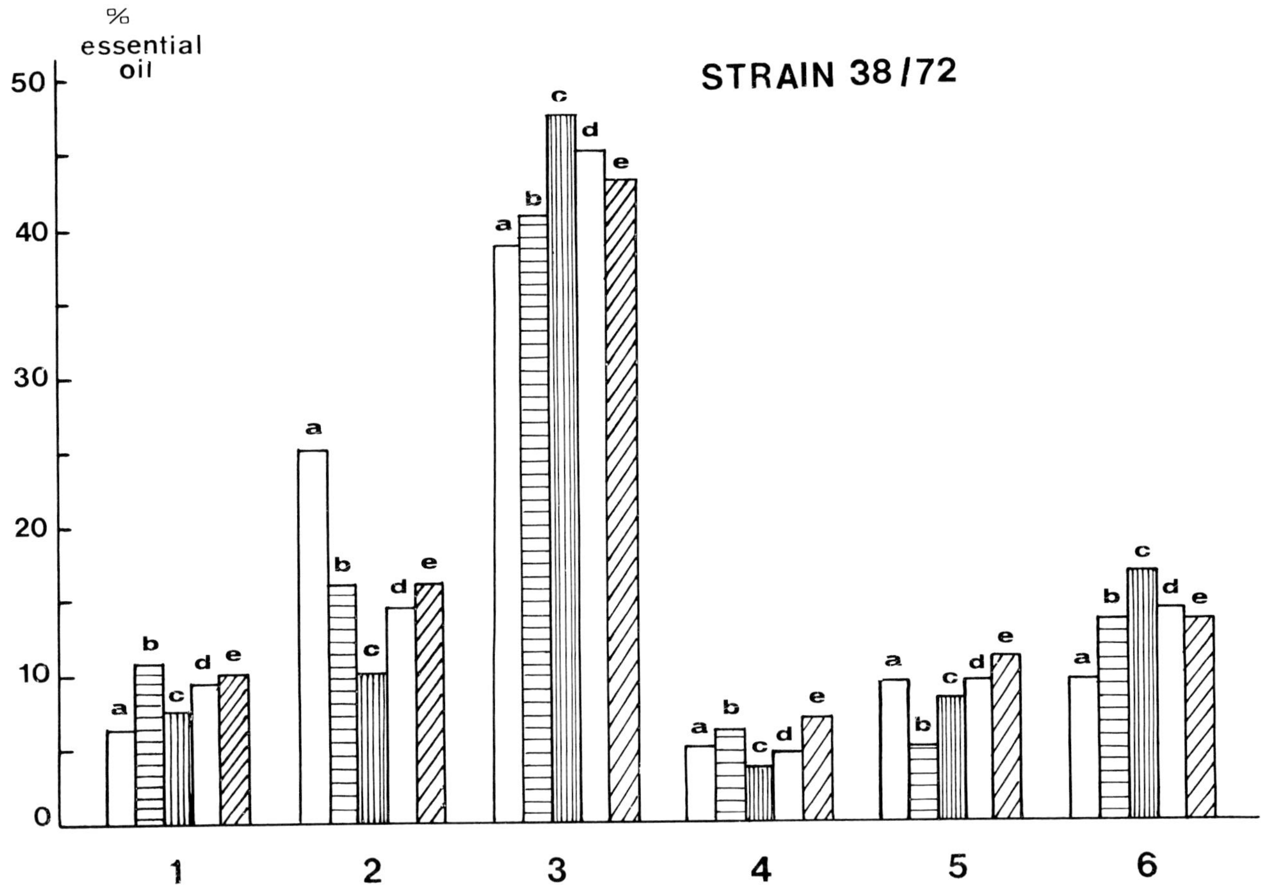
%
essential
oil
STRAIN 38/72
50
40
30
20
10
0
1
2
3
4
5
6

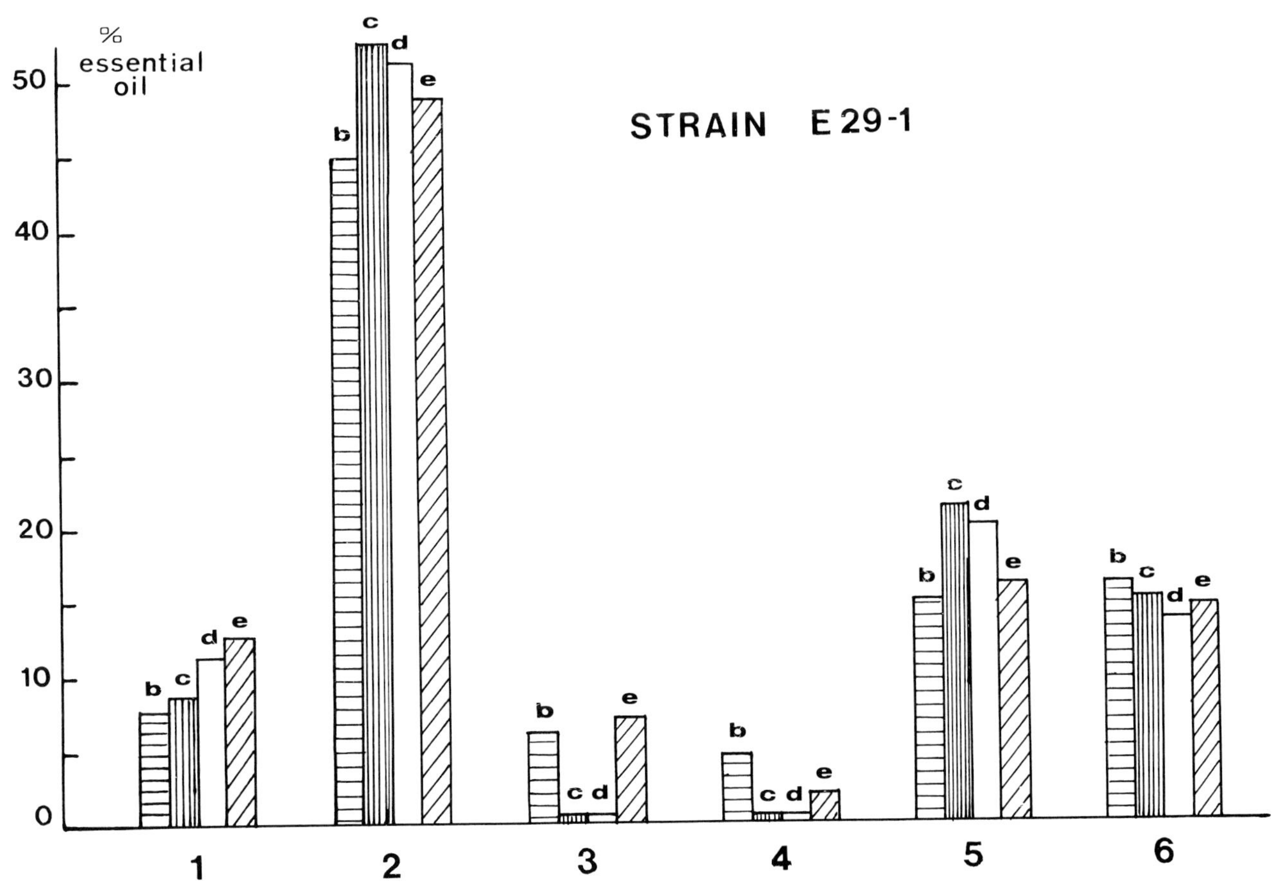
%
essential
oil
STRAIN E 29-1
50
40
30
20
10
0
1
2
3
4
5
6

REFERENCES

1. Conn, E. E., Preface to Volume 7, in *Biochemistry of Plants*, Vol. 7, Stupf, P. K. and Conn, E. E., Eds., Academic Press, New York, 1981, XIX.
2. Davis, D. H. and Heywood, V. H., *Principles of Angiosperm Taxonomy*, Van Nostrand - Reinhold, Princeton, NJ, 1963.
3. Flück, H., Intrinsic and extrinsic factors affecting the production of secondary plants products, in *Chemical Plant Taxonomy*, Swain, T., Ed., Academic Press, New York, 1963, 167.
4. Chatterjee, S. K., Nandi, R. P., and Sarkar, D. P., Effect of mineral nutrients on growth and essential oil formation in *Cymbopogon winterianus*, *Sci. Cult.*, 50, 26, 1984.
5. Brachet, J., Cosson, L., Ducourtioux, D., and Scheidecker, D., Effect of sodium chloride on the tropane alkaloid content in *Datura innoxia* Mill. cultivated in a controlled environment, *Physiol. Veg.*, 19, 77, 1981.
6. Nandi, R. P. and Chatterjee, S. K., Physiological and biomedical control of alkaloid biogenesis in *Datura innoxia* Mill. with special reference to photoperiodic and gibberellic acid treatments, *Indian J. Pharm. Sci.*, 44, 21, 1982.
7. Nandi, R. P. and Chatterjee, S. K., Physiological studies in some medicinal plants with special reference to synthesis of active principles in relation to photoperiodic and GA_3 treatments, *Environ. Physiol. Ecol. Plants*, 373, 1978.
8. Hofman, P. J. and Menary, R. C., Variations in morphine, codeine and thebaine in the capsules of *Papaver somniferum* L. during maturation, *Aust. J. Agric. Res.*, 31, 313, 1979.
9. Gheorghin, A. and Ionescu-Matiu, E., Influence of viral infection on the vincaleucoblastine content of *Vinca rosea* L. as well as on the morphological and anatomical character of this plant, *Ann. Pharm. Fr.*, 25, 807, 1967.
10. Yamada, H., Ota, Y., Tsujiwaki, Y., Ikemasu, M., Mizukami, H., Ikenaga, T., and Ohashi, H., Ecology and cultivation of Swertia plant. XII. Study on planting density in cultivation of S. japonica, *Shoyakugaku Zasshi*, 37, 115, 1983.
11. Bernáth, J., Influence of ecological factors on quantitative and qualitative production of medicinal plants, *Herba Hung.* 22, 48, 1983.
12. Alston, R. E. and Turner, B. L., *Biochemical Systematics*, Prentice-Hall, Englewood Cliffs, NJ, 1963.
13. Tétényi, P. T., *Infraspecific Chemical Taxa of Medicinal Plants*, Chemical Publishing, New York, 1970.
14. Mabry, T. J., Infraspecific cariation of sesquiterpene lactones in Ambrosia (composita), in *Applications to Evolutionary Problems at the Populational Level in Phytochemical Phylogeny*, Harbone, J. B., Ed., Academic Press, New York, 1970, 269.
15. Crawford, D. J., The morphology and flavonoid chemistry of synthetic infraspecific hybrids in Coreopsis mutica (compositae), *Taxon*, 21, 27, 1972.
16. Adams, R. P., Chemosystematic and numerical studies of natural populations of *Juniperus pinchotii* Sudw., *Taxon*, 21, 407, 1972.
17. Irving, R. S. and Adams, R. P., Genetic and biosynthetic relationships of monoterpenes, *Recent Adv. Phytochem.*, 6, 187, 1973.
18. Jones, D. A., Co-evolution and cyanogenesis, in *Taxonomy and Ecology*, Heywood, V. H., Ed., Academic Press, New York, 1973, 213.
19. Harborne, J. B., The biochemical systematics of flavonoids, in *Flavonoids*, Harborne, J. B., Mabry, T. J., and Mabry, H., Eds., Academic Press, New York, 1975, 1056.
20. Smith, P. M., *The Chemotaxonomy of Plants*, Arnold, E., Elsevier, New York, 1976.
21. Paris, R. R. and Moyse, H., *Matière Médicale*, Masson, Paris, 1976, chap. 2.
22. Adams, R. P., Diurnal variation in the terpenoids of *Juniperus scopulorum* (Cupressaceae). Summer versus winter, *Am. J. Bot.*, 66, 986, 1979.
23. Wagner, H., *Pharmazeutische Biologie*, Vol. 2, 2nd ed., Anbau von Arzneipflanzen, Gustave Fisher Verlag, Stuttgart, 1982, 4.
24. Yaniv, Z. and Palevitch, D., Effect of drought on the secondary metabolites of medicinal and aromatic plants. A review, in *Cultivation and Utilization of Medicinal Plants*, Atal, C. K., and Kapur, B. M., Eds., United Printing Press, New Delhi, 1982, 2.
25. Duke, J. A., Ecosystematic data on medicinal plants, in *Cultivation and Utilization of Medicine Plants*, Atal, C. K.,and Kapu, B. M., Eds., United Printing Press, New Delhi, 1982, 13.
26. Trease, G. E. and Evans, W. C., *Pharmacognosy*, 12th ed., *Factors Involved in the Production of Drugs*, Bailliere Tindall, London, 1983, 83.
27. Gracium, F. and Dumitrescu, C., Fundamentals to establish and apply the culture technologies to medicinal and aromatic plants, in *Medicinal and Aromatic Plants, Importance, Culture, Industrialization and Analysis*, Unido/10.578, 1982, 86.
28. Paun, E., Male, S. and Mircea, M., Culture methods of medicinal and aromatic plants, in *Medicinal and Aromatic Plants, Importance, Culture, Industrialization and Analysis*, Unido/10.578, 1982, 114.
29. Dumitrescu, A., Mihalea, A., and Paun, E., Amelioration and production of seeds of medicinal and aromatic plants, in *Medicinal and Aromatic Plants, Importance, Culture, Industrialization and Analysis*, Unido/10.578, 1982, 144.
30. Bernáth, J. and Tétényi, P., Climatron — An up-to-date tool in the research work on medicinal plants, *Herba Hung.*, 21, 49, 1982.
31. Walter, H., *Vegetation of the Earth and Ecological Systems of the Geobiosphere*, Springer-Verlag, New York, 1979.
32. Duvigneaud, P., *La Synthèse Ecologique*, Doin, Paris, 1980.
33. Hoppe, H. A., *Drogenkunde*, Vol. 1, *Angiospermen*, 8th ed., Walter de Gruyter, Berlin, New York, 1975.
34. Hoppe, H. A., *Drogenkunde*, Vol. 2, *Gymnospermen, Kryptogamen, Tierische Drogen*, Walter de Gruyter, Berlin, 1977.
35. Hoppe, H. A., *Taschenbuch der Drogenkunde*., Walter de Gruyter, Berlin, 1975.
36. Paris, R. R. and Moyse, H., *Précis de Matière Médicale*, Vol. 3, Masson, Paris, 1971.
37. Paris, R. R. and Moyse, H., *Précis de Matière Médicale*, Vol. 1, 2nd ed., Masson, Paris, 1976, 275.
38. Paris, R. R. and Moyse, H., *Précis de Matière Médicale*, Vol. 2., 2nd ed., Masson, Paris, 1981.
39. Trease, G. E. and Evans, W. C., *Pharmacognosy*, 12th ed., Bailliere Tindall, London, 1983.
40. Tyler, V. E., Brady, L. R. and Robbers, J. E., *Pharmacognosy*, 8th ed., Lea & Febiger, Philadelphia, 1981.
41. Wagner, H., *Pharmazeutische Biologies*, Vol. 2, 2nd ed., Gustav Fischer Verlag, Stuttgart, 1982.
42. Cosson, L., Chouard, P., and Paris, R., Influence of light on ontogenic variations of alkaloids in *Datura tatula*, *J. Nat. Prod.*, 29, 19, 1966.
43. Cosson, L., Effect of light on the alkaloid content in *Datura*. Effect on the biogenesis of alkaloids, *Herba Hung.*, 5, 157, 1966.
44. Cosson, L., Effect of light on the ontogenic variations in scopolamine and hyoscyamine levels of *Datura metel* leaves, *Phytochemistry*, 8, 2227, 1969.
45. Cosson, L., Importance of climatic factors and development stages on the production of tropanic alkaloids, in *Etudes de Biologie Végétale*, Hommage au Professeur P. Chouard, R. Jacques Ed., Paris, 1976, 483.
46. Cosson, L., Some aspects of the metabolism of alkaloids in *Datura tatula* L., *Plant. Med. Phytother.*, 2, 269, 1981.
47. Cosson, L., Escudero-Morales, A., and Cougoul, N., Ecophysiological regulation of the metabolism of tropane alkaloids (hyoscyamine and scopolamine), *Plant. Med. Phytother.*, 12, 319, 1978.
48. Nandi, R. P. and Chatterjee, S. K., Photoperiodic regulation of growth and alkaloid formation in *Datura innoxia* Mill., *Indian Biol.*, 7, 31, 1976.
49. Towers, G. H. N. and Chi-Kit, W., Phenylpropanoid metabolism (review), *Planta Med.*, 37, 97, 1979.
50. Cosson, L. and Vaillant, J. C., Tropane alkaloids of the leaves of neocaledonian *Duboisia myoropoides*, *Phytochemistry*, 15, 818, 1976.
51. Cosson, L. and Kuntmann-Cougoul, N., Regulation and metabolism of tropane alkaloids (hyoscyamine and scopolamine) in *Duboisia myoporoides* R.Br. and *Datura* cultivated in controlled conditions, *Herba Hung.*, 18, 135, 1979.

52. Karnick, C. R. and Saxena, M. D., On the variability of alkaloids production in *Datura* species, *Planta Med.*, 18, 266, 1970.
53. Pelt, J. M., Younos, C., and Hayon, J. C., On the alkaloid constitution of some solanaceae from Afghanistan. I., *Ann. Pharm. Fr.*, 25, 59, 1967.
54. Pelt, J. M., Younos, C., and Hayon, J. C., On the alkaloid constitution of some solanaceae from Afghanistan. II., *Ann. Pharm. Fr.*, 25, 101, 1967.
55. Gupta, S., Prabhakar, W. S., and Madau, C. L., Study on the distribution of total alkaloids and major components in the different organs of *Datura metel var. fastuosa* at various stages of growth, *Planta Med.*, 23, 370, 1974.
56. Aslanov, S. M., Dynamics of the alkaloid content in *Physochlaina orientalis* (Bieb.), G. Don Fil., *Restit. Resur.*, 19, 65, 1983.
57. Lebeau, P. and Janot, M. H., *Traité de Pharmacie Chimique*, Masson, Paris, 1956, 2765.
58. Waller, G. R. and Nowachi, E. K., *Alkaloids Biology and Metabolism in Plants*, Plenum Press, New York, 1978, 85.
59. Ikenaga, T., Saisho, Y., and Ohashi, H., Studies on the production of Duboisia in Japan. V. The effect of soil water on the growth and the alkaloid content of *Duboisia myoporoides* R. Br. and the amount of water absorption from its roots, *Nettai Nogyo*, 21, 11, 1977.
60. Ikenaga, T., Kino, S., Zensho, I., and Ohashi, H., Study on the production of *Duboisia* in Japan. VII. Comparison of the cultivation of *Duboisia myoporoides* R.Br. in Okinawa and Nagasaki, *Nettai Nogyo*, 21, 86, 1978.
61. Sarin, Y. K., Indian raw materials for the production of tropane alkaloids. A review, in *Cultivation and Utilization of Medicinal Plants*, Atal, C. K., Kapur, B. M., Eds., United Printing Press, New Delhi, 1982, 251.
62. Tétényi, P. and Vaguyfalvi, D., Changes in the alkaloid content of *Papaver somniferum*, *Herba Hung.*, 3, 317, 1964; *Pharmazie*, 20, 731, 1965.
63. Phillipson, J. D., Scutt, A., Baytop, A., Ozhatay, N., and Sariyar, G., Alkaloids from Turkish samples of *Papaver orientale* and *P. pseudo-orientale*, *Planta Med.*, 43, 261, 1981.
64. Cheng, P. C., Cultivation and Analysis of *Papaver bracteatum* Lindl., M.Sc. thesis, University of Mississippi, University, 1972.
65. Bare, C. E., Toole, V. K., and Gentner, W. A., Temperature and light effects on germination of *Papaver bracteatum* Lindl, *P. orientale* L. and *P. somniferum* L., *Planta Med.*, 34, 135, 1985.
66. Fairbain, J. W. and Wassel, G., The alkaloids of *Papaver somniferum* L. Evidence for a rapid turnover of the major alkaloids, *Phytochemistry*, 3, 253, 1964.
67. Fairbain, J. W. and Hakim, F., *Papaver bracteatum* Lindl. A new plant source of opiates, *J. Pharm. Pharmacol.*, 25, 353, 1973.
68. Mika, E. S., Growth, development and morphine content of opium poppy, *Bot. Gaz.*, 116, 323, 1955.
69. Il'Inskaya, T. N. and Yosifova, M. G., Influence of the conditions under which the poppy is grown on the alkaloid content of the opium obtained, *Bull. Narcotics*, 8, 38, 1956.
70. Tookey, H. L., Spencer, G. L., Grove, M. D., and Kwolek, W. F., codeine and morphine in *Papaver somniferum* grown in a controlled environment, *Planta Med.*, 30, 340, 1976.
71. Greppin, H., Delessert, B., Rossier, A., Klüppers, F. J. and Salemink, C. A., Etude de *Papaver bracteatum* sous diverses contraintes de milieu, *Sanssurea*, 6, 253, 1975.
72. Bernáth, J. and Tétényi, P., Alteration in compositional character of poppy chemotaxa affected by different light and temperature conditions, *Planta Med.*, 36, 230, 1979; *Herba Hung.*, 18, 91, 1979; *Acta Hort.*, 96, 91, 1980.
73. Bernáth, J. and Tétényi, P., The effect of environmental factors on growth, development and alkaloid production of poppy. I. Responses to day length and light intensity, *Biochem. Physiol. Pflanzen*, 174, 468, 1979.
74. Bernáth, J. and Tétényi, P., The effect of environmental factors on growth, development and alkaloid production of poppy. II. Interaction of light and temperature, *Biochem. Physiol. Pflanzen*, 176, 599, 1981.
75. Bernáth, J. and Tétényi, P., Characteristics of cultures productions of *Papaver somniferum* L. from different origins and vegetative periods, *Bull. Stup.*, 34, 123, 1982.
76. Gauhl, E., Physiologische und biochemische charakteristiken verschiedener ökotypen von *Solanum dulcamara* L., *Ber. Dtsch. Bot. Ges.*, *84*, 483, 1971.
77. Honváth, I., Bernáth, J., and Tétényi, P., Effect of the spectral composition of light on dry matter production in *Solanum dulcamara* L. Ecotypes of different origin, *Acta Agron. Acad. Sci. Hung.*, 26, 345, 1977.
78. Máthé, I., Jr., Toth, G., Vajda, S., and Máthé, I., Study on the effects of ecological factors on *Solanum dulcamara*, *Acta Agron. Acad. Sci. Hung.*, 24, 325, 1975.
79. Bernáth, J. Tétényi, P., Honváth, I., and Zámbó, I., Effects of light intensity and spectrum composition on the formation of steroidal alkaloids, *Herba Hung.*, 15, 43, 1976.
80. Bernáth, J. Tétényi, P., Honváth, I. and Zámbó, I., Change in the steroid aglycon content of clones of *Solanum dulcamara* L. of various origin and chemism as depending on the spectral composition of light, Herba *Hung.*, 17, 57, 1978.
81. Máthé, I., and Máthé, I., Jr., Data to the European area of the chemical taxa of *Solanum dulcamara* L., *Acta Bot. Acad. Sci. Hung.*, 19, 441, 1973.
82. Chatterjee, S. K., Nandi, R. P., Choudhuri, R., and Chatterjee, S., Effects of external and internal factors on increase of alkaloids and steroids in some commercially cultivated medicinal plants of Darjeeling Hills, West Bengal, India, *Herba Hung.*, 18, 125, 1979.
83. Bernáth, J. and Tétényi, P., Ecological factors adaptability-relationship of steroid alkaloid production based on investigation of two species, *Solanum laciniatum* Ait and *Solanum dulcamara* L., *Acta Bot. Acad. Sci. Hung.*, 24, 41, 1978.
84. Karnick, C. R., Seasonal periodicity of sapogenin production in *Dioscorea deltoidea* Wall and *Dioscorea prazeri* Prain and Burhill, *Planta Med.*, 16, 269, 1968.
85. Karnick, C. R., Effect of photoperiod on steroidal sapogenins and other constituents in *Dioscorea deltoidea*, *Ann. Bot. (London)*, 36, 605, 1972.
86. Young, J. C., Variability in the content and composition of alkaloids found in Canadian ergot. I. Rye, *J. Environ. Sci. Health*, B16(1), 83, 1981.
87. Young, J. C., Variability in the content and composition of alkaloids found in Canadian ergot. II. Wheat, *J. Environ. Sci. Health*, B16(4), 381, 1981.
88. Young, J. C. and Zhen-Jian, C., Variability in the content and composition of alkaloids found in Canadian ergot. III. Triticale and Barby, *J. Environ. Sci. Health*, B17(2), 93, 1982.
89. Chatterjee, S. K., Cultivation of *Cinchona* in Darjeeling Hills, West Bengal, in *Cultivation and Utilization of Medicinal Plants*, Atal, C. K. and Kapur, B. M., Eds., United Printing Press, New Delhi, 1982, 222.
90. Keene, A. T., Anderson, L. A., and Phillipson, J. D., Investigation of *Cinchona* leaf alkaloids by high-performance liquid chromatography, *J. Chromatogr.*, 260, 123, 1983.
91. Perera, P., Sandberg, F., and Verpoorte, R., Variation in alkaloid content in the leaves of *Tabernaemontana dichotoma* from different collection sites, *Acta Pharm. Suec.*, 21, 195, 1984.
92. Pimenova, M. E., Kogan, V. I., Mamocinka, L. F., and Surikova, E. M., Intraspecific geographic variability of the composition of triterpenoid glycosides of *Patrinia intermedia* (Horn.) Roem. et Schult., *Restit. Resur.*, 14, 160, 1978.
93. Gozin, A. A., Effect of ecological factors on the level of biologically active agents in *Vaccinium myrtillus* berries, *Restit. Resur.*, 8, 256, 1972.
94. Franz, C. and Fritz, D., Cultivation of *Gentiana lutea* and main substances of some ecotypes, *Planta Med.*, 28, 289, 1975.
95. Lebeau, P. and Janot, M. M., *Traité de Pharmacie Chimique*, Masson, Paris, 1956, 1113.
96. McCarthy, T. J. and Van Rheede Van Oudtshoorn, M. C. B., the seasonal variation of aloin in leaf juice from *Aloe ferox* and *Aloe marlothii*, *Planta Med.*, 14, 62, 1966.

97. McCarthy, T. J., The metabolism of anthracene derivatives and organic acids in selected *Aloe* species, *Planta Med.*, 16, 348, 1968.
98. McCarthy, T. J., and Price, C. H., Acid metabolism in leaves of *Aloe* species, *Planta Med.*, 14, 200, 1966.
99. Tonn, O., Digitalis culture in Java, *Pharm. Monatschr.*, 13, 8, 1932.
100. Demianets, P. F., *Avtorefrat*, Vilar, Moscow, 1947.
101. Silva, F., Vererbung und Veränderlichkeiteinigen eigenschaften bei Chemocultivars des wollingen fingerhutes (*Digitalis lanata* Ehrh.), *Planta Med.*, 14, 302, 1966.
102. Shah, C. S., Gadry, J. S., and Chauhan, M. G., Chemical races in Fennel, *Planta Med.*, 18, 285, 1970.
103. Tóth, L., Study on the essential oil from *Foeniculum vulgare*, *Planta Med.*, 15, 157, 1967.
104. Tóth, L., Studies on the essential oil from *Foeniculum vulgare*. II. Modifications of different Fennel oils before and after collection, *Planta Med.*, 15, 371, 1967.
105. Willuhn, G., Studies on the constituents of Arnica roots. II. Yield of essential oil in roots and rhizomes of different Arnica roots, *Planta Med.*, 17, 127, 1969.
106. Grahle, A. and Hoeltzel, C., Photoperiodic dependence of the formation of essential oils in *Mentha piperita*, *Naturwissenschaften*, 50, 552, 1963.
107. Franz, C., Fritz, D., and Schröder, F. J., Influence of ecological factors on the essential oils and flavonoids of two camomile cultivars, *Planta Med.*, 27, 46, 1975.
108. Hörster, H., Variability of the oils of *Juniperus communis*, *Planta Med.*, 26, 45, 1974.
109. Paris, R. and Dumerac, M. L., Teneur en résine et activité physiologique du Chanvre cultivé dans la région parisienne, *Ann. Pharm. Fr.*, 7, 47, 1949.
110. Dehay, C., Mémoires consacrés aux plantes médicinales, *Bull. Soc. Bot. Fr.*, 57, 1961.
111. Herisset, A., Besson, Ph., and Austin, C., Aperccu toxicologique sur les variétés de chanvre cultivés en Anjou, *Ann. Pharm. Fr.*, 23, 631, 1965.
112. Stephanou, E., Lawi-Berger, C., and Kapetanidis, I., *Cannabis sativa* L., *Pharm. Acta Helv.*, 59, 216, 1984.
113. Brant-Boucher, F., Fournier, G., Cosson, L., and Paris, M., Chemotypes multiplicity in *Cannabis sativa* L., *Planta Med.*, 39, 255, 1980.
114. Boucher, F., Cosson, L., Unger, L., and Paris, M. R., *Cannabis sativa* L., chemotypes or varieties, *Plant Med. Phytother.*, 8, 20, 1974.
115. Boucher, F., Paris, M., and Cosson, L., Occurrence of two chemotypes of *Cannabis sativa* L. from South Africa, *Phytochemistry*, 16, 1445, 1977.
116. Brant-Boucher, F., Cosson, L. and Paris, M., Ecophysiological behavior and biochemical relationships: the genus *Cannabis*, *Herba Hung.*, 18, 301, 1979.
117. Brant-Boucher, F., Effects of ecophysiological conditions on the growth, development and cannabinoids levels of clones corresponding to two chemical types of *Cannabis sativa* L. from South Africa, *Physiol. Veg.*, 18, 207, 1980.
118. Coffman, C. B. and Guntner, W. A., Cannabinoid profile and elemental uptake of Cannabis sativa L. as influenced by soil characteristics, *Agron. J.*, 67, 491, 1975.

Chapter 6

METHODS FOR BIOLOGICAL ASSESSMENT OF PLANT MEDICINES

B. N. Dhawan

I. INTRODUCTION

Plant medicine is a broad category of medicaments which include drugs used in traditional systems of medicine, folklore and ethnomedical products, as well as drugs discovered from plants having no documented therapeutic use. In the first two categories of plants there is a record of long clinical use (in many cases for centuries), and this is helpful in choosing appropriate biological test systems. In case of plants having no documented therapeutic use the problem is rather more difficult and very often a broad-based screening must be resorted to.

There has been a resurgence of interest in the development of new drugs from plant sources. In spite of tremendous advances in synthetic chemistry, natural products continue to play a major role in developing new drugs in many countries and in several areas of biological activity. Once a plant has been identified for investigation, biological screening is the first major step to be taken. The objectives of this screening are (1) to find new drugs, (2) to discover lead molecules (essentially novel chemical moieties) which can be modified through chemical means into new drugs, and (3) to provide a rationale for clinical use of traditional drugs even if activity is not of an order high enough to warrant development of the active moiety as a new drug.

II. THE PRIMARY SCREENING

There are a large number of ways by which a particular activity may be evaluated by a pharmacologist. The cost and availability of material constitute major constraints in selecting the appropriate test systems which are likely to be most rewarding in a particular situation. The first phase of biological assessment, therefore, involves scanning, i.e., use of test or groups of tests which will permit detection of a desired physiological activity so as to distinguish potentially useful materials from the useless materials rapidly, comprehensively, and inexpensively.[1] Such procedures, therefore, have got to be quick and dirty.[2] They have to be quick since the primary screening of a new material should not require more than a few days. They are called dirty since they do not involve in-depth studies, because the aim is to get rid of a large number of materials to pick out a few for detailed study. Finally, in most cases the results depend heavily on the skills of the investigator under very limited conditions of observation. It is, therefore, also necessary to very carefully evaluate the results of this initial screening to decide about further follow-up procedures which constitute an important part of any screening programme. For obvious reasons, no two screening programmes can be identical, even though the main objective of all such programmes is to unequivocally establish the presence of biological activity, to classify it, and to make at least a rough estimate of potency. The information is largely descriptive and qualitative, the analytical and in-depth evaluation being undertaken during secondary screening or follow-up studies. Due to constraints of space, only the general principles of developing and operating a biological screening programme for plant medicines are discussed, along with a brief description of the screening programme developed at Central Drug Research Institute (CDRI). For detailed description of individual tests the reader is referred to several available publications on the subject.[3-8]

III. TYPES OF SCREENING PROGRAMMES

The screening programmes can be broadly classified into three subtypes:[5]

1. Individual activity screening: Termed as simple screening by Turner,[5] the procedure utilises one or two tests to detect substances having a particular property, e.g., hypoglycaemic. The objective here is to use suitable, sufficiently accurate method(s) rather than a battery of tests. For example, Malone et al.[9] reported on muscarinic activity in Inocybe using a simple test procedure. Some of the earliest studies have involved screening for antibacterial,[10,11] anticancer,[12,13] antidiabetic,[14] and antimalarial[15] activities in each case using a few test systems. A detailed and critical review of the work at the National Cancer Institute on developing anticancer drugs from plant sources, one of the biggest projects covering any area of pharmacological activity of medicinal plants was undertaken by Suffness and Douros in 1979.[16] In India, there has been much effort on screening plants for fertility regulation, which was recently reviewed by Kamboj and Dhawan,[17] and certain other types of biological activities.[18,19]
2. Broad-based screening: Also known as blind screening,[5] such a procedure is generally employed with a new series of chemical compounds. It has been equally well employed for plant products where no information is available on their possible pharmacological activity. The main objective of this approach is to find out if the material is pharmacologically inert or has any exploitable potential. The strategy here also should be to employ simple and quick tests which should cover a broad area of biological activity and as such, requires careful planning. The ultimate objectives of a research programme determine the choice of the test systems. A broad-based biological screening of medicinal plants was initiated in several countries during the early 1960s,

including the U.S.S.R.,[20] the West Indies,[21-23] and the U.S.[24-26] Some reports are also available from South American countries,[27] China,[28] etc., but the programmes there still have not been well established. A similar programme was initiated by us in India in 1963 and the first report published in 1968.[29] A review of results from 2000 plants has been made[30] and the programme continues.[31] Subsequently, some other laboratories in India have also gotten involved.[32] Reports of Western Samoan plants have been published in the U.S.[33] Similar studies have been undertaken with marine products in several countries, including India.[34-39]

3. Programmed screening: The programmed screening is midway between the two approaches outlined above. The approach is more useful with synthetic compounds when a drug of a specific type is sought, but can also be used with natural products available in a fairly pure state. The tests to be used in such a programme need greater precision and capability of providing some indication of safety as well. The antistress and stimulant effects of glycosides from ginseng and eleutherococcus have been screened by Brekhman and Dardynov[40] in this manner.

IV. DEVELOPMENT OF A SCREENING PROGRAMME

As has been pointed out earlier, in most cases a broad-based screening programme is best suited for studies on plant medicines, particularly when the pharmacological activity is not well defined. Therefore, the development of such a screening programme is primarily discussed in this section. Care should be taken to select screening tests which are able to provide information about the main pharmacological activity as well as other pharmacological properties of the test material, and preferably also indicate whether activity is of sufficiently high order or warrant and justify a more detailed study.[41] In some cases, e.g., chemotherapeutic agents, adequate information may be obtained from *in vitro* studies on the infecting organisms, but in majority of the cases, additional tests have be undertaken in one or more mammalian species. The material may be administered to (1) conscious intact animals (rodents are preferred), (2) anaesthetised animals with appropriate surgical intervention (cats or dogs are the usual choice, but techniques employing rodents are being increasingly developed and used), and (3) *in vitro* studies employing isolated tissues from various mammalian species, including tissue samples obtained from slaughterhouses.

Biological screening involves studying the effects of a new product at certain arbitrarily fixed dose levels in a species of animal, prediction of its effect over the entire dosage range, and extrapolation to other species. Several parameters are utilised for the latter purpose, including the surface area ratios.[8,42] Further, to achieve these objectives the following basic premises are taken into consideration when developing a screening programme:[43]

1. All drugs operate in a dose responsive manner, the intensity and/or duration of effects increasing to a certain limit with dose. At still higher doses all agents produce serious toxicity, including death.
2. The dose response profile of each class of drug is characteristic and usually recognisable. Hence, with dose response data of various classes of drugs it is possible to predict the possible utility and toxicity of a new drug by comparing it with the available profiles.
3. The route of administration in most cases produces only quantitative change in the response.

At the start of such a programme it is important to identify the parameters to be measured. It is also important to know the correlation between the pharmacological effect being studies in an animal model and the clinical effect ultimately expected from the compound under test. Once animal effects to be studied have been decided it is necessary to select tests that can provide the most specific and reliable information about such effects. The following guidelines are useful in this context:

1. Does it measure what it is supposed to measure? The biological effect being measured should reflect the relevant therapeutic effect. Thus, it will hardly be valid to select a potential neuroleptic agent on the basis of cataleptic activity, which is an undesirable side effect of neuroleptics. Some indication must also be available about the reliability of the test system used, i.e., how consistently it measures whatever it does measure?
2. In the case of plant medicines it is important to select test systems that can be used for crude extracts, semipurified materials, and purified active constituents. One has also to be careful in selecting the vehicle, route of administration, etc.
3. The programme should have built-in safety mechanisms such as periodic testing of known standard compounds (also under unknown code numbers) or blank placebos, etc.

These factors tend to make such a programme reliable, but it becomes biased toward picking up compounds with already known types of biological activity. It is much more rewarding to pick up activity which is new, unexpected, or unique and this requires the programme to be unbiased, broad, and comprehensive. In most cases a careful balance has to be struck between these two apparently contradictory requirements. A constant evaluation of the tests is also necessary in terms of their validity, reliability, and cost, etc., as newer methods become available or limitations of selected tests are evident.

V. THE CDRI PROGRAMME ON MEDICINAL PLANTS

A broad-based programme of biological screening of medicinal plants was initiated at the Central Drug Research Institute, Lucknow, in 1963 and the first results were reported in 1968.[29] The programme continues, and results on over 3000 plant materials have been published.[29,31,43-49] In addition, results of testing 967 plants for antiviral activity,[50] 1600 plants for their spermicidal and semen coagulant activity,[48,51-52] and 644 plants for leather tanning and insecticidal activity have also been reported.[19] The programme was subsequently expanded to cover the marine flora and fauna and initial results indicate its suitability for screening these materials.[38,39] Several of the tests are being concurrently used for screening of synthetic compounds, with minor modifications of the protocols, route of administration, etc.

A. THE TEST SYSTEMS

The screening programme employs about 100 different test systems covering several broad areas of biological activity, as indicated in Table 1. The details of tests used for each of these activities have been reported elsewhere.[8,18,29,43-52] Some of the test systems have been developed at CDRI,[53,54] while others are standard tests sometimes suitably modified. A detailed description of all the tests is beyond the scope of present review, but a brief description of tests used for assessing CNS activity is given below to illustrate the operation of the programme.

B. THE CNS TESTS

The system for assessing CNS activity has the capacity of handling about 1000 materials of various grades of purity utilising five to six technicians needing the supervision of one scientist. The initial testing is done with 50% ethanol extract dried under reduced pressure and suspended in 2% gum acacia. The same procedure is followed for other test systems (*vide supra*). Once activity is detected it has to be reconfirmed in one of the primary fractions before selecting the plant for detailed study. In addition to identifying the type of CNS effects it also provides some indication of therapeutic index or safety margin. The programme essentially consists of four parts and the tests are done in albino mice at a constant room temperature of 24° to 25°C. Whenever other species are employed appropriate conversion factors are used to calculate approximate doses.[8] Mice are used in batches of five to ten each and test compounds are administered intraperitoneally.

1. Determination of Acute Toxicity

The technique of Horn[55] is followed and doses of >1500 mg/kg are generally used. The method provides a reasonably accurate estimate of LD_{50}, although it is called approximate LD_{50} (ALD_{50}). Further testing is done at $^1/_2$ ALD_{50} for crude extracts and at $^1/_5$ ALD_{50} for pure substances.

2. Effect on Gross Behaviour

Various aspects of gross behaviour, analgesia, spontaneous motor activity, and rectal temperature are monitored at regular intervals. The behavioural effects are graded as mild or severe. Based on the effects the test substance is classified as being stimulant, depressant, or with no gross effects. The additional tests are selected on the basis of gross effects.

3. Drug Interaction Studies

These evaluate reversal of drug-induced altered behaviour to baseline level with the test compound and represent better equivalent of human disease processes. Further they also help in characterising the CNS effect more precisely. The tests include:

1. Amphetamine-induced hyperactivity and toxicity in aggregated and nonaggregated mice (for detection of anxiolytic and neuroleptic activities).
2. Antireserpine and potentiation of yohimbine toxicity tests (for detection of antidepressant activity).
3. Antistrychnine test for central skeletal muscle relaxant activity.
4. Potentiation or antagonism of barbiturate narcosis (barbital may be used if indirect effect due to altered metabolism in liver is to be excluded) to pick up CNS depressants or analeptics.
5. Carrageenin-induced paw edema (test compound given orally) for anti-inflammatory compounds.
6. Phenylquinone writhing (test compound by oral route) for aspirin type of analgesic activity.

4. Additional Tests

The following additional tests are used wherever indicated:

1. Supramaximal electroshock seizure pattern test to identify anticonvulsant agents
2. Forced motor activity (rota rod)
3. Eddy's hot plate (for analgesia)
4. Anorexigenic activity
5. Catalepsy test

5. Tests in Other Species

In a limited number of cases, additional confirmatory tests are performed in rats or cats. The tests done in rats include (1) the conditioned avoidance and the secondary conditioned avoidance tests for CNS depressant compounds showing positive effect in the amphetamine toxicity and hyperactivity tests, respectively, to confirm anxiolytic or neuroleptic activity; (2) carrageenin edema, granuloma pouch, and ulcerogenic index tests for compound showing anti-inflammatory activity; and (3) antipyretic activity in the yeast pyrexia test for compounds active in the phenylquinone writhing test. Compounds exhibiting antistrychnine activity are further investigated for effect on poly- and monosynaptic reflexes in anaesthetised cats.

Malone et al.,[9] while proposing their Hippocratic screen in rats, enumerated several advantages of their procedures. The procedure developed by us meets most of their stipulated advantages such as use of a single species, constant volume of a single vehicle, single route of administration, reliance on gross behavioural changes, etc. It is also capable of picking up various types of compounds which are picked up by the procedure developed by Irwin.[1] Our procedure offers several additional advantages. The major portion of testing is done in mice, thus significantly reducing the requirements of a test sample. This is particularly important when fractions are being tested. All tests are done at a constant fraction of LD_{50}, thus automatically generating comparative data and providing indication of available safety margin. Over 10,000 synthetic compounds or natural products have already been successfully

TABLE 1
Showing Areas of Biological Activity Investigated at CDRI and the Test Systems Utilised

Activity	Tests done
A. Antiviral	1. Ranikhet disease virus, chorioallantoic membrane (CAM), chick embryo fibroblast monolayers (CEF) *in vitro*
	2. Vaccinia virus —CAM and CEF
	3. Semliki forest virus (mouse, *in vivo*)[a]
	4. Interferon inducers (mouse)[a]
B. Antibacterial *In vitro* unless mentioned otherwise	1. *Streptococcus faecalis*
	2. *Klebsiella pneumoniae*
	3. *Mycobacterium leprae (in vivo,* mouse)[a]
	4. *Mycobacterium tuberculosis (in vivo* also)[a]
	5. *Staphylococcus aureus*
	6. *Salmonella typhi*
	7. *Escherichia coli*
	8. *Pseudomonas aeruginosa*[a]
	9. *Bacillus subtilis*[a]
C. Antifungal *In vitro* unless mentioned otherwise	1. *Candida albicans (in vivo* also)
	2. *Trichophyton mentagrophytes (in vivo* also)
	3. *Aspergillus fumigatus*
	4. *Cryptococcus neoformans*
	5. *Sporotrichum schenckii*
	6. *Fusarium solani (in vivo)*[a]
	7. *Curvularia lunata (in vivo)*[a]
	8. *Acrophialophora fusispora (in vivo)*[a]
D. Antiprotozoal	
1. Antiamoebic	1. Axenic culture *in vitro*
	2. Intestinal infection in rat
	3. Hepatic infection in rat[a]
	4. Meningoencephalitis in mouse[a]
2. Antimalarial[a]	
	1. Blood schizonticide
	a. *in vitro - P. falciparum*
	b. *in vivo* - mouse, rat, mastomys, monkey
	2. Causal prophylactic - mastomys, mouse, monkey
	3. Antirelapse — monkey *in vitro* and *in vivo* (hamster)
3. Antileishmanial[a]	
E. Anthelminthic	
1. Filaria[a]	1. *Litmosoides carinii* (cotton rat, mastomys)
	2. *Brugia malayi* (cat, mastomys)
	3. *Dipetalonema viteae* (mastomys)
2. Hookworm	*Ancylostoma ceylonicum* (rat, hamster)
3. Tapeworm	*Hymenolepis nana* (mouse, rat)
F. Anticancer	*In vitro* cell line
G. Endocrinal	
1. Antifertility	
In female	1. *In vivo*
	anti-implantation (rat, hamster)
	abortifacient (rat)
	Tubal occluding (monkey)[a]
	Blastotoxic (rat)[a]
	Oxytocic (cat)[a]
	2. *In vitro*
	Abortifacient (rat uterus)

TABLE 1 (CONTINUED)
Showing Areas of Biological Activity Investigated at CDRI and the Test Systems Utilised

	Activity	Tests done
	In male	1. Antispermatogenic (rat) 2. Spermicidal - *in vitro* 3. Semen coagulant - *in vitro*
2.	Hypoglycaemic	Rabbit, rat
H.	Central nervous system (mouse and rat)	1. Gross effects on behaviour 2. Body temperature 3. Analgesia 4. Supramaximal electroshock seizure pattern 5. Barbiturate potentiation and antagonism 6. Amphetamine hyperactivity and toxicity 7. Strychnine/metrazol seizures 8. Antireserpine 9. Phenylquinone writhing 10. Rota rod 11. Learning and memory[a]
I.	Cardiovascular and autonomic effects (cat, dog)	1. Blood pressure 2. Respiration 3. E.K.G. 4. Antiarrhythmic[a] *in vitro* — guinea pig auricle *in vivo* — rat, aconitine arrhythmia
J.	Hypolipidaemic	
	1. Lipid lowering *in vivo* (rat, rabbit, monkey)	Serum cholesterol, triglycerides and lipoproteins in normal and high fat fed/triton treated animals
	2. Cholesterol biosynthesis	Rat liver slices
	3. Inhibition of platelet aggregation	*in vitro* and *in vivo*
K.	Anti-inflammatory	Carrageenin paw edema (mouse, rat) Granuloma pouch (rat)[a] Adjuvant arthritis (rat)[a] Ulcerogenic index (rat)[a]
L.	Other pharmacolgoical activities	
	1. Acute toxicity (LD_{50})	Mouse, rat (i.p., oral)
	2. Diuretic	Rat
	3. Spasmolytic	Isolated guinea pig ileum
	4. Neuromuscular conduction[a]	Cat
	5. Antiallergic[a]	PCA (mouse, rat) Haemagglutination (mouse) Cell mediate immunity (rat)
6.	Hepatoprotective[a]	Liver damage by CCl_4 Lanthanum and galactosamine in rats

[a] Restricted to selected plants only.

evaluated using this procedure and a large number of leads picked up. Moreover, data on standard compounds having various type of CNS activity have been generated and a quantitative assessment of activity of a new test material is therefore possible even during the initial screening tests. Standard work charts and record books have been developed and records are amenable to computer processing.

C. EVALUATION OF CDRI PROGRAMME

As pointed out earlier, 50% ethanolic extracts from about 3000 plant materials have been run through the programme. Table 2 lists various types of activities observed and the number of plants showing a particular activity. The analysis covers results from approximately 2700 materials. A perusal of Table 2 indicates significant biological activity in about

TABLE 2
Summary of Primary Biological Screening Results of 2700 Plant Extracts at CDRI

S. no.	Pharmacological activity	No. of active extracts
1.	Abortifacient	12
2.	Antiamoebic	1
3.	Anticancer	132
4.	Antihelminthic	2
5.	Anti-implantation	10
6.	Anti-inflammatory	31
7.	Antimicrobial (antibacterial + antifungal)	13
8.	Antiviral	13
9.	Cardiovascular effects (hypotensive, hypertensive, cardiotonic)	71
10.	CNS depressant	77
11.	Diuretic	39
12.	Hypoglycaemic	12
13.	Hypolipaedemic	5
14.	Insecticidal	21
15.	Oxytocic	14
16.	Semen coagulant	49
17.	Spasmolytic	52
18.	Spermicidal	16
Total		570

21% of the plants tested which could be confirmed in fractions. About 33% of the plants had a low (though significant) level of activity which did not merit a detailed follow-up. The programme still provided a sufficiently large number of leads for follow-up studies. Detailed results with the more promising plants are discussed in another chapter of this book.[56]

VI. CONCLUDING REMARKS

Biological screening of plant medicine serves several purposes. The primary function is to develop new drugs and new leads. The leads may represent not only new chemical moiety but also novel types of biological activity. The example of reserpine is well known. More recently the concept of drugs acting as "adaptogens" has come from experimental studies with ginsenosides obtained from *Panax ginseng*.[57] Further, a large number of plant materials of doubtful utility are being used particularly in developing countries and traditional systems of medicine. It is necessary to systemically screen them so as to throw out the inactive products. Material having even moderate activity could be used since they have a rationale and are more easily accepted by the patient population in such countries. Unfortunately most of these countries lack the necessary testing facilities and trained manpower to screen them. Second, negative results are seldom published, particularly by the drug houses. Such data, if published, will be of value for selecting plants as well as test systems for screening.

Another difficulty in screening plant medicines is availability of suitable test systems. The test systems are usually developed by large pharmaceutical houses primarily for synthetic compounds of high purity. These are often not suitable for use with crude or semipure materials and there is an urgent need for developing or modifying existing test procedures toward this end. Many sophisticated techniques have been applied to chemical studies on natural products in recent years with rewarding results, but unfortunately use of sophisticated pharmacological technique is still sporadic. The necessity of this approach will be amply clear from the following two examples. We recently investigated experimentally an Unani herbal preparation claimed to improve myocardial function. No effect was detectable in commonly used test systems but a study of the changes in regional circulation with radiolabeled microsphere clearly brought out a selective improvement in the blood supply to the myocardium without affecting other parameters of cardiovascular haemodynamics.[58] Similarly, it is now possible to screen plant materials for effects on biochemical parameters such as prostaglandin synthesis, thromboxane synthesis, etc. Such a study led to the uncovering of the antithrombotic effect of curcumin, a major active constituent of *Curcuma longa*.[59]

There are certain types of activities that cannot be satisfactorily evaluated in existing animal models. Thus, the time-honored herbal medicine valerian contains the iridoids valepotriates, which cannot be evaluated by existing procedures but are clinically useful. There are a number of plant medicines used for treatment of haemorrhoids, pregnancy toxaemias, liver disorders, etc., for which suitable animal models are not available. With carefully selected remedies an initial clinical evaluation is a quicker way of establishing activity, particularly in the developing countries, and for preparations being used clinically in traditional medicine. Thus, Antarkar et al.[60] could demonstrate the beneficial effects of an Ayurvedic preparation in infective hepatitis in a double-blind clinical trial.

In many laboratories, pharmacological studies are heavily biased toward plants used in traditional systems of medicine or having some ethnomedical data available. Such a bias is understandable in view of possible cost effectiveness and greater chances of getting an active molecule, and also the local pressures of investigating these plants on a priority basis. Our analysis of CDRI data has revealed certain interesting facts. Among the first 1975 plants screened by us some ethnomedical data were available for 536 (27.1%) plants, whereas the remaining 1439 (72.9%) were devoid of any such information. Among the plants of the first category 97 plants (18.3%) were active, whereas 272 (18.9%) plants in the second group showed activity.[61] These results clearly bring out the necessity of broad-based pharmacological evaluation of a much larger proportion of local flora to discover new drugs or leads.

REFERENCES

1. Irwin, S. Drug screening and evaluative procedure. *Science*, 136: 123, 1962.
2. Nodine, J. H. and Siegler, P. E. *Animal and Clinical Pharmacologic Techniques in Drug Evaluation,* Year Book Medical Publishers, Chicago, 1964.
3. Siegler, P. E. and Moyer, J. H. *Animal and Clinical Pharmacological Techniques in Drug Evaluation.* Vol. 2. Year Book Medical Publishers, Chicago, 1967.
4. Laurence, D. R. and Bacharach, A. L. *Evaluation of Drug Activities: Pharmacometrics.* Vol. 1 and 2. Academic Press, London, 1964.
5. Turner, R. A. *Screening Methods in Pharmacology.* Academic Press, New York, 1965.
6. Turner, R. A. and Hebborn, P. Eds. *Screening Methods in Pharmacology,* Vol. 2, Academic Press, New York, 1971.
7. Mantegazza, P. and Piccinini, F. Eds *Methods in Drug Evaluation.* North Holland, Amsterdam, 1966.
8. Dhawan, B. N. and Srimal. R. C. Eds. *The Use of Pharmacological Techniques for the Evaluation of Natural Products.* UNESCO, New Delhi, 1984.
9. Malone, M. H., Robichaud, R. C., Tylar, V. E., Jr., and Brady, L. R. A bioassay for antimuscarinic activity and its detection in certain inocybe. *Lloydia,* 24: 204, 1961.
10. Lucas, E. H. and Lewis, R. W. Antibacterial substances in organs of higher plants. *Science,* 100: 597, 1944.
11. Hays, L. Survey of higher plants for presence of antibacterial substances. *Bot. Gaz.* 108: 408, 1947.
12. Taylor, A., McKenna, G. F. and Burlage, H. M. Cancer chemotherapy experiments with plant extracts. *Tex Rep. Biol. Med.* 10: 1062, 1952.
13. Hardinge, M. G., Courville, D. A., Hardinge, M., Fujikawa, B., and Harvey, R. Action of plant extracts on transplanted mouse tumors. *Cancer Res. (Suppl.)* 21: 573, 1961.
14. Sharaf, A. A., Hussein, A. M., and Mansour, M. Y. Studies on the antidiabetic effects of some plants. Preliminary report. *Planta Med.* 11: 159, 1963.
15. Spencer, C. F., Koniuszy, F. R., Rogers, E. F., Shavel, J., Jr., Easton, N. R., Kaczka, E. A., Kuehl, F. A., Jr., Phillips, R. F., Watti, A. Flokers, K., Malanga, C., and Seeler, A. O. Survey of plants for antimalarial activity. *Lloydia,* 10: 145, 1947.
16. Suffness, M. and Douros, J. Drugs of plant origin. *Methods Cancer Res.* 16: 73, 1979.
17. Kamboj, V. P. and Dhawan, B. N. Research on plants for fertility regulation in India. *J. Ethnopharmacol.* 6: 191, 1982.
18. Atal, C. K., Srivastava, J. B., Wali, B. K., Chakravarty, R. B., Dhawan, B. N., and Rastogi, R. P. Screening of Indian plants for biological activity. VIII. *Indian J. Exp. Biol.* 16: 330, 1978.
19. Fernandes, F. and Costa Pereira, F. X. R. Antimicrobial activities of Indian medicinal plants. Proc. III Asian Medicinal Plants & Spices, Colombo, 1977, p. 93.
20. Aliev, R. K. Alkaloidal plants of the Azerbaidhzan SSR and their pharmacological preparations. *Gyogyszereszet.* 6: 86, 1962.
21. Feng, P. C., Haynes, L. J., Magnus, K. E., and Flimmer, J. R. Further pharmacological screening of some West Indian medicinal plants. *J. Pharm. Pharmacol.* 14: 556, 1962.
22. Feng, P. C. Haynes, L. J., Magnus, K. E., and Flimmer, J. R. Further pharmacological screening of some West Indian medicinal plants. *J. Pharm. Pharmacol.* 16: 115, 1964.
23. Hooper, P. A. and Leonard, B. E. Pharmacological properties of some West Indian plants. *J. Pharm. Pharmacol.* 17: 98, 1965.
24. Farnsworth, N. R., Henry, L. K., Svoboda, G. H., Blomster, R. N., Yates, M. J., and Euler, K. L. Biological and phytochemical evaluation of plants I. Biological test procedures and results from two hundred accessions. *Lloydia,* 29: 101, 1966.
25. Farnsworth, N. R., Henry, L. K., Svoboda, G. H., Blomster, R. N., Fong, H. H. S., Quimby, M. W., and Yates, M. J. Biological and phytochemical evaluation of plants. II. Test results from an additional two hundred accessions. *Lloydia,* 31: 237, 1968.
26. Fong, H. H. S., Farnsworth, N. R., Henry, L. K., Svoboda, G. H., and Yates, M. J. Biological and phytochemical evaluation of plants. X. Test results from a third two hundred accessions. *Lloydia,* 35: 35, 1972.
27. Gothlieb, O. R. The search for medicinal and aromatic plants in South America. Proc. II Int. Meeting Med. Aromat. Plants, Citta de Castello, 225, 1981.
28. Peigen, X. Traditional experience of Chinese herb medicine. Its application in drug research and new drug searching, in *Natural Products as Medicinal Agents.* J. L. Beal and E. Reinhard, Eds. Hippocrates Verlag, Stuttgart, 351, 1981.
29. Dhar, M. L., Dhar, M. M., Dhawan, B. N., Mehrotra, B. N., and Ray, C. Screening of Indian plants for biological activity. III. *Indian J. Exp. Biol.* 6: 232, 1968.
30. Rastogi, R. P. and Dhawan, B. N. Research on medicinal plants at the Central Drug Research Institute, Lucknow (India). *Indian J. Med. Res.* 76(Suppl): 27, 1982.
31. Aswal, B. S., Bhakuni, D. S., Goel, A. K., Kar, K., and Methrotra, B. N. Screening of Indian Plants for biological activity. XI. *Indian J. Exp. Biol.* 22: 487, 1984.
32. Sharma, M. L., Chandokhe, N., Ray Ghatak, B. J., Jamwal, K. S., Gupta, O. P., Singh, G. B., Mohammed Ali M., Thakur, R. S., Handa, K. L., Rao, P. R., Jamwal, P. S., and Sareen, Y. K. Pharmacological screening of Indian medicinal plants. *Indian J. Exp. Biol.* 16: 228, 1978.
33. Norton, T. R., Bristol, M. L., Read, G. W., Busbrell, O. A., Kashiwagi, M., Okinaga, C. M., and Oda, C. S. Pharmacological evaluation of medicinal plants from Western Samoa, *J. Pharm. Sci.* 62: 1677, 1973.
34. Kaul, P. N. Pharmacologically active substances of marine origin. Proc. Marine Tech. Soc. p. 229, 1972.
35. Kaul, P. N., Kulkarni, S. K., Weinheimer, A. J., Schwartz, F. J., and Karns, T. K. B., Pharmacologically active substances from the sea. II. Various cardiovascular activities found in the extracts of marine organisms. *Lloydia,* 40: 253, 1977.
36. Bhakuni, D. S. and Silva, M. Biodynamic substances from marine flora. *Bot. Mar.* 17: 40, 1974.
37. Ruggieri, G. D. Drugs from the sea. *Science* 194: 491, 1976.
38. Naqvi, S. W. A., Solimabi, Kamat, S. Y., Fernandes, L., Reddy, C. V. G., Bhakuni, D. S., and Dhawan, B. N. Screening of some marine plants from the Indian coast for biological activity. *Bot. Mar.* 34: 51, 1980.
39. Kamat, S. Y., Solimabi, Naqvi, S. W. A., Fernandez, L., and Reddy, C. V. G. Bioactivity of the extracts from some marine animals of the Indian Coast. *Mahasagar* 14: 117, 1981.
40. Brekhman, I. L. and Dardynov, I. V. Pharmacological investigation of glycosides from ginseng and eleutherococcus. *Lloydia,* 32: 26, 1969.
41. Smith, W. G. Pharmacological screening tests, in *Progress in Medicinal Chemistry.* Vol. 1, G. P. Ellis and G. B. West, Eds., Butterworths, London, 1961, 1.
42. Paget, G. E. and Barnes, J. M. Toxicity tests, in *Evaluation of Drug Activities: Pharmacometrics* Vol. 1, D. R. Laurence and A. L. Bacharach, Eds., Academic Press, London, 1964, 135.
43. Bhakuni, D. S., Dhar, M. M., Dhawan, B. N., and Mehrotra, B. N. Screening of Indian plants for biological activity. II. *Indian J. Exp. Biol.* 7: 250, 1969.
44. Bhakuni, D. S., Dhar, M. L., Dhar, M. M., Dhawan, B. N., Gupta, B., and Srimal, R. C. Screening of Indian plants for biological activity. III. *Indian J. Exp. Biol.* 9: 91, 1971.
45. Dhar, M. L., Dhar, M. M., Dhawan, B. N., Gupta, B. Srimal, R. C., and Tandon, J. S. Screening of Indian plants for biological activity. IV. *Indian J. Exp. Biol.* 11: 43, 1973.
46. Dhar, M. L., Dhawan, B. N., Prasad, C. R., Rastogi, R. P., Singh, K. K., and Tandon, J. S. Screening of Indian plants for biological activity. V. *Indian J. Exp. Biol.* 12: 208, 1974.
47. Dhawan, B. N., Patnaik, G. K., Rastogi, R. P., Singh, K. K., and Tandon, J. S. Screening of Indian plants for biological activity. VI. *Indian J. Exp. Biol.* 15: 208, 1977.
48. Setty, B. S., Kamboj, V. P., and Khanna, N. M. Screening of Indian plants for biological activity. VII. Spermicidal activity of Indian plants. *Indian J. Exp. Biol.* 15: 231, 1977.

49. Aswal, B. S., Bhakuni, D. S., Goel, A. K., Kar, K., Mehrotra, B. N., and Mukherjee, K. C. Screening of Indian plants for biological activity. X. *Indian J. Exp. Biol.* 22: 312, 1984.
50. Babbar, O. P., Joshi, M. N., and Madan, A. R. Evaluation of plants for antiviral activity. *Indian J. Med. Res.* 76 (Suppl.): 54, 1982.
51. Setty, B. S., Kamboj, V. P., Garg, H. S., and Khanna, N. M. Spermicidal potential of saponins isolated from Indian medicinal plants. *Contraception* 14: 571, 1976.
52. Kamboj, V. P. Setty, B. S., and Khanna, N. M. Semen coagulation — a potential approach to contraception. *Contraception,* 15: 601, 1977.
53. Ahmad, A. and Dhawan, B. N. Metrazol test for rapid screening of anticonvulsants. *Jpn. J. Pharmacol.* 19: 472, 1969.
54. Srimal, R. C. and Dhawan, B. N. On the suitability of mice as an experimental animal for study of anti-inflammatory agents. *Indian J. Pharmacol.* 3: 4, 1971.
55. Horn, H. J. Simplified LD50 (or ED50) calculations. *Biometrics,* 12: 312, 1956.
56. Dhawan, B. N. and Rastogi, R. P. Recent developments from Indian medicinal plants. this book, chap. 13.
57. Phillipson, J. D. The pros and cons of herbal remedies. *Pharm. J.* 22: 388, 1981.
58. Gulati, A., Srimal, R. C., and Dhawan, B. N. Experimental and clinical studies on the cardiovascular effects of Rooh Afza. Hamdard National Foundation Research Award Monograph 2: 1, 1985.
59. Srivastava, R., Dikshit, M., Srimal, R. C., and Dhawan, B. N. Anti-thrombotic effect of curcumin. *Thromb. Res.* 40: 413, 1985.
60. Antarkar, D. S., Vaidya, A. B., Doshi, J. C., Athavale, A. V., Vinchoo, K. S., Natekar, M. R., Tathed, P. S., Ramesh, V., and Kale, N. A double blind clinical trial of Arogyawardhani — an Ayurvedic drug in acute viral hepatitis. *Indian J. Med. Res.* 72: 588, 1980.
61. Dhawan, B. N. Current status of studies on medicinal and aromatic plants in India. Proc. II Int. Meeting Med. Aromat. Plants, Citta de Castello, 207, 1981.

Chapter 7

TECHNOLOGIES FOR THE PROCESSING OF MEDICINAL PLANTS

Ezio Bombardelli

I. INTRODUCTION

Medicinal plants, which have been in past centuries a main therapeutic resource for treating a wide variety of diseases, still account with their derivatives for 20% of all medical prescriptions in industrialized countries and for 80% in developing countries. The medicinal plants used as sources for extracts or pure products for therapeutic use represent a rapidly expanding area. Of ever-increasing importance are pure substances, or purified and standardized extracts, which permit better analytical characterization and so meet the quality, efficacy, and safety requirements with which every modern drug, whether natural or synthetic, must comply. In any case, many traditional remedies are still in therapeutic use, including drugs as such, or extracts prepared in accordance with the pharmacopeia of the country where they are used or according to the requirements of use in the pharmaceutical forms for which they are intended.

Obviously, the technology for obtaining these derivatives, which differ widely one from another, likewise differ, but if a material of constant quality is to be produced industrially, some criteria apply universally and they must be complied with more strictly for less sophisticated products than for pure active principles.

All these preparations, when for therapeutic use in man, whether as drugs or their derivatives, must comply with the requirements of constancy, stability, sterility, purity, a limit of residual solvents, and pesticides as laid down by the international health authorities. This is the source of manufacturing problems that are often difficult to solve — more difficult, as mentioned above, for the simpler extracts than for those with a higher technological content. The crucial stages in the process are, as a rule, the choice of raw material and the procedure of extraction. While the choice of the extraction solvents is an important factor in finished product quality, the choice of the drug is pivotal. The drug must first be thoroughly checked before extraction as far as botanical, phytopathological, and chemical aspects are concerned. In connection with the last point alone, one needs only to consider that plants normally contain several active substances, and in a certain ratio to one another. Present regulations require that this ratio be kept, within narrow limits, constant from one preparation to another, which means starting from an already standardized raw material. To attain this first objective, if the plant to be extracted is still harvested in the wild, the area of origin, time of harvesting, the conditions of drying and of storage must be specified, and at the extraction stage several batches must be combined to achieve a homogeneous mix.

In the case of the most important drugs, to overcome these difficulties, crops are grown from homogeneous genetically selected strains. The genetic selection of medicinal plants through segregation of the desired characteristics or meristematic micropropagation from cell cultures of high grade plants, already operative in many countries, is the present and medium term future solution to this aspect of the problem of medicinal plants of industrial interest. This is pending the realization on a much larger scale of tissue cultures in the liquid phase, which at the moment, apart from a few exceptions is still at the laboratory stage. While drug quality has become an essential factor in the production of extracts, in the production of active principles it is a great advantage in that it simplifies the process and reduces the cost of the finished product. On looking at the technological side of vegetal derivatives, whether in the form of extracts or pure products, one obviously finds common stages, which essentially include grinding and extraction of the drug, concentration and drying of the extracts, and so on. Each of these stages is considered separately in the paragraphs that follow both from the theoretical point of view and from the point of view of the hardware needed in practice.

II. DRUG GRINDING

A. GENERAL CONCEPTS AND PURPOSES OF GRINDING

Grinding or mincing a drug means mechanically breaking down a given vegetable material, leaves, roots, seeds, or other into very small units, ranging from a coarse fragment to a fine powder. This is normally the first stage in the preparation of any vegetal derivative, whether simple or complex. With regard to the straight use of ground drugs as finished product for putting into capsules, sachets, or other packaging, for the technical procedures we refer the reader to the principal pharmacopeias (USP XXI, BP 80, EU, Ph.), which specify the criteria of selection, details of particle size of the various products described in them, and methods of control. This review is concerned solely with the industrial use of the vegetable material, intended for extraction in large-scale installations.

In the process of grinding, irrespective of the type of equipment used, particle size homogeneity is normally a basic parameter; this governs at the extraction stage the uniform exhaustion of the drug, which depends on the rate of diffusion of a substance from the granule to the solvent, the contact time, the rate of passage of the solvent through the drug, and other aspects to be considered later. Theoretically, the finer the granule, the faster (within certain limits) the extraction should be. When there are no technical impediments, like the release of mucilage, which hinders filtration of the solvent through the drug, reasonably fine powders (diameter approaching, but not less than, 0.5 mm) are preferred. This size is normally suited to modern installations, in which continuous extraction may operate.

The foregoing applies, of course, to dried drugs. For the extraction of fresh vegetal material the problem of grinding is mainly a problem of chemical stability of the substances to be extracted. Many fresh drugs in the course of crushing undergo numerous chemical changes, hydrolyses, or oxidations, related to the release of enzymes from the plant cells. Such snags can in part be overcome by adding specific enzyme inhibitors to the drug or by using material prefrozen to temperatures below 25°C to deter reactivity by stopping the reaction in the solid liquid phase or vice versa.

In the pharmaceutical field the extraction of fresh drugs is steadily losing ground because of difficulties in checking the material to be extracted and is practised only in the case of costly products for which deep freezing for long periods is justified.

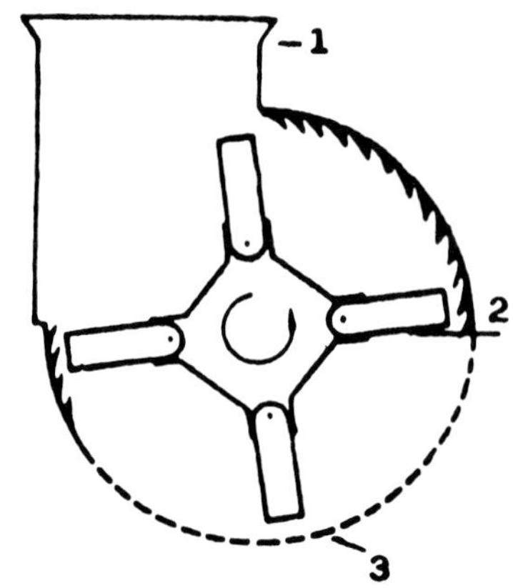

Figure 1. Diagram of hammer mill. (1) Drug feed. (2) Hammers. (3) Grind.

B. EQUIPMENT FOR GRINDING DRUGS

There are machines for prechopping or mincing drugs and grinding machines proper. In many cases mincing must precede grinding, as in the case of coarse hard woody parts, of roots of trees, of very tough fibrous materials. Because of the different nature of the materials to be pretreated, these machines are very strong and may cut in various ways, lengthwise and crosswise. The drug is cut into nonuniform pieces of 1 to 10 mm in size. If the drug has to be pulverized, there are several types of mill which are hard to classify because the various manufacturers now provide interchangeable parts for solving different problems for application to the same machine. An idea may be obtained from the literature, where the subject has been comprehensively reviewed.[1] Of these mills the standard type, still widely used, is the hammer mill. These are sturdy machines for breaking up friable substances coarsely by means of rotating hammers. The mill consists of a rotor to which pendulum beaters are attached (Figure 1). The hammers, following the rotation of the shaft to which they are applied, are arranged radially and break up the material that is introduced into the rotor chamber. On the walls of the chamber is a grid, which determines the size of the particles that pass through it. Then there are knife mills (Figure 2), which operate by cutting the material conveyed into the housing, where the blades can assume a horizontal or vertical movement. As with the hammer mill, the fineness of the chopped material depends on the grid applied. This type of mill is very suitable for grinding leaves, barks, and roots for subsequent extraction by maceration or percolation.

Another way of grinding drugs is to pass them through a system of toothed disks that can operate either horizontally or vertically. This type of grinder is suitable for grinding hard seeds or drugs that have been prechopped (Figure 3). The grinding mechanism of the mill consists of two toothed disks, one fixed and the other rotating. The teeth are arrayed on concentric rows which brush against one another during movement, and the drug is ground as it passes between the disks. Particle size is set by the distance between the two disks and by the reciprocal velocity between the disks. Very fine powders can be obtained from this type of machine. By way of conclusion to this brief review of machines for grinding drugs, it should be added that there are a wide variety of machines on the above lines that can solve any technical problem. One need only consult the manufacturers. There are also other types of mills, rotating cylinder pulverizers, and so on, that have facilities for freezing drugs containing labile compounds to –20°C, which complete the range of machines for preparing drugs for extraction.

III. DRUG EXTRACTION

A. CONCEPTS AND DEFINITIONS

The extraction of a drug is the separation by physical or chemical means of a solid or liquid material from a solid, i.e., the drug. Normally, when this operation is performed with solvents for extracting the vegetal material, which is the most common practice, it is called solid/liquid extraction. In the course of solid/liquid extraction two processes take place in parallel: the release of extractive substances from destroyed cells and the release of extractive substances from intact plant cells by a process of diffusion. The latter process is usually enhanced when the plant cell is treated with aqueous or partly aqueous solvents, which cause swelling with consequent increased permeability or rupture of the cell wall. The procedures of extraction of drugs, in light of the above remarks, may be classified in two main groups:

1. Procedures in which it is sufficient to achieve within set limits the equilibrium of concentration between drug and solution (tinctures, decoctions, macerations)
2. Procedures in which the drug is extracted until exhaustion of the soluble substances in the chosen medium

In type 1, of decreasing industrial importance, except for the preparation of tinctures, the simplest case is that of maceration, which may be static or dynamic; it also forms part of all processes, excluding countercurrent processes, in which the aim is to ensure exhaustion of the drug. In maceration, equilibrium depends on the characteristics of the drug, on its content of moisture, on the grinding, on the solvent used, and on the contact time; these parameters influence one another and the

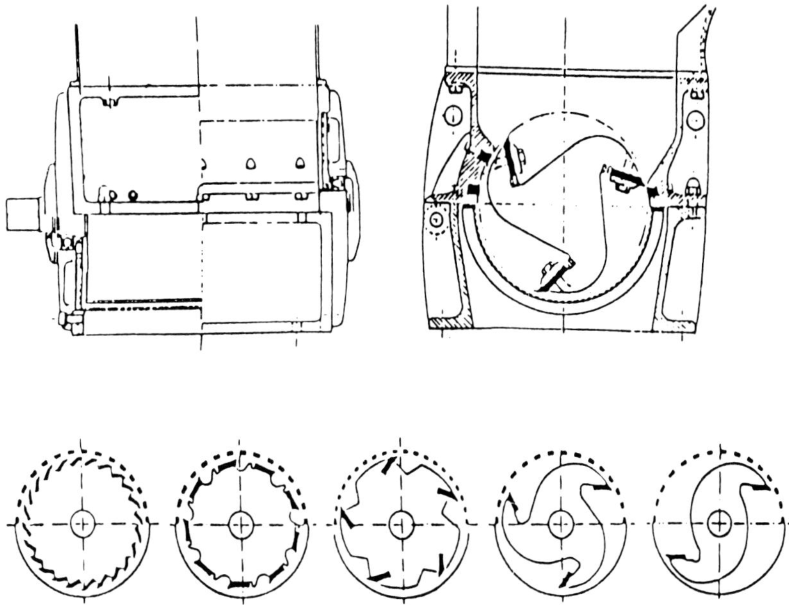

Figure 2. Different types of knife mills.

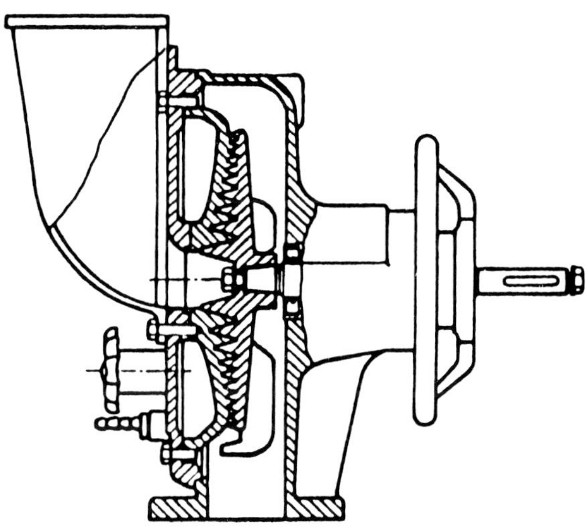

Figure 3. Diagram of teeth mill.

optimal parameters have to be sought for each drug. The theoretical and mathematical aspects and reciprocal parametric relations are well described in the literature.[4,5]

Every extractive procedure that leads to a concentration equilibrium stops when the distribution of the extractable substances between solvent and residual drug is constant. It is essential to know the value of this constant before deciding on the duration and number of extractions needed to exhaust a drug. Industrially, maceration processes are often related to those of percolation. Broadly speaking, percolation is repeated maceration conducted in a suitable container. In the following paragraphs this process is considered in detail, because it is still widely practised and is easy to handle.

B. PARAMETERS INFLUENCING EXTRACTION

1. Swelling of the Drug

In all industrial processes involving extraction by percolation or continuous extraction, pretreatment of the ground drug with solvent outside the extractor is as a rule indispensable. The main reasons for this are

1. To avoid sudden swelling of the drug in a closed container, because, if the solvent is aqueous, the drug may swell to two to three times its original volume and so burst the extractor or make percolation impossible
2. To ensure uniform moistening of the material for extraction and so prevent the formation of preferential channels, increasing the contact and passage of the solvent
3. To increase the porosity of the cell wall, thus facilitating diffusion of the extractive substances from cell to solvent or penetration of the cell by the solvent

The value of the first two points is technically self-evident. The third point is of particular importance when, for reasons of selectivity, an extraction has to be performed in a heterogeneous phase, as is the case with the extraction of many alkaloids. The ground drug is wetted with a roughly equal weight of water or aqueous solutions of Na_2CO_3 or NH_4OH to induce cell lysis and permit subsequent extraction with aliphatic or aromatic hydrocarbons. Swelling of the vegetal material in these cases ensures permeation of the solvent, and consequent removal of the substances soluble in it, simply and selectively.

2. Diffusion, pH, Particle Size, and Temperature

For the extraction of a substance from a vegetal material the solvent must diffuse into the cell and in its turn the substance must be sufficiently soluble in the solvent. In this way an equilibrium will be established between solute and solvent, the speed with which it is attained depending normally on temperature, pH, particle size, and the movement of the solvent around the particle. The pH normally plays an important role in selectivity, while temperature and movement of the liquid about the solid shift the equilibrium toward saturation of the solvent. Movement of the liquid can be ensured by keeping the drug still and cycling the liquid with pumps or with mechanical agitators, which mix solvent and drug continuously, or by means of ultrasound waves. Particle size is normally adjusted to the chemical composition of the compounds to be extracted. As a rule, a finely powdered drug is easy to extract, but for many substances, according to the solvent used, it must not be forgotten that an equilibrium is established between extraction and reabsorption of the components already dissolved in the fibrous material of the drug. Therefore, the choice must be guided by appropriate tests.[5,6] Fine particle size must be ruled out when mucilaginous substances are present, as these by swelling would prevent the solvent from passing through the drug.

3. Choice of Extraction Solvent

To obtain the complete extraction of a given active principle from a drug, the ideal solvent is obviously one that presenting maximum selectivity, has the best capacity for extraction in terms of coefficient of saturation of the product in the medium and is compatible with the properties of the material to be extracted. These requisites must as a rule be determined experimentally for each drug, since the choice often depends on the stability of the compound to be extracted and on possible interactions with other substances present. In principle, considering the above points separately, it may be said that aliphatic alcohols with up to three carbon atoms, or mixtures of them with water, are the solvents with the greatest extractive power for almost all natural substances of low molecular weight like alkaloids, saponins, and flavonoids, Ethyl alcohol, in particular, is the solvent of choice according to the pharmacopeias for obtaining classic extracts such as tinctures and soft, fluid, and dry extracts still widely used in pharmaceutical formulations. As these solvents have great extractive power, they are the least selective and can still be of use not only in the preparation of the above established extracts but also for the extraction of plants whose active fraction is not yet well known, and an extract as complete as possible is needed. Still regarding the use of this type of solvent, the ideal alcohol/water ration for the extraction of woody parts of plants or barks, roots and seeds is about 7:3 or 8:2, whereas it must be lower (compatibly with the stability of the active principles) than 1:1 for extracting leaves or aerial green parts. With an alcohol/water ratio of 1:1 it is possible to avoid the extraction of chlorophyll, of resinous or polymeric substances that are normally of no importance to the activity of the extract but that greatly complicate the subsequent stage of concentration by giving rise to gummy precipitates that are hard to eliminate. Lower strength hydroalcoholic mixtures with an alcohol/water ratio of 2:8 or 3:7 may be used in special cases not only for extraction but also under suitable temperature conditions for accomplishing target enzyme conversions in the actual course of extraction.

A case in point is the classic conversion of primary glucosides of *Digitalis lanata*, lanatosides A, B, and C, into digoxin, gitoxin, and digitoxin, respectively. This takes place at room temperature during the moistening of the drug with water only or with water containing up to 20% alcohol. In the contrary case, for the extraction of primary glucosides it is essential to operate with hydroalcoholic mixtures containing more than 50% alcohol to block hydrolase activity, and so on. Obviously, numerous other substances present in plants may behave like the glucosides of digitalis and may be modified or profoundly degraded if the extraction solvent or the temperature are not right for the products to be extracted. Mere grinding, for example, of a fresh drug containing phenolic substances can lead to their oxidative degradation. Another aspect in the extractive phase is the selectivity of a given solvent. A large number of alkaloids can be properly extracted from any plant with hydrocarbons after the drug has been moistened with water or water rendered alkaline by the addition of inorganic bases; alkaloids extracted in this way are accompanied only by other lipophilic substances, which can easily be separated off by counterextraction of the alkaloids with acid. Another large number of alkaloids can be extracted from vegetal material with acids, organic or inorganic; the alkalized extracts can be purified in a continuous process by counterextraction with hydrocarbons or other apolar solvents. Flavonoids and terpenes, on the other hand, can be selectively extracted at neutral pH with ethyl acetate or aliphatic ketones. For other classes of substances it is more difficult to give general indications of selectivity, given the heterogeneity of the components, the great difference in polarity, and their adaptability to the available machinery. For these reasons, solvent and extraction system need to be tailored for any important product.

C. EXTRACTION PROCEDURES

1. Static and Dynamic Maceration

The processes of maceration, referred to earlier when considering the parameters that influence extraction, are the simplest procedures for obtaining an extract and are describe in all the pharmacopeias (USP IX, FU IX, BP 80), since they are suitable both for small quantities of drug and for industrial quantities. The simplest process consists of pouring solvent onto the drug and, after a set time for every drug, straining off the extract and washing the drug with fresh solvent to a prescribed weight. This procedure is useful for preparing tinctures or particular extracts and sometimes is the only one that can be used for drugs rich in mucilage. However, it is wasteful because it never exhausts the vegetal material. The drug retains a considerable portion of solute which has to be recovered by pressing or by centrifugation. This final stage is a necessary step in any type of maceration, static or dynamic, which is done by turboextraction with more or less hard crushing of the drug. If the material being extracted is costly, it is normal to choose a method that leads to exhaustion of the residue by means of one of the processes described later.

2. Extraction by Simple or Continuous Percolation

In simple or continuous percolation the objective is usually to deplete the vegetal material totally of the desired components. In simple percolation the drug is always extracted to exhaustion with fresh solvent. This a long and expensive process due to the large quantities of solvent used, depending on several parameters:

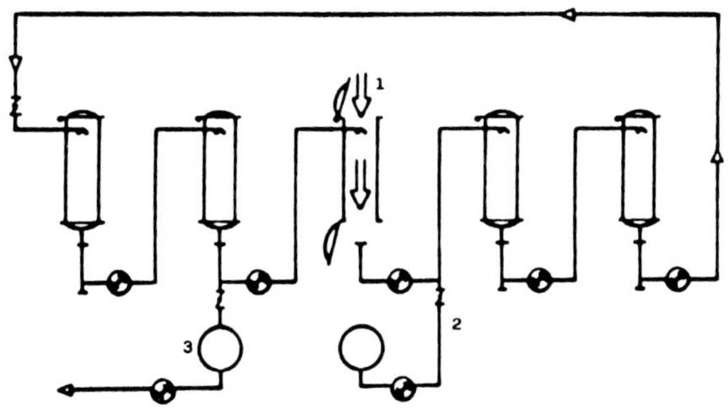

Figure 4. Battery of percolators and operation scheme. (1) Drug feed. (2) Fresh solvent. (3) Extract.

1. Time taken to reach solvent-solute equilibrium
2. Quantity of solvent needed to effect the first extraction on a reasonable industrial scale
3. Quantity of solvent needed to dilute completely the quantity of solute retained by the residual drug in the first extraction

The costly practice of always using fresh solvent can be circumvented by using more than one percolator and percolates less rich in extract for treating new. The concept is repercolation and is the prelude to continuous countercurrent extraction, in which fresh drug is placed in contact with solvent already rick in solute and new solvent is added to partially extracted drug (Figure 4).

3. Percolation and Repercolation

The mathematical approach to the process of percolation is based mainly on the concept of diffusion.[7] In the course of percolation the process of diffusion is a function of the speed of percolation, the quantity of solvent, and the drug/solvent diffusion constant. Some theoretical aspects of this process have been outlined in recent publications[8] which describe the trend of extract concentrations as a function of elution, etc.

Because it is easy to handle, percolation is one of the procedures most used for the extraction of drugs and, like maceration, it can be used both in the laboratory and industrially. In percolation, as in maceration, the drug is ground to the appropriate particle size, but not too finely so that the powder does not impede filtration of the solvent through the drug. As the drug is placed in a very thick layer in the percolator containers, it is first moistened with the extraction solvent, as a rule outside the extractor, and allowed to swell before it is loaded. The drug, conveyed to the percolator via spiral feed systems, after formation of the layer is covered with solvent. In very large installations the solvent is kept in continuous movement by pumps which recycle it from bottom to top to further the attainment of equilibrium as far as possible, and extraction can be accomplished by a weak reflux system, under pressure or at room temperature.

Hot extraction is used, as a rule, when the ratio of solubility of a given substance in the chosen solvent is improved by heat. Exhaustion can be achieved with a single percolator or, to reduce the quantity of solvent, in several percolators, using the extracts with a lower concentration for extracting a new charge of drug or a charge at a different degree of exhaustion. By appropriate calculation of the quantity of solvent and the number of percolators, some types of extracts can be prepared directly without need for concentration, etc.

This simple system of extraction has some disadvantages which include, in large installations: the formation of preferential channels with consequent nonhomogeneous extraction of the drug; the elimination of the residual solvent retained by the drug, which, in the absence of agitation, has to be either distilled in a stream of steam or recovered by more complicated processes or redrying; and difficulties of empting the exhausted material.

4. Countercurrent Extraction

In continuous countercurrent extraction the drug moves against the solvent. The drug enters the percolator meeting extract enriched solvent and leaves it encountering fresh solvent. In this way it is possible to obtain complete exhaustion by appropriate setting of the flow of the drug toward the solvent, thereby ensuring an absolute countercurrent. In main installations (discussed later) the countercurrent is relative because the drug stays still and the solvent moves. This process is similar to continuous percolation. The theoretical premises of countercurrent extraction are discussed in several publications,[9,10] which give the equations and relevant curves.

The theoretical ratios were calculated first for liquid-liquid countercurrent extraction and later adapted for solid-liquid extraction. From the practical point of view continuous countercurrent extraction can be adapted with advantage for large-scale processing and production of extractives. The equipment designed for this purpose is of large capacity and so it is suitable for high consumption products not always found in the pharmaceutical industry. Equipment of smaller capacity is now available on the market, but it requires a production cycle that runs for 24 h/day.

The most widely used continuous extractors are screw extractors, in which drug and solvent move one against the other; carousel or U extractors, in which the material, closed in percolator type subunits, moves and is extracted on meeting solvent of different saturation gradients; and turboextractors, in which the drug is continuously treated, with an excess of solvent under very strong agitation, combining a decanter in cascade for the separation of the exhausted material.

In screw extractors an absolute continuous countercurrent is achieved, U or carousel extractors achieve a relative countercurrent, and turbo/centrifuge extractors achieve a continuous countercurrent. In choosing one of these extractors for solving more than one production problem, several points need to be considered. In continuous extractors, for obvious reasons, one of the fundamental and most critical parameters is the material to be extracted and above all, its particle size. Except for extractors with combined decanter, where the quantity of solvent is normally high and subsequent filtration forced, other extractors require not too fine grinding of the drug and less than 10% of particles with a diameter of less than 0.5 mm to avoid problems in obtaining and clarifying the extract. In a continuous installation, to achieve optimal exhaustion of the vegetal material, all the parameters involved from particle size of the drug to solvent flow rate, contact times, temperature, and pressure should be defined experimentally and observed strictly. Every drug has its own characteristics and so the procedure must be tailored to the equipment, since, given the wide range of products that a producer of raw materials for the pharmaceutical industry has to offer, it is not realistic to design an ideal installation for any one of them.

With the equipment available, drug extraction is as a rule very good and once production is running, inexpensive. The drawback of these machines is that they are not easily adapted to short production programs mainly because of the idle times in changing the operative conditions between processings or the presence of large quantities of mucilage in the material to be extracted, which can give rise to unforeseeable difficulties due to swelling in the course of extraction with consequent change in other parameters. From the foregoing, the most important of all the factors governing the choice of extraction procedure that we have examined are the quantity and nature of the drug. If one has to choose between continuous and discontinuous extraction, then quantity is the main problem. In choosing between maceration and percolation, the nature of the drug material is the decisive factor.

Sufficient variety of hardware exists on the market today, both in type and in size, to enable one to solve all the most common problems connected with medicinal plants.

D. EQUIPMENT FOR EXTRACTION

1. Equipment for Static and Dynamic Maceration

Machines for static maceration are large, closed containers in which the drug is kept in contact with the solvent for a set time, after which the extract is recovered by filtration or decantation. This type of process has practically been abandoned industrially and replaced by equipment in which solvent and drug are in movement. Extractors of this type can be fitted with internal agitator or move by rotating on their axis. Figures 5 and 6 show examples of two types of extractor without agitator. In the one shown in Figure 5, which can be of small capacity, the drug is loaded with the solvent and is kept in vigorous movement for the set time for extraction. One of the two tops of the extractor is equipped with a grid for filtering the drug to a fixed container for obtaining the extract. Figure 6 shows an extractor that can reach a capacity of 3000 to 4000 L and a horizontal cylindrical container fitted with heating jacket or, if preferred, an internal heating coil which rotates on its shaft. Through this shaft the extract is discharged and fresh solvent introduced. With the use of the appropriate solvents, extraction can be effected at the desired temperature and under pressure. The drug is loaded and discharged through aperture A. At the end of the extraction the solvent retained by the drug can be removed by heating and distillation while the residue is stirred. After removal of the solvent, the exhausted drug is discharged through the same aperture by swinging the extractor several times. Drug exhaustion can be effected by suing several extractors in series. Several improvements have been made in this extractor model, including a change in the shape of the container to facilitate unloading of the exhausted material, introduction of filtering surfaces to obtain clear

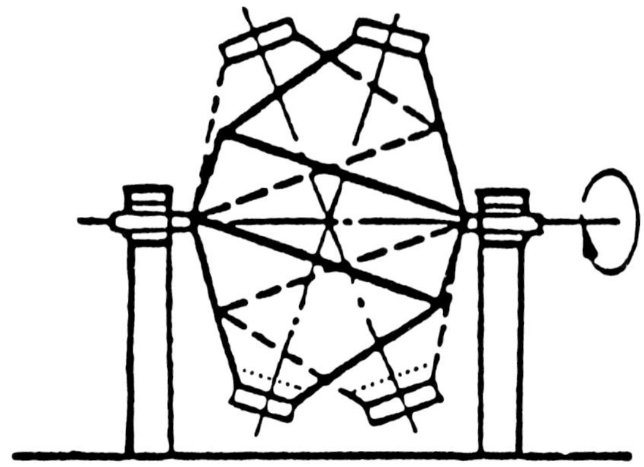

Figure 5. Scheme of dynamic extractor suited for maceration..

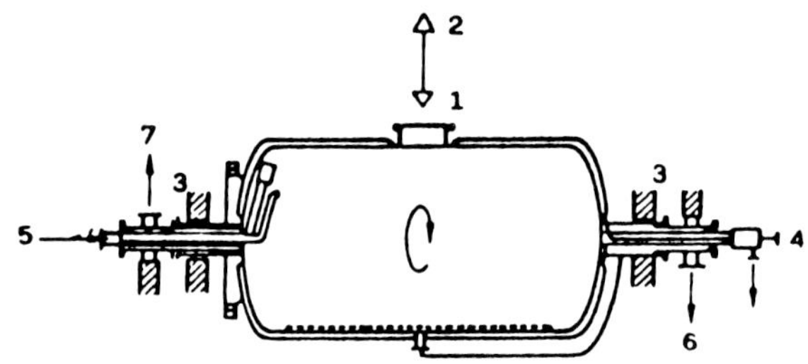

Figure 6. Horizontal extractor rotating on its shaft. (1) Solids feed. (2) Solids discharge. (3) Bearing. (4) Heating. (5) Fresh solvent. (6) Extract. (7) Vapours.

extracts, etc. The advantage of rotary extractors is that the entire process can be handled by one machine.

2. Extractors with Agitator

Extractors fitted with agitator are usually cylindrical or conical in shape and as a rule are equipped with a double heating and cooling jacket. In horizontal extractors having a central shaft fitted with arms (Figure 7), the drug is in continuous movement. The extractor may be fitted with grids at the bottom for filtering the extract. The drug is loaded from an aperture at the top and unloaded from the opposite side. Machines of this type are very useful for extracting drugs made alkaline with aqueous solutions, with solvents immiscible with water, or fresh drugs to be extracted always with solvents immiscible with water. The Nauta extractor (Figure 8) also consists of a container with double jacket for the circulation of thermoconvector fluids and inside has a screw which brushes against the wall, conveying the material from bottom to top. This type of apparatus lends itself to the extraction of all types of drugs, including those with a high mucilage content. Another extractor with agitator is the Ultra Turax, in which the drug is homogenized with rupture of the cell walls. The rate of extraction is increased, but the residue has to be separated by centrifugation.

3. Equipment for Percolation

In the percolation process, as already mentioned, the drug must as a rule be moistened with the extraction solvent before being placed in the percolator. The drug is usually moistened

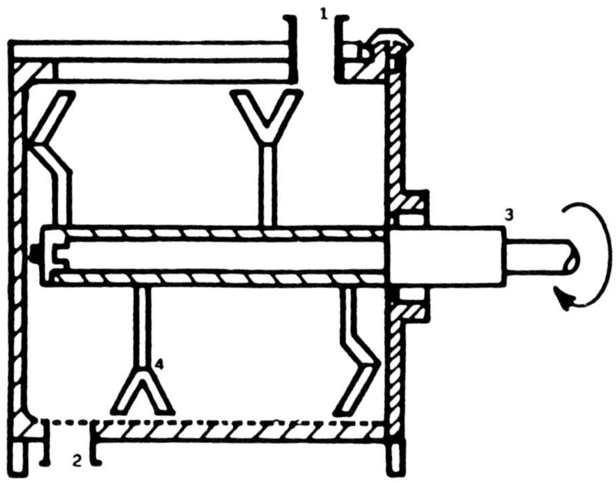

Figure 7. Scheme of extractor with agitator. (1) Drug feed. (2) Solids discharge. (3) Shaft. (4) Arm.

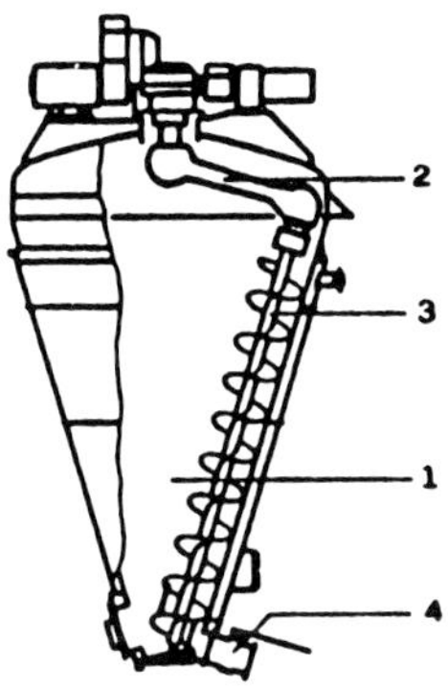

Figure 8. Nauta extractor. (1) Extractor. (2) Swinging arm. (3) Mixing screw. (4) Solids discharge.

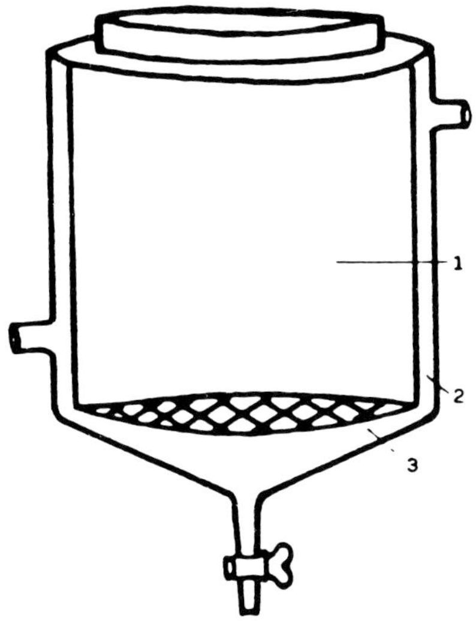

Figure 9. Scheme of traditional percolator. (1) Percolator. (2) Jacket. (3) Grid.

with from 0.3 to 1 part of solvent and left to stand for about 2 h until swelling is complete. The percolator consists of a conical or cylindrical container closed at the bottom by a filter sieve with a nonreturn valve (Figure 9). For percolation the drug must be ground coarsely (1 to 3 mm) to allow rapid drainage of the solvent through the drug. With this type of apparatus a large number of extracts are prepared on an industrial scale using a series of percolators in order to reduce solvent consumption and avoid unnecessary concentrations of percolate. In a semicontinuous installation of this type the solvent passes from one percolator to the next until saturation or the required degree of concentration is reached. Fresh solvent is used at the exhaustion stage and then reused for the extraction of fresh drug. This type of installation is very advantageous for the preparation of traditional extracts such as fluid and soft extracts, as it produces them without a lot of manipulation and with a considerable saving of solvent. The main disadvantages are that the drug has to be moistened before it goes into the percolator and that the drug, according to the height of the percolator, often gets compacted after the first few extractions so that the passage of the solvent is hindered. This means that the passage of the percolate from one extractor to the next is slowed down so that operating ceases to be smooth. Recovery of the solvent retained by the drug is often laborious and the same applies to the unloading and removal of the exhausted material. When large quantities of drug have to be handled, continuous extraction as described below is preferred to percolation.

4. Equipment for Countercurrent Extractions

As already reported in the section describing processes of extraction, in continuous extraction the types of equipment most commonly used are screw, carousel, U-, radial pressure extractors, or combinations of these. Screw extractors operate on the absolute countercurrent principle, according to which drug and solvent move in opposite directions (Figure 10). The Niro extractor consists of a long pressure-tight steel cylinder,[12] which can be heated and cooled in sections, equipped inside with a system of screw feeders. The drug reaches the extractor through a loading device at one end of the cylinder and is driven by the screw to the opposite end. The solvent enters the extractor at the opposite end to the drug entry point. Mixing of drug and solvent is a very vigorous process, the intensity of which can be regulated by the number of revolutions of the screw, the slope of the cylinder, the temperature, and the pressure. Through modification of these parameters, when a drug has been through the length of the extractor, continually meeting fresh solvent, it is practically exhausted. The carousel extractor operates on the principle of relative countercurrent extraction. It consists in practice of a battery of countercurrent percolators, in which each compartment of the extractor represents a percolator. Figure 11 illustrates well the principle on which it is based. The percolator is made up of 10 to 20 segments, on one or two planes. The plane rotates at a speed of 10 mm/s. During the movement each chamber containing drug is washed by a rain of solvent, which comes from the underlying part and has a different concentration in each compartment. The carousel which rotates clockwise is always loaded and unloaded at the same point. The exhausted drug at the end of the run is conveyed to the desolventizer. One needs to know all the "ins and outs" of this apparatus to make it work properly.

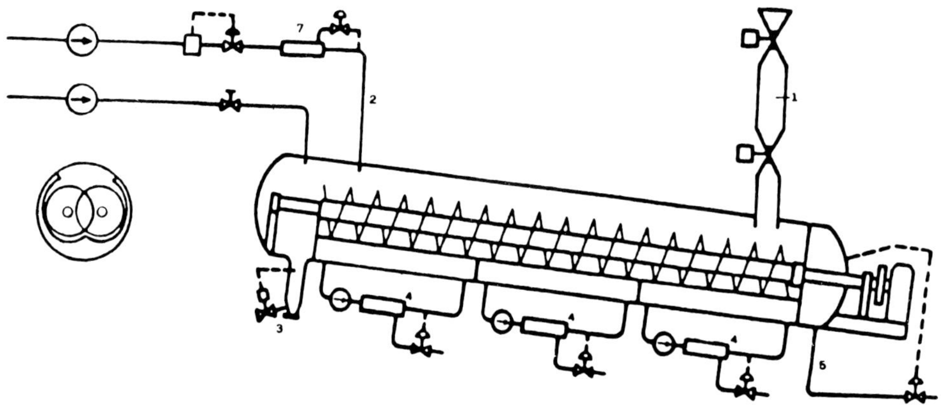

Figure 10. Countercurrent horizontal extractor. (1) Solid feed. (2) Solvent inlet. (3) Solids discharge. (4) Heat exchanger. (5) Extract. (6) Spent-solvent outlet. (7) Heater.

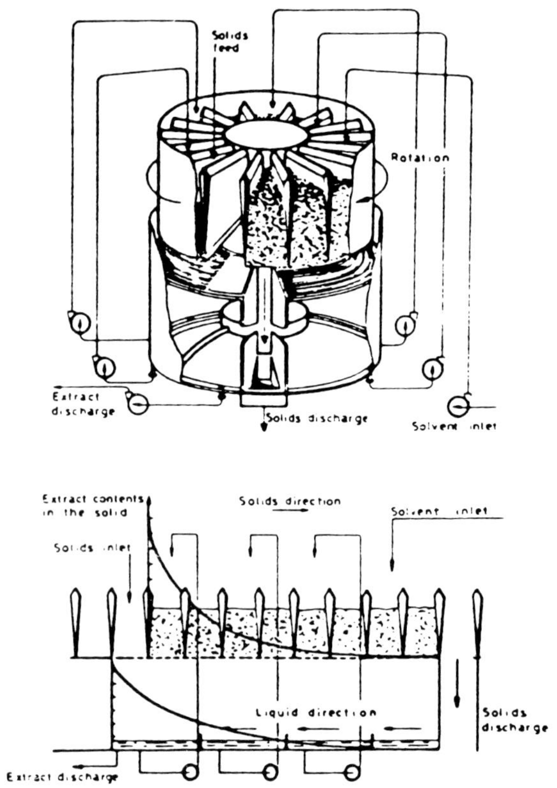

Figure 11. Diagram of carousel extractor.

As a rule, the drug has to be pretreated and particle size has to be right to get a constant speed of percolation. Grinding is best done with knife mills to prevent the formation of too fine particles, which should not exceed 10%. The carousel extractor is popular among pharmaceutical manufacturers for the extraction of numerous active principles. Like the carousel, the U-extractor operates on the principle of relative continuous extraction. This extractor likewise contains the drug in a fixed compartment, through which the solvent passes in the opposite direction. The basic difference between the two types is that in the carousel the drug moves on a horizontal plane, whereas in the U-extractor it moves in a vertical plane. The apparatus consists of a U-shaped tube (Figure 12) in which a series of filtering floors is moved by a conveyor belt.[14] The drug is conveyed on the bottom by a screw system and transported through the tube. The solvent flows in the opposite direction. At one position of the tube a vibrating device prevents the formation of preferential channels and renders extraction homogeneous. The exhausted drug at the end of the run is automatically sent to a press and then to the desolventization tower.

Other extractors to operate on the U principle are the Hildebrandt[15] extractor and moving tank[16] extractors. Inoxa type pressure extractors work on a slightly different principle.[17] The percolator is a cylinder (Figure 13) which contains a finely perforated box to hold the drug for extraction. At the center of the extractor is a tube through which the solvent is pumped continuously. A very high pressure may be attained, due to friction; extraction is completed very quickly. The duration of the operation, which is the time taken to reach equilibrium between solvent and solute, is governed by the following formula:

$$T = \frac{1}{KSV^2 (C_1 - C_2)}$$

where T = time needed to reach desired concentration, K = constant depending on the solid/solvent ratio, S = surface of contact, V = rate of passage of the solvent, and C_1-C_2 = solid/liquid concentration difference. It is clearly of interest to increase V in order to decrease the time since V is in the denominator and is squared. With this extractor a drug can be exhausted in about 2 h and drugs with a high mucilage content can be extracted. In addition to these standard types of extractor, a whole series of drug extraction equipment is on the market in which the extractor is combined with a decanter for solid/liquid separation or presses.[18,19] In these processes the drug is kept in vigorous agitation, where normally it is homogenized and disintegrated to facilitate extraction.

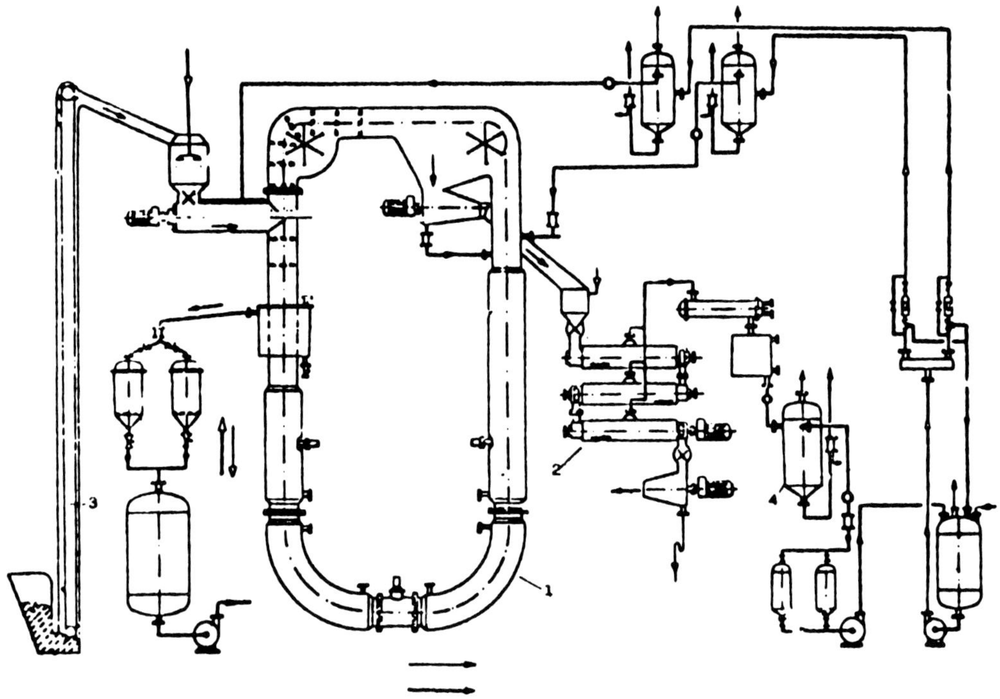

Figure 12. U-extractor. (1) Extractor. (2) Desolventizers. (3) Solids feed. (4) Concentrators.

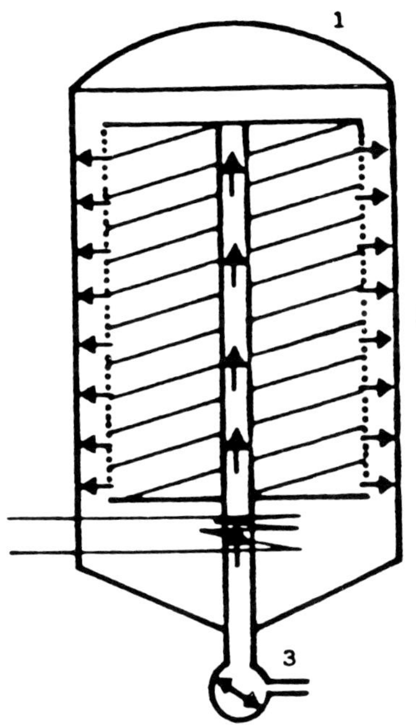

Figure 13. Inoxa extractor. (1) Extractor. (2) Basket. (3) Pump.

5. Nonconventional Extraction Systems

A recent innovation in drug extraction is the use of a gas in hypercritical phase. The extractor operates with gases as solvent under pressure.[20,21] However, such installations are very little used in the preparation of pharmaceuticals, probably because of the high running costs. They therefore are not described here.

IV. EXTRACT CONCENTRATION

A. GENERAL ASPECTS

After extraction of a drug, the resulting solutions containing the soluble fraction as a rule undergo a process of concentration or of liquid/liquid extraction, if the aim is to select a particular class of components. These extracts may be

1. Concentrated partially or totally according to whether they are to be used for preparing total fluid, soft, or dry extracts
2. Concentrated partially or counterextracted with compatible solvents for conversion into purified extracts
3. Concentrated partially and counterextracted with selective solvents for the isolation of a given active principle
4. Counterextracted right away, without concentration, for the isolation of particular products

Apart from the last case, in which concentration takes place at a later stage, extracts must be concentrated. Concentration is a tricky stage in the process in which many chemically labile products may undergo degradation, mainly due to the temperature. For this reason several types of high performance concentrators have been invented for operating at reasonably low temperatures (25° to 30°C) or at high temperatures for very short times.

B. EQUIPMENT FOR EXTRACT CONCENTRATION

The parameters that govern the choice of an evaporator are the quantity of solution to be concentrated per unit and the

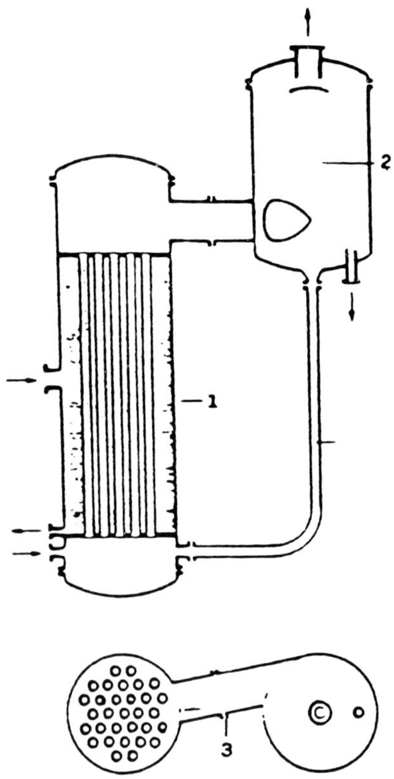

Figure 14. Traditional Robert type concentrator. (1) Concentrator. (2) Vapours chamber separation. (3) Concentrator section.

properties of the solution. Stable solutions, which do not tend to encrust, may be concentrated in simple apparatus, like ordinary reactors, in which the solutions can be evaporated at ordinary pressure or under vacuum. Industrially, for large quantities of solution one of the most widely used concentrators of the old type is the Robert concentrator.[22] This concentrator consists of a next of concentrator tubes arrayed concentrically around a central groove. The tubes are steam heated from outside and the solution to be concentrated passes through them. In its course through the tubes, evaporation takes place and the solvent is separated from the liquid in a separation chamber (Figure 14), from which it passes to the condenser. These concentrators exist in several versions and can operate with normal or forced circulation, and are very common in chemical industry.

The most up-to-date types of concentrator in the field of natural products are the descending film, thin layer or plate concentrators. The thin layer evaporator, of which there are many versions, both in design and in capacity, consists basically of a hollow cylindrical body, heated externally, in which a rotor rotates (Figure 15).[23,24] The rotor design governs the evaporation capacity, which can be adapted to the solution to be evaporated. The rotor spreads the liquid to be concentrated over the hot wall. Two versions of a rotor that can be applied to the same range of concentrators are described later. One type of concentrator has a rotor that consists of a shaft carrying a set of scoops, which spread the solution over the inner surface of the heated cylinder. Another version has a continuous lobed rotor and the product is centrifuged against the hot wall of the cylinder. The first type of rotor is normally used for viscous products, when crystals are present in suspension or when the solution is concentrated to a powdery dry residue. The second type of rotor is suitable for all the other types of solution to be concentrated.

In film evaporators, the phase change occurs in a very thin layer of liquid. The volume of solution involved is very small and so the time spent in the apparatus is very short. By operating under vacuum and hence at a fairly low evaporation temperature it is possible to ensure gentle treatment of thermosensitive products. The action of the rotor, whether mechanical, as in the case of the pan rotor, or dynamic for the centrifugal projection of particles against the wall, prevents the encrustation of the inner wall of the heated cylinder, the heat exchange characteristics therefore remaining unchanged through time, due to the high turbulence of the film inside the evaporator. As shown in Figure 16, the liquid to be concentrated is metered by pump from the extractors and introduced into the upper part of the evaporator, where a fixed ring on the rotor distributes it uniformly throughout the cylinder. The actual evaporation phase starts here, for the product, projected on to the hot wall by the rotor and in contact with the vacuum, initiates a violent separation of the vapor from the liquid. This separation is completed as the product gradually falls to the bottom. The concentrate accumulates in containers outside the evaporation tower while the vapors are conveyed to the condenser. The rate of entry of the solution to be concentrated can be regulated so that the desired concentration can be accomplished in one pass. Otherwise the concentrate is recycled.

Alongside these vertical film concentrators there are other models with a horizontal axis (Figure 17).[25] The difference between the two types of concentrator lies in the mode of regulation of the flow of solution through the evaporation chamber. In the vertical type the rate of passage of the solution is partly governed by gravity, whereas in the horizontal type the thickness of the thin layer and the stay time can be controlled by regulating the distance between the rotor shovels and the cylindrical heated surface.

Another type of evaporation, suitable for treating thermolabile substances, is the plate system, in which a wide hot contact surface promotes evaporation.[26]

In addition to these general purpose evaporators, there are numerous others based on the same principles but devised to meet particular requirements or for products to be treated for inclusion in installations in continuous operation and dealing with a specific raw material. Those mentioned above are suitable for the pharmaceutical industry producing preparations of vegetal origin, as they are available in small sizes and are fairly versatile.

V. PURIFICATION OF EXTRACTS

A. GENERAL ASPECTS

In this context this purification of extracts means the treatment of extraction liquids to remove drug residues or undesirable matter formed in the process. The clarification phase is practically indispensable, whether one intends to

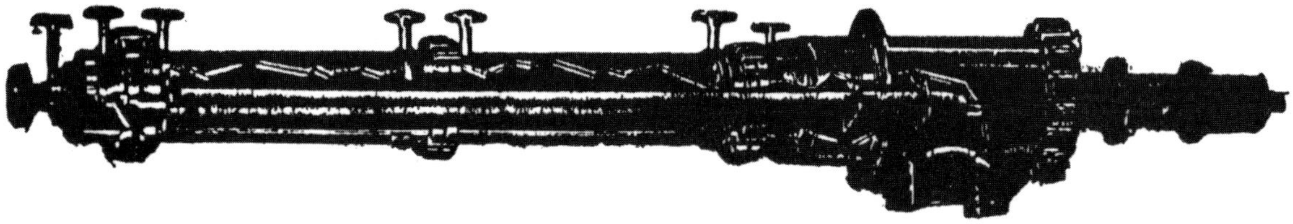

Figure 15. Diagram of descending-film evaporator.

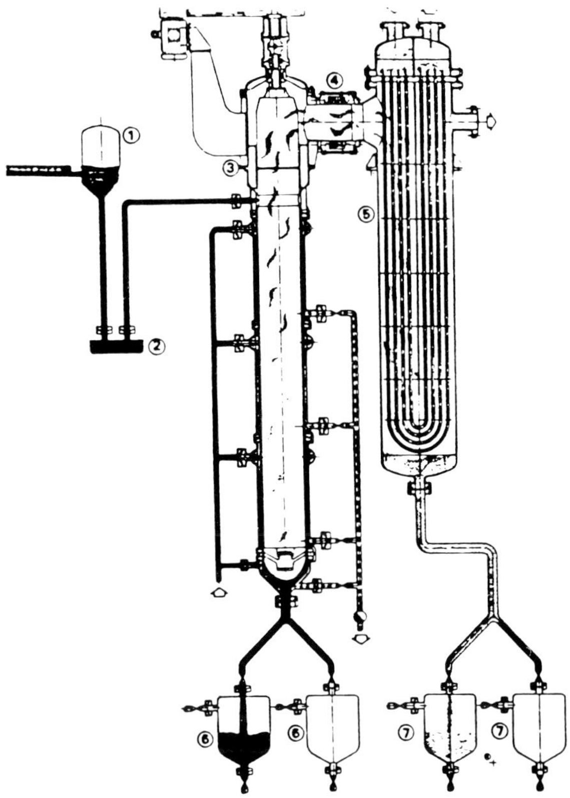

Figure 16. Diagram of concentration plant. (1) Product feed tank. (2) Dosing pump. (3) Film evaporator. (4) Valve. (5) Condenser. (6) Concentrate containers. (7) Condensate containers.

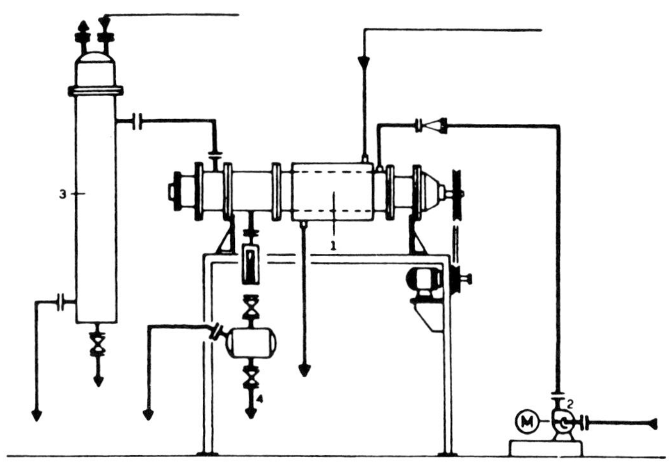

Figure 17. Horizontal thin layer concentrator. (1) Concentrator. (2) Extract feed. (3) Condenser. (4) Concentrate.

obtain the finished product by simple concentration or when the solution must be purified. In the extraction of a drug a small portion of very fine vegetal material usually manages to pass through the filtering system of the extractor. Inert substances extracted, especially in hot maceration processes, often give rise in the subsequent cooling to flocculates or precipitates that render the solution cloudy and inhomogeneous. The presence of solid particles or of thick sediment makes it impossible to carry out the normal continuous countercurrent extractions involving the use of liquid/liquid centrifugal separators, which would thus get blocked, or even the use of extractions in tower separators, because of the emulsions or formation of additional phases that such material generates. Since products intended for pharmaceutical use must not contain solid particles extraneous to the extract, clarification is obviously essential. Another aspect of the purification of extracts is the reduction of the bacterial content, which is not dealt with here as it cannot be dealt with in general terms. Sterile methods of filtration, such as sterilizing membrane filtration, are not suitable for extracts, especially when they contain mucilaginous or polymeric substances which hinder filtration. These processes therefore need to be devised case by case.

B. EQUIPMENT FOR CLARIFYING EXTRACTS

The purification of extracts includes, again, clarification. For this purpose two types of filtering apparatus are used: (1) equipment for simple or pressure filtration and (2) centrifugal separators and decanters.

Equipment for filtering extracts normally operates under pressure. The liquid is force-pumped into a series of compartments containing filters suitable for handling the material to be removed.[22] This system of filtration is used, there being no provision for continuous working when the quantity of suspended solid is small and the solid is very fine. In other cases, especially in high capacity extraction installations, centrifugal decanters are preferable for removing drug or sludge or supercentrifuges for clarifying extremely fine suspensions.[27,28] As this is a marginal operation in the production of extracts, only two types of equipment are cited, but not described in detail: horizontal decanters and Westphalia-type supercentrifuges. The ordinary perforated or closed drum centrifuges are also useful as clarifiers, as are other types of filters. As these are extremely common in the chemical industry and not specific to the production of vegetable derivatives, however, information on them can be found elsewhere.

VI. DRYING

A. GENERAL ASPECTS AND BRIEF DESCRIPTION OF EQUIPMENTS

After partial concentration of the extraction or counterextraction liquids, the dense residual solutions are in many cases evaporated to dryness. Although there are still large numbers of nondried extracts on the market, such as tinctures, fluids, and soft extracts, the tendency these days is to obtain solid extracts because of storage and transport, but above all because of their chemical stability. In our experience dry extracts are by far the most stable and if appropriately prepared

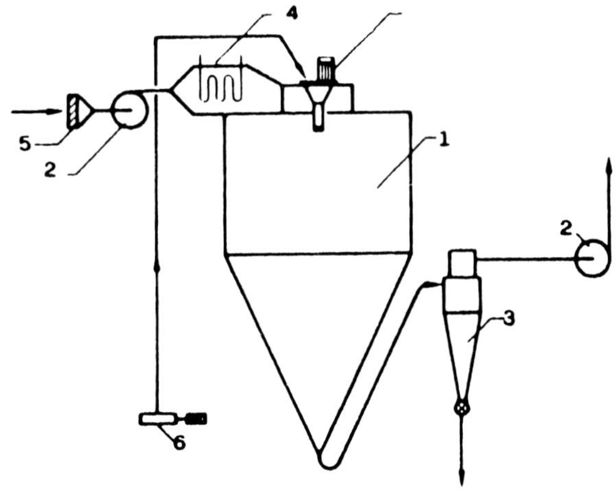

Figure 18. Principle diagram of spray dryer. (1) Drying chamber. (2) Ventilator. (3) Cyclone. (4) Air heater. (5) Filter. (6) Pump.

are perfectly suitable for pharmaceutical use. The bacterial content is, moreover, normally lower than in the other forms and in the event of contamination, the product in the solid state can be sterilized by gamma irradiation. This process is difficult to apply to solutions because of the series of radical reactions that it sets up.

There are several types of equipment for obtaining dry extracts ranging from vacuum freeze dryers for the more thermolabile products to the traditional cabinet vacuum dryers. Between these two extremes, depending on the quantity and type of liquids to be evaporated, many intermediate solutions have been found. Merely by way of example, there are drum dryers operating on a continuous basis, belt dryers operating in a stream of hot air, to those most widely used today, atomizers. Drying by atomization, as freeze-drying, is suitable for thermolabile substances but much cheaper; it is adaptable to small- and large-scale production; and it supplies products already in powder form, of fairly uniform particle size and density.

The atomizer is therefore becoming the dryer of choice, especially when the evaporization solvent is water. Figure 18 shows the diagram of an atomizer with minimum of accessories. The central body of the equipment is a drying chamber into which a stream of filtered and heated air is introduced from above. The solution to be concentrated (normally with a dry residue of around 20 to 30%) is pumped in and the dispersed by a sprayer or dispersor. The solvent in droplets of solution is evaporated instantaneously on contact with very hot air and the dried product precipitates to the bottom of the tower, where it can be collected or can be discharged together with the air to be separated in one or more cyclones. The hot air is kept circulating by a powerful fan connected to a strategic point of the dryer. Depending mainly on size, these dryers can operate with the current in the same direction, countercurrent, or mixed. There are several systems for letting the liquid into the evaporation chamber and dispersing it into droplets: the most common are rotating disk systems, in which dispersion in mainly centrifugal, and one- or two-way injection systems, in which micronization takes place under pressure in the presence of a dispersing gas (usually air). Given the versatility of these installations, it is possible to dry a large range of

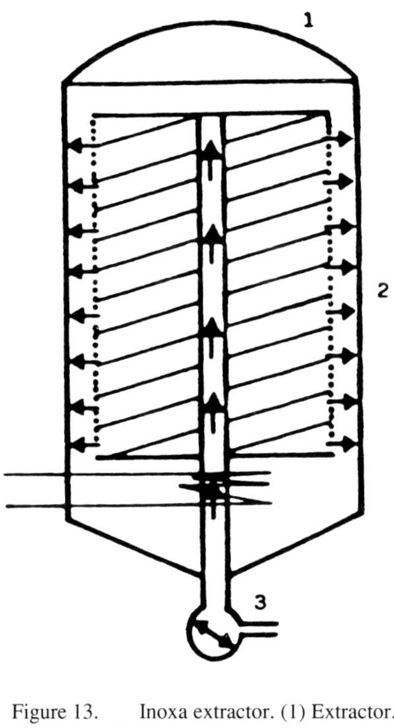

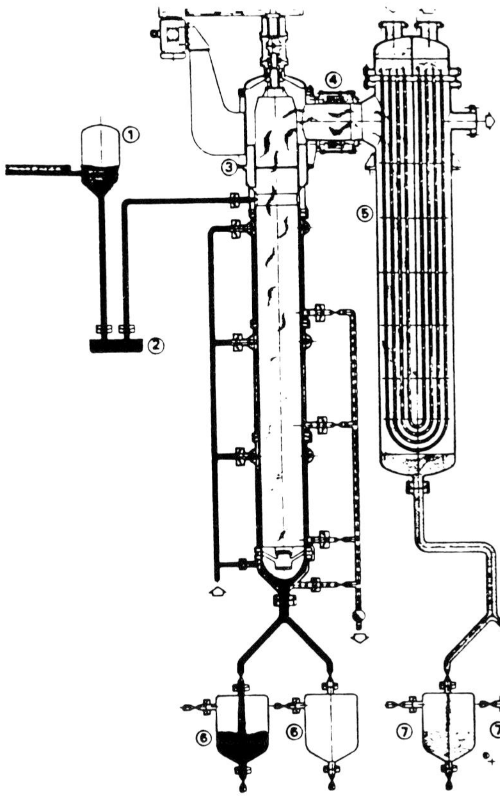

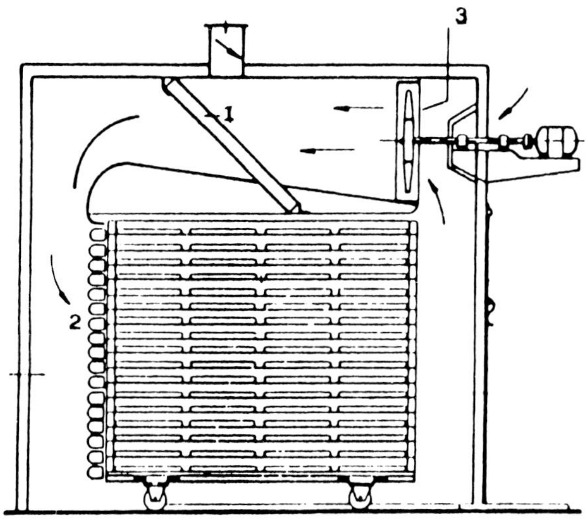

Figure 19. Example of cabinet air drier. (1) Heating device. (2) Air distribution. (3) Ventilator.

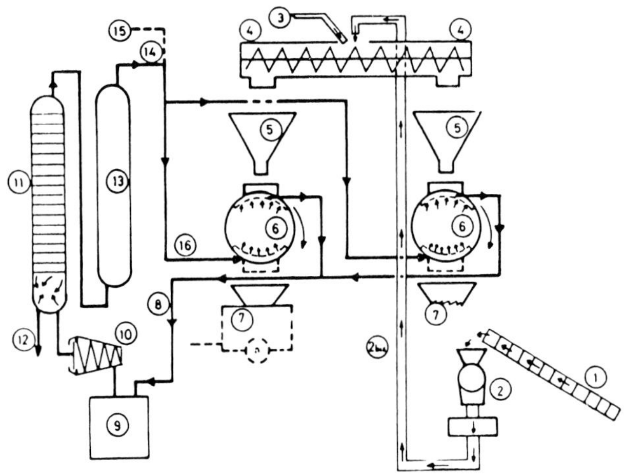

Figure 20. (1) Conveyer belt. (2) Mill. (2bis) Ground drug duct. (3) Alkaline water inlet. (4) Drug/alkaline water mixer. (5) Humidified drug silos. (6) Extractors. (7) Drug discharge. (8) Duct for extract containing alkaloids. (9) Sedimentation chamber (10) Decanters. (11) Acid counterwash tower. (12) Outlet solution containing alkaloids. (13) Counterwash exhaust solvents. (14) Exhause and washed solvents outlet. (15) Feeding washed nonrectified solvents. (16) Feeding washed and slightly rectified solvents to direct recycle.

products, including those with a low softening point, for which it is possible to introduce devices to prevent the deposition of the substance on the walls of the evaporating chamber with consequent encrustation. If sterile extracts are required, it is possible to fit a set of sterilizing filtering systems onto the installation.[30] Large-scale atomizers are equipped with a whole series of accessories for the partial recovery of the thermal energy used in the drying process, and according to whether the solutions to be evaporated, consist of water only or of solvents, the installation can be adapted to suit special requirements.

Alongside this drying system that is ideal for continuous planned operating, the traditional system of drying in cabinets in streams of air or under vacuum is still widely used, especially for small quantities of finished product. Cabinet dryers operate discontinuously. In forced air dryers the material to be dried is normally layered (Figure 19) on trays resting on trolleys or on bearing elements of the drying chamber in which heating fluid may be circulated. The operating temperature is usually between 60° and 80°C. The solvent evaporated is eliminated by convection. Although these dryers are widely used and are manageable, their disadvantage is that they can be used only for stable substances. Drying is often not uniform and so to obtain homogeneous material granular predrying must be followed by further drying. Some of the snags of air dryers are present even in vacuum dryers, which are designed on the same principles. The difference is that the trays of product rest on the heating tubes or plates that evaporate the solvent, which is removed by vacuum. To obtain products with a constant weight loss, it is necessary here again to granulate and then dry the product again. The advantage of these dryers over air dryers is that operating at lower temperatures they can be used for drying thermolabile substances. Considering that in the production of vegetal derivatives for the pharmaceutical industry one is most unlikely to be confronted with the problem of having to dry very large quantities of extracts, as might be the case in the food or chemical industries, continuous belt or drum systems of drying and even freeze drying processes are not dealt with here.

VII. CONCLUSIONS

Since it is not possible to lay down general rules for the isolation of pure substances, we now propose a flow chart that might prove useful because of its versatility and adaptability to short production runs. The proposed charge (Figure 20) begins with a battery of four extractors equipped with agitation system that can operate singly or in series. Preferably the extractor would be of the rotary type with direct desolventization at the end of extraction, in the same jacket-heated container that can handle dry drugs, drugs moistened with aqueous solutions, or even fresh drugs. Next comes a filtration system consisting of a horizontal decanter that can solve problems of elimination, whether of residual drug or of sludge formed during the process. For the concentration stage we propose a thin layer concentrator that can cope both with problems of quantity of solution to be evaporated simultaneously and with problems of instability of many active principles. As an alternative to the direct concentration of extraction liquids, to achieve advantageous reductions of solvent volume, in the case of possible countercurrent extractions, we include a tower separator with counterwashing. At the drying stage the atomizer is the best choice.

With an installation of this type it is possible to prepare all vegetal derivatives whose end product is normally a solid material or one obtainable from it. We have not dealt with the problems related to essential oils or products obtained by the expression of drugs, which would take us too far away from the complicated technology used in the industrial production of natural products of vegetal origin, of which we hope we have succeeded in giving a realistic, though rough, picture.

REFERENCES

1. Samans, H., Zerkeinerungstechnik im Bereich der Chemischen und Pharmazeutischen Industrie Cz, *Chemie-Technik,* 2, 345, 1973.
2. Schoenemann, et al., Zur berechnung von Feststoffextractionsanlagen, *Fette Seifen,* 54, 385, 1952.
3. Boucher, D. F. et al., Extraction of oil from a porus solid, *Trans. Am. Ind. Chem. Eng.,* 38, 967, 1942.
4. Müller, F., Physikalisch-chemische Deutung der Maceration, *Pharm. Zentral.,* 107, 270, 1968.
5. Büchi et al., Theoretische and Praktische Untersuchungen über das Perkolations verfarhen, *Pharm. Acta Helv.* 11, 334, 1936.
6. Melicher, M., Abhängigkeit der Ausbeute von Dragenausatz bei einstufiger Maceration, *Arzneim. Forsch.* 18, 1341, 1968.
7. Schultz, O. et al., Gesetzmässigkeiten der Mazeration un Perkolation, *Arzneim. Forsch.* 3, 471, 1953.
8. Müller, F. et al., Gesetzmässigkeiten der Mazeration un Perkolation, *Acta Pharm. Technol.* 26, 240, 1980.
9. Schoenemann et al., Zur Berechnung von Feststoffextraktionsanlagen, *Fette Seifen* 54, 385, 1952.
10. Figlmüller, J. K., Die quantitativen Grundlagen der Extraktionssysteme Fest-Flüller J. K., *Oesterr. Chem. Z.,* 56, 261, 1955.
11. Figlmüller, J. K., *Oesterr. Chem. Z.,* 56, 221, 1956.
12. Niro extractor-Niro-Atomizer A/S 305 Gladsaxevy DK-2860 Soeborg.
13. Carousel extractor-Estractionstechnik, Gesellschaft für Anlagenbau mbH D 2000 Hamburg 76.
14. U Extractor-Hungar Patent 159977.
15. Voeste, T., Hildebrand extractor, Extraktion von Feststoffen, *Ullmann's Enzyklopädie der Technischer Chemie,* Aufl. Bd. 2. 5726 Verlag Chemie, Weinheim, 1972.
16. Tank-extractors, *Ullmann's Enzyklopädie der Technischer Chemie,* Aufl Bd. 4, 5726 Verlag Chemie, Weinheim, 1972.
17. Inoxa extractor, Inoxa, Via Carriana, 9, Milano.
18. Brutner K. H., Separatoren und Dekanter für die Kontinuierliche Extraction, Westfalia, Technischwissenschaftlike Documentation, 1982.
19. Minima, S. A. et al., Investigation of extraction from plant raw material in an extractor press, *Pharm. Chem. J.,* 10, 7, 1976 (in English).
20. Coehnen, H. et al., Die Trennung von Stoffgemischen durch Extrakton mit überkritischen Gasen, *Krupp-Forsch. Ber.* 30, 1, 1982.
21. Stahl E., Hochdrurckextraction von Naturstoffen mit ü berkritischen und verflüssigten Gasen, *Parfumerie Kosmetic* 63, 117, 1982.
22. Billet, R., *Verdamfung und ihre Technischen Anwendungen,* Verlag Chemie, Weinheim, 1981.
23. Mabo evaporator-Mabo-Macchine e impianti industriali, Fiorenzuola d'Arda (PC).
24. Luwa concentrator-Luwa-Verfahrenstechnick Zürich Anemonenstrasse 40, CH 8047 Zürich.
25. Horizontal concentrator-Rotofilm Manzini SpA, 43100 Parma, Italy.
26. Plate concentrator-APV-H & K Plattenverdampfer Maschinen Fabrik Phönix GMBH Dortmund.
27. Hemming W., *Verfahrenstechnik,* Aufl 2, Vogel-Verlag, 1980.
28. Westfalia Separator A.G. OELDT, Feichthafstrasse 99, München.
29. Anhydro Spray Dryers Anhydro A/S, 8, Ostmarken DK 2860 Seborg, Denmark.
30. Nielsen, F., Trockung Pharmazeutischer Producte, Niro Atomizer.
31. Kneule F., *Das Trocknen,* Aufl 3, 1975, Verlag Sonerländer, Frankfurt, 1975.

Chapter 8

QUALITY CONTROL AND INSTRUMENTAL ANALYSIS OF PLANT EXTRACTS

Jan Karlsen

I. INTRODUCTION

The quality control of plant products is a general requirement to be fulfilled if a plant product is to be released in a market as a drug constituent. Too often a specific pharmacological action is reported for a certain plant product and when we look for the chemical characterization we find a total lack of reliable data. It is the aim of this chapter to call attention to this fact and underline the necessity of good quality control or good quality assurance when dealing with plant products which are intended to be used as drugs or as test substances in basic pharmacological experiments. This quality assurance is necessary if plant products are to fill the need for cheap and reliable medicines, or if natural products are to be used as templates for new drug molecules. It is generally accepted that the plant kingdom offers tremendous potential in this field and should be exploited and used as a supplement to the purely synthetic drugs on the market. The same is valid for the research on novel drug molecules. However, since the synthetic drugs are subject to severe quality control, the plant products must also comply with the same quality standards. In addition to these requirements, which must be considered vital to a proposed registration of a natural product as part of a drug formulation, there are also requirements to be met in research projects involving medical doctors and chemists using natural product preparations. The pharmacologist, being occupied with the biological testing procedures, would not even consider spending his time doing experiments with uncharacterized material.

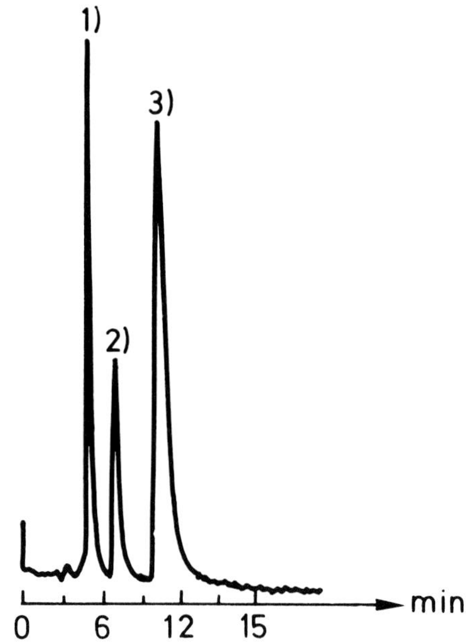

Figure 1. HPLC chromatogram of commercially available pure standard of curcumin isolated from *Curcuma longa*. Pure standards of natural products may often not be "pure". Peak number 3 is curcumin, 2 and 1 are isomers.

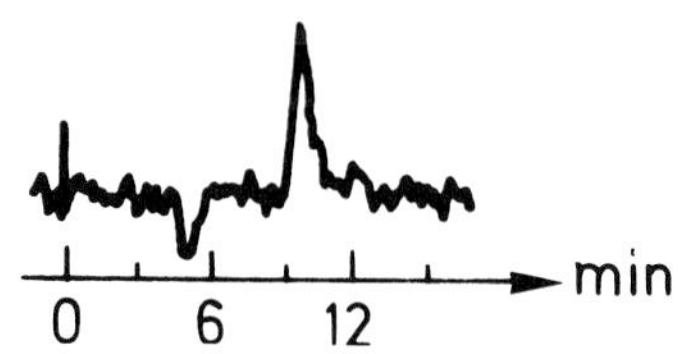

Figure 2. HPLC chromatogram of 20 pg of one of the curcumin isomers to show the detection limit of the analytical equipment. In this case we had to use fluorescence detection to get a useful signal when a plant extract was investigated.

For the pharmacological evaluation of a plant extract or a purified naturally occurring compound the chemical composition must be known if the results of the *in vivo* tests are to be reproduced. These comments may seem self-evident, but experience shows this not to be so.

The equipment necessary for plant extract control consists mainly of chromatographic instrumentation. The separation techniques were developed specifically for the analyses of complex mistures which all plant extracts are. Even though there may be a need for spectroscopic methods such as NMR and MS, they play a less important role in the phytochemical laboratory today compared to the overall importance of the chromatographic methods.

To characterize plant extracts all the different chromatographic tenchniques may have to be used for quality control. They have different advantages and shortcomings but will in combination be able to solve almost all separation problems. A great number of textbooks on the different methods exists, but their cited examples are often practically unsound as they usually show the separation of artificial mixtures. This chapter is therefore meant as a small guideline through the most important techniques used in phytochemical research laboratories. Table 1 indicates the importance of proper analytical control of plant extracts.

II. CHROMATOGRAPHY

The chemical analysis of plant extracts is very important in quality control. After the identity of the starting material has been established, the qualitative and quantitative investigation has to be started. This analysis will most certainly involve some kind of chromatographic method. The quality of the plant extract can be given as a "fingerprint" chromatogram (essential oils) in the case the active ingredients are not known or

TABLE 1
A List of the Steps Necessary for the Evaluation and Production of a Novel, Plant-Based Drug Formulation

1.	Selection of suitable plant material
2.	Botanical investigation
3.	Chemical investigation (analysis)
4.	Pharmacology (primary screening)
5.	Activity oriented fractionation of crude extracts (analysis)
6.	Isolation of active principle(s) (analysis)
7.	Structure determination (analysis)
8.	Comparison with other "look alike" compounds
9.	Detailed pharmacological evaluation
10.	Structure modification of active principle
11.	Pharmacodynamic studies (analysis)
12.	Pharmacokinetic studies (analysis)
13.	Formulation (analysis)
14.	Clinical trials (analysis)
15.	Production of drug formulation (analysis)
16.	Registration by the National Drug Authorities

Note: Use of the word "analysis" indicates where an analytical method must be applied.

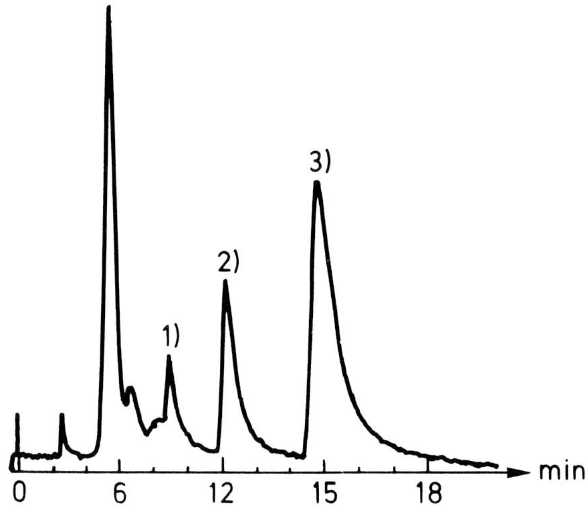

Figure 3. HPLC chromatogram of the total extract obtained from *Curcuma longa* where the content of curcumin was of interest. Peak number 3 is curcumin. Although inadvisable, we may inject the crude extract directly onto the HPLC column.

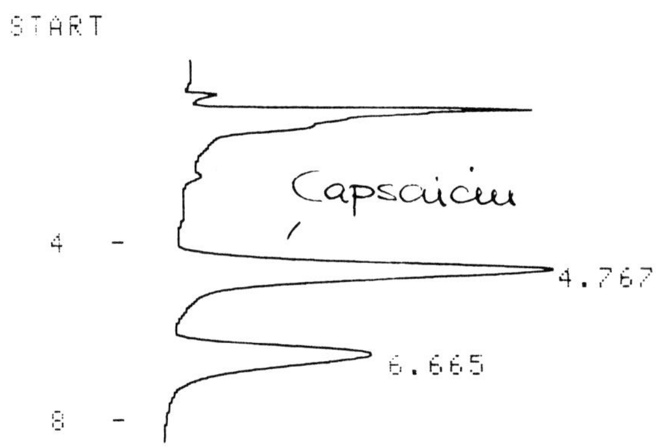

Figure 4. HPLC chromatogram showing the signal obtained when the total oleoresin of *Capsicum annuum* is analyzed.

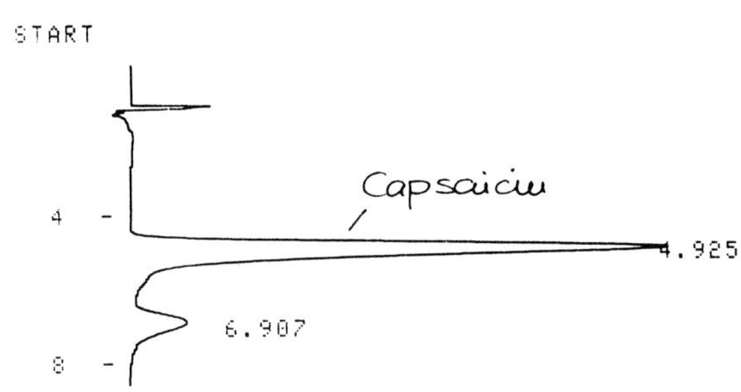

Figure 5. HPLC chromatogram of pure commercially available standard of capsaicin. This again indicates the need for making our own standards for quantitative purposes.

too complex. We then use TLC, GC, and HPLC as qualitative methods. If the active ingredient are known the same techniques must be applied to give us a quantitative result. Even though the active ingredients are not known, as in the first case, a chromatogram run with a suitable chromatographic technique will give us enough information to allow reasonable conclusions to be drawn for and from subsequent experiments. This is valid for plant extracts being used directly as medicinal agents or extracts which are to be used in further pharmacological screening. Simple physical data such as melting point, boiling point, optical rotation, and refractive index are still being used as quality criteria in most pharmacopoeias, but these requirements are changing to include chromatographic data which give much more information about the plant product.

To make the right choice of the suitable separation techniques (Table 2) may not be easy, but the detection limitations of the method combined with the complexity of the extracts, as well as the expected content of active ingredient, will give an indication to the best choice.

Chromatography is by far the most applicable technique. The separation achieved, however, may not be a complete identification of one of the compounds of interest, but by the application of specific detectors such as electrochemical, electron capture, nitrogen, phosphor detectors, etc., the detection can be made more specific. The use of simple and direct methods such as spectrophotometry and titration of an extract may be tempting and is also described in the different pharmacopoeias for quality control but not recommended where a pharmacological action is expected by the use of the preparation. We need specific methods whereby the compounds of interest are directly measured.

We may take an essential oil as an example. The essential oil may be characterized by its refractive index, optical rotation, TLC, HPLC, and GC. We know that an essential oil may contain several hundred constituents, many of which will be main constituents. Obviously, the chromatogram obtained by means of a capillary GC-column will give the best description of the oil, whereas the use of other methods may result in a "fingerprint" which will be difficult to relate to single compounds. Another example is the oleoresin of *Curcuma longa*. This oleoresin will contain a very large number of compounds, mainly of sesquiterpenic origin. However, the interesting compounds may be the coloured curcuminoids.

TABLE 2
A List of Detection Limits for the Most Widely Used Analytical Techniques in Phytochemical Research

Technique	Limits of detection
Spectrophotometry	1—0.1 μg
Spectrofluorimetry	0.01—0.001 μg
Standard TLC	0.1—0.01 μg
HPTLC	0.01 μg
HPLC	10—1 ng
GC	0.1 —0.01 ng
GC/MS	0.01 ng

Three isomers will be found, so a more simple technique than GC can be applied, even though a total separation of the constituents may be recommended. An HPLC method giving a clear separation of the three curcuminoids should suffice for the quality control. This oleoresin is used as a food colourant and even though the flavour of the oleoresin is important, it is the content of coloured compounds that matters for the price in the market. The curcuminoids are not volatile and the method of choice for nonvolatiles will be HPLC.

III. COLUMN CHROMATOGRAPHY

Column chromatography is the oldest separation technique in use in a phytochemical laboratory and in the classical sense is only used for preparative purposes. It is used for separations based on adsorption, partition, ion exchange, gel filtration, and gel permeation. As a rule of thumb we use gel filtration (Sephadex materials) for the separation of high molecular weight compounds (peptides, proteins, polysaccharides, etc.); ion exchange for alkaloids, acids, and bases; partition for nonionic compounds which are hydrophilic, and adsorption techniques for nonionic lipophilic substances such as steroids. The separation, enrichment, or fractionation of an extract can be performed by elution analysis, displacement, or partition analysis and is often effected by more than one type of interaction at the same time. It is therefore diffcult to "preprogramme" a separation. Therefore, column chromatography on a preparative scale is often carried out as repetitive or recycling chromatography. Column chromatography is therefore often used as a prefractionation technique where a crude extract is purified prior to quantitative analysis. This technique is often applied in clinical analysis and there may be valuable hints on our own separation problems regarding plant extracts to be had from publications on the analysis of body fluids.

The quality of the final separation done by GC or HPLC methods greatly depend upon a prefractionation carried out on the crude plant extract. This is a very important field where column chromatography is often applied.

IV. THIN LAYER CHROMATOGRAPHY

In contrast to column chromatography, TLC is carried out on a thin layer (open column). The principles of separation will be the same as for column chromatography. A thin layer gives us the possibility of spraying the plate and making compounds without colour visible. All the separations possible by column chromatography can be made on a much smaller scale by TLC in a much shorter time and often with better separation effect. TLC requires simple and cheap equipment for qualitative analysis and is at the moment the most versatile and popular of the separation techniques. For quantitative purposes there are available reliable and efficient densitometers which make this technique compatible with GC and HPLC.

If the main purpose is to use a separation technique as a semipreparative or as a qualitative method, TLC would often be the method of choice. TLC chromatograms can be run unidimensional, circular, wedge-strip, or two-dimensional. The main application and the standard technique is unidimensional, ascending chromatography. Gradient chromatography can now also be applied as a standard technique. At the moment all kinds of thin layer plates are available commercially and this makes TLC a very competitive technique compared to GC and HPLC.

In TLC the Rf-value is a measure of the migration rate of different substances in a given solvent. Known standards must be run together with the sample to ensure the identity or identification. This is particularly important when plant extracts are to be investigated. Due to the variation of the thin layers, the Rf-values can never be absolute values. TLC is an analytical method mainly employed as adsorption chromatography, but may also be used in reversed phase, partition, ion exchange, or gel permeation systems or for the separation of optical isomers.

Preparative TLC must be mentioned briefly as this technique is applied quite frequently to the investigation of plant extracts. For the investigation of plant products we very often need reference substances. Preparative thin layer chromatography offers a very convenient possibility of obtaining small amounts of these reference compounds. Usually thicker layers than for standard, qualitative analysis will be used, but the separation and handling of the plates remains the same. The zone with the compound in question together with the sorbent layer is removed from the plate after the chromatographic run and extracted and identified by different spectroscopic techniques.

Until 1987 the emphasis of TLC was as a qualitative and/ or as a semipreparative method. However, the instrumental development of microprocessors and the application of dedicated computers to analytical instruments has made quantiative TLC a reliable method as well. Previously, quantitative TLC was carried out by eluting the interesting compound and measure by spectroscopy the amount present in an extract. This technique, if applicable, was very time consuming, tedious, had a large error incorporated, and was therefore shunned. The advent of modern densitometry has completely changed this picture. Quantitative TLC has become a method also applicable as GC and HPTLC.

When the chromatographic run has been carried out the spot of interest is located and the absorbance or fluorescence is measured directly on the thin layer. In this way several mixtures can be measured on the same plate. For the control of plant extracts quantitative TLC has become a very convenient method for quality control. It is also strongly emphasized that

TLC as a cleaning method for crude extract is recommended prior to GC or HPLC analysis.

V. HIGH PERFORMANCE LIQUID CHROMATOGRAPHY

According to literature surveys HPLC has the largest number of applications regarding the analysis of biological extracts. The reason may be that this technique allows direct analysis of aqueous extracts. A number of available detectors will also give us compound-specific signals. It means that even though the separation achieved has not been complete, one can compensate for this by using detectors with a high degree of specificity. Since we aim at specific detection of the compounds of interest, HPLC has become the routine technique for the screening of plant extracts. For compounds such as saponins, glucosides, peptides, and carbohydrates, few other possibilities exist for adequate characterization of extracts than by using HPLC techniques.

Modern HPLC equipment also allow the use of preparative as well as microbore columns. This means that the sample size will be of minor importance in HPLC.

In principle, on HPLC we can run the same separation as that obtained by TLC and this makes TLC and HPLC supplementary techniques. With the introduction of modern computer-aided TLC densitometry this last method is gaining acceptance. Still, the quantitative HPLC analysis is the method of choice in a phytochemical laboratory. For the HPLC analysis we have an almost unlimited choice of mobile and stationary phases. Even though plant extracts, when injected directly into the chromatographic column, can seriously contaminate the system, we can take necessary precautions by using precolumns and gradient elution. In this way our separation system will remain stable and reliable.

The separations are usually carried out at room temperature. Thermolabile compounds will pass through an HPLC column without degradation and this is attractive to the phytochemist who is used to compounds which very often give artefacts during the work-up. The HPLC analysis can also be automated so that a large number of samples can be run during the night as well. Automation of the analytical procedure also leads to better reproducibility of the results.

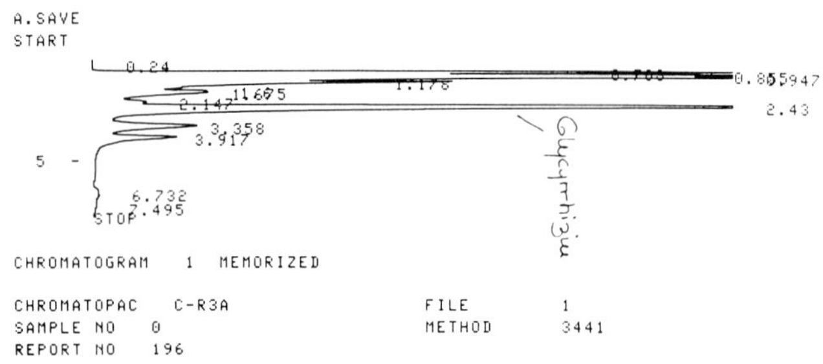

Figure 6. HPLC chromatogram of aqueous extract obtained from *Glycyrrhiza glabra*. The peak of glycyrrhizin is reasonably well separated from other compounds and can be quantitated.

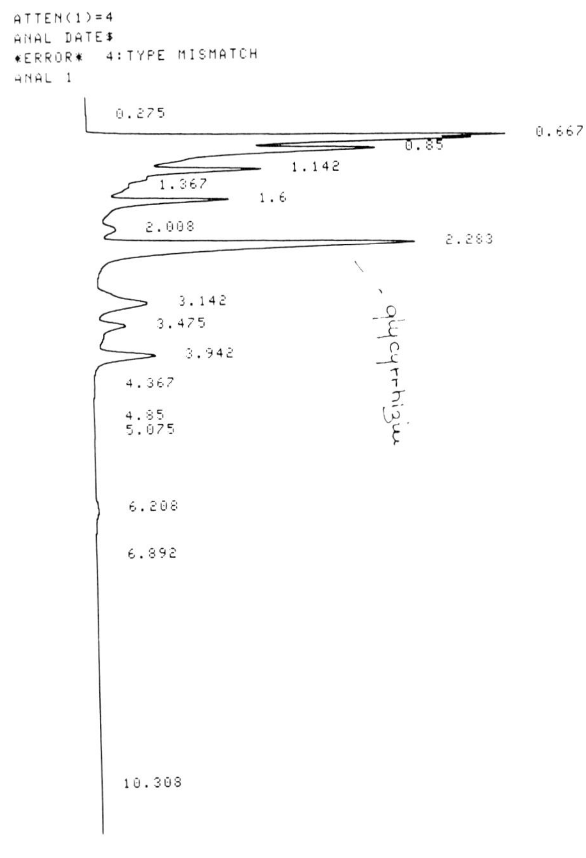

Figure 7. An optimal separation of the constituents present in an aqueous extract obtained from *Glycyrrhiza glabra* showing a clear separation of glycyrrhizin from other components. Here, a better separation is obtained, but the analysis time is considerably longer.

VI. GAS CHROMATOGRAPHY

When a plant extract contains volatile compounds of interest, GC will be the method of choice for the separation to be carried out. However, one must bear in mind that the sample has to be heated and that this procedure invariably leads to the formation of artefacts. GC, which was initially developed as an analytical tool for the petrochemical industry, has become indispensable for working with essential oils and natural products for the perfume industry. Essential oils will contain a very large number of compounds and GC is the only technique that has enough separation power to give a suitable "fingerprint" of such mixtures. The GC separation power of volatile mixtures is so large that one often needs only two capillary columns for the total separation — one polar and one apolar. The volatile compounds have specific retention times on the two columns and a tentative identification could be possible. A capillary system is regarded as necessary since all quality control laboratories will refer to capillary GC analysis in their certificates.

Nonvolatile constituents can also be analyzed by GC, but the compounds then need to be transformed into volatile derivatives. This transformation may not be quantitative and thus causes problems in a quantitative analysis. However, as in clinical analysis, when the detection limit of other techniques is too high, we have no other choice than to use derivatives. Polar groups such as hydroxyl, carbonyl, amines, etc., which

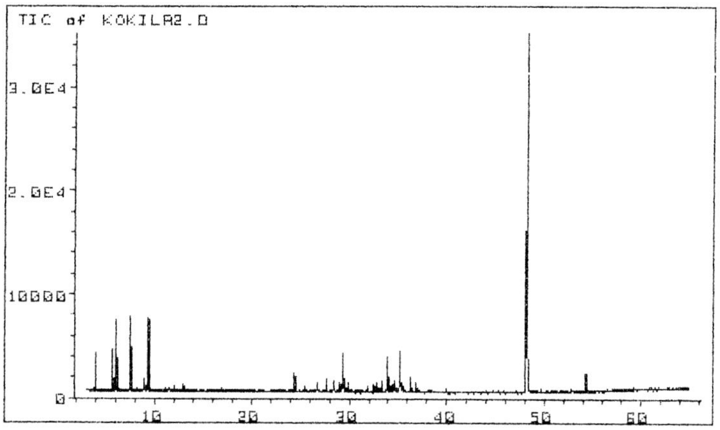

Figure 8. A total ion current GC-MS chromatogram obtained by using a capillary column coupled to a quadrupole mass spectrometer. Every 4 s a mass spectrum of the effluents of the column is measured and stored in the dedicated computer coupled to the GC-MS. When the chromatogram has been run, all the mass spectra can be recalled and recorded on a printer-plotter. The essential run is Kokila oil obtained from Royal Drugs Research Laboratories, Kathmandu, Nepal.

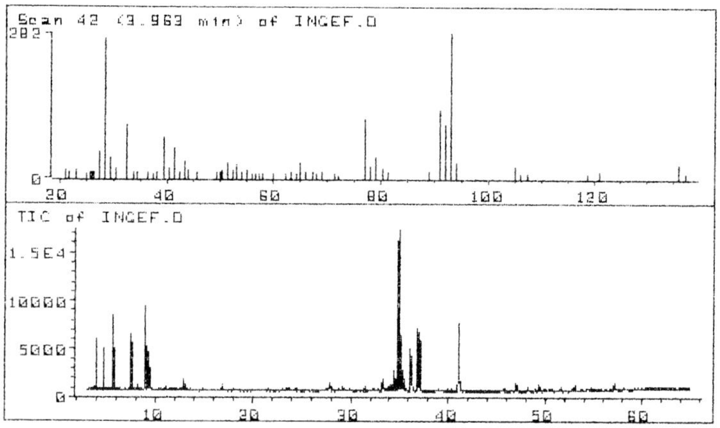

Figure 9. This figure shows the printout from the GC-MS of the total ion current chromatogram as well as the mass spectrum of a compound eluting from the column at 3963 after the run was started.

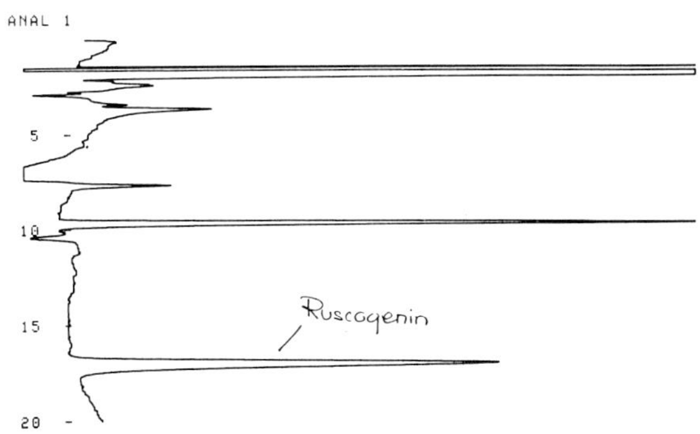

Figure 10. HPLC chromatogram of a Ruscogenin sample showing again the impurities present in isolated standards.

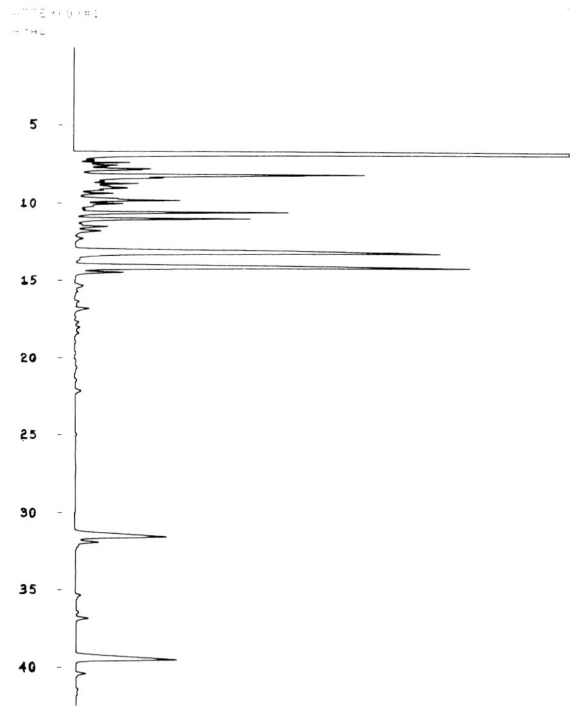

Figure 11. GC chromatogram of the essential oil of *Origanum onites* run on a capillary column. The mixture of compounds is complex and obviously needs capillary analysis for a clear separation.

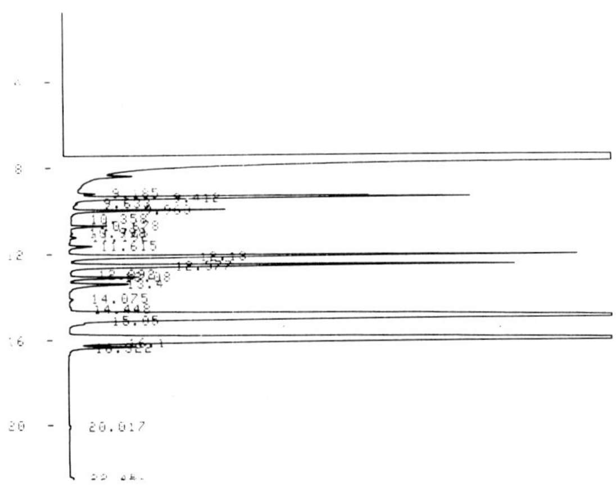

Figure 12. This GC chromatogram shows the separation of the monoterpene hydrocarbon fraction of the essential oil of *Origanum onites*. To obtain the best separation from a capillary column a prefractionation of the oil into subfractions may be necessary.

make the compounds nonvolatile, can be covered by different silylation reagents and changed into volatile derivatives.

Another well-established and popular quality of GC is the ease with which we can couple this technique to a mass spectrometer.

The modern table-top versions of quadrupole mass spectrometers are ideal instruments for the GC-MS coupling. Again, these instruments have become standard equipment for the investigation of volatiles in plant material. If a laboratory cannot afford such instruments, the GC equipment should at least be fitted with a capillary column system.

VII. LABORATORY INSTRUMENTATION FOR QUALITY CONTROL OF PLANT EXTRACTS

In view of the comments on analytical instrumentation necessary for proper control of plant extracts there are some standard equipment requirements for a phytochemical laboratory. A list to aid the purchase of analytical instrumentation is given below.

1. Thin layer chromatography (TLC)
   Glass plates and absorbent
   Developing tanks
   10 μL glass capillaries
   Storing cabinets for activated plates
   Computer-aided densitometer
2. Column chromatography (CC)
   Low pressure solvent pump
   Fraction collector
   Low dead-volume chromatographic columns
3. Gas chromatography (GC)
   Standard instrument fitted for capillary columns and temperature programming
   Flame ionization detector
   One polar and one apolar capillary column with chemically bonded stationary phase
   Electronic integrator with printer/plotter
4. High performance liquid chromatography (HPLC)
   Programmable pump unit-gradient system
   Loop injector
   Variable wavelength detector (UV-VIS)
   HPLC columns with reversed-phase stationary phase
   HPLC columns with normal phase stationary phase
   Guard columns to prolong the life of the analytical columns
   Electronic integrator
   Fluorescence detector
5. UV-VIS scanning spectrophotometer
6. Infrared scanning spectrophotometer

VIII. CONCLUDING REMARKS

There is a need for the standardization of plant extracts. This will have to follow international quality assurance regulations. The plant products available in the commercial market still lack, to a large extent, quality certificates that tell us something about the content of the active ingredients or about a necessary standardization procedure.

No one is prepared to take plant products seriously unless certain elementary quality control criteria have been fulfilled. Through phytochemical research the development of chromatographic techniques have all become valuable tools in the isolation of new compounds. One would like to see these techniques applied to the same extent to quality control purposes. The introduction of quality assurance criteria may also have the effect that the rigid quality control will have to start at the cultivation and harvesting of plant material. In this way a total quality assurance scheme for the development of plant products can be established.

That plant products from the same plant species show large variation in the content of interesting constituents was recognized long ago, but the analytical tools for quantitative measurements were lacking. Today these methods (chromatography and spectroscopy) are all part of normal laboratory equipment. A large number of textbooks describing these techniques in detail are available (see reference list for a selection). We have the means of quality control at hand.

Wherever the active ingredients of the active fraction of plant extracts are known, plant material, extraction procedure, and final product must be controlled. Whether the active ingredients are known, we must consider a plant extract as a product which has to comply with strict quality criteria. The content of "active" ingredients must be standardized in such a way that the extract alone or in a formulation can be identified and quantified following a given procedure.

The quality control and the standardization starts with a well-defined plant material through a defined extraction procedure and ends with a quantitative method for one or more active ingredients.

Only if the quality control is rigidly applied to plant products can these products really compete with synthetic ones. As mentioned before, the pharmacological screening of plant products also requires rigid quality control of the natural products, if the final results are to be taken seriously.

REFERENCES

Journal of Ethnopharmacology
Journal of Natural Products
Analytical Chemistry
Journal of Chromatography (General)
Journal of Chromatography (Biomedical Section)
Phytotherapy Research
Planta Medica
Journal of Liquid Chromatography
Journal of High Resolution Chromatography
Chromatographia
Trends in Pharmacological Sciences
Trends in Analytical Chemistry
Clinical Chemistry
Chemical and Pharmaceutical Bulletin (Tokyo)
Natural Product Reports
Medicinal and Aromatic Plants Abstract

1. C. J. O. R. Morris and P. Morris, *Separation Method in Biochemistry*, Pitman Publishing, London, 1976.
2. E. Heftmann, *Chromatography*, Van Nostrand Reinhold, New York, 1976.
3. J. M. Miller, *Separation Methods in Chemical Analysis*, John Wiley & Sons, New York.
4. D. W. Grant, *Gas Liquid Chromatography*, Van Nostrand Reinhold, New York.
5. W. Jennings, *Gas Chromatography with Capillary Columns*, Academic Press, London.
6. W. Jennings and T. Shibamoto, *Qualitative Analysis of Flavour and Fragrance Volatiles by Glass Capillary Gas Chromatography*, Academic Press, New York.
7. R. Ikan, *Chromatography in Organic Microanalysis*, Academic Press, New York.
8. D. N. Knapp, *Handbook of Analytical Derivatization Reactions*, John Wiley & Sons, New York.
9. J. C. Touchstone, *Quantitative Thin Layer Chromatography*, John Wiley & Sons, New York.
10. *Journal of Chromatography* Library Vol. 1-28, Elsevier, Amsterdam.
11. L.N. Snyder and J.J. Kirkland, *Introduction to Modern Liquid Chromatography*, John Wiley & Sons, New York.
12. R. E. Kaiser and E. Oelrich, *Optimierung in der HPLC*, Huthig Verlag, Heidelberg.
13. J. C. Touchstone and D. Rogers, *Thin Layer Chromatography*, John Wiley & Sons, New York, 1980.
14. R. Macrae, *HPLC in Food Analysis*, Academic Press, London, 1982.
15. W. W. Christie, *Lipid Analysis*, Pergamon Press, Oxford, 1982.

Chapter 9

FORMULATION OF PLANT EXTRACTS INTO DOSAGE FORMS

Attilio Bonati

I. INTRODUCTION

The formulation of plant extracts into dosage forms is a complex operation which cannot be regarded only as a problem of pharmaceutical technology. It is more demanding in that it calls for close collaboration between manufacturer and formulator of extract-based products, which must be founded on a clear grasp of and full respect for one another's problems.

Unlike pure active principles, whether synthetic or natural, extracts are raw materials that always contain, alongside variable but small amounts of the active principle(s), large quantities of secondary material (organic and inorganic salts, organic bases and acids, saponins, polyphenols, tannins, sugars, polysaccharides, etc.) that can appreciably affect the technology of preparation and the stability of the finished pharmaceutical form.

For this reason, before dealing with the subject of formulating extracts into dosage forms, some general remarks on the extracts themselves, on their preparation and properties, and on the problem of control are in order.

II. PREPARATION, TYPES, AND PROPERTIES OF EXTRACTS

The pharmacopoeial definitions of extracts are substantially similar in concept but differ considerably in detail, especially in regard to modes of preparation; some pharmacopoeias, perhaps more from habit than to meet real requirements, describe methods of preparation better suited to the pharmacist's laboratory than to industry while others more realistically include, even if in general terms, methods also suitable for production on an industrial scale.

The ninth edition of the Italian Pharmacopoeia (F.U. IX ed.) may be numbered among the pharmacopoeias that partly, because of the real local situation, show awareness of the problems arising from the industrial production of extracts. The monograph on "Extracts" in F.U. IX ed. states:

> "Extracts are liquid (fluid extracts) or solid (dry extracts) or of intermediate consistency (soft extracts) preparations obtained from vegetable raw materials, as a rule dried (drugs).
> Extracts are prepared by exhaustion with appropriate solvents, through maceration or percolation, or other suitable procedures, from drugs reduced to appropriate particle size.
> In some cases the drugs to be extracted must undergo preliminary treatment (for example: inactivation of enzymes, defatting). In other cases unwanted matter is removed after extraction.
> Concentration to the prescribed consistency is accomplished by appropriate methods, usually at low pressure and at a temperature at which deterioration of active principles is minimal."

This short monograph summarizes all the fundamental aspects of the preparation of an extract on any scale, in the pharmacist's laboratory or in the industry:

1. *The drug must be suitably ground* to ensure total extraction of the active or characteristic principles in a short time, thereby reducing the risk of alteration.
2. *Extraction must be carried out with an appropriate solvent.* "Appropriate solvent" means a solvent that not only ensures extraction without degradation of the active or characteristic principles but is selective, i.e., dissolves little of the secondary extractive material. This selectivity, always a good quality, becomes essential in the preparation of high strength extracts.
3. *The extract may be prepared by maceration, percolation, or other suitable procedure.* This is an innovative point of great importance. Traditional methods such as percolation and maceration may be replaced by other methods that the pharmacopoeia does not specify, except by means of the general but binding concept of "suitable procedure". This is of great interest industrially in that it officially releases the preparation of extracts from the traditional methods, allowing the use of more rational techniques (extraction under stirring, hypercritical CO_2 extraction, etc.).
4. *Concentration to the prescribed consistency must be accomplished by suitable methods,* in order to subject the active principles to as little heat as possible. This obvious operating specification, which is self-explanatory, might well have been accompanied by another concept of great importance to the user of extracts, first and foremost the pharmaceutical technologist: the final physical form of the extract, i.e., fluid, soft, or dry, must be such as to guarantee the stability of the active or characteristic principles in the extract. By way of example, from a drug that contains unstable substances in an aqueous or hydroalcoholic solution (anthocyanosides, iridoids, labile esters, etc.) *only* soft or (even better) dry (not fluid) extracts must be prepared, as this physical form does not guarantee stability of the active principles. From the general concept that an extract must be prepared only in the stable physical form, it results that the pharmaceutical formulator must likewise observe a similar basic rule, namely that extracts must be formulated only into dosage forms that are compatible with the stability properties of the extracts themselves.
5. *Drugs may have to undergo preliminary treatment,* such as defatting and enzyme inactivation, the purpose of which is obviously (1) to remove oily or fatty inactive substances that impede the obtaining of a dry extract and the subsequent preparation of solid pharmaceutical forms

(2) to arrest the enzyme degradation of active principles, always possible in an aqueous medium or in dilute alcohol. Even more important is the fact that F.U. IX ed. allows the *removal of "unwanted matter" after extraction.* Although USP XXI used to read, "Powdered extracts that are made from drugs that contain a considerable proportion of inactive oily or fatty matter should be freed from this by a suitable method", F.U. IX ed. does not confine the removal treatment to "inactive oily or fatty matter". Indeed, it speaks in general terms of "unwanted matter", and so these may include other inert matter such as sugars, salts, resins and so on. In other words, in allowing the removal of substances generically defined as "unwanted", F.U. IX ed. seems to pave the way for the concept of *purified extract,* i.e., of an extract freed from substances that may not only be of little or no value in terms of activity, but may be harmful in the formulation of the finished dosage form either at the time of formulation (hygroscopic substances, fatty substances, resinous substances) or later. "Later" meaning once the dosage form has been finished, since these so-called inert substances may impair the stability of the finished preparation.

It is clear from the monograph on extracts that expressions such as "exhaustion of the drug", "appropriate solvents", "suitable extraction procedures", "deterioration of the active principles", have meaning only if the composition of the drug to be extracted is fully known or known for all practical purposes. This may seem all too obvious and indeed, in general terms, it is. One would hardly think of preparing an extract without the knowledge of the drug's composition that alone can enable one to operate in conditions that prevent degradation, to use the appropriate solvents, etc.

It is only in the past few years, however, on the strength of information supplied by specific analytical techniques, principally chromatography, that the manufacturer of extracts has been able to work on material whose active principles and hence the relevant solubility ("appropriate solvents", "exhaustion of the drug") and stability ("deterioration of the active principles") characteristics are known to him. In other words, the manufacturer of extracts can now operate rationally, producing extracts that are defined by a precise content, qualitatively and quantitatively, of active principles and not by a generic extract/drug ratio. Further, chromatographic analysis techniques which have proved essential, indeed irreplaceable, have transformed ill-defined, "old" traditional preparations into rational preparations of known content and hence with constant activity.

Extracts can be classified in two main groups: total extracts and purified extracts. The terms "total" or "traditional extracts" denote extracts that contain all the extractive matter obtained by treating the drug with a solvent, as a rule aqueous or hydroalcoholic. The term "purified extracts" means preparations from which substances considered useless if not actually detrimental to the activity of the extract have been removed. A purified extract may be obtained by extracting the drug with a selective solvent and/or by removing the "inert" substances by various methods (defatting, passing over absorption resins, etc.) after the primary extraction. The term "inert substances" applies chiefly to resins, fats, and sugars, all materials that constitute a major hindrance in the preparation of pharmaceutical forms, especially solid forms, because they are hygroscopic, sticky, and so give rise to formulation problems. Traditional extracts may in turn be divided into fluid extracts, soft extracts, and dry extracts, with which all pharmaceutical technologists are familiar. Purified extracts, on the other hand, are as a rule in the dry state. Their content of active principles varies enormously: in the main we may say that the minimum content is about 20% of the entire extract, though in some cases it may be as high as 60 to 70%.

A point that does not fall specifically within the scope of this chapter but that nonetheless must be made is that a purified extract does not constitute a halfway house between the total or traditional extract and the pure product; i.e., it is not an intermediate for subsequent transformation into a pure product. Purified extracts have an identity and validity of their own because they are natural mixtures of chemically related substances, all endowed with the same or similar pharmacological activity. The isolation from these natural mixtures of a particular constituent in the pure state may be pharmacological and economic nonsense, justified only in certain cases by the need for a single more easily identifiable principle, especially in studies of pharmacokinetics.

Examples of purified extracts are extracts of senna with a 45 to 60% content of sennosides, of cascara or of frangula containing 20 to 25% hydroxyanthracene derivatives, of *Vaccinium myrtillus* containing 32 to 37% anthocyanosides, and of *Silybum marianum* containing 60 to 70% hydroxyphenylchromanones (silymarine). Many go a stage beyond the total extract and possess activity comparable with that of a pure product.

Preparation of a purified extract must be preceded by confirmation that the activity of the drug is conserved; total extracts often contain substances such as saponins and polyphenols that are regarded as inert but that may be crucial to the absorption of the active principles. One has to be sure that this property has not been lost in the purified extract.

III. CONTROL OF EXTRACTS

The problem of controlling extracts has two aspects: the control of extracts per se and the control of extracts as constituents of a finished product. The analysis of extracts per se can be undertaken and accomplished satisfactorily as long as one does not demand what the nature of the matter to be tested cannot supply, i.e., exactness of data that only chemically pure products can yield. The types of tests to be done on an extract are essentially three: for the determination of the physical characteristics, for qualitative and quantitative standardization, and for potential impurities and total aerobic microbial count. All three types of tests are relevant at the time of formulation into dosage forms.

To the traditional testing of physical properties (appearance, pH, solubility, total solids, ash) might be added, as a useful practical innovation, the determination of solubility of soft and dry extracts in solvents most commonly used in formulations, such as sugar syrup or sorbitol solution. Another interesting point is the particle size of dry extracts with tolerance limits for fine powder. The last datum is very helpful for cases in which the extract has to be used in the preparation of tablets by the direct compression technique. For the qualitative standardization of an extract the chromatographic procedures now available ensure truly satisfactory results. The purpose of this test is twofold: to ensure that the chromatographic pattern obtained tallies with that of the drug corresponding to the name of the extract and to ensure that no degradations have occurred during the extractive process. To attain these objectives qualitative testing should not be confined to a single product or family of products but should extend to several substances or groups of substances in order to have several grounds for appraisal; for example, an extract of boldo calls for the identification of boldine, total alkaloids and flavonoids, one of hawthorn that of flavonoids and catechinic polyphenols.

The specific qualitative determination based on several parameters permits the exclusion or discovery of adulterations or "reconstructions" of extracts, the latter being obtained by diluting an extract with a very high content of a given active principle with large quantities of excipients. A truly specific qualitative determination is, moreover, an essential complement for the quantitative standardization of characteristic active principles when no specific assays are as yet available. The presence of several products of like structure in an extract often necessitates global colorimetric and spectrophotometric determinations, as of total alkaloids, total hydroxyanthracene derivatives, and so on. Without questioning the validity of these determinations, it is obvious that their lack of specificity may be the source of quite serious error and inexactitude, especially in a stability study of pharmaceutical formulations. This drawback may be overcome by combining a very specific qualitative test with the nonspecific assay.

Another item of cardinal practical importance in the control of an extract is the determination of the total aerobic microbial count. Even if there are no official standards of acceptability, the absence of pathogens and a total bacterial count of between 10^3 and 10^4/g or /mL are satisfactory standards and can usually be achieved without any particular difficulty.

Although of limited importance for the purposes of formulation, testing for residual solvents and for contaminants (weed killers, pesticides) is a further datum that completes the analysis of an extract.

The analysis of an extract contained in a pharmaceutical formulation is a very complex problem because, especially in liquid formulations, the excipients may give rise to a variety of interference. There are no general rules for solving these problems, for every analysis tells its own story. There are, however, numerous technical possibilities that ensure satisfactory results: HPLC, which has the advantage of being applicable directly, for instance, to syrups, drops, and ointments, without prior operations of purification and/or concentration of the active principle(s) (Figures 1 to 4).

The importance of a specific analysis of an extract or of a formulation containing extracts does not end there: a specific method of analysis is an incomparable method for studying the stability of an extract or of formulations containing it (Figure 5).*

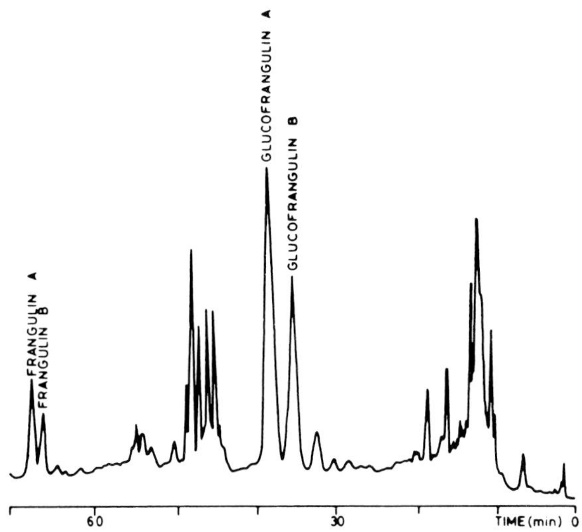

Figure 1. HPLC tracing at 280 nm of Frangoselect® (Indena-Milan), a purified extract of *Rhamnus frangula* bark.

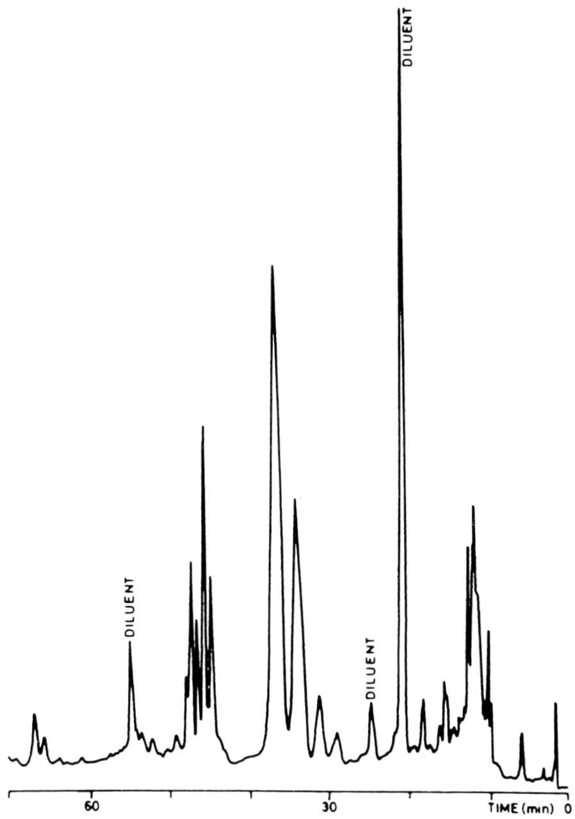

Figure 2. HPLC tracing at 280 nm of a syrup containing 0.5% of Frangoselect®.

---

* Figures 1—5: HPLC tracings concerning the analysis of a purified extract (Frangoselect®) of *Rhamnus frangula*, and of a syrup containing 0.5% of this extract. The HPLC analyses have been measured at two wavelengths, exactly at 280 and 430 nm. In the first case it can be noted, besides the active principles frangulin A and B and glucofrangulin A and B, the presence of several products; at 430 nm the HPLC tracing shows only the four anthraquinone derivatives.

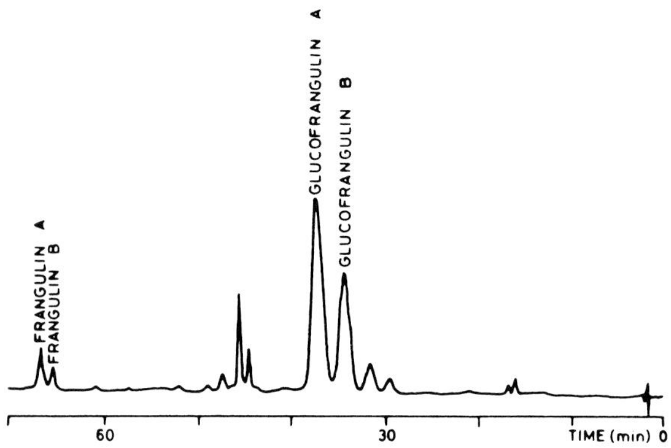

Figure 3. HPLC tracing at 430 nm of Frangoselect®.

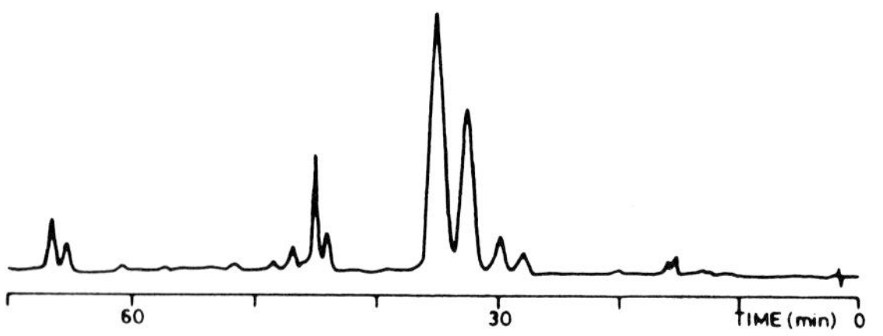

Figure 4. HPLC tracing at 430 nm of a syrup containing 0.5% of Frangoselect®.

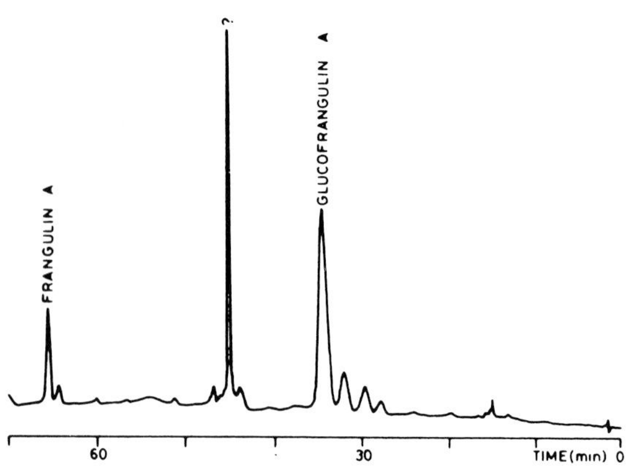

Figure 5. HPLC tracing at 430 nm of a syrup containing 0.5% of Frangoselect® kept at 50°C for 3 months. The modification in the ratios frangulin A/B and glucofrangulin A/B is evident.

The most important consequence of having specific methods of analysis is perhaps that guaranteeing the stability of composition of an extract means guaranteeing also the constancy through time of the activity of the extract.

IV. DOSAGE FORMS

The foregoing remarks on the preparation and control of extracts highlight the fact that in the past the pharmaceutical formulator, having no sure knowledge of the active or characteristic principles of an extract and no specific methods of identification and assay, might (and sometimes did) devise inappropriate, unstable formulations. Reasoning in reverse, he adapted the extract to the desired dosage form instead of adapting the dosage form to the characteristics of the extract, of which he had but a sketchy knowledge. These days the pharmaceutical technologist can rely on extracts with well-defined characteristics, which enable him to design dosage forms that are rational in terms of physical and chemical stability.

Just as the process technologist is tied in the choice of the extracting solvents and of the physical form of the extract by the nature of the active principles, and cannot prepare at will aqueous, hydroalcoholic, fluid, soft, or dry extracts from a given drug, so the pharmaceutical technologist is tied by the solubility and stability properties of the extract and cannot always prepare tablets, syrups, and ointments at will from a given extract. Predictably, it is chiefly the traditional liquid forms (syrups, drops, tinctures) that are constrained. However, this limitation is more apparent than real, since, if the usual liquid forms are not stable, it is nearly always possible to prepare granules for the user to make up into solutions, thus ensuring the availability of a "liquid" form.

With a precise knowledge of the nature of the extracts and objective methods of testing the material used, the pharmaceutical technologist can tackle and solve successfully the complex problems of bioavailability and physical and chemical stability at every stage in the life of a finished product, and thus guarantee constancy of composition and hence of activity.

While on the theme of rationalizing formulation into dosage forms, another general point is worth noting: it is always preferable to formulate simple preparations containing not more than two or three extracts. Another practice to avoid is the inclusion of several extracts having the same type of activity, as found in old preparations still in use. Typical is the example of laxatives containing extracts of cascara and frangula, cascara and aloes, senna and cascara, and cascara, frangula, and rhubarb. A simple preparation offers many advantages: the problem of control of the various active principles is simplified; the interference between different extracts is greatly reduced with obvious gains in stability, and the therapeutic activity is more well defined.

The technological problems involved in the formulation of extracts into dosage forms are more demanding than is the case with pure products, natural or synthetic. This is due to the very nature of extracts, in which the active principles are often present in low concentrations, thus making it necessary to use comparatively high doses. The pharmaceutical formulator has not only to solve problems inherent in the characteristics of solubility and stability of the active principles of the extract but also has to cope with those presented by the so-called inert substances contained in the extract, such as hygroscopicity in solid dosage forms or poor solubility with clouding in liquid dosage forms.

Practically all dosage forms can be prepared from extracts:

1. Solid dosage forms such as tablets, sugar-coated tablets, effervescent tablets, chewing tablets, retard tablets, hard gelatin capsules, and granules
2. Liquid dosage forms such as syrups, drops, solutions or suspensions for soft gelatin capsules
3. Forms for local use such as creams or ointments, gels, collyria, and suppositories

Although feasible, the formulation of injectables containing extracts is not recommended because, even if high strength purified extracts are used, they always contain a variable percentage of secondary products whose nature is not always well known and which, at least conceivably, could give rise to unforeseeable side effects.

A. PREPARATION OF SOLID DOSAGE FORMS

For the preparation of solid dosage forms mostly dry extracts are used. Except in rare cases, dry extracts are hygroscopic and so they cannot be used without prior granulation for preparing tablets, s.c. tablets or for filling soft gelatin capsules. Granulation is thus the critical stage in the process of preparing all solid dosage forms containing extracts.

To overcome the snag of hygroscopicity, silica gel, a high-porosity excipient, is widely used. Further, when operating with the traditional method of mixing, granulation and drying, it is advisable to use binders such as cellulose derivatives and polyvinylpyrrolidone in solution with organic solvents such as chloroform, acetone, or alcohol. If aqueous solutions of cellulose derivatives, gum arabic, gelatin, or aqueous gels such as starch mucilage are used, several problems may arise in the course of processing, especially if the formulation involves large quantities of extracts. Sugars or saponins, frequently present as secondary components of extracts, dissolve in the water used for mixing and form granules that are difficult to dry or form, during the drying process, very hard flakes of glassy appearance, making the granules hard to compress. Another snag found in granulation with aqueous solutions is the formation of compressed tablets that are difficult to disintegrate. The best course is therefore wet granulation with organic solvents or the use of nonhygroscopic ready-granulated extracts (containing an absorbent and possibly coated with methylcellulose). With the latter products, after dry mixing with appropriate excipients it is possible to obtain by direct compression, without further manipulation, tablets that are technologically satisfactory in terms of hardness, disintegration, and friability. The excipients most commonly used for this purpose are microcrystalline cellulose, lubricant absorbents such as silica gel, and small quantities of lubricants such as magnesium stearate.

Crippa,[1,2] who was long concerned with all the problems relating to the formulation of extracts into dosage forms, cites by way of example two formulations of tablets containing a granular extract of cascara, both prepared by dry mixing of the various constituents and direct compression (Table 1).

The substitution of powdered silica gel and microcrystalline cellulose for traditional excipients such as talc and kaolin yielded distinct improvements in hardness, friability, and disintegration, especially when a rotary tableting machine was used (Table 2).

Dry dosage forms present no particular problems of stability because in the dry state there is practically no degradation due to hydrolysis, oxidation, polymerization, etc. In formulations containing more than one extract, on the other hand, interactions may occur between the components of the different extracts with the formation of products endowed, for example, with new solubility characteristics. This may occur when formulating an extract containing alkaloids with one containing tannins or with an extract containing organic acids capable of forming stable salts with the alkaloids; in one case the formation of alkaloid-tannin derivatives may lead to sparingly soluble products (of slow bioavailability) and in the other there may be formation of soluble salts (of rapid bioavailability).[3]

These problems can easily be solved by preparing separate granules of each extract, suitably coated with resins, which can then be mixed and tableted.

For the preparation of hard gelatin capsules nonhygroscopic free-running granules are essential. Since the hard gelatin shell does not protect the granules contained in the capsule against humidity, the granules should be protected by coating with resins or methylcellulose. This coating is always advisable to ensure that the moisture filtering through the gelatin shell does not cause degradation of very labile products (*Vaccinium myrtillus* anthocyanosides) or more simply, but very frequently, transformation of initially free-running granules into a compact block that is poorly soluble.

TABLE 1
Examples of Formulations Containing a Dry, Granular Extract of Cascara (INDENA, Milan, Italy)

	A	B
Cascara dry granular extract	125 mg	125 mg
Lactose free flowing	60 mg	60 mg
Talc	6 mg	—
Kaolin	6 mg	—
Magnesium stearate	3 mg	1.5 mg
Silica gel powder	—	5 mg
Microcrystalline cellulose	—	8.5 mg

TABLE 2
Technological Data of Tablets Obtained by Direct Compression of Cascara Dry Granular Extract

	Rotary		Eccentric	
	A	B	A	B
Herberlein hardness	0.5 kg	2-3 kg	1.5 kg	3-3.5 kg
Loss in Roche friability meter (25 rps for 5′)	10%	0.6%	0.9%	0.25%
Disintegration time	8-12′	8-12′	10-15′	10-15′

B. PREPARATION OF LIQUID DOSAGE FORMS

Fluid, soft, or dry extracts may be used for preparing liquid dosage forms (syrups, drops, solutions, or suspensions for soft gelatin capsules). Of dry extracts, purified ones are the most suitable because they are free from secondary matter that may give rise to problems both of preparation and of stability.

Liquid dosage forms are undoubtedly far more difficult to prepare than solid forms. While the main problem with the latter is the hygroscopicity of the extracts used, with liquid forms there is the problem of solubility of the extracts, which obviously have to be diluted if fluid, or redissolved if soft or dry, in the solvent of syrups or drops. This is the trickiest stage in the whole operation because it may result in the formation, either immediately or later, of turbidity and precipitates due to unsatisfactory solubilization of the active principles and/or of the secondary matter.

These phenomena, which may jeopardize the stability of the formulation, can be averted by taking some or all of the following measures:

1. When redissolving a dry or a soft extract or when diluting a fluid extract, solvents of exactly the same alcoholic strength as the one used for preparing the extract should always be used. This precaution, which yields fairly satisfactory results in the case of fluid extracts, is not always so successful with dry or even soft extracts. In fact, in the concentration and the drying necessary to obtain extracts in the soft or dry state changes may occur which alter their solubility characteristics. Consequently, the dry or soft extract, prepared by extraction with a given solvent, is no longer completely soluble in a solvent with the same composition. In such cases it is necessary to add cosolvents or surfactants to the dissolution solvent.
2. In cases in which a solvent of exactly the same composition as that used in the preparation of the extract cannot be used for a liquid formulation, cosolvents or surfactants must be used. This is a fairly frequent occurrence since the solubility characteristics of the active principles or certain manufacturing exigencies require the use of extraction solvents such as high strength alcohol or organic solvents, which cannot be used in the formulation into liquid dosage forms. Cases of this kind arise, for instance, with extracts obtained from balsamic drugs (thyme, eucalyptus, mint, etc.), rich in essential oils, which present well-known solubility problems. Glycols, polyglycols, glycerin, and sorbitol are used as cosolvents; their solubilizing effect is not always satisfactory unless large quantities are used and this is not always compatible with the formulation. In cases in which cosolvents are not satisfactory, excellent results may be obtained with a nonionic surfactant such as polysorbates (polyoxyethylene derivatives of sorbitan esters) and polyoxyethylene derivatives of oleic alcohol and castor oil. An example[1] of the use of these nonionic surfactants is supplied in Table 3. In the first formulation the soft extract of *Ruscus aculeatus* is insoluble, while in the second formulation, containing 2.5% of polyoxyethylen-20-sorbitanmonostearate, a clear syrup is obtained.
3. Variations in the pH can definitely improve the solubility and stability of some extracts. Obviously this approach is feasible when the active principles are substances whose solubility increases through the formation of salts. The most typical case is that of extracts containing alkaloids. The agents most frequently used for dissolving alkaloid-containing extracts are lactic, citric, or tartaric acid. The choice of a particular pH is not only a means of improving the solubilization of an extract; in many cases it also helps to ensure the chemical stability of the alkaloids contained in the formulation. In the case of belladonna extracts containing tropane alkaloids, a pH within the 4 to 5 range not only ensures physically stable solutions but prevents racemizations or hydrolyses of the alkaloids.
4. Filtration to remove insoluble matter is obviously a very simple way of obtaining clear solutions, but it is possible only if it does not bring about a decrease in the content of active principles. This operation is feasible, for example, for clarifying solutions of extracts containing active principles that are easily soluble, such as saponins, and that therefore do not undergo losses with filtration.

TABLE 3
Examples of Use of a Nonionic Surfactant in Syrup Containing *Ruscus aculeatus* Soft Extract

Ruscus aculeatus soft extract	2 g	2 g
Sucrose	20 g	20 g
Alcohol	10 g	10 g
Polyethylenglycol 400	5 g	5 g
Mixture of *p*-hydroxybenzoates	0.1 g	0.1 g
Polyoxyethylen-20-sorbitanmonostearate	—	2.5 g
Purified water q.s. to	100 g	100 g

The stability, both chemical and physical, of a liquid dosage form is affected by other factors, for example, the possibility of fermentation and interference with other components of the finished product. The phenomenon of fermentation can be controlled either by using alcohol of appropriate concentration (although this approach needs checking every time to guard against unpleasant surprises — extracts are excellent culture media) or by adding suitable preservatives such as the methyl, ethyl, or propyl esters of *p*-hydroxybenzoic acid or mixtures of same. The quantities used are of the order of 0.1% on average. Possible interactions between other constituents of the end product are another source of instability. To mention only interference between constituents of vegetable origin, phenomena of precipitation (and hence of inactivation) can arise, for instance, from the concomitant presence of an alkaloid-containing extract and a tannin-containing extract. For such cases there are no general rules: sometimes the answer will be the addition of cosolvents or surfactants, sometimes the purification of individual extracts to remove the interfering "inert" substances.

An accessory but nonetheless important problem that arises in the preparation of liquid dosage forms containing extracts is how to ensure satisfactory organoleptic properties. Extracts often have an unpleasant, most bitter taste, which requires improvement, not always obtainable simply by sweetening. Very helpful here are the so-called flavor correctives, which may be other extracts having desirable organoleptic properties or pure products, such as ammonium glycyrrhizinate, obtained from liquorice, which is considered a potent sweetener tasting of liquorice, very helpful for correcting many bitter flavors.

The preparation of solutions of suspensions for soft gelatin capsules raises no particular problems, except that the solvents used must not be such as not to damage the gelatin shell. Those preferred are therefore oils, maybe with small quantities of glycols or polyglycols added. In our experience this is not a general finding; soft gelatin capsules made with tannin-containing extracts tend to harden with time, thereby considerably increasing their disintegration time. This phenomenon is probably due to interaction between gelatin and tannins.

C. PREPARATIONS OF DOSAGE FORMS FOR LOCAL USE

Although less commonly than in solid and liquid dosage forms, extracts are formulated into forms intended for local use such as suppositories and creams:

1. Dry or soft extracts are normally used for suppositories, with water-soluble or fatty or self-emulsifying excipients. If the extracts are not soluble in the excipients, and fatty excipients must be used for stability reasons, the extracts have to be thoroughly dispersed in the suppository mass. It is therefore helpful to absorb them on an inert material, such as silica gel or starch, to finely powder the resulting granules, and add them to the melted mass of excipients under careful stirring in order to get a homogeneous dispersion.
2. Dry or soft extracts are normally used to prepare solid emulsion-like creams. As a general rule, problems of physical stability are not encountered when extracts are formulated into oil-in-water (o/w) emulsions. On the contrary, water-in-oil (w/o) emulsions in the presence of many plant extracts are harder to make physically stable. Saponins, which are common components of extracts, make the HLB value of creams more hydrophilic and so lipophilic w/o emulsions, with an HLB value of under 10, break down with phase separation. An o/w cream with an HLB value of over 10 does not undergo any major HLB change and so remains stable.

As regards the chemical stability of creams, the high percentage of water contained in the emulsions and a pH value not carefully controlled can facilitate hydrolyses, polymerizations, racemizations, etc. A further problem to face in the preparation of extract-based creams is the microbiological stability. Extracts contain products such as sugars and amino acids which can favour a microbial proliferation. For this reason it is necessary to add to the cream antibacterial and antiyeast products such as a 20% chlorhexidine gluconate solution (0.1 to 0.2% of the cream) and an antimould mixture of *p*-hydroxybenzoates, undecylenic acid, and phenoxetol (0.3 to 0.5% of the cream).

REFERENCES

1. Crippa, F., Problems of pharmaceutical technics with plant extracts, *Fitoterapia*, 49, 257, 1978.
2. Crippa, F., Problems involved in pharmaceutical and cosmetic formulations containing extracts, *Fitoterapia*, 51, 59, 1980.
3. Longo, R., Biodisponibilita degli estratti terapeutici e dei loro principi attivi, *Boll. Chim. Farm.*, 112, 796, 1973.

Chapter 10

SOME CURRENT TRENDS IN MEDICINAL PLANT RESEARCH

N. G. Bisset, P. J. Houghton, and P. J. Hylands

I. INTRODUCTION

Within the limits imposed on the present chapter, it is clearly not possible to cover the enormous amount of chemical and pharmacological research that is currently going on in the vast field of medicinal plants. Instead, the topics chosen for discussion, albeit rather brief, are intended to show the type of work that is being carried out, and perhaps more importantly, some of the ways in which problems associated with medicinal plants research are being tackled.

Flavonoids and related substances (Section II), a major class of plant constituents, comprise different groups of phenolic compounds, many of which have been shown to bring about a wide range of effects on biological systems. The use of certain plants in traditional medicine can be understood on the basis of their content of these substances. Now that adequate techniques are available, triterpenes and particularly saponins (Section III) are receiving increasing attention, and they are another group of naturally occurring substances that is being found to elicit a broad range of biological responses. Their anti-inflammatory and adaptogenic properties are of particular interest. Plants containing cardiac glycosides, notably *Digitalis*, have long been in use for treating heart conditions (Section IV), but the dangers associated with their use have prompted a continuing search for less toxic materials. The problem has been attacked from several angles: exploration of the plant kingdom for other cardiac-glycoside-bearing plants and chemical modification of existing glycosides, as well as detailed investigation of their mechanism of action and of the receptor at which they act. Lower terpenes, especially sesquiterpene lactones, are discussed in Section V. Here again, research is uncovering new representatives at an ever-accelerating pace, and the study of their actions on biological systems is being pursued on a highly sophisticated level. Section VI deals with a single substance, the essential fatty acid γ-linolenic acid, a deficiency of which is linked with various disease states. Research is aimed at determining how it functions and its potential clinical uses, but is also directed toward a search for better-yielding natural sources of the compound. Immunostimulant substances (Section VII) are found in a wide variety of flowering plants and belong to many different chemical groups of both high and low molecular weight. The same applies to the antihepatotoxic substances considered in Section VIII. Current work in these two areas is extensive. These two sections well illustrate the necessity of having appropriate test systems or models for determining activity.

The account that follows is illustrative rather than comprehensive and, at the risk of stating the obvious, the amount of space devoted to any particular topic is not a direct measure of its importance. There is inevitably some overlap between the sections, owing to the wide-ranging actions that certain of the compounds discussed have on biological systems.

II. FLAVONOIDS AND RELATED COMPOUNDS

Flavonoids are phenyl-benzpyrone derivatives of greatly diverse structures found in higher plants and often occur as glycosides; they are thought to play a role in photosynthesis. Because of their widespread occurrence, they form a part of the dietary intake of all people, the average amount ingested being 1 g/day.

Plants containing large amounts of flavonoids have been used in the traditional medicine of many different cultures. In the late 1930s Szent-Györgi showed that extracts from citrus fruits increased capillary resistance. The therapeutic usefulness of flavonoids was disputed, but in recent years a substantial body of evidence has built up to show that many of them have potential uses in therapy, and indeed some are now commercially available. Much recent research has been aimed at elucidating the biochemical basis of their observed pharmacological actions.

Many reviews and monographs have been written on the flavonoids. Recent papers have sought to draw attention to the biological activity and therapeutic potential of these compounds.[1-8,256,262] In many instances these properties can be ascribed to the inhibitory effect of flavonoids on various enzyme systems. The areas of current interest are outlined below, except for the uses of flavonoids as liver protectants and immunostimulants which are dealt with elsewhere in this chapter.

A. FLAVONOIDS AS ANTI-INFLAMMATORY AGENTS

Several flavonoids show anti-inflammatory activity which is thought to be due to inhibition of some of the enzyme systems involved in the inflammation process.

1. Inhibition of Histamine Release

Quercetin (**1a**) and other flavonoids are effective inhibitors of the release of histamine induced by various agents.[9-12] Quercetin was originally found to be the most active inhibitor of antigen-induced histamine release from human basophils but studies on peritoneal rat mast cells have revealed that luteolin (**2a**) and amentoflavone (**3**) had higher activity.[11] Activity also varies according to the antigen used to stimulate release of histamine.[10] Structure-activity relationships have been examined.[10,11] The mechanism by which flavonoids inhibit histamine release is still not clear, although a number of theories have been put forward.[10,11,13]

Flavonoids only affect basophils activated by an antigen and not those in the "resting" state.

2. Effect on Prostaglandin and Leukotriene Synthesis

The leukotrienes and series 2 prostaglandins are produced in leukocytes by the metabolism of arachidonic acid (AA) and other polyunsaturated fatty acids (see Figure 1, located in Section VI on γ-linolenic acid) through the mediation of several enzyme systems.

Quercetin (**1a**) has been shown to reduce the release of AA by inhibition of the phospholipase system.[2] It also has an inhibitory effect on the 5-lipoxygenase system which converts AA to the leukotrienes.[14] Recent work has presented a more confused picture of the interaction between different flavonoids and the enzymes involved. Aglycones inhibit prostaglandin formation while the corresponding glycosides enhance it.[15] In studies on the effect of *Scutellaria* flavonoids it was also shown that the aglycones had a greater effect than the glucuronides on the inhibition of the AA metabolism.[16] Some components, e.g., baicalein (**4a**), inhibited the lipoxygenase but others, e.g., 2′,5,6′,7-tetrahydroxyflavanone (**5a**), had the opposite effect.

The anti-inflammatory properties ascribed to many plants used in folk medicine could well be due to the interactions between their flavonoid content and the enzyme systems indicated. An example is the recent isolation of the anti-inflammatory substance hypolaetin 8-*O*-glucoside (**2b**) from the Spanish plant *Sideritis migronensis* Borja (Lamiaceae (Labiatae)).[15]

3. Flavonoids and Capillary Permeability

Increased permeability of the blood capillaries is a feature of several disease states and is manifested in such symptoms as oedema, bleeding, and hypertension. Diseases which are usually associated with increased permeability include diabetes, chronic venous insufficiency, haemorrhoids, scurvy, varicose ulcers, and bruising.

Early research on flavonoids as plant extracts established that they reduced the permeability and fragility of capillary walls.[17] Recent developments have led to the introduction of flavonoid derivatives, e.g., venoruton (**1b**), as pharmaceuticals for treating some of the above conditions. Venoruton is a derivative of rutin found in many plants but obtained commercially from the flowers of *Sophora japonica* L. (Papilionaceae). the anthocyanin glycosides from *Vaccinium myrtillus* L. (Ericaceae) are more active than venoruton.[8]

The mechanism by which capillary permeability is decreased is not fully understood.[1,18]

Another group of flavonoid-related compounds which cause a decrease in capillary permeability are the procyanidins, substances believed to be the basis for an antiscorbutic remedy formerly employed by the Amerindians of Eastern Canada.[19] Tests have shown that the procyanidins, particularly as dimers or oligomers, stabilize collagen by forming cross-links to a much greater extent than do other flavonoids, thus leading to increased vascular resistance. Atherosclerosis is often associated with an increase in vascular permeability, so that procyanidins are of value in its treatment.[19] These compounds are also known to inhibit ascorbic-acid breakdown.

Flavonoids are potent inhibitors of cyclic nucleotide phosphodiesterase (PDE), a key enzyme in the promotion of platelet aggregation. The structural features necessary for a high inhibitory activity are a lack of sugar residues, a C-2(3) double bond, and some, but not too many, hydroxyl groups.[263]

4. Cardiac Activity

Flavone (**1c**) and other flavonoids are known to cause dilatation of coronary blood vessels.[3] However, the major interest in Western Europe in recent years has been in the use of a traditional remedy, extracts of the Hawthorn, *Crataegus* species (Rosaceae), in treating and preventing various types of heart disorder. Although some of the flavanols present, e.g., vitexin (**2c**), dilate the coronary blood vessels, the principal activity is thought to reside in the oligomeric procyanidins, such as (**6**).[18] This type of compound is also present in *Vitis vinifera* L. and *Cupressus sempervirens* L. and has been found to inhibit the angiotensin I converting enzyme (ACE) both *in vitro* and *in vivo*.[264] Oligomeric procyanidins have anti-arrhythmic, cardiotonic, and hypotensive effects. Developments in separation and assay techniques have meant that extracts can be standardized and several commercial preparations are available in Europe.

5. Anti-Ulcer Effects

Experimental systems have demonstrated that some flavonoids prevent ulcer formation,[4] and they are also useful in treating ulcers as they relieve the pain and stop the bleeding. Recent *in vivo* studies in mice showed that myricetin (**1d**) and gnaphalosid (**5b**) were the most potent of the compounds tested.[20]

Particular attention has been paid to the anti-ulcer activity of the procyanidins. (+)-Catechin (**7a**) prevents the formation of gastric lesions in rats. The related dimers and trimers from *Lindera umbellata* (Maxim.) Momiyama (Lauraceae) are more active than the monomers and tetramers.[21]

6. Treatment of Diabetes-Related Cataracts

In diabetes there is a buildup of glucose in the body fluids. In the lens of the eye, excess sorbitol or dulcitol is produced by the conversion of glucose by aldose reductase. The dulcitol cannot be further metabolized and therefore forms a hard crystalline layer in the lens, the cataract. Moreover, the high concentration of dulcitol in the eye fluid causes an increase in osmotic pressure and pain ensues.

Flavonoids are potent inhibitors of the enzyme, with quercitrin (**1e**) and myricitrin (**1f**) among the most powerful.[18] Testing of several other flavonoids has shown that methoxylation at C-6 or C-8 confers significant inhibitory properties.[265] Thus, there may be a role for flavonoids to play in counteracting cataract formation by preventing the formation of dulcitol.

7. Anti-Viral Activity

Some flavonoids, notably quercetin (**1a**), quercitrin (**1e**), catechin (**7a**), and hesperitin (**7b**), have shown anti-infective or anti-replicative activity *in vitro* against several plant and animal viruses.[2-4] A large number of flavonoids was recently tested *in vitro* against types of Herpes simplex (HS), influenza, and rhinovirus viruses.[22] Little activity was found against the

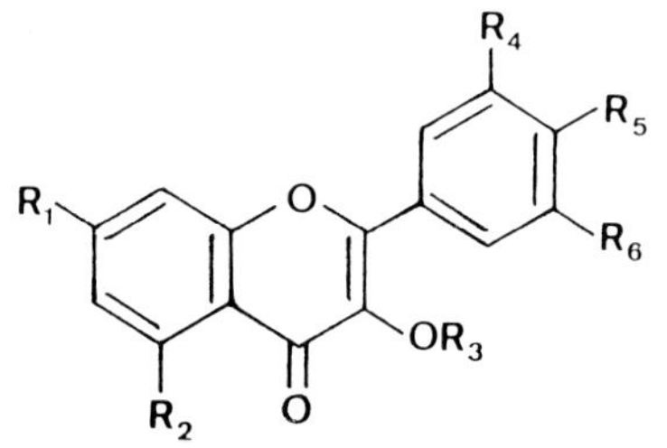

1a Quercetin $R_1 = R_2 = R_4 = R_5 = OH, R_3 = R_6 = H$
1b Venoruton $R_1 = R_4 = R_5 = OCH_2CH_2OH, R_2 = OH, R_3 =$ α-L-Rha-β-D-Glu*

*Abbreviations for the sugars are as follows: Ara = arabinose; Cym = cymarose; Dix = digitoxose; Fuc = frucose; Gal = galactose; Glu = glucose; GlA = glucuronic acid; Gul = gulomethylose; Ole = oleandrose; Rha = rhamnose; The = thevetose; Xyl = xylose.

1c Flavone $R_1 - R_6 = H$
1d Myricetin $R_1 - R_6 = OH$
1e Quercitrin $R_1 = R_2 = R_4 = R_5 = OH, R_3 = R_6 = H$
1f Myricitrin $R_1 = R_2 = R_4 = R_5 = R_6 = OH, R_3 =$ α-L-Rha-β-D-Glu

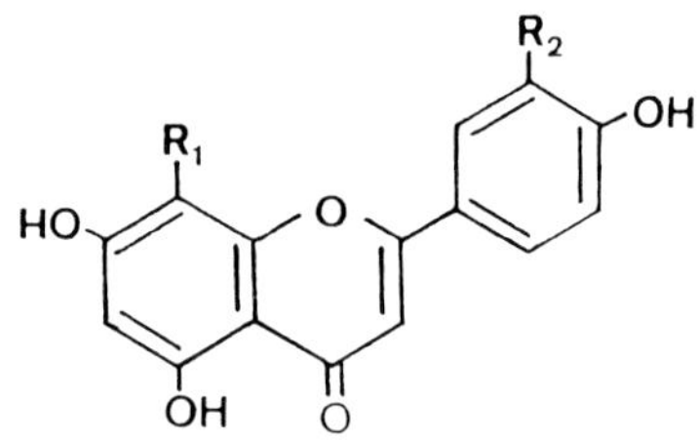

2a Luteolin $R_1 = H, R_2 = OH$
2b Hypolaetin 8-*O*-glucoside $R_1 =$ O-β-D-Glu, $R_2 = OH$
2c Vitexin $R_1 =$ O-β-D-Glu, $R_2 = H$

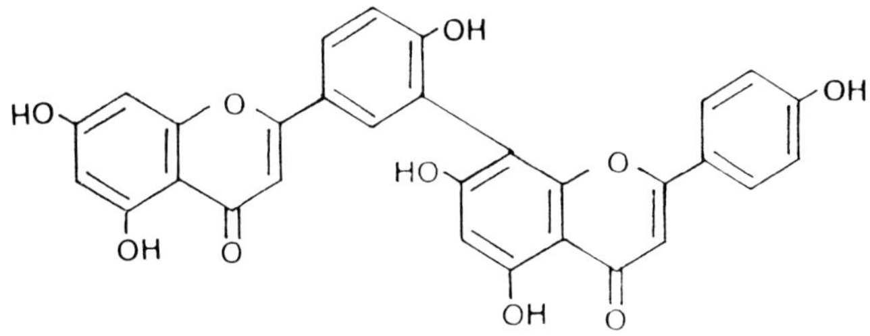

3 Amentoflavone

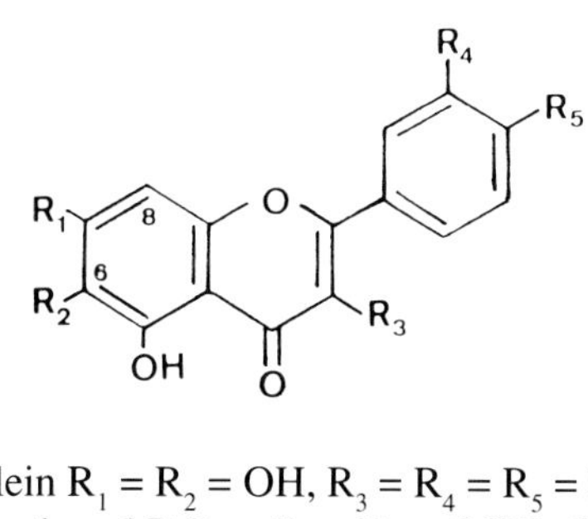

4a Baicalein $R_1 = R_2 = OH, R_3 = R_4 = R_5 = H$
4b Chrysophenol B $R_1 = R_2 = R_4 = OCH_3, R_3 = R_5 = OH$
4c Eupatin $R_1 = R_2 = R_5 = OCH_3, R_3 = R_4 = OH$

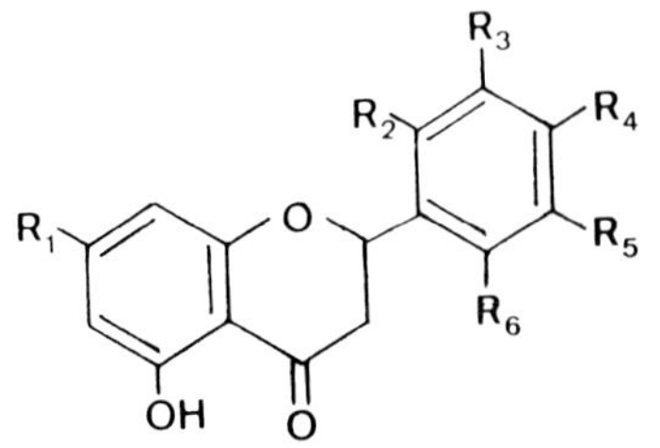

5a 2′,5,6′,7-Tetrahydroxyflavanone $R_1 = R_2 = R_6 = OH$, $R_3 = R_4 = R_5 = H$
5b Gnaphaloside R_1 = *Caffeyl*-β-D-Glu, $R_2 = R_6 = H$, $R_3 = OCH_3$, $R_4 = R_5 = OH$

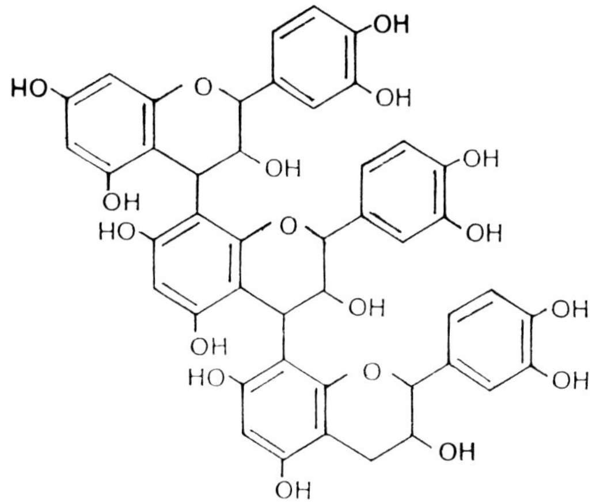

6 An oligomeric procyanidin

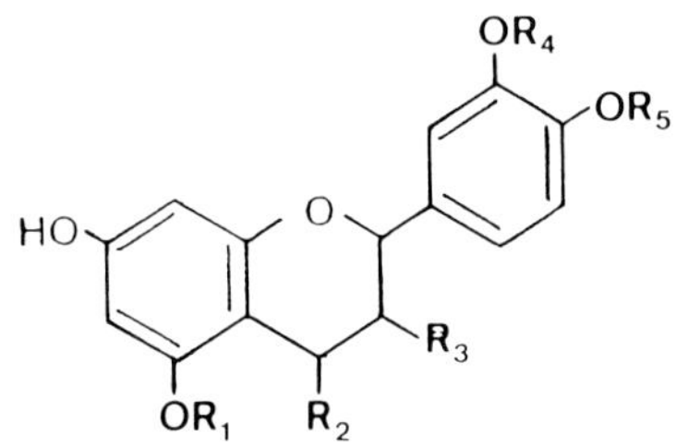

7a (+)-Catechin or (+)-3-Cyanidol $R_1 = R_2 = R_4 = R_5 = H$, R_3 = β-OH
7b Hesperitin $R_1 = R_2 = R_3 = R_4 = H$, $R_5 = CH_3$
7c Diffutin R_1 = β-D-Glu, $R_2 = R_3 = H$, $R_4 = R_5 = CH_3$

HS and influenza virus, but some 3-methoxyflavones, e.g., chrysophenol B (**4b**), were found to be active against rhinovirus. Since the rhinovirus used in the test is responsible for about 50% of common colds, an extract of *Chrysosplenium* species (Saxifragaceae), plants which contain high amounts of the active flavonoids, has been suggested as a treatment for the common cold.

Another *in vitro* study on HS and other viruses showed that quercetin (**1a**) reduced both infectivity and intracellular replication of viruses.[23] Catechin (**7a**) reduced infectivity only and hesperitin (**7b**) intracellular replication only. 3-Methylquercetin has a high level of antiviral activity but relatively little effect on the cardiovasular system.[266] The use of flavonoids in treating virus infections has not yet reached the clinical testing stage but shows potential.

It has been suggested that the antiviral activity of flavonoids is due to their inhibition of ATPase and phospholipase A_2 enzymes.[1] These enzymes aid the removal of the protein coat surrounding the nucleic acid and the preceding fusion of the viral mantle with the lysosomal membrane.

8. Adaptogenic Activity

Adaptogenicity is the term used to describe the antistress and corrective effects of drugs such as Ginseng. Some flavans, e.g., diffutin (**7c**) from *Canscora diffusa* R.Br. (Gentianaceae), have been shown to have this type of activity.[24] Interestingly, the plant is used in Indian traditional medicine for treating certain CNS disorders.

9. Flavonoids as Anti-Cancer Agents

In vitro testing against cultures of malignant cells has shown that some flavonoids, e.g., eupatin (**4c**), have moderate activity.[1,3,18] The highest activity is seen in compounds with a tri-substituted A ring and methylation at either C-6 or C-8. The mechanism of action has been investigated.[1]

Flavanoids may also be useful in inhibiting chemically induced carcinogenesis because of their enzyme-inducing effects.[1] Induction of aryl hydroxylases and epoxide hydrolases is known to occur. Aryl hydroxides and epoxides are known to be oncogenic because of their reactivity with nucleic acid bases; hence, their quick destruction by a stimulated enzyme system would be an advantage. In spite of these studies and the low toxicity of the flavonoids, none of this group of compounds has been subjected to clinical testing.

Clearly, flavonoids have a wide range of potential therapeutic usefulness. Studies of individual compounds and of structure-activity relationships have been hampered by problems in isolation procedures and structure elucidation. These have been overcome with new techniques and test systems. It is likely that further studies will validate the traditional use of many plants on the basis of their flavonoid content.

III. TRITERPENE GLYCOSIDES (SAPONINS)

In recent years, with the advent of improved methods of isolation and structural determination, there has been increasing interest in triterpene glycosides or saponins, The triterpene moieties, or aglycones, are mostly C_{29} or C_{30} compounds belonging to a variety of skeletal types built up from six isoprene units.* They occur widely in plants, usually as glycosides with one or two sugar chains (monodesmosides, e.g., **8a**, or bisdesmosides, e.g., **8e**), which may be linear or branched, comprising hexoses, pentoses, and/or uronic acids; various organic acid residues may also be present. Several reviews have been published.[25-30,257,267]

Triterpene derivatives have an important role to play in the physiology of plants, as well as in the ecological systems in which they grow.[28] Bisdesmosides tend to have weaker biological activity than the monodesmosides and they are considered rather to be a transport form.[30] Of more immediate importance are the effects these substances have in man and animals.

Interest in the adaptogenic and other properties of the triterpenoids present in ginseng root, from *Panax* species (Araliaceae), continues unabated and requires no discussion here[31,32,267] (see Sections VII.B.3 and VIII.C.3).

Triterpene glycosides affect the activity of the gastrointestinal tract, leading to a reduced intake of nutrients and hence a slower rate of growth, but there is also evidence that they affect the functions of many other organs and tissues, including the nervous system.[28] The corticoid, anti-inflammatory, and anti-ulcer properties of glycyrrhizin, a saponin from liquorice, *Glycyrrhiza* species, are well known;[27,33] see Section VII.C.1. Triterpenes from *Glycyrrhiza* (Papilionaceae) and *Antidesma* (Stilaginaceae) species, and other sources bring about a reduction in blood cholesterol levels, a property that has been extensively investigated in connection with their potential use as hypolipemics and anti-atherosclerotics.[28,34-37]

Racemoside (**9a**), from the Black Cohosh, *Cimicifuga racemosa* L. (Ranunculaceae), is another glycoside stated to have anti-ulcer activity.[38]

Through interaction with the membrane surrounding erythrocytes (red blood corpuscles), saponins cause haemolysis; the mechanism of action is still unclear, but probably involves hydrolysis of the glycoside absorbed on the erythrocyte membrane or initial hydrolysis followed by attachment of the aglycone to the membrane. Certain structure-activity relationships have been determined; acylosides and bisdesmosides show little or no activity, while acid or neutral monodesmosides are highly active. Structural features of the aglycone also affect the activity.[28,29]

Saponins have been shown to have antibacterial activity.[25,39,40] Cucumarioside, a preparation of triterpene glycosides from the sea cucumber, *Cucumaria japonica*, protected mice against *Salmonella typhimuria* infection by preventing the proliferation of salmonellae into the organs and tissues and enhancing the phagocytic activity of peritoneal macrophages.[41] Of greater interest is their fungicidal or mycocidal activity. Glycosidic derivatives of pentacyclic triterpenes with a free carboxyl function, such as oleanolic acid (**8c**) or hederagenin (**8d**), are particularly active.[28,40,42] Saponins from *Primula acaulis* (L.) Hill. (Primulaceae), although less active than nystatin and stamicin against *Candida*, are nevertheless also effective against antibiotic-resistant strains. The ester saponins derived from *Polemonium caeruleum* L. (Polemoniaceae) are active against both *Candida* and *Torulopsis*.[29]

The molluscicidal properties of monodesmoside saponins from species of *Hedera*, *Phytolacca*, *Lonicera*, and *Hedychium* have attracted attention. They have a powerful action in controlling *Biomphalaria* snails, known vectors for schistosomiasis (bilharziasis).[43-46,258]

Saponins from several plants, e.g., *Pittosporum nilghirense* Wight et Arnott (Pittosporaceae) and *Gypsophila paniculata* L. (Caryophyllaceae), have been observed to have a spermicidal activity.[29]

Of greater importance, however, are the anti-inflammatory or anti-exudative properties exhibited by certain saponins, e.g., aescin, a mixture of short-chain fatty-acid ester glycosides of various aglycones, but mainly pro-aescigenin (cf., **9b**), present to the extent of 3 to 5% in the seeds of the horse chestnut, *Aesculus hippocastanum* L. (Hippocastanaceae). It acts by normalizing the disturbed capillary permeability that characterizes the initial stage of the inflammation process. Similar properties are exhibited by the saponins from many other plants, e.g., species of *Bupleurum, Platycodum, Thea, Sanicula, Eryngium, Hedera, Polemonium, Hydrocotyle, Solidago, Sapindus,* in the rat-paw oedema and other tests.[27,29,47,48]

It is not only the triterpene derivatives of liquorice that taste sweet, but also those from the fruit of a Chinese *Momordica* species (Cucurbitaceae)[49] and from *Periandra* species (Papilionaceae);[50] the latter are proposed as useful sweetening agents.

Cytotoxic and anti-tumour properties are exhibited by various triterpenes, among them the tetracyclic cucurbitacins (cf., **10**) that occur in various Cucurbitaceae and that apparently have to bind to glucocorticoid receptors in order to take effect,[51] and the pentacyclic 3-oxolup-20(29)-en-30-al (**11**), one of the toxic principles isolated from a *Maytenus* species (Celastraceae).[52] Anti-tumour formulations containing triterpene glycosides have been patented.[53]

Clinical studies have demonstrated that the total triterpenoid fraction from *Centella asiatica* (L.) Urb. (Apiaceae (Umbelliferae)) has a favourable influence on wound healing. Experiments with cultured human embyronic fibroblasts suggest that the effect may be due to the stimulation of collagen synthesis.[54]

See Section VII.C about the anti-hepatotoxic and hepatotropic effects of triterpenoids.

IV. CARDIAC GLYCOSIDES

Ever since the publication of William Withering's *An Account of the Foxglove, and Some of its Medical Uses* in 1785, preparations of Digitalis and certain of its glycosides have been important drugs in treating the failing heart (dropsy). Their main beneficial effect is in cases of fast atrial fibrillation. Their value in treating heart failure in which the heart has

* Strictly, the plant sterols and related substances such as lanosterol are also triterpenes, but they are usually considered along with other substances of similar skeletal structure as forming part of the group of steroids.

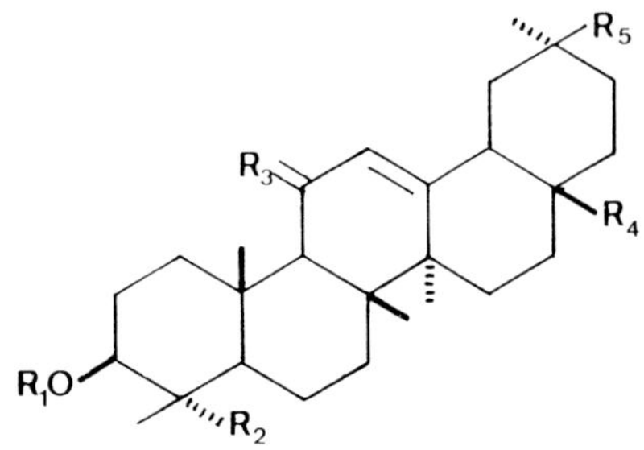

8a Glycyrrhizin R_1 = β-D-GlA(1→2)-β-D-GiA, $R_2 = R_4 = CH_3$, R_3 = O, $R_5 = CO_2H$
8b Glycerrhetinic acid R_1 = H, $R_2 = R_5 = CH_3$, R_3 = O, $R_4 = CO_2H$
8c Oleanolic acid R_1 = OH, $R_2 = R_5 = CH_3$, $R_3 = H_2$, $R_4 = CO_2H$
8d Hederagenin R_1 = OH, $R_2 = CH_2OH$, $R_3 = H_2$, $R_4 = CO_2H$, $R_5 = CH_3$
8e Bisdesmoside from *Lonicera nigra* R_1 = β-D-Glu-(1→2)-α-L-Ara, $R_2 = R_5 = CH_3$, R_3 = O, R_4 = CO-β-D-Glu-(1→6)-β-D-Glu
8f Gypsogenic acid $R_1 = R_3$ = H, $R_2 = R_4 = CO_2H$, $R_5 = CH_2OH$
8g Dianic acid $R_1 = R_3 = R_5$ = H, $R_2 = R_4 = CO_2H$

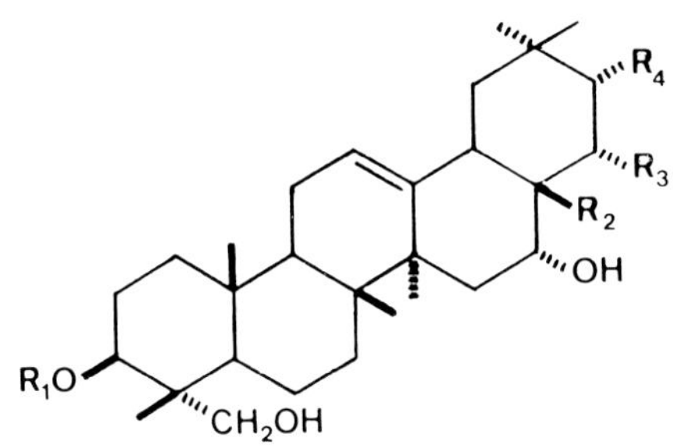

9a Racemoside R_1 = α-L-Ara, $R_2 = CO_2H$, $R_3 = R_4$ = H

9b Pro-aescigenin R_1 = [β-D-Glu(1→2), β-D-Glu(1→4)] D-GlA, $R_2 = CH_2OH$, $R_3 = OCOCH_3$, $R_4 = OCO(CH_3)\overset{+}{=}CHCH_3$

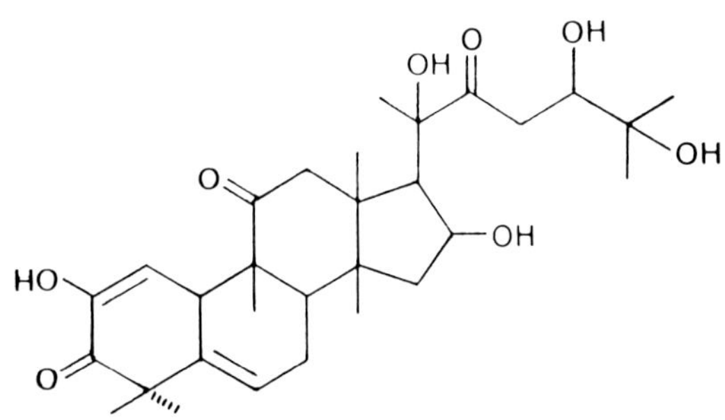

10 Cucurbitacin J

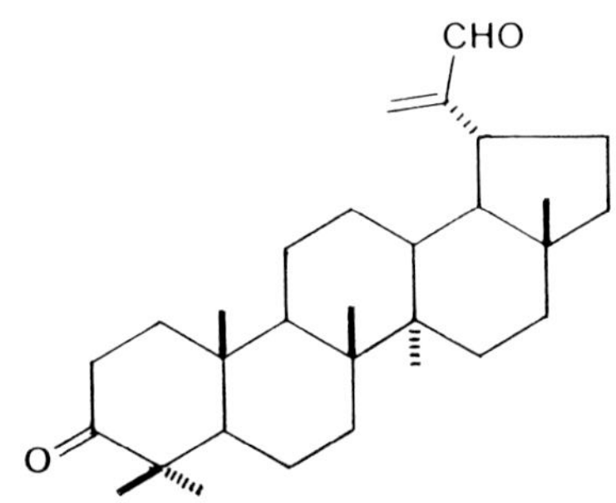

11 3-Oxolup-20(29)-en-30-al

maintained its normal rhythm has been much less spectacular, and where diuretics can be used successfully Digitalis has few, if any, advantages.[55,56] Nevertheless, as populations age, the incidence of heart disease is increasing and the use of cardiotonic drugs with it.[57]

The active principles of *Digitalis* and related drugs such as *Thevetia*, *Convallaria*, etc., are cardiac or cardiotonic glycosides of the cardenolide type; these are steroid derivatives to which a 5-membered unsaturated lactone ring is attached at the 17-position (cf., **12a**). The glycosides occurring in Squill, on the other hand, are of the scilladienolide or bufadienolide type and are also steroid derivatives but with a 6-membered doubly unsaturated lactone ring instead (cf., **19**). Besides glucose, many of the sugars present which are attached at position 3 of the steroid skeleton (cf., **12a**) are unusual and are rarely found elsewhere. They may be methylated and they may lack hydroxyl functions at positions 2 and/or 6.

It is generally accepted that the Digitalis receptor is a

membrane bound, Mg^{2+}-dependent Na^+/K^+-ATPase, one important function of which is to pump Na^+ ions out of and K^+ ions into the cell in order to maintain the high K^+/Na^+ ratio characteristic of intracellular fluid. The binding of cardiac glycosides to the enzyme prevents the conformational changes that occur during the normal activity of the pump. The resultant increase in Na^+ concentration in the cell causes exchange between Na^+ and Ca^{2+} ions and the greater concentration of Ca^{2+} brings about stronger contractions. When too many Na^+/K^+ pumps are put out of action by inhibition of the enzyme, toxic effects such as arrhythmias ensue, so that the cardiotonic and cardiotoxic effects are due to differing degrees of inhibition of the same system.[58] A lengthy monograph on various aspects of the pharmacology of the cardiac glycosides has recently been published.[59]

Two factors have rendered Digitalis and its glycosides difficult drugs to use clinically: (1) their tendency to accumulate with repeated doses, and (2) the low therapeutic ratio, i.e., the appearance of toxic symptoms with doses only a little greater than those which produce a satisfactory therapeutic response. Even under hospital conditions, toxic symptoms have been observed in up to 29% of patients being treated with Digitalis.[55] Much research has therefore been carried out in attempts to obtain cardiac drugs with improved properties. This work has taken two main directions: (1) a search in the plant kingdom for other, less toxic cardial glycosides and (2) chemical modification of glycosides already in use. Both these approaches have had limited success, and the following sections outline some of the more recent work in these fields.

A. *THEVETIA* ADANS. (APOCYNACEAE)

The yellow oleander, *Thevetia peruviana* (Pers.) Schumann (= *T. neriifolia* Juss.), is a small tree originally from tropical South America which has now become widespread as a garden plant throughout the tropics of the Old and New Worlds. The yellow-flowered variety is the most common, but plants with orange and white flowers are also known.

Seeds of *T. peruviana* contain thevetin, a mixture of the triglycosides thevetin A (**12e**) and cerberoside (thevetin B) (**12c**), which are based on the cardenolide aglycones digitoxigenin (**12a**) and cannogenin (**12b**), respectively. These two glycosides are also major components of the cardenolide mixtures present in the latex and bark.[61] Enzymatic degradation (fermentation) of thevetin A yields peruvoside (**12f**) and in the same way thevetin B affords neriifolin (**12d**). Both of these monoglycosides occur naturally in the plant; peruvoside is found principally in the seeds, while neriifolin is an important component in the seeds and latex.[61,62,77] Yields of up to 0.6% peruvoside can be obtained from defatted and fermented seeds.[63,64] The little information available suggests that the cardenolide content and composition of the seeds from the three colour varieties of *T. peruviana* is very similar.[65]

Thevetin has had some use clinically as a fast-acting cardiotonic by the intravenous rather than oral route, but it does give rise to side effects in a considerable number of cases and its failure rate is significant.[60,66] None of the *Thevetia* glycosides appears to offer any advantages over the current range of clinically used cardiotonic glycosides, with the sole exception of peruvoside (**12f**); a small monograph dealing with the chemistry, pharmacology, and clinical study of this glycoside has been published.[60]

Perhaps the most important point regarding peruvoside is that according to animal experiments, as compared with drugs such as ouabain, proscillaridin, strophanthin K, etc., its therapeutic dose affords a margin of safety that is at least twice as great.[60,67-69] The glycoside is effective by both the intravenous and oral routes, being well absorbed and having a short duration of action. Clinical reports indicate that it is well tolerated, with a rapid action, and that it has low acute and cumulative toxicity; however, occasionally, symptoms of gastrointestinal irritability have been noted.[60,70]

Chinese workers have prepared a glycoside mixture, thevetoside, from fermented seeds of Hainan *T. peruviana*. Although neriifolin and its 2′-monoacetate (cerberin) are the main constituents, up to ca. 15% of peruvoside may be present as well.[71] The pharmacology and clinical application of the plant have been studied.[72]

With the widespread availability of the yellow oleander in tropical countries, it may be that greater use could be made of peruvoside in treating heart conditions.

Several other *Thevetia* species have been investigated, but in less detail.[73-78]

B. *NERIUM* L. (APOCYNACEAE)

Nowadays, the genus *Nerium* L. is usually considered to comprise one rather polymorphic species, *N. oleander* L. (= *N. indicum* Miller = *N. odorum* Sol.), the oleander,[79] whose distribution ranges from the Mediterranean through the Middle East to India, China, and Japan. However, it is now widespread throughout the warmer parts of the globe in a variety of horticultural forms with single and double flowers varying in colour from pure white through pink to deep red; plants with yellow flowers are also known.[79]

Extracts from the leaves have been used medicinally since ancient times and they are still a source of cardiotonic drugs.[66,80] Extensive research has shown that the plant contains a complex mixture of glycosides, with those quantitatively most important being derived from the following four aglycones: oleandrigenin (**13a**), digitoxigenin (**12a**), adynerigenin (**14**), and oleagenin (**15**); derivatives of the latter two aglycones are largely devoid of activity.[79,81,82] In the fresh or air-dried leaves it is chiefly the tri- and diglycosides, including (**13c**) and (**13d**), that are present,[81,82] whereas in the oven-dried leaves enzymatic hydrolysis takes place and the monoglycosides predominate.[79]

Clinical interest has centered around the monoglycoside oleandrin (**13b**).[66,80] Air- and oven-dried leaves from various forms of oleander have been analysed; up to 0.425% oleandrin may be present.[65,79] In commercial leaf samples there was between 0.013 and 0.12% oleandrin; the tri- (and di-) glycosides usually dominated, but the varying ration of tri-/di- to monoglycosides points to differences in the drying and storage conditions.[81] There appears to be no correlation between morphology and chemistry. The considerable variation in the oleandrin content suggests the possibility of improving the yield through a breeding programme.[79]

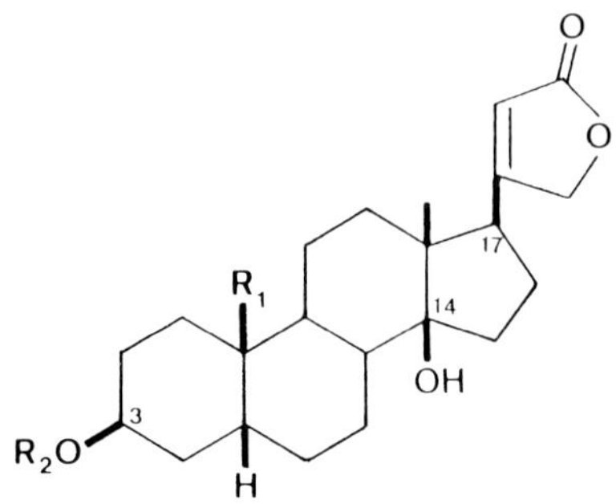

12a Digitoxigenin $R_1 = CH_3$, $R_2 = H$
12b Cannogenin $R_1 = CHO$, $R_2 = H$
12c Cerberoside $R_1 = CH_3$, $R_2 = \alpha$-L-The-(β-D-Glu)$_2$
12d Neriifolin $R_1 = CH_3$, $R_2 = \alpha$-L-The
12e Thevetin A $R_1 = CHO$, $R_2 = \alpha$-L-The-(β-D-Glu)$_2$
12f Peruvoside $R_1 = CHO$, $R_2 = \alpha$-L-The
12g Digitoxin $R_1 = CH_3$, R_2 = (β-D-Dix)$_3$
12h Purpureaglycoside A $R_1 = CH_3$, R_2 = (β-D-Dix)$_3$-β-D-Glu
12i β-Methyldigitoxin $R_1 = CH_3$, R_2 = (β-D-Dix)$_2$-4‴-Methyl-β-D-Dix
12j ASI-222 $R_1 = CH_3$, R_2 = 4′-Amino-4′,6′-Dideoxy-β-D-Gal
12k ASI-253 $R_1 = CH_3$, R_2 = β-D-Gal
12l ASI-254 $R_1 = CH_3$, R_2 = 4′-Amino-4′,6′-Dideoxy-β-D-Gal

Other parts of the oleander, stem and root bark and seeds, contain cardiac glycosides, but oleandrin and the corresponding triglycoside are not among them.[83]

C. *CONVALLARIA* L. (LILIACEAE)

Convallaria majalis L., the Lily-of-the-Valley, has long been used as a medicinal plant. The flowers (inflorescences), herb, and rhizomes have all been included in various pharmacopoeias, and the aerial parts are still an official drug in several European countries.[66,84,85] About 40 different cardenolide glycosides have been isolated from *C. majalis*, of which only about six are major components (see also below): convallatoxin (**16b**), convalloside (**16c**), and desglucocheirotoxin (**16d**), which are derivatives of strophanthidin (**16a**); convallatoxol (**16f**) and convallatoxoloside (**16g**), which are strophanthidol (**16e**) glycosides; and lokundjoside (**17b**), which has bipindogenin (**17a**) as aglycone.[86]

Bleier et. al.[87] analysed about 100 samples of leaves and flowers originating from localities scattered throughout most of Europe. The total glycoside content of the leaves ranged from about 0.1 to 0.6% with an average of 0.2 to 0.3%; the flowers usually had 2 to 3 times as much glycoside as the corresponding leaves. Convallatoxin (**16b**) formed between 10 and 50%, convalloside (**16c**) between 4 and 46%, convallatoxol (**16f**) between 12 and 44%, and lokundjoside (**17b**) between 4 and 46% of the total glycosides. Plants from northwestern Europe tend to be distinguished by a very high convallatoxin and a high convallatoxol content, while those from eastern and southeastern Europe have a high proportion of convalloside. In a small part of central Europe (southeastern Austria and northern Yugoslavia) lokundjoside is the dominant glycoside, but elsewhere in central Europe (Austria, Czechoslovakia, eastern Germany) the composition is intermediate with roughly equal proportions of the four glycosides (**16b**), (**16c**), (**16f**), and (**17b**). The glucoside composition of the flowers is similar to that of the corresponding leaves, but usually there is a slight increase in the convallatoxin content at the expense of convallatoxol.

The variations in glucoside content and composition could be due to the existence of geographical races, differences in soil and climate, and/or the existence of chemical races.[87] However, the situation may be rather more complicated, since within a given plant population there are large differences in glycoside content between one individual and another.[87,88] There are, in addition, daily and seasonal changes.[89]

The best time for collection of the drug is just before flowering, when the cardenolide content is at its highest.[90] Cultivation of the plant brings about a considerable increase in the proportion of convallatoxin in the total glucoside complex — in some cases by as much as 100%; the reason for this is unclear.[91]

Among the major glycosides of the dried flowers and aerial parts of the Japanese lily-of-the-valley, *C. keiskei* Miq., are convallatoxin (**16b**), desglucocheirotoxin (**16d**), and rhodexin A (**17c**).[92-94] The main glycoside of *C. transcaucasica* Utkin ex Grossh. is also convallatoxin.[95]

The clinical pharmacology of *Convallaria* has been reviewed.[96]

D. *ADONIS* L. (RANUNCULACEAE)

This is a small genus of herbaceous plants distributed mainly throughout the temperate regions of the northern hemisphere. One of the species, *A. vernalis* L., has been included in several pharmacopoeias, extracts of the flowering herb being recommended as a cardiotonic.[67,86] Paper chromatographic studies show that the leaves and flowers have very

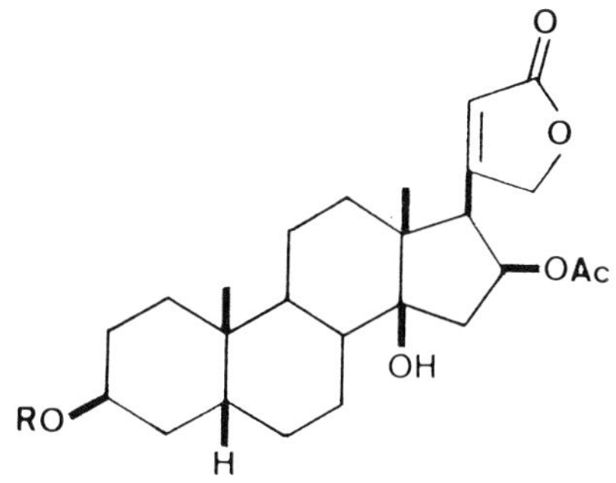

13a Oleandrigenin R = H
13b Oleandrin R = α-L-Ole
13c Glucosyloleandrin R = a-L-Ole-β-D-Glu
13d Gentiobiosyloleandrin R = α-L-Ole-(β-D-Glu)$_2$

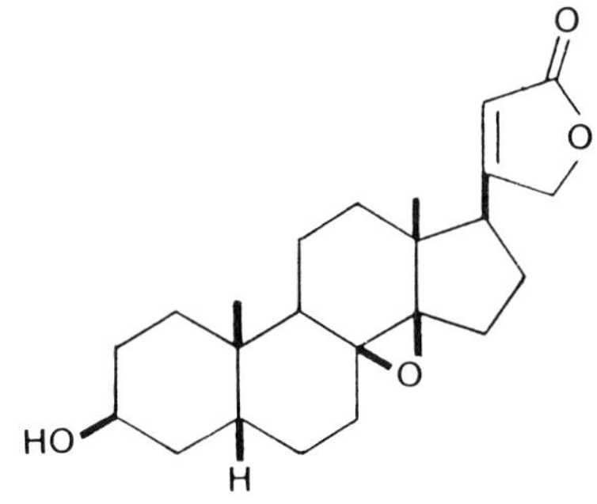

14 Adynerigenin

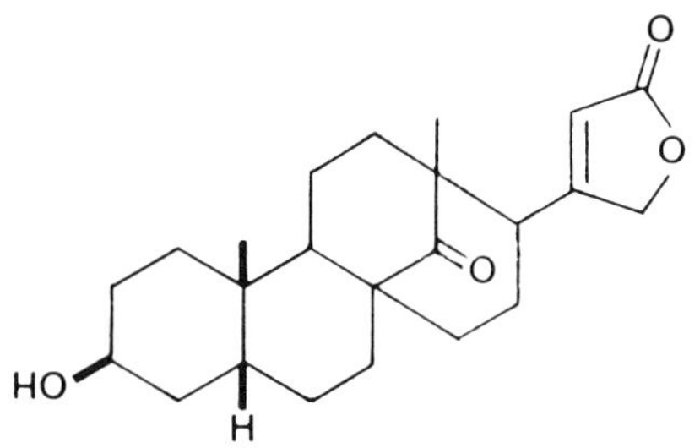

15 Oleagenin

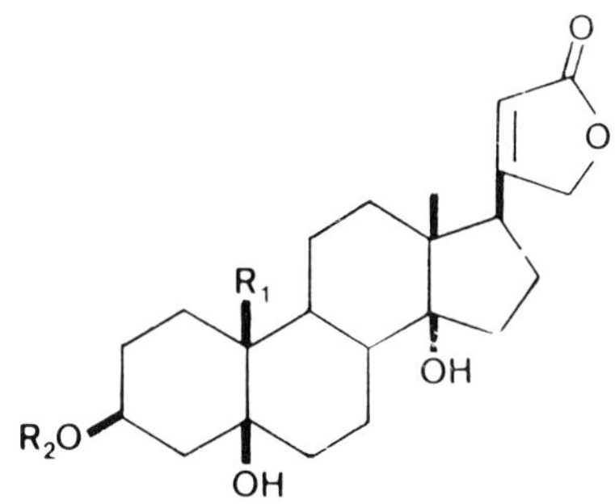

16a Strophanthidin R_1 = CHO, R_2 = H
16b Convallatoxin R_1 = CHO, R_2 = α-L-Rha
16c Convalloside R_1 = CHO, R_2 = α-L-Rha-β-D-Glu
16d Desglucocheirotoxin R_1 = CHO, R_2 = β-D-Glu
16e Strophanthidol R_1 = CH_2OH, R_2 = H
16f Convallotoxol R_1 = CH_2OH, R_2 = α-L-Rha
16g Convallotoxoloside R_1 = CH_2OH, R_2 = α-L-Rha-β-D-Glu
16h Cymarin R_1 = CHO, R_2 = β-D-Cym
16i k-Strophanthin-β R_1 = CHO, R_2 = β-D-Cym-β-D-Glu
16j k-Strophanthoside R_1 = CHO, R_2 = β-D-Cym-(β-D-Glu)$_2$
16k Cymarol R_1 = CH_2OH, R_2 = β-D-Cym
16l Erysimin (Helveticoside) R_1 = CHO, R_2 = β-D-Dix
16m Erysimoside R_1 = CHO, R_2 = β-D-Dix-β-D-Glu

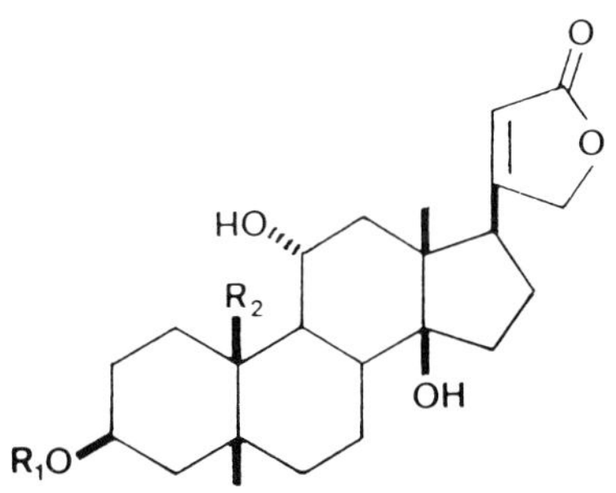

17a Bipindogenin R_1 = H, R_2 = OH
17b Lokundjoside R_1 = α-L-Rha, R_2 = OH
17c Rhodexin A R_1 = α-L-Rha, R_2 = H

complex mixtures of glycosides.[97,98] Among the principal components are the strophanthidin (**16a**) derivatives cymarin (**16h**), k-strophanthin-β (**16i**), and k-strophanthoside (**16j**); the strophanthidol (**16c**) derivative cymarol (**16k**) is also present.[99] The aglycone adonitoxigenin (**18a**) is isomeric with strophanthidin, and numerous glucosides derived from it also occur in the plant; these include adonitoxin (**18b**) and its 2′- and 3′-acetyl derivatives, and recently several diglycosides with adonitoxigenin as aglycone have been isolated.[100] The aglycone adonitoxigenol (**18c**), which, like strophanthidol, has a 19-hydroxymethyl function, is present in adonitoxol (**18d**). With the isolation of the new aglycone 3-*epi*-periplogenin,[99] it would appear that there is a stepwise oxidation of the methyl group a C-19 to a hydroxymethyl and then aldehyde function, similar to that observed in *Convallaria* (q.v.). Glycosides of several other aglycones have also been obtained.

The glycoside content of the aerial parts appears to be highest at the flowering and fruiting stages, and as much as 0.69% strophanthidin glucosides has been found at this period.

A. aestivalis L. and *A. autumnalis* L. have also been official drugs.[86] While so far only cymarin (**16h**) has been isolated from *A. aestivalis*,[101] several strophanthidin and adonitoxigenin glycosides are found in *A. autumnalis*.[102]

Many other species, primarily of Russian origin, including *A. wolgensis* Stev. ex DC. and *A. mongolica* Simonovich, have been examined, and they all contain strophanthidin and/or k-strophanthin-β and/or cymarin; in several of them adonitoxigenin or its 3′-*O*-acetyl derivative have also been found.[103-108] Yang and Qin[109] briefly reviewed the cardenolide constituents of *Adonis* species and their occurrence and spectroscopic properties.

Evdokimov et al.[110] propose *A. tianshanicus* (Adolf) Lipschitz as a new medicinal plant. The biological activity of the roots appears to be higher than that of the leaves which have about the same activity as those of *A. vernalis*.[111,112] In the case of *A. turkestanica* Adolf, the highest biological activity is again during the flowering and fruiting period, with the greatest concentration of cardenolides in the roots, green fruits, and the upper part of the stems,[113] The most appropriate conditions for storage of the drug have been determined.[114]

E. OTHER GENERA

Seeds from certain species of the European and Asian genus *Erysimum* L. (Brasicaceae (Cruciferae)) are a rich source, especially of the strophanthidin-based glycosides erysimin (helveticoside) (**16l**) and erysimoside (**16m**); in *E. canescens* Roth as much as 2% erysimosides has been found.[115] Both of these compounds have been introduced into medicine,[66,116] and along with them a number of standardized products, including coreside, an extract from *E. cheiranthoids* L., and cardiovalen from the fresh juice of *E. canescens*.[116,117]

Since the detailed investigations of the 1950/1960s, when the cardenolide glycosides of *Strophanthus* DC. (Apocynaceae) were seen as a potential source of raw material for the steroid industry, its species have received little attention. The African *S. hispidus* DC. and *S. kombe* Oliver are still official in several European pharmacopoeias; among the strophanthidin-based glycosides obtained from them are k-strophanthin, a standardized mixture, and cymarin (**16h**).[66] *S. kombe* seeds have large amounts of k-strophanthoside (**16j**), k-strophanthin-β (**16i**), and erysimoside (**16m**).[118,119] Enzymatic hydrolysis (fermentation) yields up to 3.5% cymarin.[120] Asian species have been less well examined.[121]

Bulbs from two species of *Drimia* Jacq. (*Urginea* Steinh.) (Liliaceae) have been official drugs — the aggregate species *Dr. maritima* (L.) Stearn from North Africa and southern Europe and *Dr. indica* (Roxb.) J. P. Jessop from India. Unlike the genera discussed so far, they contain cardiac glycosides of the bufadienolide type. The glycoside content (higher in autumn than in spring) and composition of various European forms of *Dr. maritima* have been determined.[122] Triploid forms of *Dr. indica* have the highest glycoside content.[123] Of the known *Drimia* glycosides,[124] only proscillaridin (**19**) is in current medical use.[66]

H. *DIGITALIS* L. (SCROPHULARIACEAE)

While preparations of *Digitalis*, from biologically standardized leaf powder of *D. purpurea* L., continue to be official drugs, nowadays the emphasis clinically is on the use of purified individual glycosides and some of their semi-synthetic derivatives (see below). Of the many glycosides in *D. purpurea*, digitoxin (**12g**) is still being used. More and more, however, it is the glycosides from *D. lanata* Ehrh., based on digoxigenin (**20a**), that are being prescribed, especially digoxin (**20b**) and to a lesser extent also deacetyl-lanatoside C (**20c**) and lanatoside C (**20d**). Recent clinical, pharmacokinetic, and metabolic studies with gitoxin (**18e**) suggest that this glycoside might have advantages over digoxin and digitoxin in cases of renal insufficiency.[125,259]

Digitalis is indigenous to central and southeastern Europe. It is cultivated not only in central Europe,[57] but also in the Netherlands, Ecuador, and the U.S.[126] The plant is a biennial and the leaves are harvested at the end of the first year before flowering, which would lead to a reduction in the digoxin content. High yielding strains of *D. lanata* have been obtained by two-stage selection involving the radioimmunoassay of over 10,000 individual plants. The digoxin content was raised 2- to 3-fold and several strains with an average yield of 0.6% were isolated; some individual plants had up to 1% of the glycoside.[127] Genetic selection can also be practised, but with the resulting inbred strains continuous selection is still required in order to prevent reversion of the digoxin content to normal levels.

1. Cell Cultures

The use of cell cultures as a means of producing *Digitalis* glycosides has been extensively investigated.[128-130] Undifferentiated cells produce little or no cardenolides or lose the ability to do so on subculturing. Under suitable conditions, however, regeneration of embryos and plantlets (shoots) does trigger cardenolide production. The extent of embyro development and cardiac-glycoside formation depends on appropriate hormone treatment and on the carbon and nitrogen sources in the medium. *D. lanata* strains have been developed, regenerated plantlets of which accumulate up to 0.15 μg/mg dry weight of glycosides (Table 1). With strain VII, although at the end of the first vegetative period the original plant formed a high proportion of digoxin (**20b**) and derivatives, the *in vitro* adventive embryos and plantlets contained mainly digitoxin (**12g**); the latter finding agrees with the fact that the cardenolides present in seedlings and young plants of *D. lanata* mostly have little or no digoxin derivatives. Embryoid structures grown from *D. lanata* cells recultured after long-term cryopreservation at the temperatures of liquid nitrogen maintain their capacity for cardenolide biosynthesis and plant regeneration.[131,132]

In spite of the interest of this work, it is questionable whether it can be exploited as a practicable alternative means of obtaining digoxin and related compounds.

2. Biotransformation of Digitalis Glycosides

Since current emphasis in therapy is on the utilization of digoxin, stocks of digitoxin (**12g**) have been accumulating as a byproduct. Another approach to the problem of increasing the supply of digoxin has been to investigate the bioconversion of digitoxin. Glucosylation, acetylation, 12β- and 16β-hydroxylation, and 16β-formylation are among the reactions that take place on addition of digitoxin to *D. lanata* cell cultures.[133,134] Digitoxin (**12g**) added to suspension cultures and cells immobilized in alginate gel is converted mainly to purpureaglycoside A (**12h**), with only minute amounts of digoxin (**20b**) and deacetyl-lanatoside C (**20c**) being formed.[135] Applied to β-methyldigitoxin (**12i**), suspension cultures of a

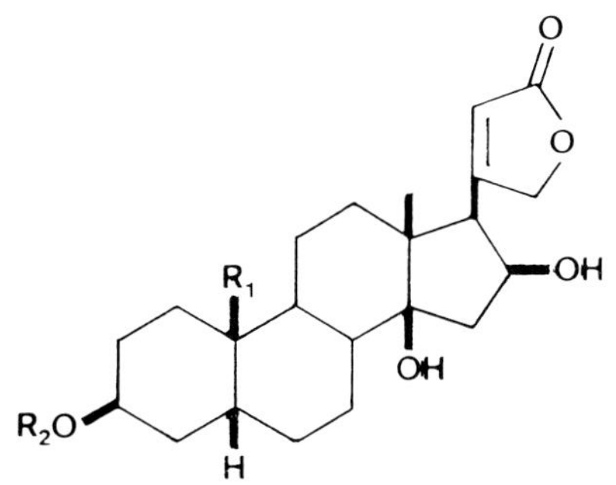

18a Adonitoxigenin R_1 = CHO, R_2 = H
18b Adonitoxin R_1 = CHO, R_2 = α-L-Rha
18c Adonitoxigenol R_1 = CH_2OH, R_2 = H
18d Adonitoxol R_1 = CH_2OH, R_2 = α-L-Rha
18e Gitoxin R_1 = CH_3, R_2 = $(\beta$-D-Dix$)_3$

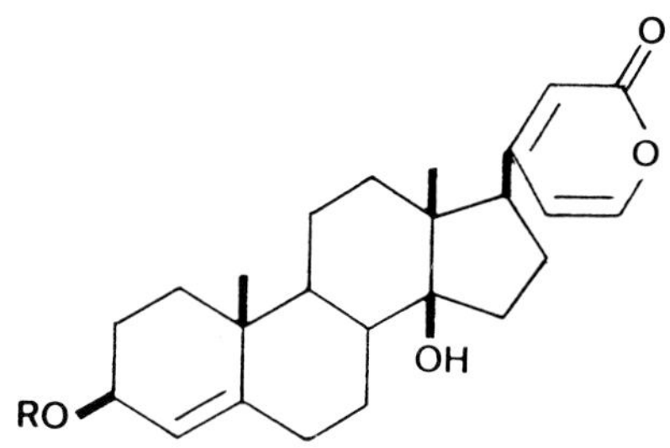

19 Proscillaridin R = α-L-Rha

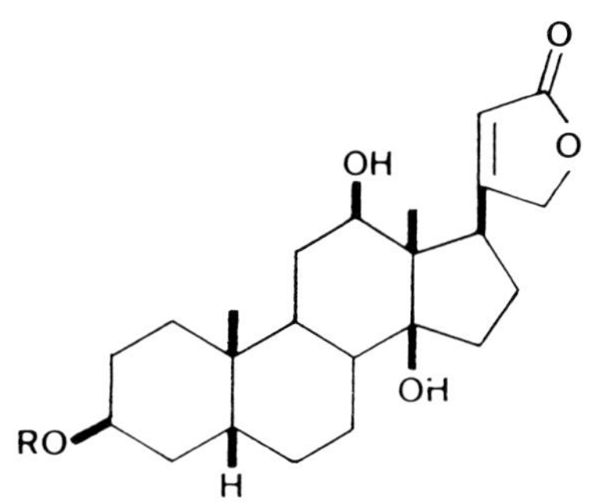

20a Digoxigenin R = H
20b Digoxin R = $(\beta$-D-Dix$)_3$
20c Deacetyl-lanatoside C R = $(\beta$-D-Dix$)_3$-β-D-Glu
20d Lanatoside C R = $(\beta$-D-Dix$)_2$-3‴-Acetyl-β-D-Dix-β-D-Glu
20e β-Methyldigoxin R = $(\beta$-D-Dix$)_2$-4‴-Methyl-β-D-Dix

TABLE 1
Cardenolide Content of *Digitalis In Vitro* Cultures[131]

Plant/strain	Material cultivated	Cardenolide content (µg/mg dry weight)
D. lanata		
Strain VII	Plantlets	0.15
Strain VII	Adventive embryos	0.1
Strain V	Adventive embryos	0.1
—	Shoots	0.1
D. purpurea	Shoots	0.05
D. grandiflora	Adventive embryos	0.05

selected *D. lanata* strain, no. 291, in air-lift bioreactors were able to convert considerable amounts of the glycoside almost quantitatively to the corresponding digoxin derivative, medigoxin or β-methyldigoxin (**20e**).[136] Immobilized cells have about half the activity of suspended cells, but can be used for more than 60 days.[135] Cryostorage of suspension cultures at the temperature of liquid nitrogen does not affect the capacity of the cultures subsequently produced to carry out the 12β-hydroxylation.[132] The commercial production of 12-hydroxy compounds using tissue cultures of *D. lanata* is currently being carried out in Germany.[58]

It is evident that the biotransformation of β-methyldigitoxin offers a greater prospect of exploitation, but it remains to be seen whether it can be applied with success to digitoxin itself. Be that as it may, medigoxin is an effective cardiac agent, which in the body is demethylated to digoxin. Its onset of action is rapid and its duration of action about the same as that of digoxin. It is, however, better absorbed and the effective dose is smaller.[66]

3. Chemical Modification of Digitalis Glycosides

Attempts have been made to design drugs that are as effective as the natural compounds but with a greater margin of safety.

Toxicity experiments with whole animals indicated that a 17β-unsaturated lactone, a 14β-hydroxyl group, and consequently a *cis* C/D ring junction and a *cis* A/B ring junction were essential for activity, with variations in the 3β-sugar moiety having at best a minor effect.

The introduction of improved *in vitro* test systems such as the isolated guinea pig left atrium for measuring inotropy and purified Na^+/K^+-ATPase for measuring relative potency, has allowed pharmacokinetic effects to be avoided and enabled the required structural parameters to be defined more precisely.

Replacement of the unsaturated lactone ring at C-17 in the cardenolide aglycone by several different unsaturated side chains[137-139] indicates that isosteric conjugated 4-centre π-electron systems, with the negative charge at the far end away from the steroid nucleus, likewise have inotropic activity.[58,138] This structural requirement is incorporated into the model of the *Digitalis* receptor discussed by Thomas et al.[140] in which the site of interaction between the receptor and cardiac glycoside lies in a cleft. Long-range interaction is initiated at the lactone or unsaturated end of the molecule, while Van der Waals and hydrophobic short-range interactions ensure binding of the steroid and sugar moieties, at the same time bringing about conformational rearrangement, and thereby causing inhibition of the Na^+/K^+-ATPase.

A 14β-hydroxyl group, hence a *cis* C/D-ring junction, is indeed essential for activity, but the introduction of other oxygen functions in the steroid part of the molecule tends to reduce potency, presumably through destabilization of the drug-receptor complex by reducing hydrophobic bonding. The glycosides are more potent than the aglycones, and variation in the sugar part of the molecule likewise affects the potency, again through changes in the bonding of the sugar(s) to the hydrophobic binding site. In spite of all the work carried out so far, it cannot be said that the structure-activity relationships are fully understood.

Randimbivololona et al.[125] have reviewed some of the work on semisynthetic cardiac glycosides related to those from *Digitalis*. By and large, polyformyl, mono-isopropylidene, methyl, acetyl, and halogen derivatives of digitoxin, digoxin, and gitoxin do not show any major improvement in properties over the parent compounds.

Digitoxigenin derivatives with sugar moieties having an amino function (**12j–l**) have attracted some attention, owing to their improved solubility properties.[125] In the isolated rabbit heart, ASI-222 (**12j**) has a greater inotropic effect than digitoxin and digoxin, and in the dog it increases the contractile effect at smaller doses than do these two glycosides. Inhibition of ATPase by ASI-222 and ASI-254 is ten times that of ASI-253 and digoxin. Moreover, ASI-222 and ASI-254 have a greater therapeutic index than digoxin (7.2 and 2.3, respectively).

SC-4453 or Plentil, a digitoxin derivative with a 17-(3′-pyridazyl) function (**12b**, but with **21** instead of the lactone ring at C-17), while having a positive inotropic effect comparable with that of digoxin, is more rapidly eliminated from the body.[125]

Compared with the parent compounds, inclusion complexes of digitoxin or digoxin with various cyclodextrins have better stability and higher solubility, as well as greatly improved dissolution rates, and given in this form may provide a new lease on life for these glycosides.[141,142]

V. LOWER TERPENES

Naturally occurring terpenes are a class of natural compounds which have become of increasing importance in recent years, largely due to development in instrumentation, particularly high-field nuclear magnetic resonance spectroscopy, which has facilitated their structural elucidation. Good examples are provided by the sesquiterpene derivatives, particularly the lactones. In 1960, the structures of barely a dozen naturally occurring compounds of this type had been fully elucidated, but by 1985 nearly 1000 compounds were known. The pace of their discovery is quickening all the time.[143] Most sesquiterpene lactones are found in the Asteraceae (Compositae), with certain genera, such as *Ambrosia*, *Artemisia*, and *Helenium*, having been particularly closely examined. Significantly, their occurrence harmonizes, in part at least, with divisions laid down by classical taxonomical methods, especially if the presence of *biosynthetic routes* rather than individual compounds is considered.[144]

These compounds are produced by cyclization of C_{15} precursors, principally *trans,trans*-farnesyl pyrophosphate. Because of the degree and location of their substitution, chiefly with olefins and their derived epoxides, further cyclizations and modifications lead to an impressive array of different skeletal types. Oxidation of one of the terminal methyl groups of the starting isopentenyl moiety to the corresponding carboxylic acid followed by lactonization with an adjacent hydroxyl leads to α-methyl-γ-lactones. Oxidation, either before or after lactonization, of the remaining methyl function gives rise to compounds containing the typical α-methylene-butyrolactones (**22**). It is this class of compound which has been shown to exhibit wide biological activity, some examples of which are discussed below.

The Asteraceae have a wide and varied history of use in traditional medicine and it is therefore not surprising that as more and more *in vitro* and *in vivo* methods of assessing biological activity become available, which mirrors the increase in the level of sophistication in the understanding of biological processes at the subcellular and even biochemical level, many plants, plant extracts, and isolated substances are the subject of biological evaluation.

Leaves and roots of the butterbur, *Penasites hybridus* (L.) P.Gaertner, B.Meyer, et Scherb., have been used since the Middle Ages for their anticonvulsive activity in asthma and for disturbances of the alimentary canal. As early as 1958, it was postulated[145] that the agents responsible were the angelate or the unusual sulfur-containing methylmercapto-acrylate esters of the eremophilane derivatives petasol (**23**) or its allylic isomer (**24**).

A. SESQUITERPENE LACTONES

Some sesquiterpene lactones have been shown to be antimicrobial, which may go part way to explaining their role in the plant. High concentrations of them are often found in association with glandular structures on the leaves. Components of the seeds and leaves of feverfew, *Tanacetum parthenium* (L.)

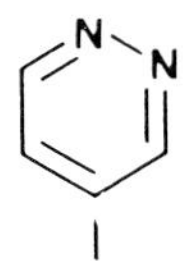

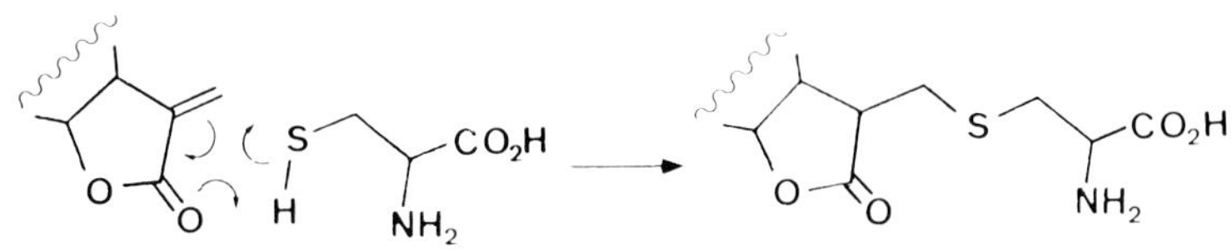

21 **22** Michael-type addition of α-methylene-γ-lactone with cysteine

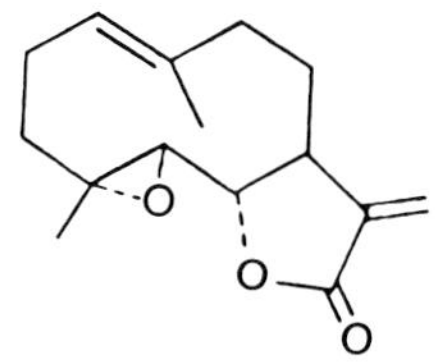

23 Petasol

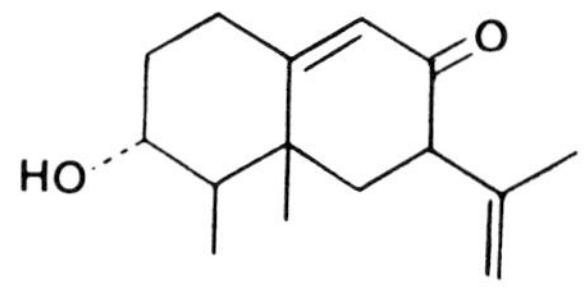

24 Petasol, allylic isomer

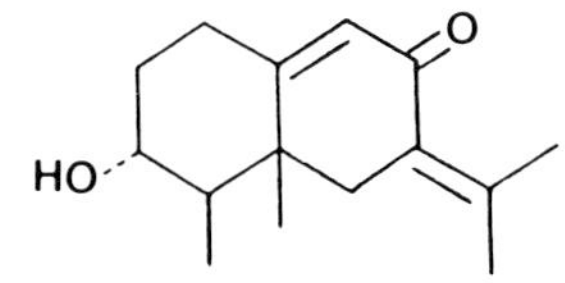

25 Parthenolide

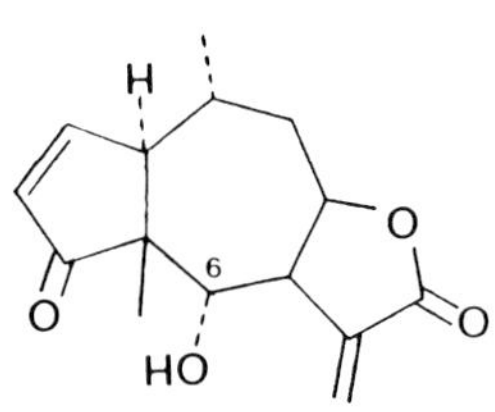

26 Helenalin

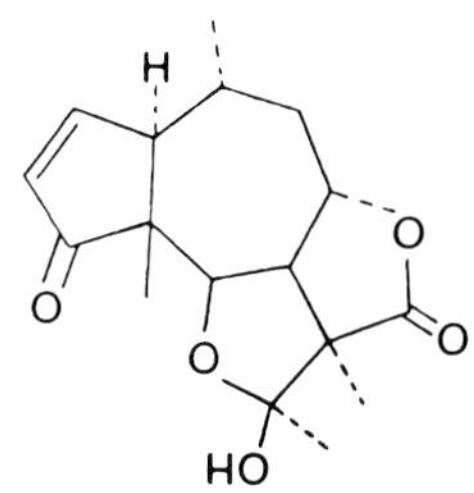

27 Tenulin

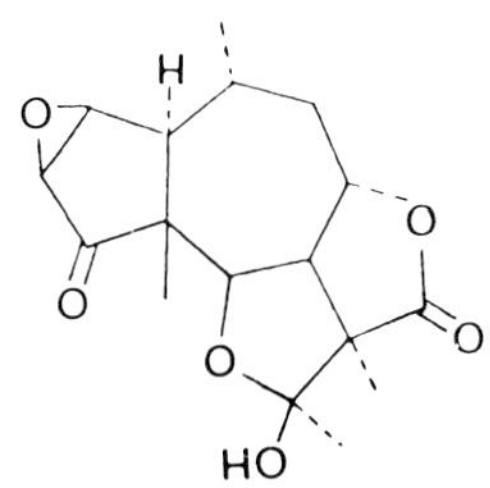

28 Tenulin 2,3-epoxide

Schultz-Bip., inhibit the growth of Gram-positive bacteria and other microorganisms. This has been related to the presence of the major sesquiterpene lactone present in the plant, the germacranolide parthenolide (**25**), and has led to the suggestion that the function of the glandular trichromes in the Asteraceae, usually rich in essential oil, may be as a protection against pathogenic bacteria, yeast, and filamentous fungi.[146]

More recently, Hall et al.[147] have reported on the antihyperlipidaemic activity of a number of guaianolides and germacranolides. The test was an *in vivo* study in mice. For example, helenalin (**26**), tenulin (**27**), and its 2,3-epoxide (**28**), deoxyelephantopin (**29**), and eupahyssopin (**30**), all at daily doses of 20 mg/kg, resulted in a lowering of the serum cholesterol level by more than 30% and that of serum triglycerides by 25%. It is interesting to note that Clofibrate, a commercially available drug indicated in hypercholesterolaemia and hypertriglyceridaemia, had no effect in rodents at this dose level. It has been suggested that the sesquiterpene lactones manifest their effect by an inhibition of the thiol-bearing enzymes responsible for lipid synthesis such as acetyl-coenzyme A and acetyl-coenzyme A synthetase.

1. *Matricaria chamomilla*

The anti-inflammatory and papaverine-type spasmolytic effects of German Chamomile, *Matricaria chamomilla* L., have been investigated.[148] The plant contains matricin (**31**), but on steam distillation (the traditional means of obtaining oil of chamomile) this suffers elimination and decarboxylation to yield chamazulene (**32**). The activity is not, however, confined to the sesquiterpene lactones. Spasmolytic activity has been associated with the unsaturated monocyclic sesquiterpene alcohol α-bisabolol (**33**),[149] of which ether and ester derivatives have been prepared which show enhanced activity,[149] and the flavonoid apigenin and its 7-glucoside.

Using the notoriously variable carrageenan-induced oedema and chronic adjuvant-induced arthritis screens, a number of sesquiterpene lactones have been examined.[150] The most potent anti-inflammatory agent of those tested was helenalin (**26**). Certain structural requirements for anti-inflammatory activity have been proposed[151] which correlate well with those derived from the very detailed work that has been carried out on cytotoxicity.[152]

This interest in cytotoxicity is not surprising in view of its potential application in the therapy of cancer, but it must be emphasized that the terms "anti-cancer" and "cytotoxic" are far from being synonymous and most compounds which perform well in screens for cytotoxicity fail in later trials because of their lack of selectivity for the abnormal cells. Nonetheless, a range of sesquiterpene lactones of diverse structures has been tested with varying success. It was, however, as early as 1943 that the first suggestion regarding the mode of action of similarly substituted compounds was made

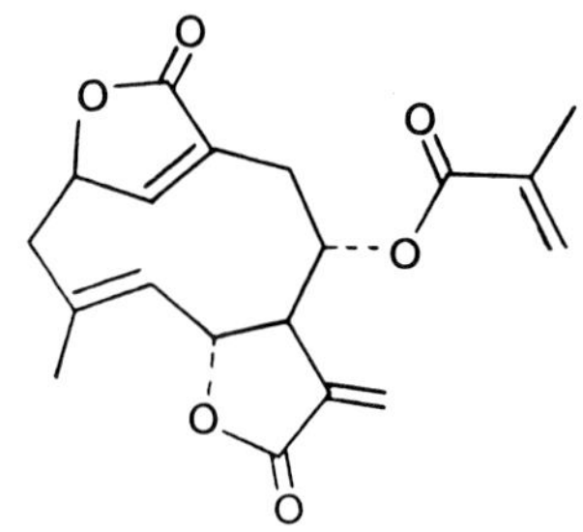

29 Deoxyelephantopin

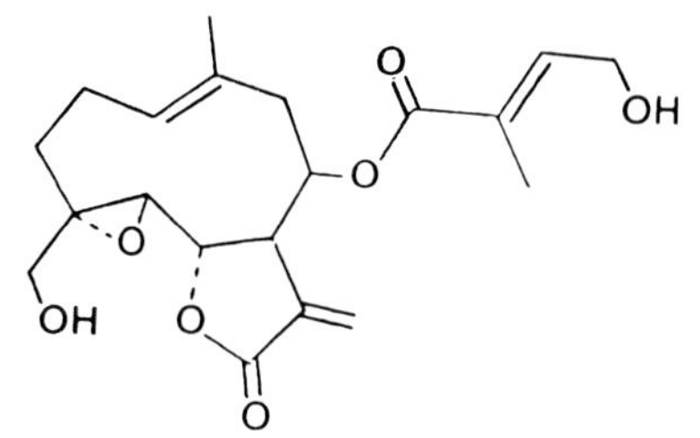

30 Eupahyssopin

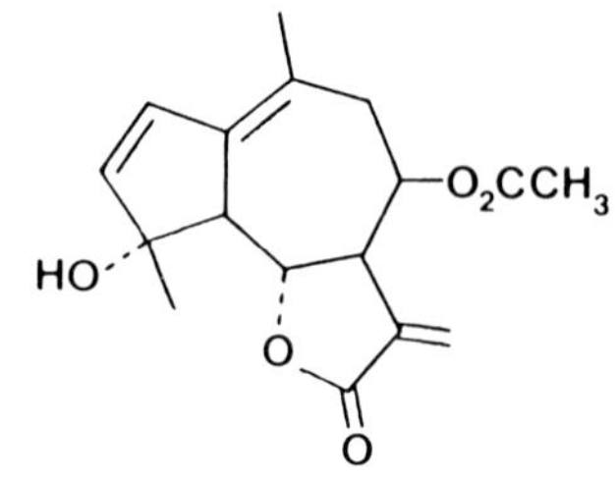

31 Matricin

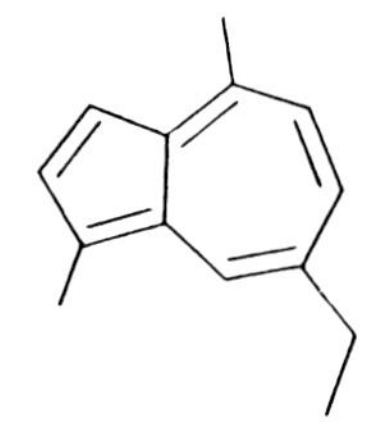

32 Chamazulene

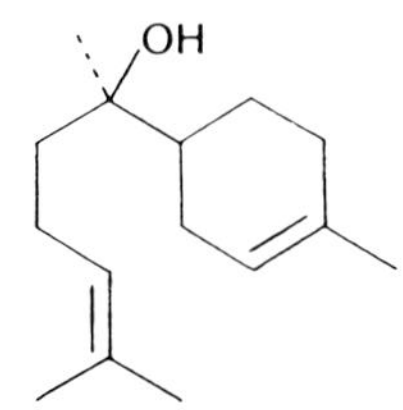

33 α-Bisabolol

by Medawar et al.[153] They demonstrated biological activity in unsaturated lactones. Subsequent studies seem to suggest that the potential anti-tumour actions of sesquiterpene lactones is due to the presence of the related α-methylene group on the γ-lactone, the precise location and potency of the action being determined by the presence of additional modifying groups such as epoxides, unsaturated esters, or ketones or other lactones. The demonstrated reactivity of unsaturated lactones toward thiols and amines suggests that perhaps all the demonstrated biological activities (including the cytotoxicity) results from irreversible alkylation of nucleophilic centres in enzymes, as exemplified by the Michael-type addition of α-methylene-γ-lactones with cysteine (cf., **22**).

a. Mode of Anti-Inflammatory Action

Hladon and Twardowski[154] have investigated the mode of action of some sesquiterpene lactones at a cellular level. At the (very low) subtoxic concentrations, they demonstrated arrest of division of the cell in interphase (in the G_1, S, or G_2 phases). At higher concentrations, complete and irreversible cytotoxic effects were seen: pykosis (chromosome condensation leading to degeneration of cell nuclei) and karyorrhexis (breaking of the nuclei with disintegration of the chromatin into shapeless granularity). Synthesis of protein and RNA was inhibited, but that of DNA was apparently not affected.

More recently, attempts have been made to explain the biological activity, particularly the anti-inflammatory actions, at a subcellular level.[151] It is worth considering some of this work in more detail because it illustrates well recent innovations.

Hall et al.[150] in 1979 compared the action of known anti-inflammatory agents such as ethacrynic acid and *N*-ethylmaleimide with sesquiterpene lactones, since they both bind sulfhydryl groups in the manner previously described. Substantial generalized anti-inflammatory actions were shown by several compounds when assessed by a number of anti-arthritis, anti-pleurisy, and related screening procedures in rodents. There was no doubt about the importance of the α-methylene-γ-lactone moiety. The most active compound was again helenalin (**26**), which contains an additional cyclopentenone ring. Significantly, the latter is β-unsubstituted in this compound and so can undergo a second Michael-type 1,4-addition. At a more subtle level, other groups may modify the overall effect. Thus, the potent helenalin contains a 6-hydroxy group while its analogous esters showed dramatically reduced anti-inflammatory activity, indicating that peripheral functions may play an important function in receptor binding. Conversely, the presence of epoxides occasionally improves activity.

In the same study,[150] of great relevance in the development of *therapeutically useful drugs* as opposed to the demonstration of activity in laboratory tests, adverse reactions (CNS, ulcerogenic, and haematological studies assessed by standard means) were also monitored. No deleterious side effects in rodents at the doses employed, usually 2.5 mg/kg/day, were noted.

In further work, in an attempt to elucidate the mode of action of the anti-inflammatory effects, Hall et al.[151] studied oxidative phosphorylation in rat and mouse liver cells or human polymorphonuclear neutrophils, lysosomal hydrolytic enzymatic activity in rodent liver preparations, cyclic adenosine monophosphate levels in isolated rat polymorphonuclear neutrophils, and prostaglandin synthetase activity (in commercial bovine semen preparations). Several sesquiterpene lactones affected a number of biochemical parameters which influence the inflammation process. The results showed that the mode of action of sesquiterpene lactones as anti-inflammatory agents resembled that of currently used drugs in that they were potent inhibitors of neutrophil migration, lysosomal rupture, enzymatic activity, and prostaglandin synthesis, which was linked to elevated cyclic adenosine monophosphate levels.

b. Mechanism of Spasmolytic Action

More detailed understanding of the physiology of the mechanisms of muscle contraction has allowed differentiation between the modes of action of the varied constituents of a number of plant extracts. It was previously mentioned that chamomile shows spasmolytic activity due, among other compounds, to α-bisabolol and apigenin. Using a recently developed technique for the specific evaluation of calcium antagonism based on the depolarization-induced contraction of rabbit aortic rings, it has been possible to differentiate between the types of activity of apigenin, a flavonoid which shows no activity in the test, and bisabolol, which was a potent inhibitor of KCl-induced contraction, thus showing it, but not apigenin, to be a calcium antagonist.[155]

2. *Tanacetum parthenium*

Another interesting instance is provided by consideration of recent biochemical experiments with extracts of feverfew, *Tanacetum parthenium*. This plant has gained considerable publicity in the U.K. over the last few years as a self-medication in treating migraine, and the results of two trials have been published which demonstrated a significant level of clinical effectiveness.[156,268] Importantly, preliminary clinical toxicological evaluation (using sister chromatid exchanges in lymphocytes and the Ames Salmonella mutagenicity test on urine) has not revealed any significant adverse reactions to the clinical consumption of feverfew in the prophylaxis of migraine.[269] The plant in known to contain a number of sesquiterpene lactones of the germacranolide, guaianolide, and eudesmanolide classes, some of which are very complex[157] and all of which are probably derived from costunolide. The relatively simple epoxycostunolide or parthenolide (**23**) is the major lactone of the plant, being present in the rather high concentrations of 0.4 to 1.0%.

Using a nonspecific spasmolytic screen in the isolated guinea pig ileum preparation, some extracts of the plant showed complete, irreversible inhibition of effects due to acetylcholine, histamine, 5-hydroxytryptamine, bradykinin, and prostaglandin E_2, all agonists implicated in the aetiology of migraine.[157] These fractions were associated with the presence of high levels of sesquiterpene lactones. This directed approach allowed the isolation of a number of new active derivatives of parthenolide. Subsequently, extracts of the plant proved to inhibit prostaglandin biosynthesis, not at the level of cyclooxygenation (as do many other nonsteroidal anti-inflammatory drugs such as aspirin),[158] but rather at the level of phospholipase A_2.[159]

Very recent work has shown that in addition to this, feverfew extracts inhibit secretory activity in blood platelets and polymorphonuclear leukocytes. It also prevents the release of vitamin B_{12}-binding protein from polymorphonuclear leukocytes under the influence of certain secretagogues,[160] and it has been suggested that perhaps the materials are acting on protein kinase C.[160] Interestingly, the pattern of inhibitory effects differs from that observed with other inhibitors of platelet aggregation and the effect on polymorphonuclear leukocytes is more pronounced than with high levels of certain commercially used nonsteroidal anti-inflammatory agents.[160,270] With its anti-platelet action, it is possible that feverfew may also have anti-thrombotic potential. These observations are leading to the development, even if not of a therapeutically useful drug, of a pharmacological tool with a unique mode and site of action. It is appropriate to mention here that these discoveries were only possible with a rapidly increased understanding of the biochemistry itself.

3. *Artemisia annua*

An example of a plant material which has already entered the therapeutic armamentarium and at the same time provided a completely novel class of therapeutic agent is a sesquiterpene lactone derived from the Chinese drug *ginghao, Artemisia annua* L.[161]

As part of a search for effective antimalarials, water and alcohol extracts of the herb were examined for antimalarial activity, but despite the long history of the plant being used in the alleviation of the chills and fever associated with malaria, curative properties of the plant could not be confirmed. It was not until *ether* extracts were re-tested *in vivo* using mice infected with *Plasmodium berghei* and monkeys infected with *P. cynomolgi* that good activity was indeed demonstrated. The active material was eventually isolated and its structure determined as (**34**), known as *ginghaosu* or artemisinin.[162] It is an unusual sesquiterpene in being not only a peroxide but also possessing a 6-membered ring lactone which is an α-methyl-δ-lactone. Its activity is thus expected to be more specific than the generalised cytotoxicity previously described for the α-methylene-γ-lactones, and this is found to be the case. Reduction of the peroxide removes activity, however, and since this is apparently readily achieved by reluxing in alcohol, it goes some way to explaining the lack of activity of the early ethanol extracts.[162]

A multidisciplinary approach was used in the evaluation of this Chinese medicinal plant and a coordinating research group was set up for its authentication. Considerable pharmacological, metabolic, toxicological, and clinical results have been obtained, and it is reported as being effective against the chloroquine-resistant *P. falciparum* and cerebral malaria.[161] Because of its unique structure, it has provided an entry into a new class of therapeutically useful drugs and well illustrates the important role of plants in providing just such materials. Thus, artemisinin has served as the template for the synthesis of a number of derivatives having improved action[162] and it has also recently been synthesised.[163] Detailed structure-activity studies are underway in an attempt to prepare more readily synthesizable potent materials.[162]

B. DITERPENES

From the foregoing discussion it is becoming increasingly clear that the biological effects of sesquiterpene lactones may be due to generalized cytotoxic actions rather than specific ones which might be exploited in the development of new therapeutic agents. However, as knowledge of the details of the biochemical transformations which control the activities of cells and of the underlying and controlling biochemistry is increasing, so novel natural compounds which have a more or less specific action on a particular enzyme or group of enzymes are becoming more important as *biological tools*. These substances are allowing more rigorous interpretation not only of the growth and control of tumour cells, but also of the mechanisms by which *normal* cell growth is governed.

Good examples are provided by some natural diterpenes. Thus, esters of polyhydric alcohols, such as phorbol (**35**), have been used in studies of tumour development as well as for investigation of other biological processes.[164]

1. *Coleus forskohlii*

Most recently, a plant, *Coleus forskohlii* (Willd.) Briq. (Lamiaceae (Labiatae)), used in Ayurvedic medicine for heart diseases and spasmodic pain, among other conditions, has been investigated. A novel labdane-type diterpene named forskolin (**36**) has been isolated and shown to have potent cardiovascular effects,[165,166] activating cyclic adenosine monophosphate-generating systems in a number of mammalian tissues in a rapid and reversible fashion. This has been interpreted in terms of the compound inducing an increase in the Ca^{2+}-uptake of heart-muscle cells and lowering the sensitivity of the smooth muscle contractile systems to Ca^{2+}.[167] Further studies indicate that forskolin activates adenyl cyclase in the intact cell,[168] accounting for its dramatic effect of increasing acid secretion and pepsinogen formation in isolated gastric glands.[169] Some structure-activity relationships have been determined.[170] Its potential as an agent to prevent cancer metastasis is demonstrated by work which has shown that forskolin strongly inhibits the aggregation of human platelets induced by melanoma cells.[171] An *in vivo* study involving pre-injection of mice with forskolin (82 μg per mouse), followed by tail-vein injection of cultured B16-F10 cells, reduced tumour colonization in the lungs by more than 70%.[171] In view of its ability to raise cyclic adenosine monophosphate levels, suggestions have been made that forskolin may therefore be of value as a bronchodilator and anti-allergy agent[172] and as an antiglaucoma agent.[173]

C. OTHER TERPENES

1. Pyrethrum

Consideration of structure-activity relationships exhibited by the ester constituents of pyrethrum flowers, from *Tanacetum cinerariaefolium* (Trev.) Schultz-Bip. has played an important role in the rational development of more potent insecticides. Pyrethrum dust and extracts have been used as insecticides for many years, their principal advantages being a spectacularly rapid "knock-down" effect on flying insects coupled with low mammalian toxicity. The major active ingredients are the natural pyrethroids, including pyrethrins I (**37a**) and II (**37b**), jasmolins I (**37c**), and II (**37d**), and cinerins I (**37e**), and II (**37f**). They are not sesquiterpene lactones but closely related, highly oxidized esters of simpler terpenic acids and alcohols, the acids being characterized by the presence of a cyclopropane ring. They are, however, unstable compounds, particularly to air and light. This makes them desirable from the environmental point of view (because of their low persistence), but at the same time and for the same reason of little value in long-term crop protection. Intensive efforts to improve this situation by increasing their stability and hence persistence, while maintaining rapid activity and low toxicity to warm-blooded animals, have included detailed synthetic studies which demonstrated that among the stereochemical requirements for the highest activity is the presence of the (+)-*trans* derivative of the acid residue.[174]

Preparation of synthetic ester derivatives of chrysanthemic acid has resulted in major successes, e.g., the compound bioresmethrin (**38**) produced by esterifying chrysanthemic acid with 5-benzylfuryl-3-methanol. It has a toxicity to houseflies about 50 times greater than that of the natural pyrethrins. When a related chlorinated derivative of chrysanthemic acid was used to prepare a series of esters, there was a dramatic increase in resistance to light and oxygen degradation. Thus, the ester of the comparatively simple 3-phenoxybenzyl alcohol with the dichlorovinyl analogue of chrysanthemic acid, biopermethrin (**39**), although more expensive than the original, is a remarkably improved material.[175] Analogues without the cyclopropane ring, produced later, still have some activity but increased mammalian toxicity.[176] The most successful one, deltamethrin (**40**), is brominated (rather then chlorinated) for improved light stability and has as its alcohol portion a nitrile-substituted 3-phenoxybenzyl alcohol.[177]

VI. γ-LINOLENIC ACID

The oils from plant seeds are largely composed of a mixture of triglycerides, compounds comprising glyceryl esters of long-chain aliphatic acids. These acids, usually known as fatty acids, vary according to chain length and presence and position of double bonds. In some plants the acids are nearly all unsaturated compounds.

γ-Linolenic acid (GLA), 6,9,12-octadecatienoic acid (**41**), is one of the group of naturally occurring fatty acids known as essential fatty acids (EFA) since they constitute a necessary ingredient in the diet (about 1 g/day) and a deficiency in intake seems to be linked with various disease states — skin rashes, lack of wound healing, irritability, and an increase in calorie intake are common effects in animals and man.

GLA is classed in the n6 series of EFAs, since its first double bond is at C-6. The other important group of EFAs is the n3 series. The importance of the EFAs lies in their roles as constituents of cell membranes and as precursors.[178] Figure 1 shows the biochemical pathways by which EFAs are transformed to prostaglandins (PG) and leukotrienes. These substances have a variety of functions and effects ranging from highly desirable to harmful.[179] The series 1 prostaglandins, e.g., PGE_1 (**42**), have generally desirable effects and are derived from GLA. PGE_1 activates the formation of cyclic adenosine monophosphate (cAMP), inhibits platelet aggregation, is a vasodilator, an inhibitor of inflammation reactions, and activates T-lymphocytes. On the other hand, the arachidonic acid (AA) derivatives, the leukotrienes and series 2 prostaglandins have less desirable effects as they are very pro-inflammatory and some promote vasospasm and thrombosis.

In humans most of the DGLA formed from GLA is metabolized to the beneficial PG 1 series compounds, the conversion of DGLA to AA being very slow. Most of the AA in humans is obtained from meat in the diet, while the GLA is obtained

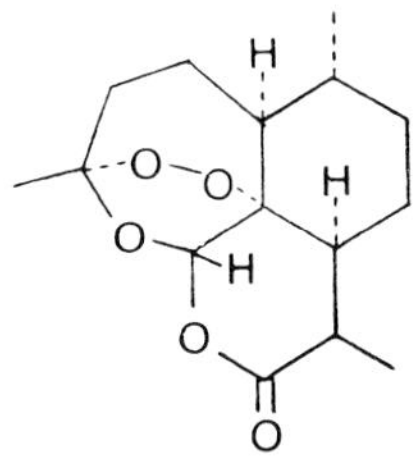

34 *Qinghaosu* or Artemisinin

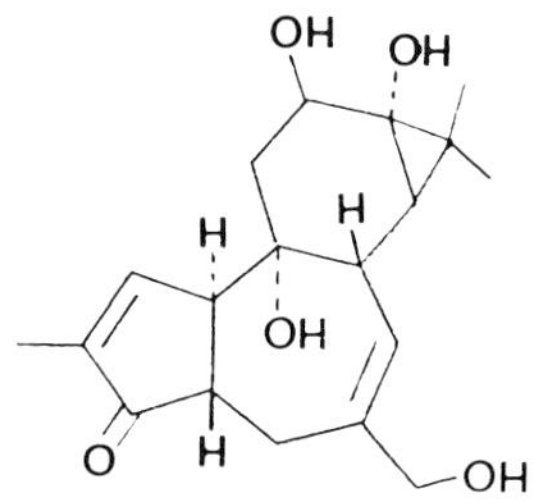

35 Phorbol

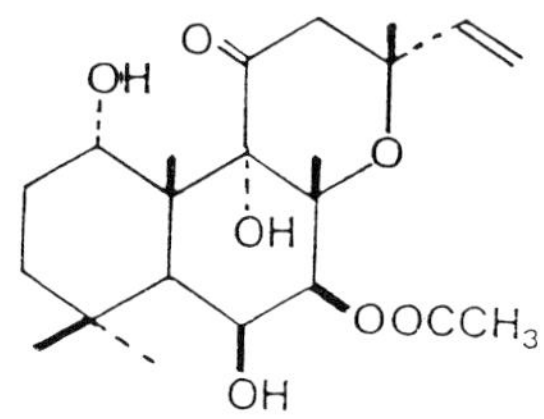

36 Forskolin

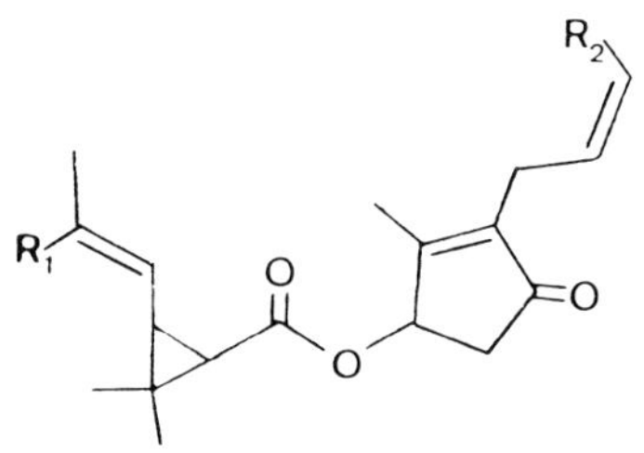

37a Pyrethrin I $R_1 = CH_3$, $R_2 = CH{=}CH_2$
37b Pyrethrin II $R_1 = CO_2CH_3$, $R_2 = CH{=}CH_2$
37c Jasmolin I $R_1 = CH_3$, $R_2 = CH_2CH_3$
37d Jasmolin II $R_1 = CO_2CH_3$, $R_2 = CH_2CH_3$
37e Cinerin I $R_1 = R_2 = CH_3$
37f Cinerin II $R_1 = CO_2CH_3$, $R_2 = CH_3$

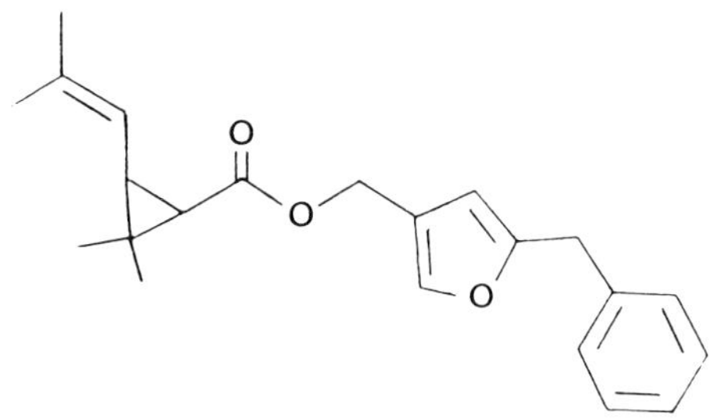

38 Bioresmethrin

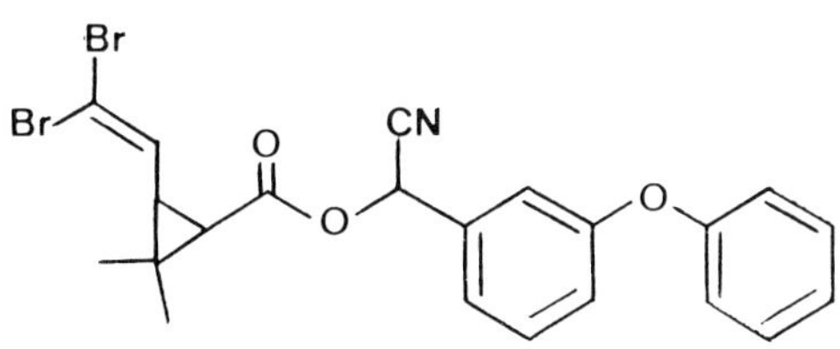

40 Deltamethrin

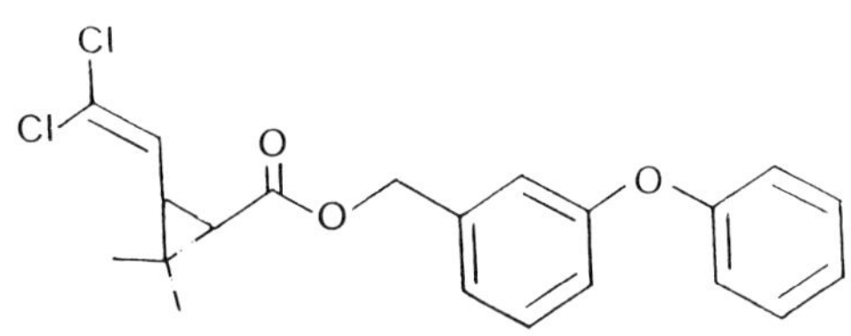

39 Biopermethrin

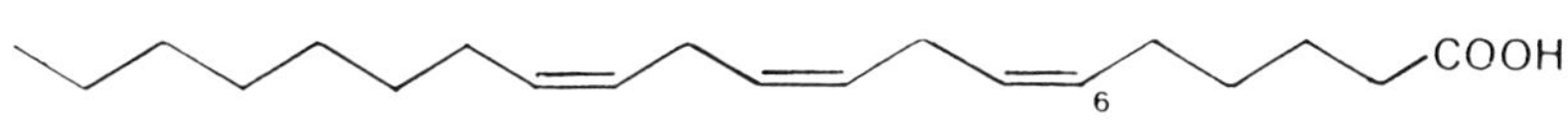

41 γ-Linolenic acid (GLA) or 6,9,12-Octadecatrienoic acid

from vegetable oils in the diet. It should be noted, however, that very few plants contain any GLA and that it is produced in the body by the action of Δ^6-desaturase on the much more common linoleic acid. The level of Δ^6-desaturase is found to be low in some individuals, which is possibly linked with aging or genetic factors. The activity of Δ^6-desaturase is inhibited in some disease states as a result of dietary substances such as saturated or *trans*-unsaturated fats and also by some drugs, e.g., glucocorticoids.[179] If the enzyme is either inhibited or lacking, the net effect is a shortage of GLA production and consequent lack of the beneficial series 1 PGs. This lack may also be brought about by a shortage of EFAs in the diet, i.e., high meat content, low vegetable content, even if sufficient enzyme is present. In recent years trials have been designed on the basis that symptoms of various diseases are caused by such a lack of PGs and thus alleviation should be observed if direct GLA intake is boosted by incorporating it in the diet.

A. SOURCES OF GLA

As mentioned above, GLA is not a common constituent of the diet. It is found in human milk in a greater amount than some other animal milks, but the main commercial source for the pure compound is from the limited number of plants where it is found in reasonable quantities in the seed oil. The major source has been the evening primrose *Oenothera biennis* L. (Onagraceae), which contains about 9% GLA in its oil. Recent reports have indicated richer sources of GLA in the seed oils of members of the Boraginaceae, especially *Nonnea macrosperma* Boiss. et Heldr. (5.1% total weight of seed), Scrophulariaceae, e.g., *Scrophularia marilandica* L. (37% oil in the seed, of which 9% is GLA),[180] and Grossulariaceae, where the

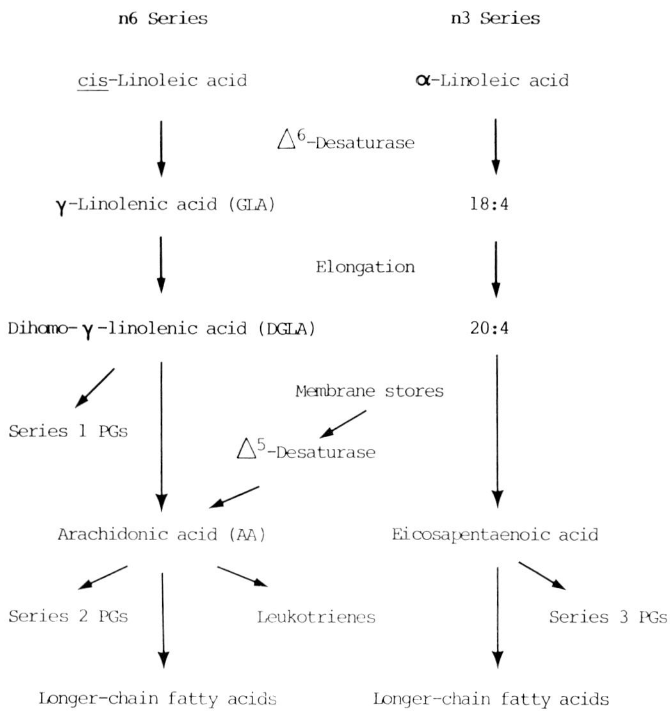

Figure 1. Biochemical pathways involving essential fatty acids (EFAs).

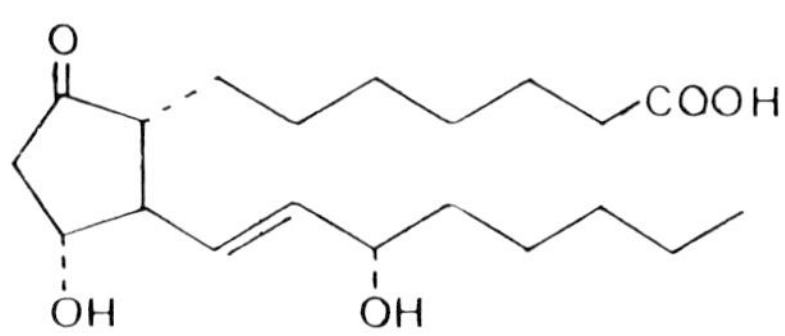

42 Prostaglandin E_1

Blackcurrant, *Ribes nigrum* L., has been found to contain 19% total weight of GLA in its seeds.[181]

Although these yields from seeds are promising for commercial exploitation, it should be noted that in Japan the production of GLA by fermentation culture of the fungus *Mortierella* is being examined.[182] This process gives yields of 50 to 90%.

B. POTENTIAL CLINICAL USES OF GLA

Two recent reviews indicate the growing interest in the use of evening primrose oil for treating a variety of conditions.[261,271]

1. Coronary Heart Disease

Coronary heart disease is linked with several factors such as high serum cholesterol levels, high blood pressure, platelet aggregation, and vascular obstruction.

An increase in dietary unsaturated fat intake at the expense of saturated fats lowers the serum cholesterol levels, but very high quantities are necessary to bring about significant reductions. Relatively low amounts of evening primrose oil have been found to exert a much greater effect, implying that GLA may play an important part in the process.[178]

EFAs lower blood pressure and GLA inhibits platelet aggregation.[183,184] The latter effect is thought to be due to the enhanced levels of PGE_1 formed from the GLA. The PGE_1 is also a peripheral vasodilator, and increased levels from an intake of GLA is of benefit in treating angina and vascular spasms due to Raynaud's syndrome.[178]

2. Diabetes

Diabetes is associated with inhibition of Δ^6-desaturase, thus leading to deficiency of GLA and series 1 PGs. PGE_1 has insulin-like activities and also potentiates the effects of insulin, so a shortage of this prostaglandin lessens the effect of insulin given in therapy. Supplementation of PGE_1 levels by administration of GLA should improve this situation. Clinical trials using GLA-rich oils are in their infancy, but encouraging results have been obtained in trials using generally EFA-rich oils in diabetes.[185]

3. Breast and Menstrual Cycle Disorders

Epidemiological studies imply a link between a high saturated fat intake and disorders in the breast and in the menstrual cycle. It is possible that this is due to inhibition of the Δ^6-desaturase by the saturated fats. Trials show that GLA supplementation in the form of evening primrose oil attenuates mastalgia, breast tenderness, cyclical nodularity, and premenstrual syndrome.[186] Evening primrose oil is now marketed as a treatment for premenstrual syndrome.

4. Inflammation and the Immune Response

Inflammation is accompanied by elevated levels of free AA and its metabolites, the series 2 PGs, and the leukotrienes. The common anti-inflammatory drugs act by interference with the release of these compounds. PGE_1, however, may be a potent anti-inflammatory, since it raises cAMP levels, thereby inhibiting the release of AA from stores and activating T-lymphocytes. Increase in PGE_1 levels by extra GLA intake may thus reduce inflammation. This has been demonstrated in studies on animals and also in the clinic.[178,187]

Atopic eczema has been seen to be associated with low levels of EFAs in the diet. Trials have shown the usefulness of evening primrose oil in treating this condition and it has recently become classified as a prescribable medicine in the U.K.[188,272] A theoretical basis for its effectiveness is related to the action of PGE_1 in stimulating T-lymphocyte cells which are defective in cases of atopic eczema. Extra PGE_1 supplied by the GLA in evening primrose oil would cause the necessary stimulation of the cells.

5. Multiple Sclerosis

There is an epidemiological connection between the incidence of multiple sclerosis and a high consumption of saturated fats. Several trials have been carried out on the use of high levels of unsaturated fats, e.g., sunflower oil, in treating the condition, but with unconvincing results. Recent considerations involving the role of PGE_1 and T-lymphocytes have led to the suggestion that GLA-rich oils might be a more useful treatment.[178,189] No trials have been reported so far.

6. Alcoholism

Alcohol stimulates the formation of PGE_1 from DGLA and

thus depletes its stores. Subsequent levels of PGE_1 are therefore considerably reduced. In addition, alcohol blocks Δ^6-desaturase and thus further contributes to low levels of the series 1 PGs. Replacement therapy with GLA might therefore counteract the effects of excessive alcohol intake due to chronic PGE_1 shortage.[178]

7. Hyperactivity

Hyperactivity in children is often associated with abnormal immune function, known as atopy. Some of the characteristics of atopy, e.g., increased water permeability of the skin and eczema, are the same as those shown in conditions of EFA deficiency. The recent suggestions of a link between hyperactivity and some food additives are noteworthy, since some of the dyes implicated are known to block the conversion of EFAs to PGs. Current trials in which hyperactive children are being treated with GLA-rich oil show encouraging results.[178]

GLA is being investigated as a therapeutic agent in several other areas,[260,261] its importance being closely linked with its role as a precursor of the series 1 PGs. This role has been discussed more fully, particularly as regards interactions with anti-inflammatory drugs.[190] Continued investigation into the links between disease states, EFA and PG levels is likely to reveal more potential uses for GLA as a replacement therapy. More clinical trials are needed to establish the validity of the theories thus far put forward and to consolidate the encouraging preliminary evidence.

VII. IMMUNOSTIMULANTS FROM HIGHER PLANTS

Immunostimulants, also called mitogens, are substances that lead to a nonspecific, usually short-lived, stimulation of the immune system. The main effect of immunostimulants is to cause increased phagocytosis by increased activity of the macrophages and granulocytes, the effect on the lymphocytes being secondary. The "memory" of the immune system is not generally affected, so the immunostimulants have to be administered frequently.

Several of the investigations mentioned below have shown that the use of some plants in folk medicine as "tonics" may be explained in terms of the immunostimulating compounds which they contain. Over the last few years the realised and potential importance of immunology in medicine has attracted much interest; the availability and development of immunostimulants could add to this, particularly from a prophylactic point of view. Some synthetic substances are known to be immunostimulants but a large variety of naturally occurring compounds also has this activity and the most important ones are dealt with below. Reviews on this topic have recently been published.[191-193,273]

The discovery of many immunostimulants over the last few years has been due to the development of systems for testing for this activity. The three most important methods are the carbon clearance test,[194] the granulocyte test,[195] and the chemiluminescence test.[196,197] These test systems have been used for both plant extracts and isolated products. It is interesting to note that many of the pure substances tested appear to have an optimum concentration for immunostimulant activity;[198] little stimulation occurs if the concentration is too high as well as if it is too low. The optimum concentration is often quite small, on the order of $M \times 10^{-3}$.

A. HIGH-MOLECULAR WEIGHT COMPOUNDS

1. Peptides and Related Compounds

Several members of the group of peptides known as lectins have been shown to stimulate T-lymphocytes, but they also have serious side effects, which precludes much in the way of development as therapeutic agents.[192] The following lectins from flowering plants have been shown to have an immunostimulating effect: concanavalin A from *Canavalia ensiformis* (L.) DC., phytohaemagglutinin (PHA) from *Phaseolus vulgaris* L., lentil lectin from *Lens culinaris* Med., abrin from *Abrus precatorius* L., and pokeweed mitogen (PWM) from *Phytolacca americana* L.[191]

2. Polysaccharides

The aqueous extracts of a large number of plants used in traditional medicine as "tonics" have shown enhancement of phagocytosis when subjected to the tests indicated above.[192,193,196] Previous phytochemical analysis of many of these plants had revealed little to support their reputation. Screening of extracts and fractions obtained from them for immunostimulant activity has resulted in the identification of complex polysaccharides as the active constituents of several drugs. These active polysaccharides have been found to be branched, water-soluble, acidic heteroglycans ranging in molecular weight from 25,000 to 500,000+. Table 2 lists those plants shown to contain immunostimulant polysaccharides.

The detailed structure of two of these polysaccharides from *Echinacea purpurea* Moench. has been elucidated.[198] Polysaccharide 1 (**43**) has a molecular weight of 35,000 and consists of a 1→4-linked xylose backbone with various other sugars, in the ratio shown, attached at each fifth xylose unit. Polysaccharide 2 (**44**) has a molecular weight of 450,000 and consists of a backbone of alternate rhamnose and galactose units with 1→4 linked side chains at every other rhamnose unit. The side chains consist of mixtures of arabinose, rhamnose, galactose, and glucuronic acid. This compound resembles the polysaccharide constituents of some gums and mucilages.

Although their structural details have not been elucidate, the relative composition of sugars in some other immunostimulant polysaccharides has been measured,[198] notably those from *Eupatorium perfoliatum* L., *Acanthopanax senticosus* (Rupr. et Maxim.) Harms, and *Chamomilla recutita* (L.) Rauscher., which all have 1→4 linked xylose chains as the core of the molecule.

The complex and anionic structure of these compounds enables them to have a higher binding affinity to membranes of immunocompetent cells as compared with simple glucan chains. The degree of complexity means that there are more immunodeterminant regions in the molecule.

At present there are difficulties in standardizing extraction and isolation procedures for this group of compounds, but in view of their apparent lack of toxicity they may well furnish useful agents in the future.

TABLE 2
Plants Containing Immunostimulant Polysaccharides

Plant	Part used
Echinacea purpurea (L). Moench.	Herb, root[198]
E. angustifolia DC.	Herb, root[198]
Eupatorium cannabium L.	Herb[198]
E. perfoliatum L.	Herb[198]
Chamomilla recutita (L). Rauscher	Flower[198]
Arnica montana L.	Flower[198]
Calendula officinalis L.	Flower[198]
Achyrocline satureoides DC.	Herb[198]
Sabal serrulata Roem. et Schult.	Fruit[198]
Eleutherococcus senticosus Maxim.	Root[198]
Viscum album L.	Fruit[199]
Astragalus gummifer Labill.	Gum[198]
Angelica acutiloba Kitagawa	Root[200]
Carthamus tinctorius L.	Flower[201]
Baptisia tinctoria (L.) R.Br.	Root[198,202]

B. LOW-MOLECULAR WEIGHT COMPOUNDS

A wide variety of small molecules have been shown to have some degree of immunostimulant activity. Those of major importance which are constituents of higher plants are discussed below.

1. Sesquiterpenes

Zexbrevin A (**45**), from a Mexican member of the Asteraceae, *Zexmenia brevifolia* A.Gray, exhibits immunostimulant activity,[203] as does another Asteraceous sesquiterpene, alantolactone (**46**), from *Inula helenium* L.[204]

2. Diterpenes

In the last few years there has been interest in forskolin (**36**), from *Coleus forskohlii*, because of its pharmacological properties (see Section V.B.1). A recent report describes the immunostimulant effect of this substance.[205]

3. Triterpenes

Although triterpenes have a vast array of biological activities (cf. Section III), no member of this group has been shown to exhibit immunostimulant activity on its own. However, there are reports that extracts of Ginseng, from *Panax ginseng* C.A.Meyer (Araliaceae), a triterpene-containing plant, enhance various humoral and cell-mediated immune responses in animals.[206]

4. Alkaloids and Related Compounds

Aristolochic acid (**47**), from *Aristolochia* species (Aristolochiaceae), has been known since 1963 to stimulate phagocytosis.[207] Its activity has been demonstrated in a variety of systems both *in vivo* (including man) and *in vitro*. A recent *in vitro* test indicated considerable immunostimulant activity with concentrations as low as 10^{-2} mg/ml.[208]

The oxindole alkaloids isopteropodine (**48a**), pteropodine (**49**), isomitraphylline (**48b**), and isorhynchophylline (**50**) are found in some species of *Mitragyna* and *Uncaria* (Rubiaceae). In addition to the other known biological activities of these compounds, their enhancement of phagocytosis has recently been reported.[196,208]

Some alkaloids show a stimulation of phagocytosis but are also cytotoxic. The isoquinoline alkaloid emetine (**51**) is an example of this.[196]

The benzylisoquinoline alkaloids berberine (**52**) and coralyne (**53**) have also been shown to be potent macrophage activators when tested *in vitro* with tumour cells.[209] Workers in Japan have isolated a bisbenzylisoquinoline alkaloid, cepharanthine (**54**), from *Stephania cepharantha* Hayata (Menispermaceae). It stimulates T-lymphocytes and thus enhances the immune defences.[210]

Another alkaloid, 1,2-β-epoxyambelline (**55**), from *Crinum latifolium* L. (Liliaceae), has recently been found to have an immunostimulant activity comparable to that of concanavalin A.[211]

5. Phenolic Substances

There is some indication that certain flavonoids may promote the production of antibodies.[191] The procyanidin (+)-3-cyanidol (**7a**) and its methoxy derivative have been shown to enhance the induced synthesis of immunoglobulins in T- and B-lymphocytes.[212] A related type of phenolic compound showing immunostimulant activity is catechin and its gallocatechin derivatives. Investigations on extracts from the bark of *Okoubaka aubrevillei* Pellegrin et Normand (Octoknemaceae), a West African tree, showed that (-)-*epi*gallocatechin gallate (**56**) had a high activity, as did the gallic and protocatechuic acids present.[213]

Plumbagin (**57**), from *Plumbago* species (Plumbaginaceae), at very low concentrations causes an increase in phagocytosis *in vitro*, although at higher concentrations it is cytotoxic.[196] Cleistanthin (**58**) from *Cleistanthus collinus* (Roxb.) Benth. (Euphorbiaceae) induces neutrophilic granulocytosis and can therefore be classed as an immunostimulant.[214]

There are thus a variety of compounds which cause stimulation of the immune system at some level. Extracts from several plants have exhibited activity but the compounds responsible have not yet been isolated. It is especially noteworthy that plants with a history of use in folk medicine are now being shown to be immunostimulant, and it is quite possible that the immunological approach to treatment, a somewhat modern concept, may be given extra impetus through the development of immunostimulant substances from plants.

VIII. ANTIHEPATOTOXIC AND HEPATOTROPIC CONSTITUENTS OF HIGHER PLANTS

The liver is an important organ in the body since it assists in general digestion and is also responsible for the metabolism of many toxins and other substances in the blood. Inflammation of the liver, hepatitis, can be caused by a wide range of chemicals and by virus infection. If untreated, hepatitis often leads to cirrhosis, the degeneration of the liver tissue due to lesions and death of the hepatocytes.

An antihepatotoxic substance antagonizes the effects of

```
[→4)-Xylose-(1→4)-Xylose-(1→4)-Xylose-(1→4)-Xylose-(1→4)-Xylose-(-]n
                  |
              1→2 link
                  |
     4-O-Methylglucuronic acid-(1→6)-Galactose-(1→5)-Arabinose-Arabinose-(1→4)-Xylose-Xylose-(1→3)-Glucose
```

43 *Echinacea* polysaccharide 1

```
[→4)-Galactose-(1→2)-Rhamnose-(1→2)-Rhamnose-(1→4)-Galactose-(1-]n
                     |
                 1→4 link
                     |
        Glucuronic acid-(1→5)-Arabinose-(1→4)-Galactose-(1→3)-Glucose-(1→4)-Xylose-(1→2)-Rhamnose
```

44 *Echinacea* polysaccharide 2

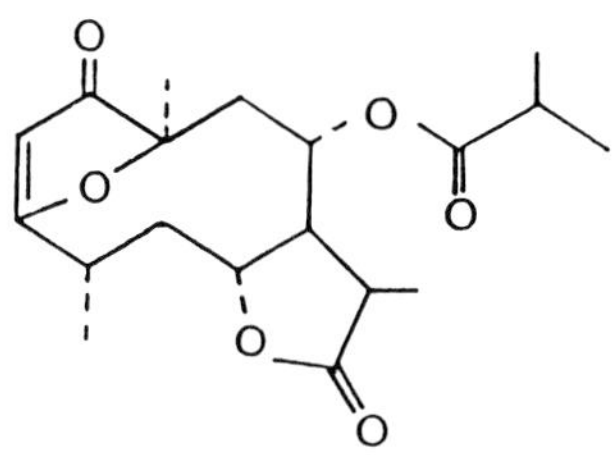

45 Zexbrevin A

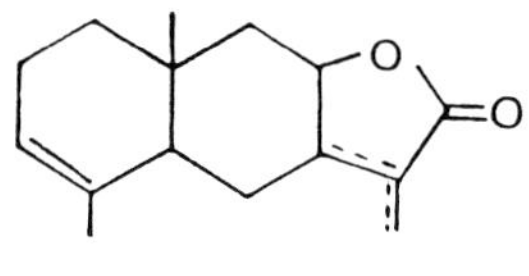

46 Alantolactone

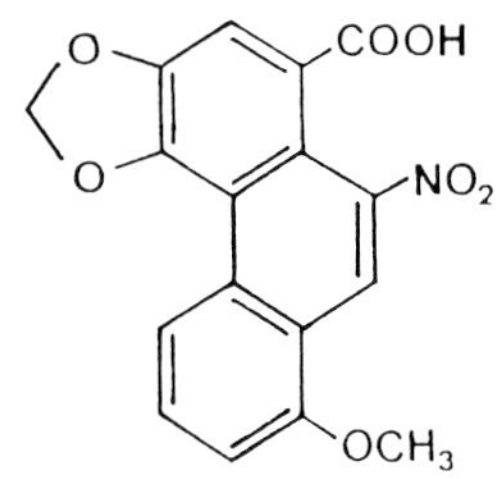

47 Aristolochic acid

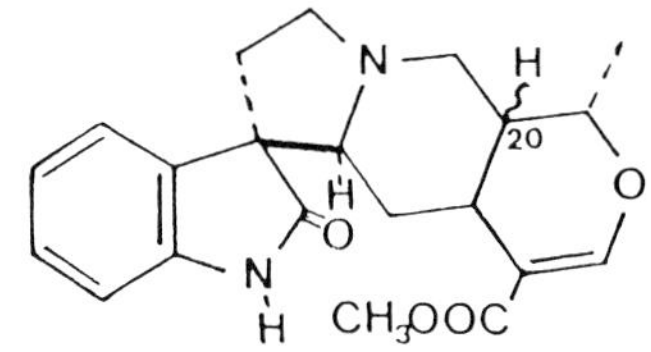

48a Isopteropodine H-20 α
48b Isomitraphylline H-20 β

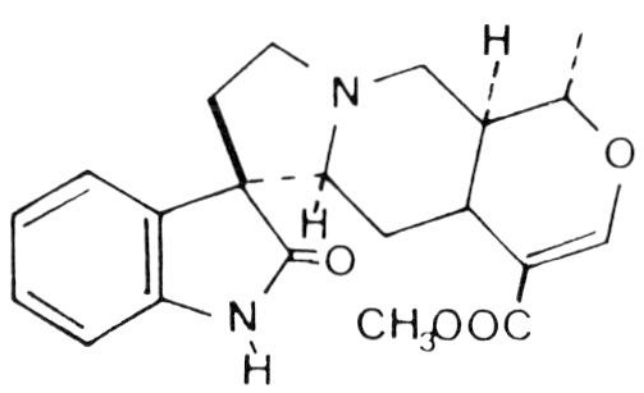

49 Pteropodine

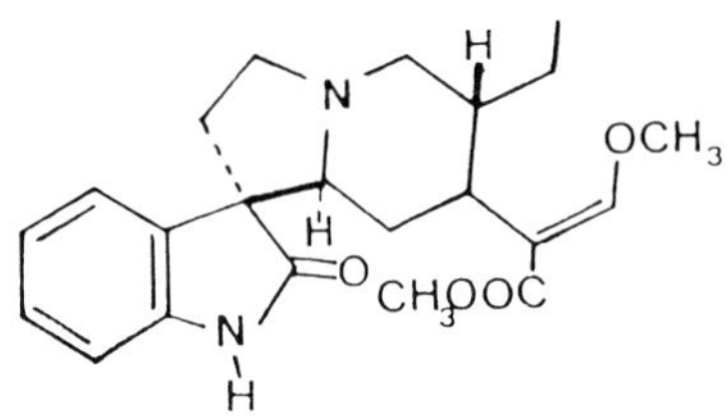

50 Isorhynchophylline

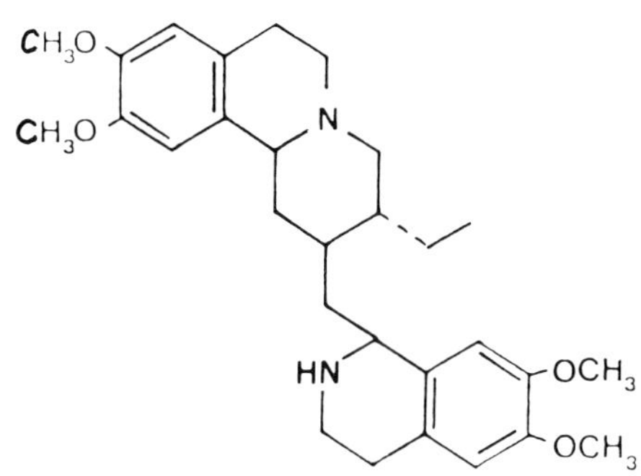

51 Emetine

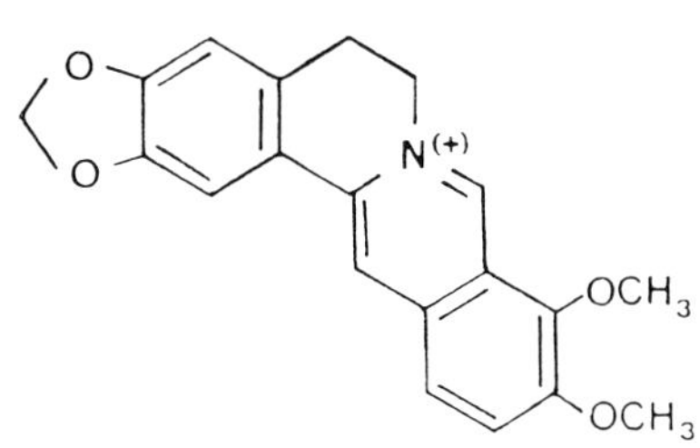

52 Berberine

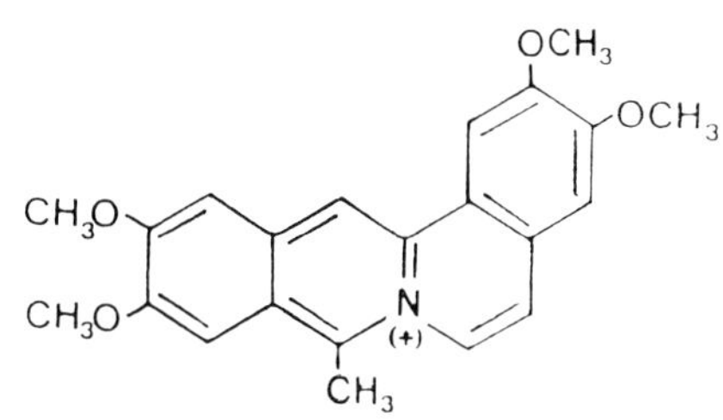

53 Coralyne

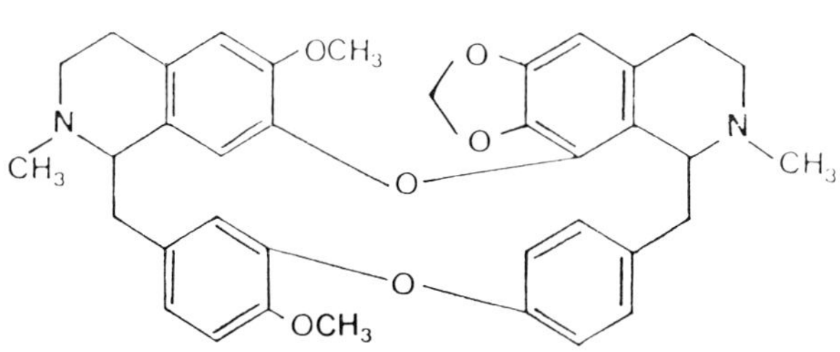

54 Cepharanthine

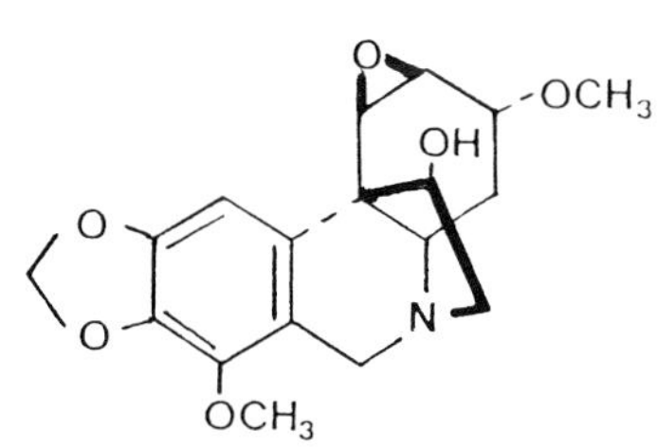

55 1,2-β-Epoxyambelline

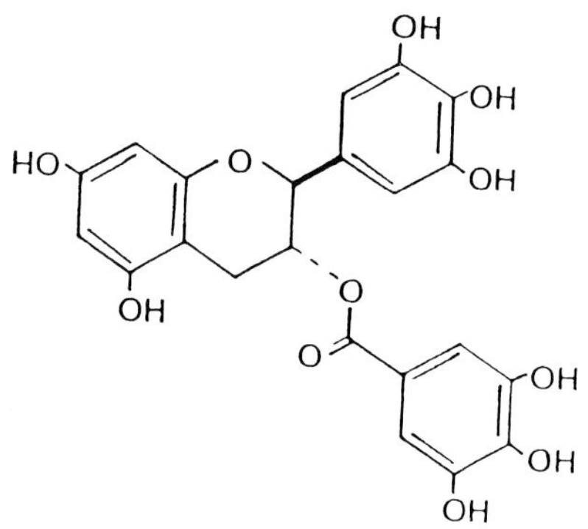

56 (–)-*Epi*gallocatechin gallate

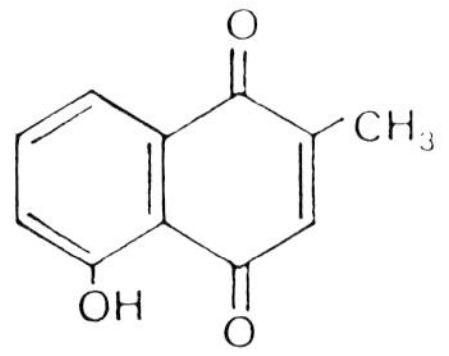

57 Plumbagin

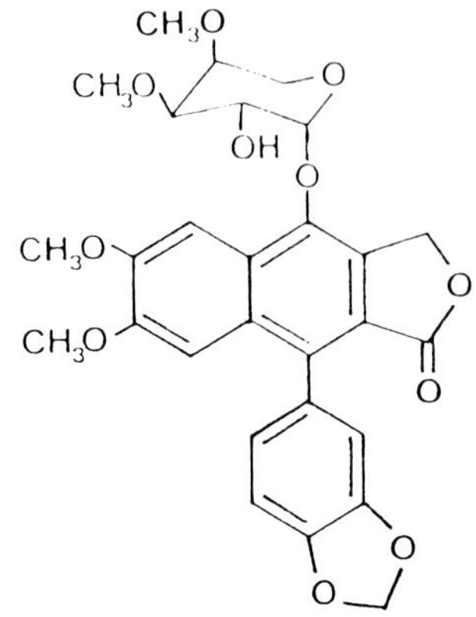

58 Cleistanthin

any compound causing hepatitis, while hepatotropic activity is exhibited by those chemicals which support or promote the regeneration and healing processes in the liver. In practice, these two activities cannot be easily distinguished from each other.

Viral hepatitis is a debilitating disease, particularly prevalent in tropical countries, but hepatitis induced by chemical factors, e.g., ingested toxins, alcohol, environmental toxins, is more of a problem in the industrialized parts of the world. There is no standard effective treatment for hepatitis, although there are many suggested treatments. Treatment for liver complaints is found in the traditional medicine of many cultures and often uses plant extracts. During the past two decades, experimental testing of several of these plants has shown them to have antihepatotoxic activity. The structure of their active constituents and modes of activity have been elucidated in many cases and these are described below.

Mechanisms of hepatotoxicity and the protective action of drugs: Damage to liver cells is believed to occur by the peroxidation of polyunsaturated fatty acids in the endoplasmic reticulum, brought about by free radicals generated by enzymes such as cytochrome P450; it may be accompanied by an influx of extracellular Ca^{2+} ions.[215] Many toxic substances are metabolised to free radicals. The antihepatotoxic effect of a compound may be due to interference at any stage in these processes.

The traditional *in vivo* test systems have been rats and mice in which liver damage has been induced by a hepatotoxic agent such as carbon tetrachloride or galactosamine.[215,216] These methods attract some criticism because the hepatotoxics used are not those normally encountered as agents causing damage. Because of this, there has been a trend in recent years to use products known to cause liver damage when ingested, e.g., the toxic peptide phalloidin from the Deathcap mushroom *Amanita phalloides* (Vaill.) Secr.

In vivo methods are costly and subject to biological variability in the test animals used. Consequently, *in vitro* methods using primary cultured hepatocytes have been introduced.[217] Hepatotoxicity is induced by agents such as those mentioned above and a decrease in the release of the enzyme glutamate-pyruvate transaminase (GPT) indicates antihepatoxic activity. Peroxides may be used instead to induce liver damage; this is a model for hepatitis caused by the peroxidized lipids present in foods and now regarded as a health problem in many countries.[218]

The following sections discuss some of the natural products known to be antihepatotoxic.

A. FLAVONOIDS AND RELATED COMPOUNDS

1. Flavonolignans from *Silybum marianum*

The Milk Thistle, *Silybum marianum* (L.) Gaertner (Asteraceae), has a tradition of use in central Europe for liver disease. In 1968, extracts of the fruits of this plant were shown to have a highly significant protective effect on the liver in a large number of model liver-damage systems.[216] The active constituents, a mixture called silymarin, have been shown to consist of three lignan-flavone isomers. The major substance present is silybin (**59**) and the two others are silydianin (**60**) and silychristin (**61a**).[216] Silybin has also been synthesized.[219] Other compounds of the same type have recently been isolated from a white-flowering variety of the plant.[220]

Silybin has been shown to be very effective against phalloidin intoxication in mice when given prophylactically. The toxic effects were even countered when silybin was given up to 20 min after the phalloidin.[216] Similar results have been seen for these flavonolignans using the *in vitro* test system.[221]

The *in vivo* tests implied that silybin acts by competitive inhibition of phalloidin in binding to the receptors on the cell membranes. It also inhibits the penetration of another toxic peptide, α-amanitin, to the cell nucleus.[216] Tests on cultured hepatocytes have shown that silybin stimulates protein synthesis and hence an increase in the production of new liver cells. Thus, the action of silybin and related compounds is due both to the protection of the membrane of intact, undamaged cells and also to stimulation of the repair of damaged cells and formation of new ones.

After absorption from intestine, silymarin is eliminated in the bile as glucuronates and sulfates. These conjugates are cleaved in the intestine and re-enter the enterohepatic circulation; hence the pronounced liver specificity of this drug.

A few structure-activity studies have been carried out on these compounds. The parent flavonoid, taxifolin (**62a**), and the 2,3-dehydro analogue have no *in vivo* activity.[216] More recent *in vitro* testing has shown that the 3-deoxy compounds,

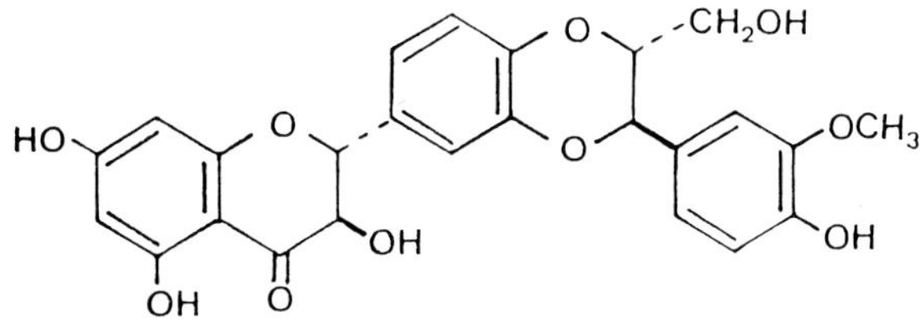

59 Silybin

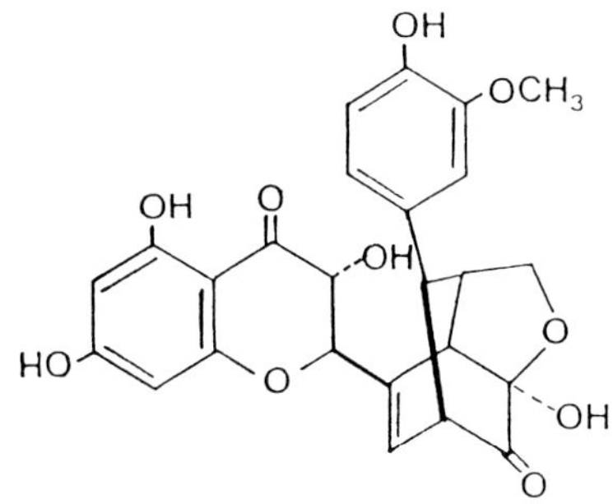

60 Silydianin

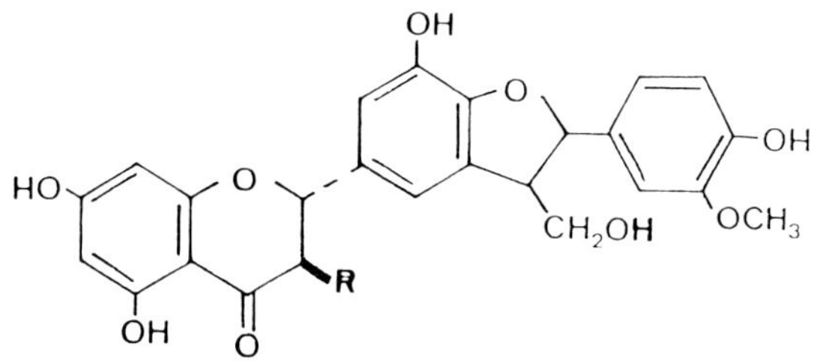

61a Silychristin R = OH
61b 3-Deoxysilychristin R = H

e.g., 3-deoxysilychristin (**61b**), are more powerful than the parent compound.[222] The same set of experiments showed that molecules with a benzodioxan link between the flavonoid and lignan moieties were the most active.

Extracts of *Silybum* are marketed in several European countries.

2. Flavonoids of *Artemisia capillaris*

The buds of *Artemisia capillaris* Thunb. (Asteraceae), the crude drug *inchinko,* have been used in oriental medicine for liver disease.[223] Initial *in vivo* tests showed great variation in activity between different batches of the drug. The greatest activity was observed in samples collected when the plant was in bud. Phytochemical analysis of the buds resulted in the isolation of a coumarin and 13 flavonoids; several of the isolated flavonoids exhibited antihepatotoxic activity, with capillarisin (**62b**) having the greatest effect. It is thought, however, that the major substance responsible for the action of the drug is the coumarin aesculetin 6,7-dimethyl ether (**63**), which is present in much greater amounts and also has appreciable antihepatotoxic activity. Tests with related coumarins indicates that alkylation at C-7 tended to decrease activity, while at C-6 it had little effect.[223]

3. Biflavonoids from *Garcinia kola*

Tests in Nigerian hospitals on patients with hepatitis resulted in an improvement in the condition of those treated with whole seeds or seed extracts from *Garcinia kola* Heckel (Clusiaceae (Guttiferae)). Subsequent screening of extracts on phalloidin-poisoned mice revealed that the active constituents were the biflavanones (**64a–d**).[224]

4. Procyanidins

The procyanidins are structurally related to the flavonoids but lack the carbonyl function in ring B. 3-Cyanidol, also known as (+)-catechin (**7a**), has only recently been shown to exhibit hepatotropic activity *in vivo*, using animals with paracetamol-induced liver damage.[225] 3-Cyanidol is the only active member of the four possible catechin isomers. Polymers of these compounds occur widely in plants as components of condensed tannins, but the monomers are not usually found in large amounts. The polymers are not thought to be active against liver damage.

3-Cyanidol may act by a similar mechanism to silybin, but it does not have such a high specificity for the liver and it is rapidly excreted by the kidneys.[216] 3-Cyanidol may be of use in alcohol-induced cirrhosis, since it normalizes various biochemical functions that are affected adversely by alcohol.[226] 3-Cyanidol is available commercially in some countries.

B. PHENYLPROPANOID DERIVATIVES

1. Constituents of *Cynara scolymus*

Decoctions of the Globe Artichoke, *Cynara scolymus* L. (Asteraceae), were used in the Middle Ages for the treatment of jaundice, a condition caused by malfunction of the liver. Tests have demonstrated that extracts of the plant exercise a beneficial effect on the liver. The active ingredient appears to be cynarin (**65**),[227] which in the gastrointestinal tract produces free caffeic acid (**66a**), a compound with hepatotropic activity.[228]

2. Constituents of *Curcuma longa*

The rhizomes of *Curcuma longa* L. (Zingiberaceae), commonly known as turmeric, have been used in Oriental medicine to treat jaundice. The main constituent is a diphenylpropide derivative, curcumin (**67**). This is known to have a choleretic effect, but recent work using both whole animals and cultured hepatocytes indicates that curcumin does indeed exert a clear-

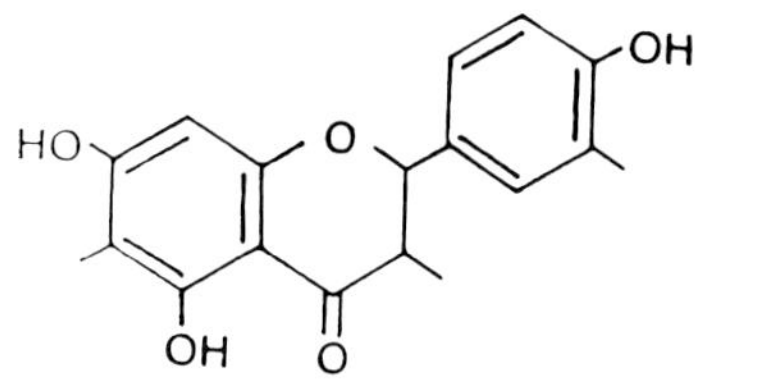

62a Taxifolin $R_1 = H, R_2 = R_3 = OH$
62b Capillarisin $R_1 = OCH_3, R_2 = R_3 = H$

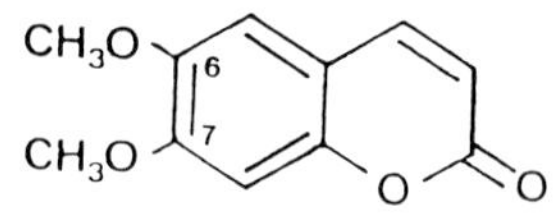

63 6,7-Dimethylaesculetin

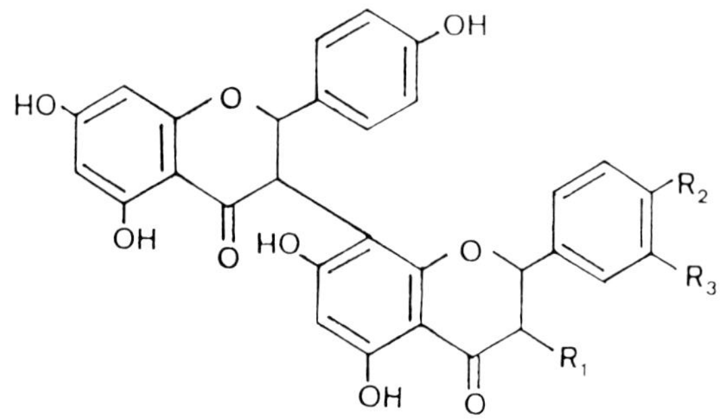

64 Biflavanones from *Garcinia kola* R_1, R_2, R_3 = H, OH, or OCH_3

cut protective effect against carbon-tetrachloride- and galactosamine-induced cytotoxicity.[229] At one time, it was thought that this activity might be due to the phenylpropanoids formed by *in vivo* cleavage of the curcurminoids to feruli and *p*-coumaric acids (**66b,c**). However, experimental investigation showed that only caffeic acid, and not the other phenylpropide acids, had any activity.[229] The curcuminoids do not afford caffeic acid on cleavage, and since they showed strong activity per se they are reckoned to be the actual active components.

3. Lignoids from *Schizandra* species

Fruits of *Schizandra chinensis* Baill. and *S. sphenanthera* Rehder et Wilson (Magnoliaceae) have been used in Chinese medicine as a tonic and sedative. Chemical investigation has disclosed the presence of over 30 lignans possessing the dibenzcyclooctane skeleton (**68**). Preliminary studies indicated that some of these compounds suppressed chemically induced liver damage *in vivo*.[216] Schizandrin B (**69**), when administered before or after a hepatotoxic substance to test animals, was found to cause rapid regeneration of liver tissue and to diminish GPT levels. More recent work with *in vitro* techniques supports the hepatotropic action of these substances, especially wuweizisu C (**70**) and schisantherin D (**71**).[230] These studies also allowed some conjecture about structure-activity relationships. A methylenedioxy substituent on the aromatic ring appears to contribute to activity, a 7-hydroxyl group tends to decrease activity, and an oxygen function at the allyl position of the phenyl rings and methylenedioxy group also tends to decrease activity as compared with the nonoxygenated analogue.

The mechanism of action of these compounds has been shown to be due to inhibition of lipid peroxidation rather than scavenging of free radicals.[231]

4. Lignans from *Thujopsis dolabrata*

The leaves of *Thujopsis dolabrata* (Thunb.) Siebold et Zucc. (Cupressaceae) are another Oriental remedy for jaundice. *In vivo* studies using mice resulted in deoxypodophyllotoxin (**72**) being isolated as the major antihepatotoxic constituent present.[232] Recent studies have concentrated on the structure-activity relationships of related compounds.[233]

5. Neolignans

Americanin A (**73**), from *Phytolacca americana* L. (Phytolaccaceae) has been reported to have antihepatotoxic activity.[234]

C. TRITERPENES

1. Constituents of *Glycyrriza* Species

Glycyrrhiza glabra L. and related species (Fabaceae) comprise the drug Liquorice, which is well known in medicine and is used for several different purposes.[33] Triterpenoids, especially glycyrrhizin (**8a**), which is hydrolysed by the gut flora to its aglycone glycerrhetinic acid (**8b**) and the 3-*epi*- and 3-dehydro analogues, are its major active ingredients.

Both glycyrrhizin and glycerrhetinic acid have shown hepatoprotective activity *in vitro*.[235] Further studies have revealed that the aglycone is a much more potent inhibitor of lipid peroxidation and free-radical generation than the glycoside.[217] Glycyrrhetinic acid may arise not only from hydrolysis in the gut but possibly also by the action of glucuronidase in the liver cells. Glycyrrhetinic acid has a stronger inhibitory effect than vitamin E, and it also seems to act by a different mechanism.

2. Saikosaponins

The saikosaponins are found in *Bupleurum falcatum* L. (Apiaceae), which has been used as a remedy in China for inflammatory disease for many centuries. The saikosaponins, like saikosaponin F (**74**), have a variety of pharmacological effects, including the prevention of liver injury induced by D-galactosamine.[236]

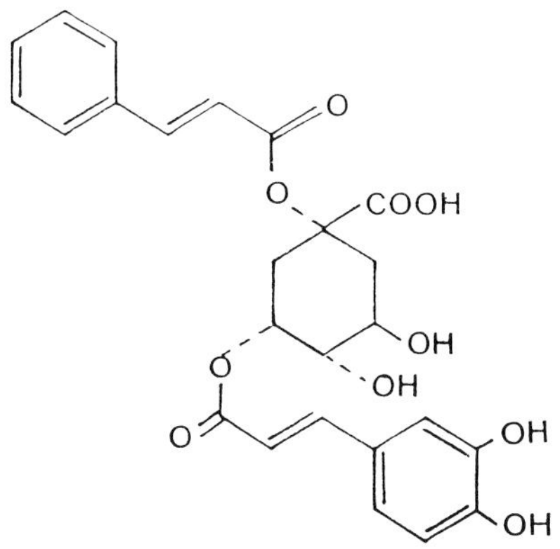

65 Cynarin

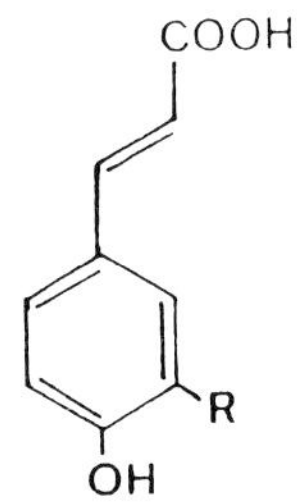

66a Caffeic acid R = OH
66b Ferulic acid R = H
66c *p*-Coumaric acid R = OCH_3

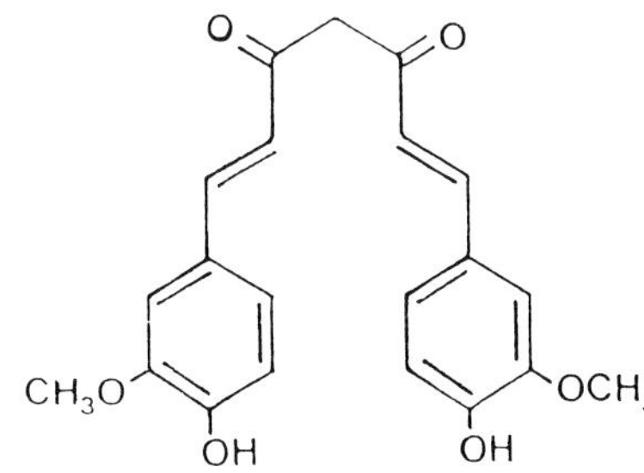

67 Curcumin

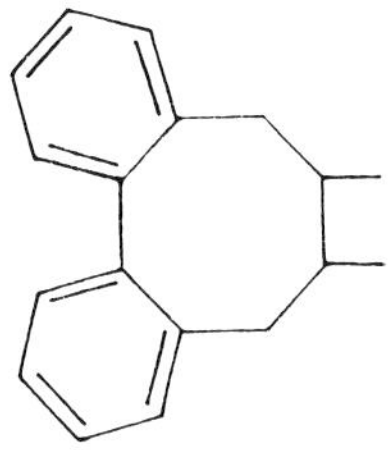

68 Dibenzcyclooctane skeleton

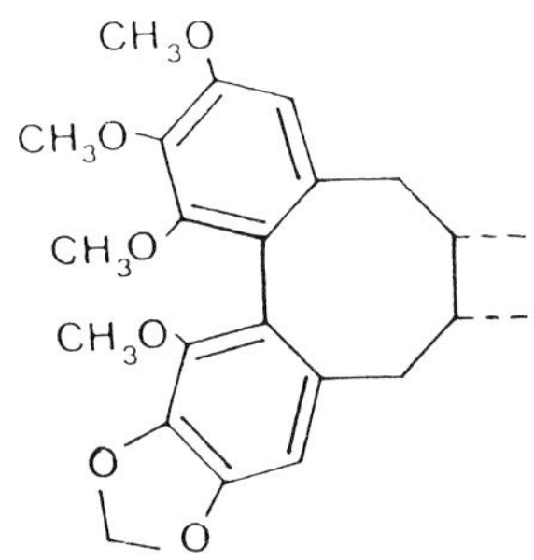

69 Schizandrin B

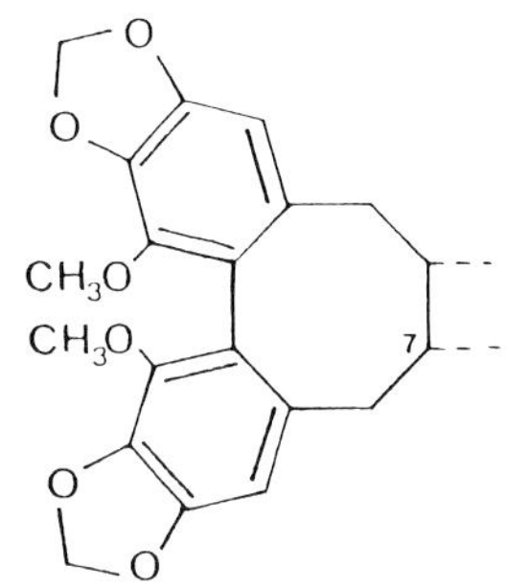

70 Wuweizisu C

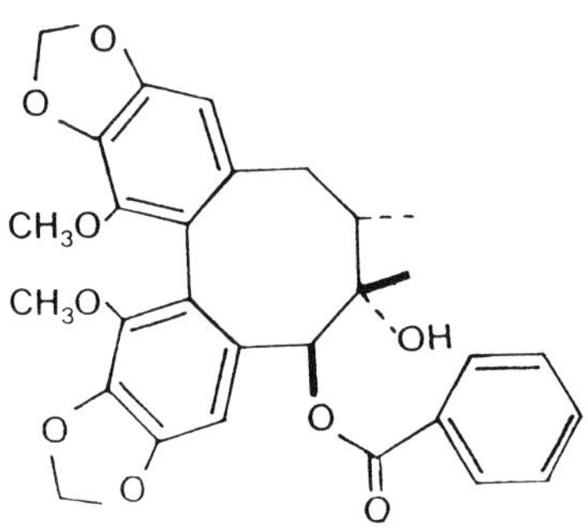

71 Schisantherin D

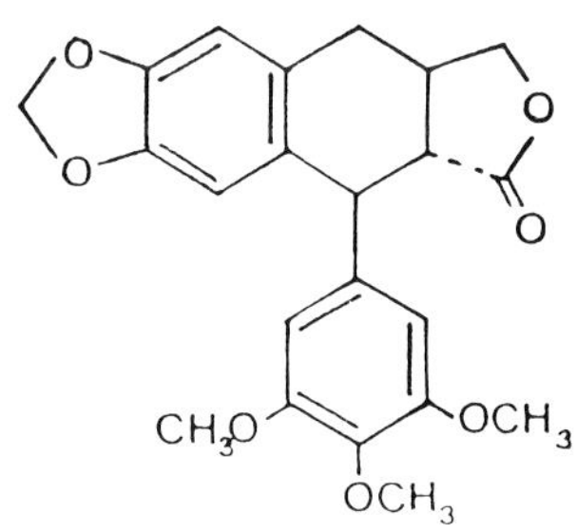

72 Deoxypodophyllotoxin

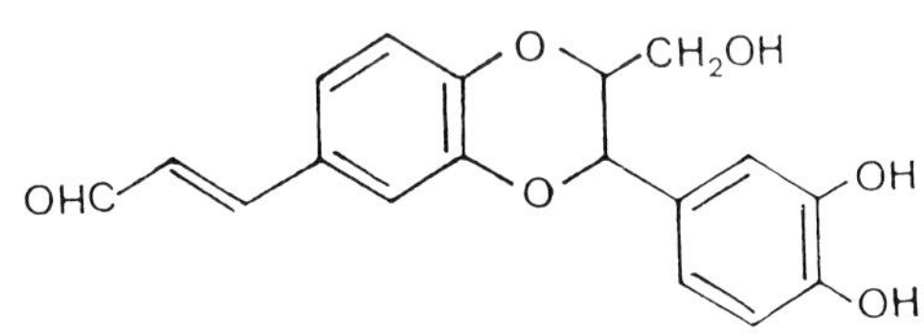

73 Americanin A

3. Ginseng Saponins

The saponins of Ginseng, *Panax ginseng* C.A.Meyer (Araliaceae), are thought to be the constituents mainly responsible for the pharmacological activity underlying the various uses of this drug. Antihepatotoxic properties for some of the saponins present have been reported on the basis of *in vivo* and *in vitro* test systems.[237] For most of the sapogenins, the (20*R*) series had a greater effect than the corresponding (20*S*) compounds. As a general rule, the antihepatotoxic activity increased with a reduction in the number of sugar moieties in the molecule. Some of the saponins were found to exhibit a cytotoxic effect as well.

Of interest is the reported difference in activity between the two commercial forms, white and red, of ginseng. The red ginseng, which is prepared by steaming and drying the white form of the drug, had a greater effect. Chemical investigation revealed that the red ginseng contains saponins not present in the white form.

Ginseng saponins have recently been shown to have a protective effect in man against viral B hepatitis.[238]

Studies on the saponins of another *Panax* species, *P. japonicus* C.A.Meyer, have also indicated that they possess hepatoprotective activity.[238] The triterpenes belong to a different type from those in *P. ginseng*, viz., the oleanane and dammarane series. The oleanane compounds appear to be responsible for the activity. In contrast to the ginsenosides and

Glycyrrhiza constituents, the number of sugars attached to the triterpene seem to have little effect on the anti-hepatotoxic properties.

4. Saponins from *Dianthus superbus* L. *var. longicalycinus* Williams

The above-ground parts of this plant have a reputation in oriental medicine as a diuretic and anti-inflammatory agent. Phytochemical examination of the plant showed the presence of oleanane triterpenes, and *in vitro* investigation indicated that although they had some antihepatotoxic activity it was not as great as that shown by oleanolic acid (**8c**).[239] The most active compounds present were found to be gypsogenic acid (**8f**) and its glycosides and the glycosides of dianic acid (**8g**); relative potencies varied according to the test system used.

Oleanolic acid (**8c**), a simple and commonly occurring triterpene, has recently been shown to have antihepatotoxic activity in *in vitro* tests.[239]

C. MISCELLANEOUS COMPOUNDS

1. Atractylon and Other Constituents of *Atractylodes* Rhizomes

The rhizomes of *Atractyylodes macrocephala* Koidzumi and *A. japonica* (Asteraceae) are used in oriental medicine as a diuretic, analgesic, and stomachic.

In vitro screening for antihepatotoxic properties showed that extracts of these plants gave positive results but that the degree of activity varied considerably between batches.[240] The active components were found to be the sesquiterpenes present, atractylon (**75**), β-eudesmol (**76**), and hinesol (**77**), the first of these being the most potent. Structure-activity studies involving related compounds established that hydrogenation causes no change in activity. However, introduction of a hydroxyl function at C-3 of dihydroeudesmol caused an increase in activity, while a ketone function at the same position resulted in a decrease.

Atractylon scavenges free radicals and inhibits enzymatic lipid peroxidation. It seems that atractylon has to be metabolized to a free radical itself before it can exert its protecting effect. Thus, as far as the liver is concerned, there appear to be both "good" and "bad" free radicals.[241]

2. Other Terpenoids

The iridoid glycoside picroside I (**78**) from the roots of *Picrorrhiza kurrooa* Benth. (Scrophulariaceae) has been shown to have hepatotropic activity both *in vitro* and *in vivo*.[220,245]

A diterpenoid kaurenoic acid from *Wedelia chinensis* Merrill (Asteraceae) is responsible for the liver-protectant effect of this plant.[274] Likewise, one of the steroid-related withanolides is an active constituent of *Withania coagulans* Dunal (Solanaceae).[275]

3. Other Compounds

The coumestans from *Eclipta alba* (L.) Hassk. (Asteraceae) have *in vitro* hepatotropic activity.[276] Investigations on garlic oil, from *Allium sativum* L. (Alliaceae), have demostrated that some of the sulfur-containing compounds also have this effect.[277] In China, sarmentosine (**79**), an alkaloid from *Sedum sarmentosum* Bunge (Crassulaceae), is used in the treatment of hepatitis; it lowers GPT levels in serum.[243]

The recent introduction of an *in vitro* test should enable the verification of more traditional remedies to be accomplished and should also result in the isolation of more antihepatotoxic compounds from plants. The compounds so far known to exert this effect encompass a wide range of chemical structures. It is likely that other types of molecules will be shown to be useful in protecting and treating the liver, and they may open up the way for the development of drugs which will be of value in treating hepatitis.

IX. CONCLUSION

Much of the work described in this chapter has only been possible through an integrated multidisciplinary approach.

Investigative procedures are being specifically adapted to facilitate the screening of plant extracts. For example, the widely utilized turbidometric method for studying platelet aggregation[244,245] is not easily applicable to the study of crude plant extracts. Often these are complex, intractable tars furnishing coloured, opaque aqueous suspensions yielding results that are difficult to interpret.[246] To overcome this, a method of assessing potential platelet aggregation inhibition activity based on a smear technique has been adapted so as to be particularly suitable for the screening of plant extracts.[247]

Great effort has been devoted to the study of the mode of action and metabolism of extracts and individual chemical substances (as well as chemically modified analogues) obtained from medicinal and other plants. One of the points emerging from the foregoing discussions is the importance of having appropriate models for testing whether extracts or individual substances do indeed have activity. However, establishing whether a model is suitable requires adequate understanding of the biochemical mechanisms responsible for the activity being investigated. This emphasizes the need to be able to follow events in normal and diseased cells, not only at the subcellular level but even more essentially at the molecular level where specific enzyme-mediated reactions are taking place. New pharmacological tools need to be developed in order to improve our understanding of receptors, recognition sites of transport mechanisms, and mechanisms of action, all of which are required for the design of better drugs.

A recently published, excellent review[248] shows very clearly how natural sources are acting as reservoirs for the discovery of new and unique substances that are finding a place in the pharmacology research laboratory. Among other things, it highlights the action of plant products on hormone and neurotransmitter receptors,[249] receptor-associated ion channels,[250] and protein-activated transduction pathways (e.g., ryanodine[251] from *Ryania speciosa* Vahl).

More and more the concept of isosterism, i.e., the correlation of molecular geometry, is being applied, rather than the repetitious synthesis of simple chemical analogues. It is becoming increasingly evident that it is the *shape, flexibility*, and *electron density*, as much as the precise nature of the chemical substituents, that are all-important in determining the mode and site of action of drug molecules.[252,253] The presence of

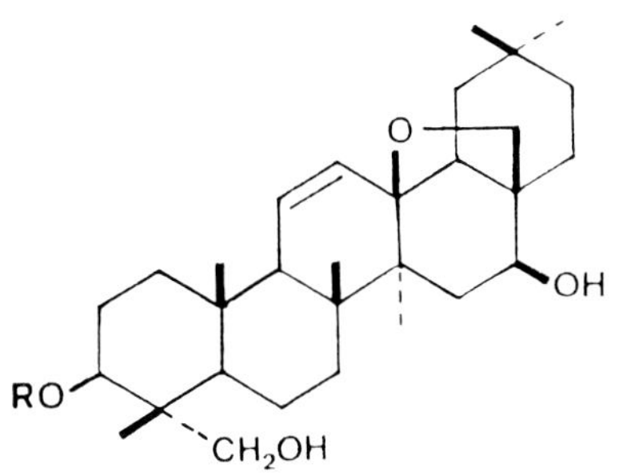

74 Saikosaponin F R = β-D-Glu-(1→3)-β-D-Fuc

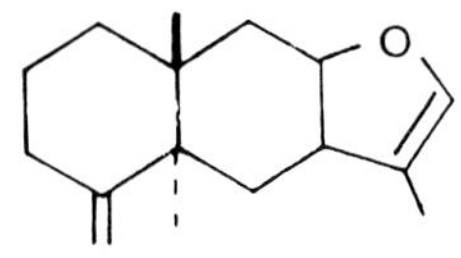

75 Atractylon

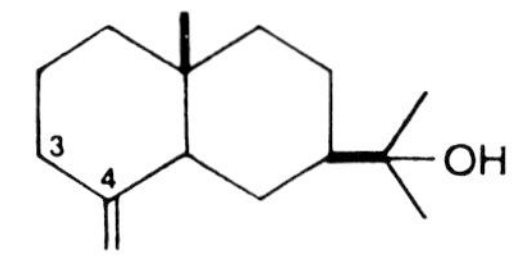

76 β-Eudesmol

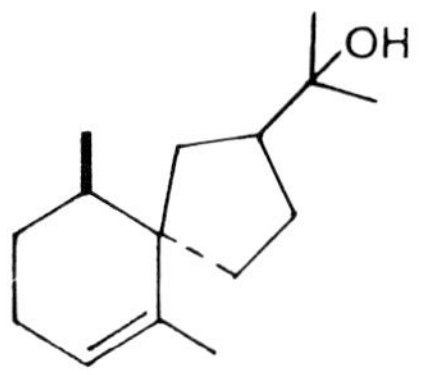

77 Hinesol

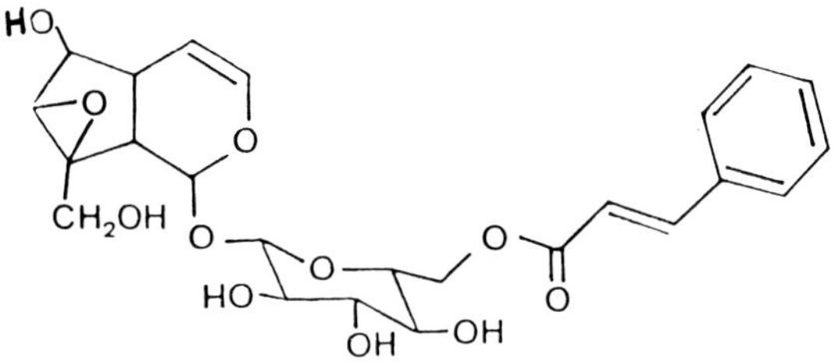

78 Picroside I

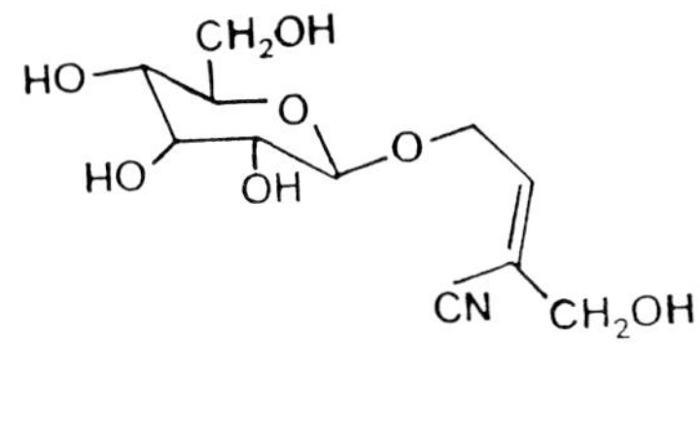

79 Sarmentosine

modifying functional groups may increase or decrease activity by directing the molecule to a particular binding site, thus inducing an appropriate conformational change in the receptor and eliciting a particular drug response.

Medicinal and other plants have proved to be a rich source of novel structures, which as a result of probing their pharmacology and biochemistry, have led either to useful drugs themselves or have acted as templates for synthetic improvements. It is a role that plants will continue to play.[254,255]

REFERENCES

1. **Havsteen, B.,** Flavonoids. A class of natural products of high pharmacological potency, *Biochem. Pharmacol.*, 32, 1141, 1983.
2. **Middleton, E.,** The flavonoids, *Trneds Pharmacol. Sci.*, 335, 1984.
3. **Tripathi, V.D. and Rastogi, R.P.,** Flavonoids in biology and medicine, *J. Sci. Ind. Res.*, 40, 116, 1981.
4. **Farkas, L., Gabor, M., Kallay, F., and Wagner, H.,** Flavonoids and Bioflavonoids 1981, *Proc. Int. Bioflav. Symp., Munich*, Elsevier, Amsterdam, 1981.
5. **Mabry, T. J. and Ulubelen, A.,** Chemistry and utilisation of phenylpropanoids including flavonoids, coumarins and lignans, *J. Agric. Food Chem.*, 28, 188, 1980.
6. **Harborne, J.B. and Mabry, T.J.,** *The flavonoids: Advances in Research*, Chapman & Hall, London, 1982.
7. **Gabor, M.,** Flavonoids, in *Handbook of Experimental Pharmacology*, Vol. 50, *Antiinflammatory Drugs*, Vane, J.R. and Ferreira, S.H., Eds., Springer-Verlag, New York, 1979, 698.
8. **Wagner, H.,** New plant phenolics of pharmaceutical interest, in *Ann. Proc. Phytochem. Soc. Europe*, Vol. 25, *The Biochemistry of Plant Phenolics*, Van Sumere, C.F. and Lea, P.J., Eds., Clarendon Press, Oxford, 1985, 401.
9. **Fewtrell, C.M.S. and Gomperts, B.D.,** Quercetin: a novel inhibitor of calcium ion influx and exocytosis in rat peritoneal mast cells, *Nature*, 265, 635, 1977.
10. **Middleton, E. and Drzewiecki, G.,** Flavonoid inhibition of human basophil histamine release stimulated by various agents, *Biochem. Pharmacol.*, 33, 3333, 1984.
11. **Amellal, M., Bronner, C., Briancon, F., Haag, M., Anton, R., and Landry, Y.,** Inhibition of mast cell histamine release by flavonoids and biflavonoids, *Planta Med.*, 16, 1985.
12. **Middleton, E. and Drzewiecki, G.,** Effects of flavonoids and transitional metal cations on antigen-induced histamine release from human basophils, *Biochem. Pharmacol.*, 31, 1449, 1982.
13. **Fewtrell, C.M.S. and Gomperts, B.D.,** Effect of flavone inhibitors of transport ATPases on histamine secretion from rat mast cells, *Biochem. Biophys, Acta*, 469, 52, 1977.
14. **Hope, W.C., Welton, A.F., Fiedler-Nagy, C., Batilla-Bernado, C., and Coffey, J.W.,** *In vitro* inhibition of the biosynthesis of slow reacting substance of anaphylaxis SRS-A and lipoxygenase activity by quercetin, *Biochem. Pharmacol.*, 32, 367, 1983.
15. **Alcaraz, A.J. and Hoult, J.R.S.,** Actions of flavonoids and the novel anti-inflammatory flavone, hypolaetin-8-glucoside, on prostaglandin biosynthesis and inactivation, *Biochem. Pharmacol.*, 34, 2477, 1985.
16. **Kimura, Y., Okuda, H. and Arichi, S.,** Studies on *Scutellariae radix*. XIII. Effects of various flavonoids on arachidonate metabolism in leukocytes, *Planta Med.*, 132, 1985.
17. **Rusznyak, S. and Sznet-Györgi, A.,** Vitamin P — flavonols as vitamins, *Nature*, 138, 27, 1936.
18. **Wagner, H.,** Phenolic compounds of pharmaceutical interest, in *Recent Advances in Phytochemistry*, Vol. 12, *Biochemistry of Plant Phenolics*, 1977, 589.
19. **Masquelier, J.,** Pycnogenols: Recent advances in the therapeutic activity of procyanidins, in *Natural Products as Medicinal Agents*, Beal, J.L. and Reinhard, E., Eds., Hippokrates, Stuttgart, 1981, 243.
20. **Barnaulov, O.D., Manicheva, O.A., Shelyuta, V.L., Konopleva, M.M., and Glyzin, V.I.,** Effects of flavonoids on the development of experimental dystrophy of the stomach in mice, *Khim. Farm. Zh.*, 18, 935, 1984; cited from *Chem. Abstr.*, 102, 72395, 1985.
21. **Ezaki, N., Kato, M., Takizawa, N., Morimoto, S., Nonaka, G., and Nishioka, I.,** Pharmacological studies on *Linderae umbellatae* ramus. IV. Effects of condensed tannin related compounds on peptic activity and stress-induced gastric lesions in mice, *Planta Med.*, 34, 1985.
22. **Tshuiya, Y., Shimizu, M., Hiyama, Y., Itoh, K., Hashimoto, Y., Nagayama, M., Horie, T., and Morita, M.,** ANtiviral activity of naturally occurring flavonoids *in vitro*, *Chem. Pharm. Bull.*, 33, 3881, 1985.
23. **Kaul, T.N., Middleton, E., and Ogra, P.L.,** Antiviral effect of flavonoids on human viruses, *J. Med. Virol.*, 15, 71, 1985.
24. **Saini, S. and Ghosal, S.,** Naturally occurring flavans unsubstituted in the heterocyclic ring, *Phytochemistry*, 23, 2415, 1984.

25. **Tschesche, R. and Wulff, R.**, Chemie und Biologie der Saponine, *Fortschr. Chem. Org. Naturst.*, 30, 461, 1973.
26. **Hiller, K. and Voigt, G.**, Neue Ergebnisse in der Erforschung der Triterpensaponine, *Pharmazie*, 32, 365, 1977.
27. **Shibata, S.**, Saponins with biological and pharmacological activity, in *New Natural Products and Plant Drugs with Pharmacological, Biological or Therapeutical Activity*, Wagner, H.K. and Wolff, R.M., Eds., Springer-Verlag, Berlin, 1977, 177.
28. **Anisimov, M.M. and Cirva, V.Ja.**, Die biologische Bewertung von Triterpenglykosiden, *Pharmazie*, 35, 731, 1980.
29. **Hiller, K. and Adler, C.**, Neue Ergebnisse über Triterpensaponine, *Pharmazie*, 37, 619, 1982.
30. **Adler, C. and Hiller, K.**, Bisdesmosidische Triterpensaponine, *Pharmazie*, 40, 676, 1985.
31. **Ginseng**, *Advances in Chinese Medicinal Materials Research*, Chang, H.M., Yeung, H.W., Tso, W.W. and Koo, A., Eds., World Scientific, Singapore, 1985, 453.
32. **Shibata, S., Tanaka, O., Shoji, J., and Saito, H.**, Chemistry and pharmacology of *Panax*, in *Economic and Medicinal Plant Research.*, Vol. 1, Wagner, H., Hikino, H., and Farnsworth, N.R., Eds., Academic Press, London, 1985, 217.
33. **Lutomski, J.**, Chemie und therapeutische Verwendung von Süssholz (*Glycyrrhiza glabra* L.), *Pharm. Unser. Z.*, 12(2), 49, 1983.
34. **Vasilenko, Yu.K., Ponomarev, V.D., and Oganesyan, E.T.**, Comparative study of the hypolipemic properties of triterpenoids, *Khni. Farm. Zh.*, 15(5), 50, 1981.
35. **Vasilenko, Yu.K., Lisevitskaya, L.I., Frolova, L.M., Parfent'eva, E.P., and Voskanyan, V.L.**, Hypolipemic properties of triterpenoids, *Farmakol. Toksikol. (Moscow)*, 45, 66, 1982.
36. Fujisawa Pharmaceutical Co. Ltd., Japan Kokai Tokkyo Kohn 58, 106000, 1983; cited from *Chem. Abstr.*, 99, 158684, 1983.
37. Society for the study of provisions, Japan Kokai Tokkyo Koho 58, 116415, 1983; cited from *Chem. Abstr.*, 99, 164017, 1983.
38. Suntory Ltd., Japan Kokai Tokkyo Koho 59, 20298, 1984; cited from *Chem. Abstr.*, 101, 43572, 1984.
39. **Valterova, I, Klinot, J., and Vystrcil, A.**, Preparation and antibacterial activity of di-, tri-, and tetraoic acids derived from 3,4-secolupane, *Coll. Czech. Chem. Commun.*, 48, 649, 1983.
40. **Shcheglov, V.V., Baranova, S.I., Anisimov, M.M., Antonov, A.S., Afiyatullov, Sh.Sh., Levina, E.V., Sharypov, V.F., Stonik, V.A., and Elyakov, G.B.**, Antimicrobial spectrum of some triterpene and steroid glycosides, *Antibiotiki (Moscow)*, 24, 270, 1979.
41. **Sedov, A.M., Elkina, S.I., Sergeev, V.V., Kalina, N.G., Sakandelidze, O.G., Batrakov, S.G., and Girshovich, S.E.**, Capacity of triterpene glycosides obtained from sea cucumbers to stimulate antibacterial resistance in experimental salmonellosis in mice. *Zh. Mikrobiol. Epidemiol. Immunobiol.*, 5, 55, 1984.
42. **Anisimov, M.M., Shcheglov, V.V., Strigina, L.I., Chetyrina, N.S., Uvarova, N.I., Oshitok, G.I., Alad'nia, N.G., Vecherko, L.P., and Zorina, A.D.**, Chemical structure and antifungal activity of some triterpenoids, *Izv. Akad. Nauk S.S.S.R., Ser. Biol.*, 4, 570, 1979.
43. **Lugt, C.B.**, Development of molluscicidal potency in the short and long staminate racemes of *Phytolacca decandra*, *Planta Med.*, 38, 68, 1980.
44. **Hostettmann, K.**, Saponins with molluscicidal activity from *Hedera helix* L., *Helv. Chim. Acta*, 63, 606, 1980.
45. **Saleh, M.M., Shabana, M.K.S., Guindi, E.R., and Kamal, R.M.**, Phytochemical and molluscicidal study of *Hedychium gardnerianum* Wall. as a method for control of Schistosoma, *Proc. 1st Int. Conf. Chem. Biotechnol. Biol. Act. Nat. Prod.*, 3, 293, 1981; cited from *Chem. Abstr.*, 97, 67752, 1982.
46. **Domon, B. and Hostettmann, K.**, Saponins with molluscicidal properties from *Lonicera nigra* L., *Helv. Chim. Acta*, 66, 422, 1983.
47. **Jacker, H.-J., Voigt, G., and Hiller, K.**, Zum antiexsudativen Verhalten einiger Triterpensaponine, *Pharmazie*, 37, 380, 1982.
48. **Poizot, A. and Dumez, D.**, Modification de la cinétique de cicatrisation après exérèses itératives chez le rat. Action d'un triterpénoide et de ses dérivés sur les délais de cicatrisation. *C. R. Acad. Sci., Paris, Ser. D*, 286, 789, 1978.
49. **Takemoto, T., Arihara, S., Nakajima, T., and Okuhira, M.**, Studies on the constituents of *Fructus momordicae*. I. On the sweet principle, *Yakugaku Zasshi*, 103, 1151, 1983.
50. **Hashimoto, Y., Ogura, M., and Ishizone, H.**, U.K. Patent Appl. 2,019,407, 1979.
51. **Witkowski, A. and Koopa, J.**, Binding of the cytotoxic and antitumor trieterpenes, cucurbiracins, to glucocorticoid receptors of HeLa cells, *Biochim. Biophys. Acta*, 674, 246, 1981.
52. **Fang, S.D., Berry, D.E., Lynn, D.G., Hecht, S.M., Campbell, J., and Lynn, W.S.**, The chemistry of toxic principles from *Maytenus nemorosa*, *Phytochemistry*, 23, 631, 1984.
53. Kokan, Takuo Tsumura Juntendo Inc., Japan Kokai Tokkyo Koho 60, 08224, 1985; cited from *Chem. Abstr.*, 102, 191182, 1985.
54. **Del Vecchio, A., Senni, I, Cossu, G., and Molinaro, M.**, Effetti della *Centella asiatica* sull'attività biosyntetica di fibroblasti in coltura, *Farmaco (Ed. Prac.)*, 39, 354, 1984.
55. **Aronson, J.K.**, *An Account of the Foxglove and its Medical Uses 1785-1985*, Oxford University Press, London, 1985, 213.
56. **Opie, L.H.**, Digitalis and sympathomimetic stimulants, *Lancet*, 1, 912, 1980.
57. **Luckner, M. and Diettrich, B.**, Herzglykoside — Struktur, Gewinnung, Verwendung in der Therapie, *Pharmazie*, 34, 477, 1979.
58. **Watson, T.R., Cheung, H.T.A., and Thomas, R.E.**, Structure-activity studies in cardiac glycosides, in *Natural Product and Drug Development*, Alfred Benson Symposium 20, Krogsgaard-Larsen, P., Christensen, S.B., and Kofod, H., Eds., Munksgaard, Copenhagen, 1984, 337.
59. Cardiac glycosides, in *Handbook of Experimental Pharmacology*, Greeff, Ed., Vol. 56 I/II, Springer-Verlag, Berlin, 1981.
60. **Arora, R.B. and Rangaswami, S.**, *Peruvoside and Other Cardiotonic Glycosides of Thevetia Neriifolia Juss.*, Thomson (India), New Delhi, 1972.
61. **Bisset, N.G.**, Cardiac glycosides. IV. Apocynaceae: A preliminary paper-chromatographic study of the glycosides from *Thevetia peruviana* (Pers.) K.Schum., *Ann. Bogor.*, 4, 145, 1961.
62. **Venkata Rao, E., Devi., V., and Krishna Rao, R.V.**, Chemical examination of the bark of *Thevetia nereifolia*, *Indian J. Pharm.*, 29, 44, 1967.
63. **Seitz, G. and Ripphahn, J.**, German Offen. 1,959,039, 1971.
64. **Balsam, G. and Kuffner, O.**, Peruvoside, German Offen. 1,959,039, 1971.
65. **De, M. and Sharma, A.K.**, Cardenolide contents in different genotypes of *Thevetia neriifolia* and *Nerium odorum*, *Nucleus (Calcutta)*, 23, 213, 1980.
66. *Martindale. The Extra Pharmacopoeia*, 28th ed., Reynolds, J.E.F., Ed., Pharmaceutical Press, London, 1982, 531.
67. **Schorscher, E., Sommer, S., Wild, A.J.N., and Block, A.**, Untersuchungen über die Wirkung von Peruvosid auf die Herz-funktionen im Tierversuch, *Arzneimittelforschung*, 18, 1582, 1968.
68. **Imai, S., Katano, Y., Takeda, K., Kumada, M., Iizuke, H., Shiraki, Y., and Sakai, K.**, Quantitative comparison of the cardiac actions of peruvoside with those of G-strophanthin and proscillaridin, *Tohoku J. Exp. Med.*, 108, 389, 1972.
69. **Gao, Shijia and Zeng, Guiyun**, Cardiotonic and toxic effects of peruvoside and neriifolin, *Yaoxue Xuebao*, 18, 572, 1983.
70. **Hundeshagen, H., Zeidler, U., Fabel, H., Garbe, A., and Nowak, H.**, Zur Pharmakokinetik des Peruvosid, *Arzneimittelforschung*, 18, 1592, 1968.
71. **Wang, Muzou, Gao Fengying, and Li Bailong**, Analysis of thevetoside, a mixture of cardiac glycosides extracted from *Thevetia peruviana* Merr., *Yaoxue Xuebao*, 16, 379, 1981.
72. **Gao, Shijia**, Pharmacological and clinical studies of *Thevetia neriifolia*, *Yaoxue Xuebao*, 18, 875, 1983.
73. **Cruz, A., García, I., Iriarte, J., Muchowski, J.M., and Regla, I.**, Seeds of *Thevetia* species as an alternative source of digitoxigenin, *J. Org. Chem.*, 42, 3580, 1977.
74. **Cruz, A., Guzmán, A., Iriarte, J., Medina, R., and Muchowski, J.M.**, 18,20-Oxido-20,22-dihydroneriifolin, an unusual oxygenated cardanolide, *J. Org. Chem.*, 44, 3511, 1979.

75. **Seelkopf, C.,** Isolation of cardiac glycosides from *Aouia nitida, Rev. Fac. Farm., Univ. Los Andes*, 15, 147, 1974; cited from *Chem. Abstr.* 83, 75351, 1975.
76. **Jolad, S.D., Hoffmann, J.J., and Bates, R.B.,** 3′-*O*-Methyl-evomonoside: a new cytotoxic cardiac glycoside from *Thevetia ahouia* A.DC. (Apocynaceae), *J. Org. Chem.*, 46, 1946, 1981.
77. **Kyerematen, G., Hagos, M., Weeratunga, G., and Sandberg, F.,** The cardiac glycosides of *Thevetia ovata* A.DC. and *Thevetia nereifolia* Juss. ex Steud., *Acta Pharm. Suec.*, 22, 37, 1985.
78. **Sáenz, J.A. and Nassar, M.,** Glicósidos de *Thevetia plumeriaefolia* Benth. I., *Rev. Biol. Trop.* 9, 233, 1961.
79. **Yamauchi, T., Abe, F., Tachibana, Y., Atal, C.K., Sharma, B.M., and Imre, Z.,** Quantitative variations in the cardiac glycosides of oleander, *Phytochemistry*, 22, 2211, 1983.
80. **Janiak, P.St., Weiss, Ek., von Euw, J., and Reichstein, T.,** Die Konstitution von Adynerin, *Helv. Chim. Acta*, 46, 374, 1963.
81. **Tittel, G. and Wagner, H.,** Qualitative und quantitative Analyse von Herzglykosiddrogen durch HPLC-Verfahren. III. Mitteilung: qualitative und quantitative HPLC der Cardenolide aus Nerium oleander-Drogen und deren Zubereitungen, *Planta Med.*, 43, 251, 1981.
82. **Yamauchi, T., Takata, N., and Mimura, T.,** Cardiac glycosides of the leaves of *Nerium odorum, Phytochemistry*, 14, 1379, 1975.
83. **Yamauchi, T., Takahashi, M., and Abe, F.,** Cardiac glycosides of the root bark of *Nerium odorum, Phytochemistry*, 15, 1275, 1976.
84. **Imbesi, A.,** *Index Platarum quae in Omnium Populorum Pharmacopoeis sunt Adhuc Receptae*, Messina, 1964.
85. **Jurenitsch, J., Kopp, B., Bamberg-Kubelka, E., Kern, B., and Kubelka, W.,** Hochleistungsflüssigchromatographische Trennung der herzwirksamen Glykoside von *Convallaria majalis* L. durch Serienschaltung verschieden polarer "Reserved Phase" Säulen, *J. Chromatogr.*, 2409, 125, 1982.
86. **Koppe, B. and Kubelka, W.,** Neue Cardenolide aus Convallaria majalis, XIV. Mitteilung über Convallaria-Glykoside, *Planta Med.*, 45, 87, 1982.
87. **Bleier, W., Kaiser, S., Kubelka, W., and Wichtl, M.,** Beziehungen zwischen Standort und Glykosid-Zusammensetzung bei *Convallaria majalis* L., *Pharm. Acta Helv.*, 42, 423, 1967.
88. **Máthé, A., Máthé, I., Jr., and Kaposi, P.,** Angaben zur Frage der geographischen Rassen und ontogenetischen Entwicklung von *Convallaria majalis* L., *Herb. Hung.*, 17(2), 7, 1978.
89. **Máthé, A., Jr. and Máthé, I.,** Study of the cardiac glycoside production of *Convallaria majalis* L., *Acta Pharm. Hung.*, 47, 195, 1977.
90. **Borisova, N.A., Stankevich, N.B., and Davydov, V.V.,** Biological activity of the [drug] raw material of *Convallaria majalis* procured in the Leningrad region, *Farmatsiya* (Moscow), 33(5), 9, 1984; cited from *Chem. Abstr.*, 101, 216262, 1984.
91. **Schenk, B., Junio, P., and Wichtl, M.,** Cannogenol-3-*O*-α-L-Rhamnosid und Cannogenol-3-*O*-β-D-allomethyloside, zwei neue Cardenolidglykoside aus Convallaria majalis, *Planta Med.*, 40, 1, 1980.
92. **Komissarenko, N.F., Chernobai, V.T., and Kolesnikov, D.G.,** Cardiac glycosides of Far Eastern COnvallaria keiskei, *Med. Prom. S.S.S.R.*, 15(1), 12, 1961; cited from *Chem. Abstr.*, 55, 21484, 1961.
93. **Kommisarenko, N.F. and Stupakova, E.P.,** Rhodexin A from *Convallaria keiskei, Khim. Prir. Soedin.*, 6, 790, 1983.
94. **Kimura, M., Tohma, M., and Yoshizawa, I.,** Constituents of Convallaria. VI. Isolation of convallatoxol, *Chem. Pharm. Bull.*, 15, 226, 1967.
95. **Dzynba, N.P., Vorob'ev, N.E., Kazarinov, N.A., Puchkova, E.I., and Sokolova, A.I.,** Chromotographic and spectrophotometric methods for analyzing phytochemical preparations and plant raw materials containing cardiac glycosides, *Postep. Dziedzinie Leku Rosl., Pr. Ref. Dosw. Wygloszone Symp. 1970* (publ. 1972), Kaczmarek, F., Eds., 53; cited from *Chem. Abstr.*, 78, 102051, 1973.
96. **Chekman, I.S., Gorchakova, N.O., and Dombrovska, A.M.,** Clinical pharmacology of Convallaria drugs, *Farm. Zh. (Kiev)*, 2, 56, 1977.
97. **Pusz, A. and Büchner, S.,** Papierchromatographische Untersuchung der Blüten von *Adonis vernalis* L., *Arzneimittelforschung*, 12, 932, 1962.
98. **Pusz, A. and Büchner, S.,** The chromatographic examination of *Adonis vernalis* L. II. The cardenolides of pheasant's eye leaf, *Arzneimittelforschung*, 13, 409, 1963.
99. **Junior, P. and Wichtl, M.,** 3-*Epi*-periplogenin: ein neues Cardenolid aus *Adonis vernalis, Phytochemistry*, 19, 2193, 1980.
100. **Winkler, C. and Wichtl, M.,** Neue Cardenolide aus *Adonis vernalis, Planta Med.*, 68, 1986.
101. **Kavalali, G.,** Isolation of cymarin from *Adonis aestivalis, Lloydia*, 36, 426, 1973.
102. **El Kiey, M.A., Darwish Sayed, M., Abd El Wahab, S.M., and Soliman, F.M.,** Chromatographic study of the cardenolides in the overground part of flowering-fruiting herbs of *Adonis autumnalis* L. and *Adonis dentata* Del., *J. Pharm. Sci., U.A.R.*, 10, 81, 1969.
103. **Yatsyuk, V.Ya., Komissarenko, N.F., and Gella, E.V.,** Chemical composition of the above-ground part of *Adonis wolgensis, Rastit. Resurs.*, 11, 515, 1975.
104. **Yatsyuk, V.Ya., Komissarenko, N.F., and Gella, E.V.,** Cardenolides from *Adonis wolgenis, Khim. Prir. Soedin.*, (5), 672, 1976.
105. **Komissarenko, N.F., Korzennikova, E.P., and Yatsyuk, V.Ya.,** Cardenolides of *Adonis wolgensis, Khim. Prir. Soedin.*, (3), 433, 1973.
106. **Goryaev, M.I., Murzagulova, Z.G., Yurina, K.L., and Shishkina, N.V.,** Chemical composition of *Adonis vernalis* and *Adonis wolgensis, Izv. Akad. Nauk. Kaz. S.S.R., Ser. Khim.*, 21(6), 31, 1971.
107. **Thieme, H. and Lamchav, A.,** Phytochemical study of Mongolian medicinal plants. II. Cardenolides of *Adonis mongolica, Pharmazie*, 25, 202, 1970.
108. **Thieme, H. and Lamchav, A.,** Über die Cardenolidglykoside von *Adonis mongolica* Sim., *Pharmazie*, 31, 565, 1976.
109. **Yang Xuemin and Qin Guowei,** Cardiac glycosides from Adonis plants and their chemistry, *Zhongcaoyao*, 14, 139, 1983.
110. **Evdokimov, P.E., Mamdov, Yu.M., and Yusupov, M.Zh.,** Tien-Shan Adonis — a new medicinal plant, *Med. Zh. Uzb.*, 2, 56, 1980.
111. **Evdokimov, P.E.,** Chemical composition of *Adonis tianshanica* (Adolf) Lipschitz and pharmacological properties of galenicals produced from it, *Rastit. Resurs.*, 15, 565, 1979.
112. **Evdokimov, P.E. and Khalmatov, Kh.Kh.,** Effect of drying on the biological activity and cardenolide composition of Golden and Tian-Shan Adonis, *Farmatsiya (Moscow)*, 28(6), 44, 1979.
113. **Evdokimov, P.E. and Nazirov, Z.N.,** Study of *Adonis turkestanica* (Korsch) Adolf, *Med. Zh. Uzb.*, 8, 44, 1977.
114. **Evdokimov, P.E. and Khalmatov, Kh.Kh.,** Shelf life and storage conditions of the herb *Adonis turkestanica* (Korsch) Adolf, *Farmatsiya (Moscow)*, 27(2), 43, 1978.
115. **Kolarova, B. and Boyadzhieva, M.,** Cardiac glycosides of *Erysimum* (Cruciferae), *Probl. Farm.*, 5, 9, 1977.
116. **Kolarova, B. and Boyadzhieva, M.,** Cardiac glycosides of *Erysimum* (Cruciferae), *Probl. Farm.*, 7, 70, 1979.
117. **Tai, Baoshan,** Study on injection of cardiac glycosides from *Erysimum cheiranthoides* L., *Yaoxue Tongbao*, 17, 325, 1982.
118. **Puchkova, E.I., Musyanovich, V.M., Kolomiitseva, T.N., Kazarinov, N.A., and Georgievskii, V.P.,** Quantitative analysis of cardiac glycosides in *Strophanthus kombe* seeds, *Rastit. Resurs.*, 11, 268, 1975.
119. **Makarevich, I.F.,** Cardenolides of *Strophanthus kombe, Khim. Prir. Soedin.*, (2), 180, 1972.
120. **Wolfgang, H.W.,** Ger. Patent 2,050,457, 1972.
121. **Hou, C.Y., Liu, Y.L., and Sun, N.C.,** Isolation, separation, and identification of lipophilic glycosides from *Strophanthus divaricatus* (Lour.) Hook. et Arn., *Zhongcaoyao*, 11, 289, 1980.
122. **Kopp, B.,** Bufadienolides. II. Bufadienolide content, composition, and distribution in some samples of *Urginea maritima, Sci. Pharm.*, 51, 238, 1983.
123. **Patil, M.M. and Torne, S.G.,** Chemical analysis of *Urginea indica* (Roxb.) Kunth cytotypes, *Curr. Sci.*, 50, 956, 1981.
124. **Kopp, B. and Danner, M.,** Bufadienolides. I. Structure of the bufadienolides of *Urginea indica* (Roxb.) Kunth, *Sci. Pharm.*, 51, 227, 1983.
125. **Randimbivololona, F., Pellegrin, P., and Lesne, M.,** Les cardiotoniques semi-synthétiques apparentés à la digitale et les nouveaux agents inotropes positifs de synthèse, *J. Pharm. Belg.*, 39, 225, 1984.

126. **Trease, G.E. and Evans, W.C.,** *Pharmacognosy*, 12th ed., Bailliere Tindall, London, 1983, 505.
127. **Weiler, E.W. and Westekemper, P.,** Rapid selection of strains of *Digitalis lanata* Ehrh. with high digoxin content, *Planta Med.*, 35, 316, 1979.
128. **Garve, M., Luckner, M., Vogel, E., Tewes, A., and Nover, L.,** Growth, morphogenesis and cardenolide formation in long-term cultures of *Digitalis lanata*, *Planta Med.*, 40, 92, 1980, and references cited therein.
129. **Kuberski, C., Scheibner, H., Steup, C., Diettrich, B., and Luckner, M.,** Embryogenesis and cardenolide formation in tissue cultures of *Digitalis lanata*, *Phytochemistry*, 23, 1407, 1984, and references cited therein.
130. **Hagimori, M., Matsumoto, T., and Mikami, Y.,** Digitoxin biosynthesis in isolated mesophyll cells and cultured cells of Digitalis, *Plant Cell Physiol.*, 25, 947, 1984.
131. **Diettrich, B., Haack, U., Popov, A.S., Butenko, R.G., and Luckner, M.,** Long-term storage in liquid nitrogen of an embryogenic cell strain (VII) of *Digitalis lanata*, *Biochem. Biophys. Pflanz.*, 180, 33, 1985.
132. **Seitz, U., Alfermann, A.W., and Reinhard, E.,** Stability of biotransformation capacity in *Digitalis lanata* cell cultures after cryogenic storage, *Plant Cell Rep.*, 2, 273, 1983.
133. **Linley, P.A., Harkiss, K.J., and Mesbah, M.K.M.,** A radio-tracer study of the biotransformation of cardiac glycosides in *Digitalis lanata*, *Fitoterapia*, 56, 267, 1985, and references cited therein.
134. **Döller, P.C. and Reinhard, E.,** Biotransformation of cardenolides. Comparative studies with cell cultures of *Thevetia neriifolia* and *Digitalis lanata*, *Planta Med.*, 37, 277, 1979.
135. **Alfermann, A.W., Schuller, I., and Reinhard, E.,** Biotransformation of cardiac glycosides by immobilized cells of *Digitalis lanata*, *Planta Med.*, 40, 218, 1980.
136. **Heins, M., Wahl, J., Lerch, H., Kaiser, F., and Reinhard, E.,** Preparation of β-methyldigoxin by hydroxylation of β-methyl-digitoxin in fermenter cultures of *Digitalis lanata*, *Planta Med.*, 33, 57, 1978.
137. **Eberlein, W., Nickl, J., Heider, J., Dahms, G., and Machleidt, H.,** Einführung von Substituenten in den Butenolid-Ring von Herzglykosiden, *Chem. Ber.*, 105, 3686, 1972.
138. **Eberlein, W., Heider, J., and Machleidt, H.,** Substitution des Butenolidringes von Herzglycosiden durch unverzweigte offenkettige π-Elektronensysteme, *Chem. Ber.*, 107, 1275, 1974.
139. **Thomas, R., Boutagy, J., and Gelbart, A.,** Synthesis and biological activity of semisynthetic *Digitalis* analogs, *J. Pharm. Sci.*, 63, 1649, 1974.
140. **Thomas, R., Brown, L., Boutagy, J., and Gelbart, A.,** The *Digitalis* receptor. Inferences from structure-activity relationship studies, *Circ. Res.*, 46, Suppl. I, I-176, 1980.
141. **Jones, H.P.,** European Patent Application EP 56,995, 1982; cited from *Chem. Abstr.*, 97, 20323, 1982.
142. Mitsubishi Yuka Pharmaceutical Co. Ltd., Japan Kokai Tokkyo Koho JP 82 62,297, 1982; cited from *Chem. Abstr.*, 97, 110336, 1982.
143. **Hegnauer, R.,** The chemistry of the Compositae, in *The Biology and Chemistry of the Compositae*, Vol. 1, Heywood, V.H., Harborne, J.B., and Turner, B.L., Eds., Academic Press, London, 1977, 283.
144. **Mabry, T.J. and Bohlmann, F.,** Summary of the chemistry of the Compositae, in *The Biology and Chemistry of the Compositae*, Vol. 2, Heywood, V.H., Harborne, J.B., and Turner, B.L., Eds., Academic Press, London, 1977, 1097.
145. **Aebi, A., Waaler, T., and Buchi, J.,** Petasin and S-petasin, the spasmolytic active principles in *Petasites officinalis*, *Pharm. Weekbl.*, 93, 397, 1958.
146. **Blakeman, J.P. and Atkinson, P.,** Antimicrobial properties and possible role in host-pathogen interactions of parthenolide, a sesquiterpene lactone isolated from *Chrysanthemum parthenium*, *Physiol. Plant Pathol.*, 15, 183, 1979.
147. **Hall, I.H., Lee, K.H., Starnes, O.C., Muraoka, O., Sumida, Y., and Waddell, T.G.,** Antihyperlipidaemic activity of sesquiterpene lactones and related compounds, *J. Pharm. Sci.*, 69, 694, 1980.
148. **Herz, W.,** Asterae — chemical review, in *The Biology and Chemistry of the Compositae*, Vol. 1, Heywood, V.H., Harborne, J.B., and Turner, B.L., Eds., Academic Press, London, 1977, 567.
149. **Wagner, H.,** Pharmaceutical and economic uses of the Compositae, in *The Biology and Chemistry of the Compositae*, Vol. 1, Heywood, V.H., Harborne, J.B., and Turner, B.L., Eds., Academic Press, London, 1977, 411.
150. **Hall, I.H., Lee, K.H., Starnes, C.O., Sumida, Y., Wu, R.Y., Waddell, T.G., Cochran, J.W., and Gerhart, K.G.,** Anti-inflammatory activity of sesquiterpene lactones and related compounds, *J. Pharm. Sci.*, 68, 537, 1979.
151. **Hall, I.H., Starnes, C.O., Lee, K.H., and Waddell, T.G.,** Mode of action of sesquiterpene lactones as anti-inflammatory agents, *J. Pharm. Sci.*, 69, 537, 1980.
152. **Lee, K.H., Huang, E.S., Piantadozi, C., Pagano, J.S., and Geissman, T.A.,** *Cancer Res.*, 31, 1649, 1971.
153. **Medawar, P.B., Robinson, G.M., and Robinson, R.,** A synthetic differential growth inhibitor, *Nature*, 151, 195, 1943.
154. **Hladon, B. and Twardowski, T.,** Sesquiterpene lactones. XXV. Studies on the mode of action. The cellular and molecular basis of cytostatic action, *Pol. J. Pharmacol. Pharm.*, 31, 35, 1979.
155. **Vuorela, H., Cook, N., and Hittunen, R.,** The calcium antagonism of some chamomile compounds tested on depolarized rabbit aortic rings, *Acta Agron.*, 34(Suppl.), 95, 1985.
156. **Johnson, E.S., Kadam, N.P., Hylands, D.M., and Hylands, P.J.,** Efficacy of feverfew as prophylactic treatment of migraine, *Br. Med. J.*, 291, 569, 1985.
157. **Johnson, E.S., Hylands, D.M., and Hylands, P.J.,** U.K. Patent GB 2,124,486B, 1986.
158. **Collier, H.O.J., Butt, N.M., McDonald-Gibson, W.J. and Saeed, S.A.,** Extract of fverfew inhibits prostaglandin biosynthesis, *Lancet*, 2, 922, 1980.
159. **Makheja, A.N. and Bailey, J.M.,** A platelet phospholipase inhibitor from the medicinal herb feverfew (*Tanacetum parthenium*), *Prostagl. Leukotr. Med.*, 8, 653, 1982.
160. **Heptinstall, S., White, A., Williamson, L., and Mitchell, J.R.A.,** Extracts of feverfew inhibit granule secretion in blood platelets and polymorphonuclear leucocytes, *Lancet*, 1, 1071, 1985.
161. Quinghaosu antimalaria co-ordinating research group, Antimalaria studies on quinghaosu, *Chin. Med. J.*, 92, 811, 1979.
162. **Klayman, D.L.,** Qinghaosu (artemesinin): an antimalarial drug from China, *Science*, 228, 1049, 1985, and references cited therein.
163. **Schmid, G. and Hofheinz, W.,** Total synthesis of qinghaosu, *J. Am. Chem. Soc.*, 105, 624, 1983.
164. **Evans, F.J. and Soper, C.J.,** The tigliane, daphnane and ingenane diterpenes, their chemistry, distribution and biological activities. A review, *Lloydia*, 41, 193, 1978.
165. **Lindner, E., Dohadwalla, A.N., and Bhattacharya, B.K.,** Positive inotropic and blood pressure lowering activity of a diterpene derivative isolated from *Coleus forskohli*: forskolin, *Arzneimittelforschung*, 28, 284, 1978.
166. **Metzger, H. and Lindner, E.,** The positive inotropic-acting forskolin, a potent adenylate cyclase activator, *Arzneimittelforschung*, 31, 1248, 1981.
167. **Lindner, E. and Metzger, H.,** The action of forskolin of muscle cells is modified by hormones, calcium ions and calcium antagonists, *Arzneimittelforschung*, 33, 1436, 1983.
168. **Ealey, P.A., Kohn, L.D., Marshall, N.J., and Ekins, R.P.,** Forskolin stimulation of naphthylamidase in quinea pig thyroid sections detected with a cytochemical bioassay, *Acta Endocrinol.*, 108, 367, 1985.
169. **Hersey, S.J., Owirodu, A. and Miller, M.,** Forskolin stimulation of acid and pepsinogen secretion by gastric glands, *Biochim. Biophys. Acta*, 755, 293, 1983.
170. **Seamon, K.B., Daly, J.W., Metzger, H., de Souza, N.J., and Reden, J.,** Structure-activity relationships for activation of adenylate cyclase by the diterpene forskolin and its derivatives, *J. Med. Chem.*, 26, 436, 1983.
171. **Agarwal, K.C. and Parks, R.E.,** Forskolin: a potential antimetastatic agent, *Int. J. Cancer*, 32, 801, 1983.

172. **Kreutner, W., Chapman, R.W., Gulbenkian, A., and Tozzi, S.,** Bronchodilator and antiallergy activity of forskolin, *Eur. J. Pharmacol.*, 111, 1, 1985.
173. **Caprioli, J., Sears, M., Bausher, L., Gregory, D., and Mead, A.,** Forskolin lowers intraocular pressure by reducing aqueous inflow, *Invest. Ophthalmol. Vis. Sci.*, 25, 268, 1984.
174. **Elliott, M. and Janes, N.F.,** Synthetic pyrethroids — a new class of insecticide, *Chem. Soc. Rev.*, 7, 473, 1978.
175. **Elliott, M., Farnham, A.W., Janes, N.F., Needham, P.H., Pulman, D.A., and Stevenson, J.H.,** A photostable pyrethroid, *Nature*, 246, 169, 1973.
176. **Ohno, N., Fujimoto, K., Okuno, Y., Mizutani, T., Hirano, M., Itaya, N., Honda, Y., and Yoshioka, H.,** 2-Arylalkanoates, a new group of synthetic pyrethroid esters not containing cyclopropanecarboxylates, *Pest. Sci.*, 7, 241, 1976.
177. Deltamethrin monograph, Roussel Uclaf, Paris, 1982.
178. **Horrobin, D.F.,** *Clinical Uses of Essential Fatty Acids*, Eden Press, Montreal, 1982, 3.
179. **Brenner, P.R.,** Nutritional and hormonal factors influencing desaturation of essential fatty acids, *Progr. Lipid Res.*, 20, 41, 1982.
180. **Wolf, R.B., Kleiman, R., and England, R.E.,** New sources of γ-linolenic acid, *J. Am. Oil Chem. Soc.*, 60, 1858, 1983.
181. **Traitler, H., Winter, H., and Richli, U.,** Characterisation of γ-linolenic acid in *Ribes* seed, *Lipids*, 19, 923, 1984.
182. **Suzuki, O.,** Production of γ-linolenic acid by filamentous fungi. Probe of lipid production by microorganisms. Advantages of concentration, *Kagaku to Seibutsu*, 23, 11, 1985; cited from *Chem. Abstr.* 102, 219549, 1985.
183. **Jacono, J.M., Judd, J.T., and Marshall, W.M.,** The role of dietary essential fatty acids and prostaglandins in reducing blood pressure, *Progr. Lipid Res.*, 20, 349, 1982.
184. **Fisher, J.M., Donegan, D., and Leon, H.,** Effects of prostaglandins and their precursors in some tests of hemostatic function, *Progr. Lipid Res.*, 20, 799, 1982.
185. **Houtsmuller, A.J., van Hal-Ferwerda, J., and Zahan, K.J.,** Favourable influences of linoleic acid on the pregression of diabetic micro- and macro-angiopathy, *Nutr. Metab.*, 24(Suppl. 1), 105, 1980.
186. **Brush, M.G.,** in *Clinical Uses of Essential Fatty Acids*, Horrobin, D.F., Ed., Eden Press, Montreal, 1982, 155.
187. **McCormick, J.N., Neill, W.A., and Sim, A.K.,** Immunosuppressive effect of linolenic acid, *Lancet*, 2, 508, 1977.
188. **Wright, S. and Burton, J.L.,** in *Clinical Uses of Essential Fatty Acids*, Horrobin, D.F., Ed., Eden Press, Montreal, 1982, 73.
189. **Horrobin, D.F.,** Multiple sclerosis: the rational basis for treatment with colchicine and evening primrose oil, *Med. Hypotheses*, 5, 365, 1979.
190. **Horrobin, D.F.,** The regulation of prostaglandin biosynthesis by the manipulation of essential fatty acid metabolism, *Rev. Pure Appl. Pharmacol. Sci.*, 4, 339, 1983.
191. **Lindequist, U. and Teuscher, E.,** Pflanzliche und mikrobielle Wirkstoff als Immunomodulatoren, *Pharmazie*, 40, 10, 1985.
192. **Wagner, H.,** Immunostimulants of fungi and higher plants, in *Natural Products and Drug Development*, Alfred Benson Symposium 20, Krogsgaard-Larsen, P., Christensen, S.B., and Kofod, H., Eds., Munksgaard, Copenhagen, 1984, 391.
193. **Wagner, H. and Proksch, A.,** Immunostimulatory drugs of fungi and higher plants, in *Economic and Medicinal Plant Research.*, Vol. 1, Wagner, H., Hikino, H., and Farnsworth, N.R., Academic Press, London, 1985, 113.
194. **Biozzi, G., Benacerraf, B. and Halpern, B.N.,** Granulopectic activity of the reticuloendothelial system. II. Kinetics of the granulopectic of the relation between the weight of organs and the activity of the reticuloendothelial system in relation to the dose of C injected, *Br. J. Exp. Pathol.*, 34, 441, 1953.
195. **Brandt, L.,** Studies on the phagocytic activity of neutrophilic leukocytes with special reference to chronic myeloproliferative conditions and megaloblastic anemia, *Scand. J. Haematol.*, Suppl. 2, 5, 1967.
196. **Wagner, H., Proksch, A., Vollmar, A., Kreutzkamp, B., and Bauer, J.,** *In vitro*-Phagozytose-Stimulierung durch isolierte Pflanzen-stoffe gemessen im Phagozytose-Chemolumineszenz-(CL)-Modell, *Planta Med.*, 139, 1985.
197. **Allen, R.C.,** in *Bioluminescence and Chemoluminescence: Basic Chemistry and Analytical Application*, de Luca and McElroy, Eds., Academic Press, New York, 1981, 63.
198. **Wagner, H., Proksch, A., Reiss-Maurer, I., Vollmar, A., Odenthall, S., Stuppner, H., Jurcic, K., le Turdu, M., and Fang, J.N.,** Immunostimulierend wirkende Polysaccharide (Heteroglykane) aus hoheren Pflanzen, *Arzneimittelforschung*, 35, 1069, 1985.
199. **Bloksma, N., Schiermann, P., de Reuver, M., van Dyk, H., and Willers, J.,** Stimulation of humoral and cellular immunity by *Viscum* preparations, *Planta Med.*, 46, 221, 1982.
200. **Kumazawa, Y., Mizunoe, K., and Otsuka, Y.,** Immunostimulant polysaccharides from *Angelica acutiloba*, *Immunology*, 47, 75, 1982.
201. **Caldes, G., Prescott, B., Thomas, C.A., and Barker, P.L.,** Characterisation of a polysaccharide from *Carthamus tinctorius* that cross-reacts with type III pneumococcal polysaccharide, *J. Gen. Appl. Microbiol.*, 27, 157, 1981.
202. **Beuscher, N. and Kopanski, L.,** Stimulation der Immunantwort durch Inhaltstoffe aus *Baptisia tinctoria*, *Planta Med.*, 381, 1985.
203. **Valdes, R., and Cordoba, F.,** Effect of zexbrevins A and B, two new sesquiterpene lactones, on the immune response of mice, *Agents Actions*, 5, 64, 1975.
204. **Dupuis, G. and Brisson, J.,** Toxic effect of alantolactone and dihydroalantolactone in *in vitro* cultures of leukocytes, *Chem.-Biol. Interact.*, 15, 205, 1976.
205. **Schorlemmer, H.U., Dickneite, G., Sedlacek, H.H., de Souza, N.J., and Dohadwalla, A.N.,** Forskolin for immune stimulation, Eur. Patent Appl. EP 126,313, 1985; cited from *Chem. Abstr.*, 102, 100797, 1985.
206. **Singh, V.K., Agarwal, S.S., and Gupta, P.M.,** Immunomodulatory activity of *Panax ginseng* extract, *Planta Med.*, 462, 1984.
207. **Mose, J.L.,** Versuche über die Wirksamkeit von Aristolochiasäure, *Planta Med.*, 11, 73, 1963.
208. **Wagner, H., Kreutzkamp, B., and Jurcic, K.,** Die Alkaloide von *Uncaria tomentosa* und ihre Phagozytose-steigernde Wirkung, *Planta Med.*, 419, 1985.
209. **Kumuzawa, Y., Hageki, A., Fukumoto, M., Fujisawa, H., Nishimura, C., and Nomoto, K.,** Activation of peritoneal macrophages by berberine-type alkaloids in terms of induction of cytostatic activity, *Int. J. Immunol.*, 6, 587, 1984.
210. **Kasajima, T.,** Effect of cepharanthine in promoting antibody production, *Jpn. J. Reticuloendothelial Soc.*, 14, 1 1974.
211. **Ghosal, S., Saini, K.S., and Arora, V.K.,** 1,2-β-Epoxy-ambelline, an immunostimulant alkaloid from *Crinum latifolium*, *J. Chem. Res. (Synop.)*, 1984.
212. **Brattig, N.W., Diao, G.J., and Berg, P.A.,** Immunoenhancing effect of flavonoid comounds on lymphocyte proliferation and immunoglobulin synthesis, *Int. J. Immunopharm.*, 6, 205, 1984.
213. **Wagner, H., Kreutzkamp, B., and Jurcic, K.,** Inhaltstoffe und Pharmakologie der *Okoubaka aubrevillei*-Rinde, *Planta Med.*, 404, 1985.
214. **Rao, R.R. and Nair, T.B.,** Investigations on induction of neutrophilic granulocytosis and toxicity of cleistanthin — CIBA GO, 4350 — a new glycoside from *Cleistanthus collinus*, *Pharmacology*, 4, 347, 1970.
215. **Hikino, H.,** Antihepatotoxic principles in oriental plants, in *Natural Products and Drug Development*, Alfred Benson Symposium 20, Krogsgaard-Larsen, P., Christensen, S.B., and Kofod, H., Eds., Munksgaard, Copenhagen, 1984, 374.
216. **Wagner, H.,** Plant constituents with antihepatotoxic activity, in *Natural Products as Medicinal Agents*, Beal, J.L. and Reinhard, E., Eds., Hippokrates, Stuttgart, 1981, 217.
217. **Kiso, Y., Tohkin, M., and Hikino, H.,** Assay method for antihepatotoxic activity using carbon tetrachloride induced cytotoxicity in primary cultured hepatocytes, *Planta Med.*, 49, 222, 1983.
218. **Kiso, Y., Kato, O., and Hikino, H.,** Assay methods for antihepatotoxic activity using peroxide-induced cytotoxicity in primary cultured hepatocytes, *Planta Med.*, 50, 1985.
219. **Merlini, L., Zanrotti, A., Pelter, A., Rochefort, M.P.,and Hänsel, R.,** Benzodioxans by oxidative phenyl coupling. Synthesis of silybin, *J. Chem. Soc., Perkin Trans. I*, 775, 1980.

220. **Fiebig, M. and Wagner, H.,** Neue antihepatotoxisch wirksame Flavolignane einer weissblühenden *Silybum*-Varietät, *Planta Med.*, 310, 1984.
221. **Hikino, H., Kiso, Y., Wagner, H., and Fiebig, M.,** Antihepatotoxic actions of flavonolignans from *Silybum marianum* fruits, *Planta Med.*, 248, 1984.
222. **Machicao, F. and Sonnenbichler, H.,** Mechanism of the stimulation of RNA synthesis in rat liver nuclei by silybin, *Z. Physiol. Chem.*, 358, 141, 1977.
223. **Kiso, Y., Ogasawara, S., Hirota, K., Watanabe, N., Oshima, Y., Konno, C., and Hikino, H.,** Antihepatotoxic principles of *Artemisia capillaris* buds, *Planta Med.*, 81, 1984.
224. **Iwu, M.M.,** Antihepatotoxic constituents of *Garcinia cola* seeds, *Experientia*, 41, 699, 1985.
225. **Gajdos, A., Gajdos-Torok, M., and Horn, R.,** The effect of (+)-catechin on the hepatic level of ATP and the lipid content of liver during experimental steatosis, *Biochem. Pharmacol.*, 21, 595, 1972.
226. **Gajdos, A., Gajdos-Torok, M., and Horn, R.,** Action of (+)-catechin on the hepatic tissue of the white rat subjected to ethanol intoxification, *C. R. Soc. Biol., Paris*, 164, 2187, 1970.
227. **Panizzi, L., Scarpati, M.L., and Scarpati, R.,** Constitution of cynarin, the active principle of artichoke, *Gazz. Chim. Ital.*, 84, 792, 1954.
228. **Maros, T., Racz, G., Katonai, B., and Kovacs, V.,** Effect of *Cynara scolymus* extracts on the regeneration of rat livers, *Arzneimittelforschung*, 16, 127, 1966.
229. **Kiso, Y., Suzuki, Y., Watanabe, N., Oshima, Y., and Hikino, H.,** Antihepatotoxic principles of *Curcuma longa* rhizomes, *Planta Med.*, 49, 185, 1983.
230. **Hikino, H., Kiso, Y., Taguchi, H., and Ikega, Y.,** Antihepatotoxic actions of lignoids from *Schizandra chinensis* fruits, *Planta Med.*, 213, 1984.
231. **Kiso, Y., Tohkin, M., Hikino, H., Ikega, Y., and Taguchi, H.,** Mechanism of antihepatotoxic activity of wuweizisu C and gomisin A, *Planta Med.*, 331, 1985.
232. **Hikino, H., Sugai, T., Konno, C., Hashimoto, I, Terasaki, S., and Hirono, I.,** Liver-protective principle of *Thujopsis dolabrata* leaves, *Planta Med.*, 36, 156, 1979.
233. **Kiso, Y., Konno, C., Hikino, H., Yagi, Y., and Hashimoto, I.,** Liver-protective actions of desoxypodophyllotoxin and its analogs, *J. Pharmaco-Biodynam.*, 5, 638, 1982.
234. **Woo, W.S., Kang, S.S., Seligmann, O., Chari, V.M., and Wagner, H.,** The structure of new lignans from seeds of *Phytolacca americana*, *Tetrahedron Lett.*, 21, 4255, 1980.
235. **Kiso, Y., Tohkin, M., Hikino, H., Hattri, M., Sakamoto, T., and Namba, T.,** Mechanism of antihepatotoxic activity of glycyrrhizin. I. Effect on free radical generation and lipid peroxidation, *Planta Med.*, 298, 1984.
236. **Hashimoto, M., Inada, K., Ohminami, H., Kimura, Y., Okuda, H., and Arichi, S.,** Effects of saikosaponins on liver tyrosine aminotransferase activity induced by cortisone in adrenalectomised rats, *Planta Med.*, 401, 1985.
237. **Hikino, H., Kiso, Y., Kinouchi, S., Sanada, S., and Shoji, J.,** Antihepatotoxic actions of ginsenosides from *Panax ginseng* roots, *Planta Med.*, 62, 1985.
238. **Joo, C.N.,** The effect of ginseng saponins on biochemical reactions, in *Surfactants Soln., Proc. 4th Int. Symp.*, Mittal, K.L. and Lindman, B., Eds., Plenum Press, New York, 1982, 2093.
239. **Hikino, H., Ohsawa, T., Kiso, Y., and Oshima, Y.,** Analgesic and antihepatotoxic actions of dianosides, triterpenoid saponins of *Dianthus superbus* var. *longicalycinus* herbs, *Planta Med.*, 50, 353, 1984.
240. **Kiso, Y., Tohkin, Y., and Hikino, H.,** Mechanism of antihepatotoxic activity of atractylon, I: Effect on free radical generation and lipid peroxidation, *Planta Med.*, 97, 1985.
241. **Kiso, Y., Tohkin, Y., and Hikino, H.,** Mechanism of antihepatotoxic activity of atractylon. I. Effect on free radical generation and lipid peroxidation, *Planta Med.*, 97, 1985.
242. **Pandey, V.N. and Chaturvedi, G.N.,** Effect of the indigenous drug Kutaki (*Picrorhiza kurroa*) on bile after producing biliary fistula in dogs, *Ind. J. Med. Res.*, 57, 503, 1969.
243. **Shen, Mei-Ling,** Advances in natural products in China, *Trends Pharmacol. Sci.*, 6, 496, 1983.
244. **Born, G.V.R. and Cross, M.I.,** The aggregation of blood platelets, *J. Physiol.*, 168, 178, 1961.
245. **Beretz, A., Roth-Georger, A., Corre, G., Kuballa, B., Anton, R., and Cazenave, J.-P.,** Polyindolic alkaloids from *Psychotria fosteriana*. Potent inhibitors of the aggregation of human platelets, *Planta Med.*, 300, 1985.
246. **Stuffness, M. and Douros, J.,** Current status of N.C.I. plant and animal product program, *J. Nat. Prod.*, 45, 1, 1982.
247. **Yun-Choi, H.S., Kim, S.O., Kim, J.H., Lee, J.R., and Cho, H.I.,** Modified smear method for screening potential inhibitors of platelet aggregation from plant sources, *J. Nat. Prod.*, 48, 363, 1985.
248. **Drummond, A.H. and Hughes, P.J.,** The interaction of natural products with cellular signalling mechanisms, *Phytother. Res.*, 1, 1, 1987.
249. **Braquet, P. and Godfroid, J.J.,** PAF-acether specific binding sites. II. Design of specific antagonists, *Trends Pharmacol. Sci.*, 7, 397, 1986.
250. **Nowycky, M.C., Fox, A.P., and Tsien, R.W.,** Three types of neuronal calcium channels with different calcium agonist sensitivity, *Nature*, 316, 440, 1986.
251. **Sutko, J.L., Itto, K., and Kenyon, J.L.,** Ryanodine: a modifier of sarcoplasmic reticulum calcium release in striated mescle, *Fed. Proc.*, 44, 2984, 1985.
252. **Thornber, C.W.,** Isosterism and molecular modification in drug design, *Chem. Soc. Rev.*, 8, 563, 1979.
253. **Ariens, E.J., Soudijn, W., and Timmermans, P.B.M.W.M., Eds.,** *Stereochemistry and Biological Activity of Drugs*, Blackwell Scientific, Oxford, 1983.
254. **Barton, D. and Ollis, W.D., Eds.,** *Advances in Medicinal Phytochemistry*, J. Libbey, London, 1986.
255. **Leeuwenberg, A.J.M., Ed.,** *Medicinal and Poisonous Plants of the Tropics*, Proc. Symp. 5-35, 14th Int. Bot. Congr., Berlin, 1987.
256. **Cody, V., Middleton, E., and Harborne, J.B., Eds.,** *Plant Flavonoids in Biology and Medicine*, Alan R. Liss, New York, 1985.
257. **Bader, G. and Hiller, K.,** Neue Ergebnisse zur Struktur und Wirkungsweise von Triterpensaponinen, *Pharmazie*, 42, 577, 1987.
258. **Mott, K.E., Ed.,** *Plant Molluscicides*, UNDP/WORLD BANK/WHO Special Programme for Research and Training in Tropical Diseases, John Wiley & Sons, Chichester, 1987.
259. **Pellegrin, P. and Lesne, M.,** A new alternative in the field of cardiac glycosides: Gitoxin, *J. Pharm. Belg.*, 41, 39, 1986.
260. **Hassam, A.G.,** The role of evening primrose oil in nutrition and disease, in *The Role of Fats in Human Nutrition*, Padley, F.B. and Podmore, J., Eds., Ellis Horwood, Chichester, 1985, 84.
261. **Shreeve, C.M.,** Evening primrose oil, *Pharm. Update*, 308, 1987.
262. **Cody, V., Middleton, E., and Harborne, J.B., Eds.,** *Plant Flavonoids in Biology and Medicine — Biochemical, Pharmacological and Structure-Activity Relationships*, Alan R. Liss, New York, 1986.
263. **Beretz, A., Stoclet, J.C., and Anton, R.,** Inhibition of isolated rat aorta contraction by flavonoids. Possible correlation with cAMP phosphodiesterase inhibition, *Planta Med.*, 39, 236, 1980.
264. **Meunier, M.-T., Villié, F., Jonadet, M., Bastide, J., and Bastide, P.,** Inhibition of angiotensin I converting enzyme by flavanolic compounds. *In vitro* and *in vivo* studies, *Planta Med.*, 12, 1987.
265. **Tomás-Barberán, F.A., Lopéz-Gómez, C., Villar, A., and Tomás-Lorente, F.,** Inhibition of lens aldose reductase by *Labiatae* flavonoids, *Planta Med.*, 239, 1986.
266. **Laekeman, G.M., Glaeys, M., Rwangabo, P.C., Herman, A.G., and Vlietinck, A.J.,** Cardiovascular effects of 3-methylquercetin, *Planta Med.*, 433, 1986.
267. **Mahato, S.B., Sarkar, S.K., and Poddar, G.,** Triterpenoid saponins, *Phytochemistry*, 27, 3037, 1988.

268. **Murphy, J.J., Hepinstall, S., and Mitchell, J.R.A.,** Randomised double-blind placebo-controlled trial of feverfew in migraine prevention, *Lancet*, 1, 189, 1988.
269. **Anderson, D., Jenkinson, P.C., Dewdney, R.S., Blowers, S.D., Johnson, E.S., and Kadam, N.P.,** Chromosomal aberrations and sister chromatid exchanges in lymphocytes and urine mutagenicity of migraine patients: a comparison of chronic feverfew users and matched non-users, *Hum. Toxicol.*, 7, 145, 1988.
270. **Losche, W., Mazurov, A.V., Heptinstall, S., Repin, V.S., and Till, U.,** An extract of feverfew inhibits interactions of human platelets with collagen substrates, *Thromb. Res.*, 48, 511, 1987.
271. **Barber, A.J.,** Evening primrose oil: a panacea?, *Pharm. J.*, 240, 723, 1988.
272. EPO for atopic eczema, *Pharm. J.*, 241, 545, 1988.
273. **Wagner, H.,** Immunostimulants from higher plants, in *Biologically Active Natural Products*, Hostettmann K. and Lea, P.J., Eds., Blackwell Scientific, Oxford, 1986, 127.
274. **Yang, L.-L., Yen, K.-Y., Konno, C., Oshima, Y., Kiso, Y., and Hikino, H.,** Anti-hepatotoxic principles of *Wedelia chinensis* herbs, *Planta Med.*, 499, 1986.
275. **Budhiraja, R.D., Garg, K.N., Sudhur, S., and Arora, B.,** Protective effect of 3β-hydroxy-2,3-dihydrowithanolide against CCl_4-induced hepatotoxicity, *Planta Med.*, 28, 1986.
276. **Wagner, H., Geyer, B., Kiso, Y., Hikino, H., and Rao, G.S.,** Coumestans as the main active principles of the liver drugs *Eclipta alba* and *Wedelia calendulacea*, *Planta Med.*, 370, 1986.
277. **Hikino, H., Tohkin, M., Kiso, Y., Namiki, T., Nishimura, S., and Takeyama, K.,** Antihepatotoxic actions of *Allium sativum* bulbs, *Planta Med.*, 163, 1986.

Chapter 11

TRADITIONAL REMEDIES AND MODERN ASSESSMENT: THE CASE OF GINSENG

Hiroshi Hikino

I. INTRODUCTION

Human beings have used, for medicinal purposes, natural resources (plants, animals, and minerals) which are termed "crude drugs". These crude drugs have contributed to some extent to the preservation of health and healing of human diseases since early periods when other remedies were not available. Knowledge concerning crude drugs, indigenous to every area in the world, has accumulated in parallel with the development of civilization. In certain areas, it has been systematized as traditional medicines.

During the development of traditional medicines, the striking therapeutic effects of crude drugs have gradually been revealed. At the same time a number of characteristic procedures have sometimes been devised in collection, processing, and mixing of the natural resources, in formulation of dosage forms, and in the administration of remedies which often involve mysterious procedures and ritualistic ceremonies. Once these procedures were accepted within the established system of traditional medicines, deviation from them was invariably prohibited whether rational or not.

The weakest point of crude drugs is their uncertainty as regards quality (ignorance of active principles, and if some active principles were known, variation of their contents). Due to the unreliable quality of crude drugs, their utilization has decreased, particularly in modern medicine where generally remedies whose quality is well defined are employed. Furthermore, philosophical and sometimes ritualistic procedures in traditional medicines which cannot be scientifically rationalized have tended to estrange physicians practising traditional medicine, from those who are carrying out therapy according to modern medicine. However, because of the limitations of modern medicine and the undesirable side effects of certain synthetic remedies, people have recently become more interested in traditional medicines utilizing natural resoures.

Although crude drugs are still not universally accepted, trials have been recently performed to evaluate traditional medicines in the light of modern science. These have revealed that traditional medicines are actually efficacious, at least in certain diseases which cannot be effectively cured by modern medicine.

However, simple accumulation of such proven examples of efficacy only adds a relatively small degree of evidence to that of thousands of years of experience by traditional doctors. This does not lead to the eventual understanding of traditional medicines unless proof of the efficacy in the levels of active constituents is accumulated.

Recently, in order to prove the validity of traditional medicines, numerous scientists have tried to isolate the active constituents from so-called effective crude drugs and to elucidate mechanisms of their physiological actions.

Isolation of active constituents from crude drugs was initiated in modern times, as in the case of morphine from opium, indicating that active principles can be isolated from natural resources, as the alchemists dreamed of but could not achieve in olden times. In the wake of such developments came the successful isolation of a number of active constituents from crude drugs. Currently, a variety of chemicals from natural resources are recorded in various pharmacopoeias and utilized as remedies. Apparently, however, most drugs of natural origin were those discovered in the last century and the relative pace of discovery of natural drugs has recently slowed down. It should be noted that those natural drugs were mostly isolated for primarily phytochemical interests and later revealed to have biological activity, and those which were isolated from efficacious crude drugs by monitoring the activity were rather exceptional. Thus, investigations have so far led to the discovery of the active principles of certain crude drugs, enabled the estimation of the contents of certain active principles in crude drugs, and clarified mechanisms of action of certain active principles. As a result, folkloric usages have been substantiated to be correct in some cases and have not been proven to be correct in other cases. The interesting fact is that clinical effects, which have hitherto not been recognized in traditional medicines, have been revealed by certain crude drug constituents. On the whole, however, it can be said that the validity of crude drugs has, until now, not been assessed satisfactorily.

The purpose of this chapter is to record recent achievements which are cornerstones for bringing the whole aspect of the validity of traditional medicines into perspective, taking as an example the most important of crude drugs in Oriental medicine — the traditional medicine best known to the modern world.

II. SOME RECENT RESEARCH CONCERNING THE ACTIVITY OF GINSENG

The most important elixirs in the Orient have been "ninjin" (ginseng), the roots of *Panax ginseng* C. A. Meyer (Araliaceae) (Figure 1), and "reishi", the fruit bodies of *Ganoderma lucidum* (Polyporaceae). Because the latter is a mushroom that has been relatively scarce until its recently successful cultivation, its scientific investigation has only just been initiated. The original species of ginseng is indigenous to Korea, China, and Siberia, and is rare and difficult to collect in the wild state. However, it has been cultivated in Korea, China, and Japan, and though it is expensive it is now freely available (Figure 2). In terms of its clinical effects ginseng has been claimed to possess analeptic, antihypothermal, anti-anaemic, aphrodisiac, stomachic, antifatigue, tranquilizing, and sedative effects.

Figure 1. *Panax ginseng* (Korea).

Figure 2. Harvest of *Panax ginseng* (Japan).

Figure 3. Preparation of red ginseng (Japan).

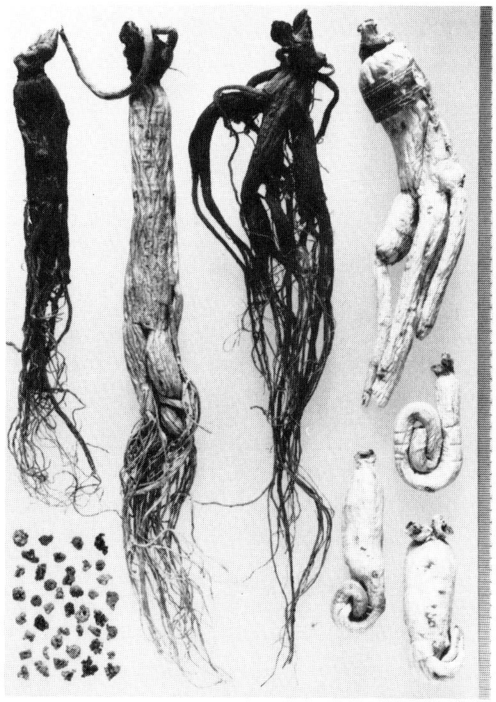

Figure 4. From left to right (top row) red ginseng (Japan), white ginseng (China), red ginseng (China), white ginseng (Korea); (bottom row) white ginseng, rolled (Korea), ginseng callus (Japan).

Fresh ginseng roots are air-dried to obtain white ginseng, steamed for a few hours and dried yielding red ginseng (Figures 3 and 4). As delivery forms, whole roots, slices, rough chops, fine chops, and powders are available in the markets.

Ginseng is one the two exceptions utilized singularly to treat illness in Oriental medicine. In most cases, ginseng is prescribed as one of the components of multi-item herbal formulas like other crude drugs. The most authentic dosage form of the crude drugs in Oriental medicine is decoction. However, decoctions are nowadays condensed *in vacuo* in factories to furnish ready-made powders, granules, and tablets of multi-item herbal preparations. In the Occident, people's interest in ginseng has been increasing, and it is now imported from the Orient and also employed as a drug (ethical and OTC) primarily in countries where ginseng is authorized (e.g., Switzerland, France, and Spain). In this case ginseng is administered in such forms as powders, granules, tablets, and capsules of extracts, those preparations being standardized by contents of the active principles, ginsenosides Rb_1 and Rg_1 (described below). In addition, ginseng is utilized in health foods or herbal teas worldwide and even in countries where it is not authorized as a remedy (e.g., Germany, the U.K., and the U.S.). In this case, ginseng is sold as powders, granules, tablets, and capsules, as well as drinks made from extracts, but lack of quality control in the health food industry brings about remarkable variations in the ginseng content of various products estimated by determination of the saponins as the marker. Indeed, ginseng products are often revealed to contain no ginseng whatsoever.

At first research on the constituents of ginseng roots was dictated solely by phyochemical interests. The most thoroughly examined principles are saponins and a number of ginsenosides and related constituents have been isolated (Figure 5).[1-3] Besides the constituents that are common to white ginseng (except for the malonyl derivatives), red ginseng was found to contain some other ginsenosides which are generated during processing (Figure 5).[1-3] The physiological actions of the ginsenosides were investigated following their isolation and these have been shown to be varied. Some other substances have been found as the active constituents; among those worth noting for their pharmacological activity are poly-

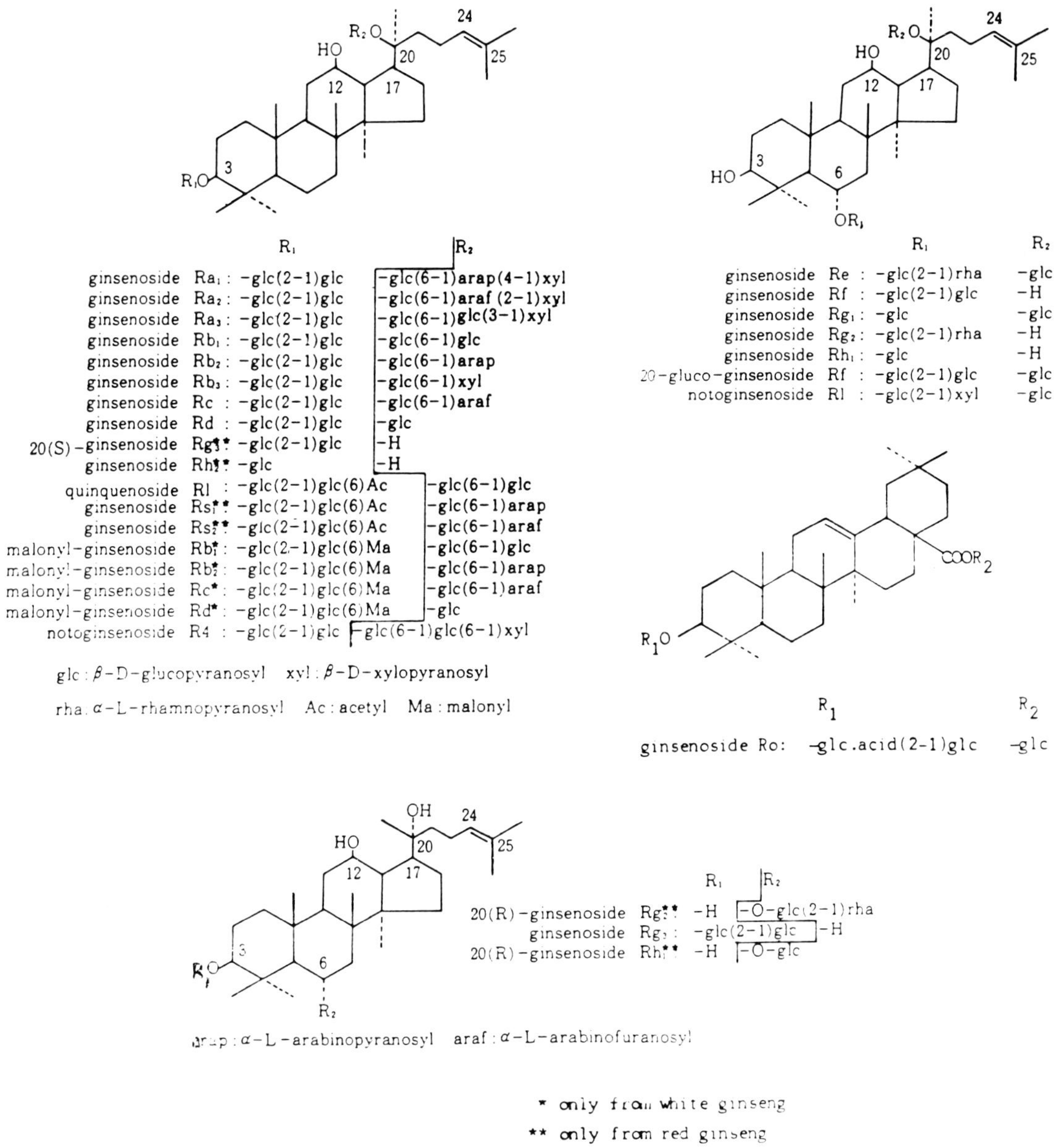

Figure 5. Saponins of ginseng roots.

saccharides.

In the following sections, the results of investigations on the pharmacological and biochemical properties of ginseng are outlined. However, the main purpose of this chapter is to introduce the achievements of recent research so that previous results are described only briefly and reference to the literature published before 1980 is excluded.

A. PHARMACOLOGY

Early research on the pharmacological actions of ginseng was previously reviewed.[3-5] Pioneering work was initiated as early as the 1910s; the first discovered was its hypoglycemic activity. Since then, numerous investigations have been carried out, but there are a number of discrepancies among the pharmacological properties of ginseng extracts and fractions reported by different investigators which may be due to differences of sources, preparative procedures, animal species, experimental conditions, etc.

Thus, improvement in stamina in humans and antifatigue activity, analgesic activity, antipyretic activity, stimulative activity on the central nervous system (CNS) (inhibitory and stimulating effects in the cerebral cortex), tranquilizing activity, hypertensive activity, hypotensive activity, stimulative activity on respiration, cholinergic activity, potentiating activity of acetylcholine, histamine-like activity, antihistamine-like activity, serotonin-like activity, ganglion stimulating activity, papaverine-like activity, contracting activity of skeletal muscle, anti-inflammatory activity, hypoglycemic activity, potentiating activity of insulin, and erythropoietic activity in animals have been reported. One interesting observation was that a water extract exhibited CNS-stimulating, cholinergic, histaminergic, and hypertensive effects while a methanol extract elicited a CNS-depressant effect. Later, these pharmacological actions came to be explained by the constituents in the two extracts. Thus, CNS depressant, antipsychotic, tranquilizing, hypotensive, antistress, anti-inflammatory, gastrointestinal motility activating, and antihemolytic actions along with stimulating actions on glycolysis and cholesterol, nuclear RNA, and protein synthesis were identified as due to the presence of ginsenoside Rb_1, while CNS stimulating, behavioral, hypertensive, antifatigue, and stress ulcer aggravating actions were attributed to ginsenoside Rg_1. Further detailed examinations of the pharmacological properties of pure ginseng saponins were performed to reveal that they exhibited suppressive actions on behavior, fighting and arousal response, biphasic blood pressure action (fall preceded by slight rise), antifatigue action, and anticholinergic action, the potency varying depending on the structure. A structure-activity related study indicated a tendency toward CNS activity, the saponins of the panaxadiol series being stimulatory and the saponins of the panaxatriol series being inhibitory; the effect of the former was stronger in toxicity, but weaker in vasodiation, electroshock avoidance reaction, and antifatigue effects than the latter.

1. Activity on Behavior

Behavioral pharmacological experiments using rats and mice indicated that ginseng saponins elicited a stimulating effect at lower doses and a sedative effect at higher doses. Among the saponins, ginsenoside Rg_1 exhibited similar effects. Although a single dosing of ginsenosides Rf, Re, and Rd suppressed the conditioned avoidance response, a repeated dosing facilitated an induced response. Further, ginsenoside Rg_1 stimulated discrimination between two different sounds in the pole climbing test but had no effect on spontaneous motor activity. In performance of exploratory behavior of rats in the Y-maze test, ginsenoside Rg_1 shortened running time and response latency. Ginsenoside Rg_1 also increased the correct response of rats in a reversal learning experiment. In a study to examine the effect on the acquisition of conditioned avoidance response by the pole climbing test in the chronic intraventricularily canulated rats, acceleration in acquisition of learning was observed in rats which had received ginsenoside Rg_1.

Biochemically, it was found that in the brain tissue ginseng saponins exerted an influence on the conversion of glucose into lipids, on the transformation of amino acids into proteins, and on the cyclic AMP (cAMP) and monoaminergic systems.[6] These observations together with the findings that they reduce the sleeping time and inhibited the extrapyramidal effect of chlorpromazine suggested that ginseng saponins may participate in the regulation of neurotransmission. In support of this supposition, a total ginseng saponin fraction inhibited the uptake of γ-aminobutyric acid (GABA), glutamate, dopamine, noradrenaline, and serotonin into rat brain synaptosomes. Among the ginsenoside fractions, a fraction containing mainly ginsenoside Rd was most active and ginsenoside Rc was also effective. Because the ginsenosides most specifically inhibited the uptake of GABA, it was claimed that ginseng may elicit its action(s) on the CNS by affecting the removal of neurotransmitters in synaptic regions, especially at the GABAnergic neurons.[7]

Recently a blue fluorescent substance was separated from ginseng and found to enhance the activity of adenylate cyclase from the rat brain at very low concentrations (50 pg to 50 ng),[8] though the structure has not been elucidated.

2. Activity on Stress

Because ginseng potentiated nonspecific resistance against stress, it was proposed that ginseng was to be designated an adaptogen. The adaptogenic effects were later defined more precisely to be (1) innocuous properties causing minimal disorders in the physiological functions, (2) antifatigue action against physical and mental loads, (3) nonspecific resistance against stress, and (4) normalization from morbid states. More recently, scientific evidence has accumulated on the adaptogenic effects of ginseng. It was reported that ginseng saponins had a weak analgesic action and might improve some symptoms of vegetative stigmatism caused by specific repeatedly cold stress and restraint/water immersion stress.[9] Further, chronically administered ginseng provided significant protection against electroshock stress and heat stress, and provided effective adaptation to fatigue in mice.[10] The effect of red ginseng on mice exposed to various stresses was examined to reveal that ginseng potentiated the performance of the forced exercise, delayed the extinction of passive avoidance response, and protected the stressed mice against decrease of sex behavior and increase in the failure of performance of passive avoidance response.[11] Many of the crude drugs used as tonics such as ginseng were shown to exhibit protective effects against the decrease in sexual behavior.[12] On the other hand, however, it was recently claimed that ginseng infusions exerted no adaptogenic response in mice.[13]

The adaptogenic actions are very much associated with the hypothalamic-pituitary-adrenal system and the immune system which is described below.

Ginseng has often been suggested to be a geriatric restorative. A considerable amount of data on this aspect is available.[14-17]

3. Activity on Stomach Ulcer

In order to verify the traditional usage of ginseng to stimulate the digestive tract, several investigations have been carried out. Thus, a ginseng extract was shown to elevate suppressed gastric, pancreatic, and biliary secretion in oscillation stress-loaded rats.[18] Moreover, ginseng saponins (p.o.) prevented and healed gastric ulceration induced by stress (water immersion), by acetic acid injection into the submucosal layer, and by local freezing of the gastric wall in rats. The combination of ginsenoside Rg and Rb fractions produced by a potentiated inhibition of ulceration in the former two models.[19] Recently, it was also reported that a ginseng extract showed inhibitory effects on the Shay ulcer by pylorus-ligation and on serotonin- and endotoxin-induced ulcer in rats, a mechanism shown to be the inhibitor of decrease in gastric mucosal blood flow induced by serotonin and endotoxin.[20]

4. Activity on Blood Circulation

Although the effects of ginseng on the cardiovascular system have been frequently examined, results are not always compatible. However, decrease of heart rate by extracts and saponins of ginseng were always observed in animals,[21,22] while the effect of ginseng and its saponins of the cardiac contractile force was variable, depending on conditions.[21-25] In one instance, a total ginseng saponin slightly, and ginsenoside Rb_1 markedly, increased the contractile force. The positive inotropic effect of ginsenoside Rb_1 was suggested to be due to catecholamine release and/or inhibition of Na^+, K^+-ATPase activity.[23] A recent investigation revealed that at doses which counteractedmyocardial ischemia in rabbits, ginsenosides decreased cyclic GMP but had no effect on cAMP levels in the myocardium.[26]

Blood pressure was for the most part decreased but sometimes increased by ginseng and its saponins in normal animals.[21,22] The hypotension was not affected by propranolol, phenoxybenzamine, or angiotensin, and seemed to have no correlation with histamine liberating activity. Although chronic antihypertensive effects of Korean ginseng in spontaneous hypertensive rats (SHR) and patients with essential hypertension were reported,[27] there is another report that Korean red ginseng exhibited no remarkable antihypertensive actions on SHR, stroke-prone SHR, two kidney-one clip hypertensive rats, and deoxycorticosterone/salt-induced hypertensive rats.[28,29] It was believed that the hypotensive mechanism involved a decrease of plasma renin activity,[27] although there is also a report to the contrary,[29] and that vasodilatation was closely related to blood pressure in most cases (but there was negative evidence for that in certain cases).[21] Due to conflicting opinions with respect to the effects of ginseng on blood pressure, further clarification is required.

Concerning the effects on the vascular smooth muscle, ginseng saponins caused relaxation of the aorta, decreased the Ca^{++}uptake of sarcoplasmic reticulum in the presence of ATP, and inhibited the Ca^{++} binding to the cellular membrane, the effects being stronger in the triol saponins than in the diol saponins.[30] Quite recently, the vasodilatory effects of ginsenosides were examined to show that ginsenoside Rg_1 of the protopanaxatriol series mediated vasodilation only in Ca^{++}-induced contraction while oleanolic acid series exhibited nonselective vasodilation, the activity being the strongest in ginsenoside Rb_1 in both cases.[31] Further, ginseng increased peripheral blood flow at the fingertip.[32] In particular, ginseng was shown to improve blood supply to the brain,[33] probably because of dilatation of vessels, and to elicit an action on vessels of the brain and fundus oculi, mainly through sympathetic innervation.

5. Activity on Nerve Growth

The nerve growth factor (NGF) is a protein which plays an important role in the development and maintenance of the functioning of the sympathetic neuron. It was found that ginsenosides Rb_1 and Rd, the 20(*S*)-protopanaxadiol oligosides, potentiated NGF-induced nerve fiber production in organ cultures of chick embryonic dorsal root ganglia and sympathetic lumbal ganglia, but promoted no nerve fiber production. A study on the structure-activity relationship indicated that removal of the glucosyl group or hydroxylation at the side chain diminished the activity.[34] Although the mechanism of potentiation of NFG-induced nerve fiber production by the ginsenosides has not been completely elucidated, it is probable that it results in an accelerating effect in the process of nerve fiber production. When ginsenoside Rb_1 was daily administered to mice after an X-ray irradiation, tyrosine hydroxylase activity in the submandibular gland remained unchanged while the nerve growth factor content decreased. It was considered that ginsenoside Rb_1 may prevent gradual destruction of the sympathetic nerve terminals in the submandibular gland by exposure to irradiation.[35]

6. Anti-Inflammatory Activity

Anti-inflammatory activity of ginseng has previously been reported. It was observed by inhibition of the vascular permeability accelerated by egg whites in rats. The degree of edema evoked by immersing a hind leg into hot water was found to vary with the time of the year. The effect of a ginseng extract on edema induced by heat and formalin in normal and adrenalectomized mice also exhibited seasonal variations. Later, it was revealed that certain ginseng saponins exerted anti-inflammatory actions in mice and rats,[25,26] and panaxynol and linoleic acid from ginseng elicited anti-inflammatory effects in the fertile egg and cotton pellet tests.[36]

7. Toxicity

Acute toxicity of ginseng and its saponins in dogs and mice was very low. Ginseng saponins exhibited low acute toxicity (ginsenosides Rb_2>Rd>Rc>Rb_1>Rg_1>Rf), ginsenosides Rf, Rg_1, and Rb_1 containing glucose being lower in their toxicities than the rest containing arabinose and/or rhamnose. In subacute toxicity studies in rats, body weight was decreased when ginseng was ingested with a diet (2 to 8%), probably because of decrease of the consumption and the efficiency of feed and protein, while body weight was increased when a ginseng saponin mixture was administered s.c. Increase of s.c. dose of the saponin mixture to rats caused an increase of liver weight/body weight ratio, a decrease of albumin/globulin ratio, and an increase of urea nitrogen in serum.[37] Daily feeding of chicks with a ginseng crude saponin gave no significant differences in body weight gain, organ weights, feed efficiency, and retention of nutrients. When beagle dogs were fed a ginseng extract (up to 15 mg/kg/day), no effect was noticed in body weight gain, food consumption, and biochemical and pathological findings.[38] The safety of ginseng was again claimed recently in mice, rats, rabbits, beagle dogs and minipigs by short-, medium-, and long-term studies.[39,40]

Although ginseng is alleged to be an elixir of life, it is not always innocuous in humans. Thus, the literature contains examples where the abuse of ginseng caused harmful effects such as glow, irritability, uneasiness, even destruction of personality and mental derangement.[41,42] These symptoms were produced by overdosage or use in diseases for which ginseng should not have been prescribed. Therefore, ginseng must be utilized for feeble patients or analogous conditions but not for stout patients or conditions where Oriental medicine contraindicates its use.

B. ACTIVITY ON BLOOD

The ability of ginseng to increase the erythrocyte count and hemoglobin was previously noticed. Further, it was found that ginseng fractions injected into rabbits induced erythropoietin activity in the liver, kidney, spleen, and bone marrow, while a ginseng total saponin enhanced harringtonin- and cytosine arabinose-produced damage to the hematopoietic precursor cells in mice.[43] Further, ginseng fractions doubled the number of mitoses in both myeloid and erythroid, and increased the numbers of nucleated cells in bone marrow and reticulocytes in peripheral blood.[44] However, no clinical effect was noticed on human anemias.[45] A crude ginseng saponin mixture inhibited 5′-nucleotidase activity in erythrocyte ghosts and mediated a biphasic effect on insulin binding by erythrocyte cells and ghosts. Another ginseng saponin mixure increased the Donnon chloride ratio of human erythrocytes.[46] These observations suggest that ginseng exerts its various pharmacological actions by interactions of its saponins with cell surface membrane.

Biochemically, it was found that incubation of human erythrocytes with ginseng saponins increased the ATP, ATP/ADP, pyruvate, and lactate contents, and stimulated glucose utilization and the efflux of Na^+ and declined ADP, while Na^+, K^+-ATPase activities exhibited a biphasic response to saponins, increasing in lower concentrations and decreasing in higher concentrations of the saponins.[47] Further, ginseng saponins ($10^{-4}\%$) stimulated glucose 6-phosphate dehydrogenase activity of the erythrocytes and stimulated protein biosynthesis by reticulocytes.[44]

Platelet adhesiveness was significantly inhibited by ginseng in high cholesterol diet rats.[48-53] Ginseng was also shown to have inhibitory activity against platelet aggregation produced by ADP, adrenalin, and collagen in human blood.[54] This was substantiated experimentally in rats.[52] A ginseng extract prevented changes of the blood platelet count, the fibrinogen content, and the prothrombin time in disseminated intravascular coagulation induced by endotoxin or thrombin, inhibited formation of fibrin thrombi in the renal glomeruli on the thrombin-produced disseminated intravascular coagulation, and exerted fibrinolytic activity in rats. The ginseng extract also promoted activation of urokinase action in the plasminogen-containing fibrin plate *in vitro*.[55] Similar actions of ginsenosides Rg_1 and Rg_2 *in vitro* were also noticed.[56,57] Further it was found that in human platelets, formation of thromboxane (TX) B_2 (a stable metabolite of TXA_2 which accelerates platelet aggregation) was inhibited by ginsenoside Rg_1 but not by ginsenosides Rb_1, Rb_2, and Rc, but formation of 6-keto-prostaglandin (PG) F_1 (a stable metabolite of prostacyclin which inhibits platelet aggregation) was not affected by these ginsenosides *in vitro*,[58] while concentration of TXB_2 was decreased and that of 6-keto-PGF_1 was increased by ginseng or ginsenosides, especially Rb_2, in cholesterol-cholic acid diet-fed rats *in vivo*.[59] After active components in the blood coagulative and fibrinolytic systems *in vitro* were revealed to be ginsenosides Ro, Rb_1, Rb_2, Rc, Rg_1, and Rg_2,[60] it was also shown that 20(*S*)- and 20(*R*)-ginsenosides Rg_3 inhibited collagen- and ADP-induced blood platelet aggregation, and 20(*S*)-ginsenoside Rg_3 inhibited the thrombin-induced conversion of fibrinogen to fibrin.[61] The changes of these parameters by ginseng or ginsenosides are considered to contribute to the improvement of atherosclerosis.

Although saponins generally possess hemolytic activity, total ginseng saponins have been revealed to exhibit no hemolytic activity, although a crude ginseng saponin mixture was reported to be hemolytic at higher concentrations. However, it was later found that fractions containing saponins of the 20(*S*)-protopanaxatriol series elicited hemolyic actions while those containing saponins of the 20(*S*)-protopanaxadiol series displayed protective actions against hemolysis by other hemolytic reagents, both actions being apparently masked in crude preparations due to their balanced counteractivity. The hemolytic effects of pure saponins were revealed to be much less than that of the crude saponins and this was ascribed to participation of unknown factors in the latter. Later, the potency of hemolytic activity was claimed to be in the order Rd>Rb_2>Re>Rc=Rf>Rb_1>Rg_1, which was different from the above result but similar to that in toxicity, indicating that the panaxatriols and the panaxadiols cannot be classified clearly with respect to hemolytic activity. Further, based on a study on structure-activity relationships, it was claimed that a bulky and hydrophlic group at C-3 of a triterpenoid or steroid nucleus was necessary for hemolytic activity and that such a group around ring D ws unfavorable for hemolytic activity.[62] However, this structural requirement is not satisfied in the ginseng saponins.

C. ACTIVITY ON LIVER

It has been previously reported that ginseng extracts elicited inhibitory actions on carbon tetrachloride (CCl_4)-induced liver damage in rabbits and on thioacetamide-induced liver damage in rats. Among the constituents, ginsenosides Rb_1, Rg_1 and Re were revealed to exhibit antihepatotoxic effects on CCl_4-induced liver lesions in rabbits, their potencies being similar. Quite recently, a more detailed comparable study on the antihepatotoxic actions of ginsenosides and their aglycones was conducted using using CCl_4- and galactosamine-induced cytotoxicity in primary cultured rat hepatocytes and showed that significant protective effects were found with 20(*S*)-ginsenoside Rh_2, 20(*R*)-ginsenoside Rg_3, and prosapogenins of ginsenoside Ro, 20(*R*)- and 20(*S*)-ginsenoside Rs in CCl_4-induced cytotoxicity, and with 20(*S*)-ginsenoside Rh_1 and prosapogenin of 20(*S*)-ginsenoside Rs in galactosamine-induced cytotoxicity, revealing that like the triterpenoids of the oleanane series, those of the dammarane series also exert similar antihepatotoxic activity. In CCl_4-produced cytotoxicity assay, the ones in the 20(*R*) series exhibited more intenst activity as compared to the 20(*S*) series with the exception of the pair of ginsenosides Rh_1. Generally, the ones possessing less sugar moieties were stronger in activity than those possessing more sugar moieties.[63] It is worthy of note that most of the ginsenoside-artifacts formed during processing (steaming) to prepare red ginseng showed stronger antihepatotoxic effects, suggesting red ginseng is more intense in antihepatotoxic activity than white ginseng. As mentioned below, ginsenosides were shown to bring about reverse transformation of cultured Morris hepatoma cells. It is likely that the action on the cell membrane participates in the mechanism of preventive activity on cytotoxicity in hepatocytes and of reverse transformation of cultured Morris hepatoma cells. It is likely that the action on the cell membrane participates in the mechanism of preventive actvity on on cytotoxicity in hepatocytes and of reverse transformation of hepatoma cells.

It was recently reported that ginseng saponins were administered to patients suffering from viral (B-type) hepatits in an attempt to understand the mechanism of the protective action of the ginseng saponins against the development of the disease.[64] Further, ginseng polysaccharides, when orally dosed to patients of chronic hepatitis, were found to improve subjective symptoms, lowered the GPT level, decreased circulating immune complex, and restored the action of T-cells.[65]

Sensitivity to ginseng and its saponins has been noticed in various hepatic enzymes such as alcohol dehydrogenase, aldehyde dehydrogenase, glutamate dehydrogenase, lactate dehydrogenase (LDH), pyruvate oxidase, alkaline phosphatase, and ATPase. Examinations of effects of ginseng saponins on hepatic enzymes were also continued recently.[66-68]

The effects of ginseng and its saponins on ethanol-induced intoxication have been frequently investigated. Thus, a number of reports indicated that they accelerated recovery from acute and chronic intoxication with ethanol. A number of other symptoms of ethanol intoxication were counteracted by ginseng and its saponins: the duration of anesthesia was shortened and so was the time of recovery; the elimination of ethanol from the blood was accelerated and the lethal dose was increased; the swelling of hepatocytes was masked and recovery from hepatic injury facilitated; an inhibiting effect on the decrease of protein was evident as well as an increase of lipid in the liver; and the effect of decrease of body weight was counteracted although the experimental evidence was inconclusive as regards the action on liver alcohol dehydrogenase and the blood ethanol level.

Recent reports about the action of ginseng on ethanol intoxication indicated that ginseng prevented various disorders, including memory inhibition induced by ethanol in mice,[69] and a ginseng fraction increased alcohol dehydrogenase activity and the microsomal ethanol-oxidizing system at a small dose and decreased it at larger doses. Recovery from ethanol intoxication could be accelerated by treatment with ginseng, although reduction of the ethanol concentration in blood is dependent upon increased activity of the above enzyme system.[70]

Treatment of rats with ethanol induced histological changes and elevated glutamic-oxaloacetic transaminase (GOT), glutamic-pyruvic transaminase (GPT), alkaline phosphatase, LDH, and bilirubin levels in liver and serum which were reduced by ginseng saponins,[71,72] although ginseng saponins ingested to mice and rats significantly increased LDH activity in liver homogenate.[66,67] Further, because stimulation of alcohol dehydrogenase, aldehyde dehydrogenase, and the microsomal ethanol oxidizing system was confirmed *in vitro* and *in vivo*, it was concluded that toxic aldehyde generated during ethanol oxidation might be removed quickly from the liver and the excess hydrogen might be utilized for the biosynthetic work in the presence of ginseng saponins, resulting in protection of the liver from ethanol intoxication. This was substantiated by histological observation.[73]

Salicyclic and vanillic acids in ginseng strongly inhibited lipid peroxidation in the liver of ethanol-intoxicated mice.[74] Recently ginseng saponins were shown to inhibit lipoperoxide formation *in vitro* and *in vivo* and the potencies were more marked in the liver than in the kidney or blood, suggesting that ginseng may elicit an anti-aging action through the liver functions.[75]

D. ACTIVITY ON NUCLEIC ACID METABOLISM

Effects of ginseng on nucleic acid metabolism have been given attention since the early 1960s. Thus, it was found that the RNA and DNA contents in the liver were slightly decreased and those in the kidney were increased by ginseng in Ehrlich ascites tumor-bearing mice. Next, ginseng was shown to affect the incorporation of radioactive thymidine into DNA of lymphocytes in the lymph node, spleen, and thymus in normal and stressed mice. Further, RNA, DNA, and the RNA/DNA ratio in adrenal gland, liver, spleen, pancrease, mandibular gland, testes, and bone marrow were influenced by ginseng but the effects varied. Meanwhile, ginseng saponins were claimed to be capable of normalizing nucleic acid formation during stress. Recently, ginseng was also found to increase visceral nucleic acids and inhibited ACTH-induced decrease in tissue nucleic acid in rats and mice.[76]

In a series of investigations, it was shown that a ginseng fraction accelerated incorporation of orotic acid into liver nuclear RNA and into cytoplasmic polysomal RNA in rats; the increase of nuclear RNA was not due to change in the pool size of endogenous soluble nucleotides and the ratio of RNA and DNA in the nuclei was constant at various periods. This activating action on nuclear RNA biosynthesis was induced by ginseng fractions containing saponins and was followed by the promotion of the biosynthesis of ribosomal RNA and messenger RNA, the activation of RNA-polymerase, the increase of polysome, the enhancement of protein biosynthesis, and indeed the activation of every step of biochemical reactions leading to protein biosynthesis. Acceleration of syntheses of RNA and protein by the ginseng fraction was observed in kidney cells as in liver cells. In another study, reconfirmation of the promoting activity of ginsenosides in incorporation of orotic acid into rat liver nuclear RNA led to clarification that incorporation of orotic acid into liver nuclear RNA in rats was accelerated by ginsenoside Rb_1 but was reduced by ginsenoside Rc and was not affected by ginsenoside Rg_1. Consistent with the incorporation study, ginsenoside Rb_1 enhanced while ginsenoside Rc repressed the RNA polymerase activity *in vivo*, while similar results were also obtained in *in vitro* experiments using RNA polymerases I and II. Further, blocking experiments suggested that the stimulation of RNA polymerase activity by ginsenoside Rb_1 occurred at the transcriptional stage. As a result of similar experiments with the bone marrow (the most active organ in mitosis), it was found that ginseng fractions promoted the synthesis of DNA, RNA, proteins, and lipids in rat bone marrow *in vivo* and *in vitro*. Active constituents were revealed to be saponins among which ginsenosides Rb_2, Rc, Re, and Rg_1 were especially effective. These saponins also lowered the cAMP level and elevated the cGMP level in bone marrow, these changes in contents of the cyclic nucleotides being comparable with those occurring at cell division.

In a related study, it was found that ginseng reduced hepatic xanthine oxidase activity and the blood urea concertrations in ethanol-induced hyperuremic mice[77] which was supported in patients with hyperuricemia.[45]

E. ACTIVITY ON PROTEIN METABOLISM

Early invesigations indicated that ginseng saponins enhanced the rate of incorporation of radioactive leucine into liver protein and increased blood serum protein formation in mice. Ginseng extracts were found to increase the rates of incorporation of glucose and amino acids into brain protein, although causing no significant increase of the protein content of the liver in mice.

Later, incorporation of leucine into mouse serum protein was found to be accelerated by ginsenosides Rb_2, Rc, Rd, Re, and Rg_1, among which ginsenoside Rd was most active, but ginsenoside Rb_1 displayed no effect. However, subsequent investigations indicated that ginsenosides Rb_1 and Rc enhanced incorporation of leucine into rat serum protein while ginsenoside Rg_1 gave no significant result. Although ginsenoside Rc reduced synthesis of liver nuclear RNA as described above, it promoted synthesis of protein and the reason for this is so far unknown. Analysis of the synthesized protein demonstrated leucine to be incorporated more in the globulin fraction than the albumin fraction. Intraperitoneal injection of a ginseng saponin fraction to mice remarkably increased the concentration of protein, the incorporation of leucine into protein, and the concentration of free amino acids (mainly nonessential amino acids such as alanine, aspartate, glutamate, glycine and serine) also in the liver. Acceleration of synthesis of DNA and protein by ginseng saponins was observed also in testis which is the second most active site of cell division next to bone marrow.

Recently it was found that aqueous extracts of ginseng inhibited intracellular protein degradation in confluent caltures of IMR-90 human diploid fibroblasts, inhibited proteolysis of long-lived proteins selectively, and stimulated protein synthesis in human fibroblasts, indicating that ginseng components are able to act directly on human cells to promote protein accumulation.[78]

F. ACTIVITY ON CHOLESTEROL METABOLISM

It is now well understood that in hyperlipidemia penetration of low density lipoprotein (LDL), rich in cholesteryl ester, into the affected arterial wall and increase of platelet aggregation followed by blood coagulation on the surface of the arterial intima results in the development of arteriosclerosis, especially atherosclerosis. Interests have been directed toward assessing the effect of ginseng on cholesterol metabolism since the 1960s. Early reports alleged that its anticholesterolemic activity was variable in rabbits, although it accelerated cholesterol metabolism in the rat liver. However, later reports pointed to reduction of cholesterol content in the serum, liver, and heart.

When rabbits were fed cholesterol, administration of ginseng saponins reduced the serum cholesterol level, lowered the ratio of cholesterol to phospholipids, elicited less fatty infiltration of the liver, diminished the penetration of cholesterol into aortic tissue, resulted in less atherosclerotic alteration, and prevented the occurrence of atheroma in the aorta. These observations suggested that ginseng saponins might exert cholesterol-lowering activity by stimulating cholesterol transport and an enzyme involved in cholesterol metabolism.[79-82] However, there was another observation that although serum cholesterol was decreased by ginseng total saponins in normal rats, the effect of the saponins on serum cholesterol was variable depending on the dose in high-cholesterol administered rats.[83]

Prior to these investigations, it was found that when rats were fed with a high cholesterol-cholic acid diet for 2 weeks, plasma cholesterol and triglycerides (TG) were increased, and this effect was inhibited by intramuscular administration of ginseng saponins. On oral administration, ginseng also inhibited increase of cholesterol and TG, increased high density lipoprotein (HDL)-cholesterol and phospholipids, decreased the nonesterified fatty acid (NEFA) level in blood, and reduced the cholesterol content in the liver of rats fed with high cholesterol diets.[48-53,84] Thus, ginseng was shown to lower the atherogenic index (total cholesterol-HDL-cholesterol/HDL-cholesterol). Among the active ingredients, saponins, ginsenoside Rb_1 notably reduced the total cholesterol, free cholesterol, NEFA, and LDL-cholesterol and increased HDL-cholesterol in blood and decreased total cholesterol in the liver in rats fed a high cholesterol diet. Such changes were not observed in ginsenoside Rg_1-treated rats fed the same diet and the ginsenoside Rb_1-treated rats fed an ordinary diet.[85] Further, in the ginsenosides tested, ginsenosides Rb_2 and Rc decreased the total cholesterol, free cholesterol, LDL-cholesterol, and TG levels and increased the HDL-cholesterol level in the serum in rats fed a cholesterol diet. Reduction in the serum total cholesterol, free cholesterol and LDL-cholesterol concentrations was also found by the 20(*R*)-prosapogenin treatment.[86] Recently ginsenoside Rb_2 was revealed also to reduce serum total cholesterol in streptozotocin-induced hyperglycemic rats.[87] It was futher revealed that a ginseng extract and ginsenosides Rb_2 Rc, Rb_1, and Rg_1, especially Rb_2, improved in the levels of cholesterol, HDL-cholesterol, VLDL (very low density lipoprotein), LDL, HDL, and apoproteins in serum and that of cholesterol in the liver in cholesterol-cholic acid-diet fed rats.[59,88] Similar results were obtained by ginsenosides Rb_2 in rats fed a high cholesterol diet.[89] On the other hand, it was shown that i.p. injection of the ginseng saponins, particularly ginsenoside Rb_1, to rats increased the concentrations of serum and liver cholesterol, and the incorporation of acetate into serum and liver cholesterol. Promotion of the incorporation of acetate into cholesterol was also observed in liver slices from rats treated with ginsenosides in which ginsenoside Rb_1 was most active. These observations demonstrated that biosynthesis of cholesterol was stimulated by ginseng saponins. Therefore, the decrease of the plasma cholesterol level was explained by acceleration of the turnover of cholesterol into bile acids and the excretion of cholesterol, thereby resulting in reduction of the cholesterol concentration in blood.[48-53] In radioactive cholesterol-fed rats, ginseng saponins accelerated appearance of radioactivity in blood and liver lipids, indicating ginseng saponins stimulated cholesterol metabolism.[90]

When rats were fed with a high fat diet, the cholesterol level in the liver was elevated and incorporation of acetate into liver cholesterol suppressed. Concomitantly, the activity of mevalonate:NADP oxidoreductase, which catalyzes conversion of 3-hydroxy-3-methyl glutaryl coenzyme A into mevalonate and is a rate-limiting enzyme in the biosynthetic pathway of cholesterol, was suppressed, confirming biosynthesis of cholesterol to be subjected to feedback control of the final product, cholesterol. Ginsenoside Rb_1 was found to reverse the effects of high fat diets. The observed effects on administration of Korean red ginseng caused elevation of serum HDL-cholesterol and reduction of serum total cholesterol, atherogenic index, TG, NEFA, and lipoperoxide, which along with inhibition of platelet adhesiveness[48-53,91] and suppression of the coagulative and fibrinolytic systems mentioned above[55,60,61] indicated ginseng to exhibit an anti-atherogenic effect in humans.

G. ACTIVITY ON LIPID METABOLISM

Much attention has been paid to the effects of ginseng on lipid metabolism which, however, again exhibited variation depending on the conditions employed in these early studies.

A report claimed that although ginseng saponins slightly reduced serum TG, no correlation was found between the saponin dose and the serum TG level in normal rats.[81] Ginseng saponins activated hepatic enzymes involved in phospholipid biosynthesis in normal and ethanol-fed rats.[82,92] Further, ginseng saponins decreased the TG concentration and lipase activity in plasmids in adrenaline-induced hyperlipidemia, and seemed to exert an inhibitory effect on TG mobilization into blood and a stimulating effect on esterification of free fatty acids (FFA) in the liver and adipose tissue.[93] Recently, it was found that ginseng selectively inhibited lipolysis induced by toxohormone-L, a lipolytic factor which releases fatty acid from rat epididymal adipose tissue, but did not affect ACTH-induced lipolysis.[94]

The effects of ginseng and its saponins on lipid metabolism were evaluated in animals fed with fats or cholesterol. Thus, the serum lipids were increased in rats fed with fats with and without ginseng saponins (0.05 ng per rat), but the absorption of fats was greatly stimulated by ginseng saponins.[90] A recent study demonstrated that ginseng saponin when administered to high lipid diet-fed rabbits reversed histopathological changes in the liver, increased blood levels of transaminases, alkaline phosphatase, bilirubin, cholesterol, phospholipids, TG, and VL- and LD-lipoproteins, and decreased blood levels of HD-lipoproteins.[95] Improvement by ginseng in serum TG and NEFA, and in hepatic TG, phospholipid, and lipoperoxide in hyperlipidemia of rats produced by cholesterol-cholic acid diet was also observed.[59]

Reduction of TG and HEFA in blood by ginseng in rats fed with high cholesterol diets was mentioned above. Beneficial effects of ginseng were also observed on fatty liver, decrease of the liver cholesterol, TG, NEFA, and lipid peroxide levels and increase of the liver phospholipid level being found in rats fed with a high cholesterol diet.[48-53,84] Nevertheless, it was found that lipid synthesis was stimulated by ginseng saponins. Thus, a ginseng saponin fraction on a single i.p. administration to rats was shown to increase the incorporation of acetate into total lipid in the liver and epididymal adipose tissue. Concurrently, increase of the lipid content in the adipose tissue and slight decrease of the TG concentration in the serum were observed. Later, it was revealed that lipid metabolism was stimulated by ginsenosides Rb_1 and Rc in the adipose tissue as well as in the liver and intestine; these organs were alleged target organs in which the liver and intestine may be sites for apolipoprotein synthesis. Lipid synthesis was also promoted in bone marrow cells by a ginseng saponin fraction *in vitro*. Further, ginsenosides Rb_2, Rc, Re, and Rg_1 when injected i.p. increased lipid synthesis in bone marrow cells. Ginseng saponins were shown to inhibit lipolysis by ACTH-like insulin. The mechanism of reduction of TG by ginseng was considered to involve (1) acceleration of treatment of TG in the liver and blood because of concurrent decrease of the TG content in the liver[48-53] and (2) elevation of lipoprotein lipase activity.[96] Ginsenoside Rb_1 was shown to inhibit lipolysis catalyzed by ACTH-sensitive lipase in the presence of Ca^{++} and to restore the reduction of TG synthesis by ACTH, a fact which partly explains reduction of the NEFA level in blood. Recently, a hyperlipemia-improving effect of ginsenoside Rb_2 in rats fed on a high cholesterol diet was reported.[89]

Metabolism of lipid is closely linked to that of sugar which is described below.

H. ACTIVITY ON CARBOHYDRATE METABOLISM

Ginseng has been alleged to be clinically effective for diabetes mellitus. In this connection, ginseng has been examined in relation to carbohydrate metabolism in animals. Thus, a number of reports on the effects of ginseng on concentrations of sugar in blood and those of glycogen in the liver and muscles have appeared, but the results are completely contradictory (hyperglycemic, noneffective, and hypoglycemic) probably due to differences in the experimental conditions employed. Quite recently, ginseng was alleged to normalize the depressed peripheral glucose utilization and insulin response, leading to an improvement of abnormal glucose tolerance in rats poisoned by a rodenticide, Vacor.[97]

It was found that an acute reduction of the blood glucose concentration was induced by ginseng fractions in adrenalectomized rats which are quite sensitive to insulin because of a lack of secretion of the antagonizing hormone, adrenaline. In normal rats, however, such an acute hypoglycemia was not observed but instead, slight reduction of the blood sugar level and a marked decrease of glycogen stores in the liver were noticed. Examination of factors which influence the effect of the ginseng fraction on the liver demonstrated that the lipogenic capacity of ginseng altered remarkably by varying nutritional states. The most striking observation was an acceleration of lipogenesis and a decrease of the liver glycogen content in rats fed with a fat-free diet. These observations suggested that hormonal regulation participates in the actions of the ginseng fraction. Later, it was confirmed that the above reduction of the hepatic glycogen content by ginseng in rats was actually caused by its component, saponins, although the liver glycogen level was not affected by ginseng saponins in exercised rats.[98] It has also been reported that ginseng saponins lowered the blood glucose level in mice[99] and rats.[98] The mechanism of the hypoglycemic activity of ginseng saponins was alleged to involve neither the enhancement of insulin secretion nor the antagonism of the action of alloxan.[99] Among the saponins, it was claimed that the protopanaxatriol group had a potentiating action on adrenalin-induced hyperglycemia while the protopanaxadiol group had an inhibitory action.[93]

Ginseng saponins dose-dependently elevated the activity of hepatic pyruvate kinase, the rate-limiting enzyme in glycolysis, and, on the other hand, inhibited the activity of hepatic serine dehydratase, the rate-limiting enzyme in the synthesis of glucose from the glycogenic amino acids. These effects suggested that ginseng saponins were capable of the acceleration of glycolysis and the inhibition of glyconeogenesis. It was further observed that glucose-6-phosphatase (G-6-Pase) exhibited a trend toward an increase in the liver and a significant decrease in the kidney in the case of ginsenoside Rb_1-treated rats fed with a high cholesterol diet. A decrease of G-6-Pase in the kidney was also noticed in ginsenoside Rg_1-treated rats fed with the same diet.[85] Among ginsenoside Rb_1, Rb_2, Rc, Rd, Re, and Rg_1, ginsenoside Rb_2 was found to cause a transient increase of hepatic glucose-6-phosphate dehydrogenase activity, a decrease in hepatic glycogen content, and an increase in the level of hepatic glucose-6-phosphate followed by an accumulation of lipid in the adipose tissue as well as an increase of the phosphofructokinase activity; however, it brought about no significant changes in the amounts of liver total lipid, TG, total cholesterol, phospholipid, glucose, pyruvate, and lactate in rats.[100] Examination of the mechanism of action of ginsenoside Rb_2 demonstrated that a series of enzyme reactions in the glycolytic pathway, lipogenetic pathway, and the dihydronicotinamide adenine dinucleotide-generating system were accelerated, with a decrease of liver glycogen; at the same time the neutral fat formed in the liver was converted into VLDL and transferred through the blood vessel system to the adipose tissue where, owing to activated lipoprotein lipase, an increase of neutral fat resulted.[101]

Although no hypoglycemic effect was observed by i.p. administration of ginseng saponins (ginsenoside Rg_1),[102] it was later found that a single administration of ginsenoside Rb_2, which showed intense effects on lipid and sugar metabolism, caused no change in the blood sugar concentration in normoglycemic rats but slightly reduced the blood sugar concentration in streptozotocin-induced diabetic rats.[87] Hence, repeated administration of ginsenoside Rb_2 to streptozotocin-produced diabetic rats was carried out, resulting in a remarkable decrease in the blood glucose level. In the ginsenoside Rb_2-treated group, a decline of G-6-Pase activity and a rise of glucokinase activity occurred as compared to the control group, showing improvement of the G-6-Pase/glucokinase ratio or acceleration of the sugar utilizing system. At the same time, a trend toward an increase of hepatic glycogen was observed.[103] Further, decreases of the TG and VLDL concentrations in blood, an increase of the TG content in the adipose tissue, an increase of lipoprotein lipase activity, and a decrease of hormone-sensitive lipase activity by ginsenoside Rb_2 were observed.[104] Ginsenoside Rb_2, when successively administered i.p. to streptozotocin-induced diabetic rats, also markedly decreased blood glucose level, remarkably increased body weight, and notably improved diabetic symptoms such as overeating, polyuria, and glycosuria.[105] These results were explained to indicate that in the liver, ginsenoide Rb_2 may increase the utilization of glucose and synthesis of fatty acids which may be transferred to the adipose tissue via VLDL in the blood and stored as TG. Effects of ginseng saponins on lipolysis and lipogenesis were further examined using the adipose tissue of normal rats to show that although the ginseng saponins exhibited no effect on the lipolytic reaction of adrenaline, ginsenosides Rb_1, Rb_2, Rc, Rd, Re, Rg_1, and Rg_2, but not Rh_1 inhibited lipolysis induced by ACTH in fat cells, while ginsenosides Rb_2, Rc, Rg_1, Rg_2, and Rh_1, but not Rb_1 and Re, reduced insulin-stimulated lipogenesis in adipose tissue slices.[106,107]

Taking advantage of the observation that a ginseng extract lowered concentrations of blood glucose and ketone bodies, purification for isolating the active components was performed to afford a hypoglycemic fraction,[102] DPG-3-2. This fraction was shown to increase the blood glucose level in alloxan-induced diabetic mice, to lose its activity by the anti-insulin serum, to stimulate insulin release from perfused rat pancreas, to enhance glucose-stimulated insulin release from the pancreas,[108] to decrease the humping effect on the glucose tolerance curves in genetically diabetic KK-CA^y mice,[109] to stimulate insulin biosynthesis in different preparations of pancreas from animals with hyperglycemia,[110] to inhibit the activity of phosphodiesterase originating from islets,[111] and to accelerate the uptake of Ca^{++} into islets in parallel with the increase of insulin release.[112] Although this fraction was subjected to frequent biological examinations, as above, it was reported to still be a mixture of ginsenosides Re and Rg_1, glucose, fructose, amino acids, uracil, etc. Although this fraction intensely released insulin from isolated islet cells of normal and diabetic mice *in vitro*,[113] this may be attributed to cytotoxicity by ginseng saponins present in this fraction. Quite recently, this fraction was further purified to give the active fraction (D-O-AN-a) from which succinic acid, malic acid and citric acid were isolated[114] and shown to potentiate glucose-induced insulin released by the pancreas perfusion method.[115]

In the course of further effort in seeking an insulin-like substance, a water extract of ginseng was found to inhibit adrenaline-induced lipolysis in isolated fat cells of rat epididymal adipose tissue which led to an acidic "peptide" with antilipolytic activity.[106] Although this compound was claimed to be a peptide with a molecular weight of approximately 1400 consisting of 14 amino acids,[106,116] the elucidation of its whole structure has not yet been reported.

During purification of the above insulin-like peptide, another fraction was found to have antilipolytic activity. The active component of the fraction was purified and revealed to be adenosine,[117] whose antilypolytic activity was already known.

Recently, an ethanol/water extract of Korean ginseng, when administered to mice, was found to significantly decrease the plasma glucose concentration in normal mice. The extract was fractionated by monitoring the hypoglycemic activity to yield eight polysaccharides containing some peptide moieties, panaxans A, B, C, D, E, F, G, and H. These glycans, when injected i.p., exhibited the ability to lower the blood sugar level in normal mice. Reduction of the blood glucose concentration in alloxan-induced hyperglycemic mice was also observed.[118,119] Although panaxans A and B possess quite different molecular weights (1.4×10^4 and 1.8×10^6, respectively), their basic structures are the same, an α-1→6 linked D-glucopyranose residue with branching points at C-3.[120,121] In seeking to obtain further quantities of material, another lot of Korean ginseng was extracted and fractionated and this yielded other hypoglycemic polysaccharides, panaxans I, J, K, and L.[122] Because polysaccharide components of ginseng produced in other countries came to be the subject of interest, ginseng of Chinese and Japanese origin was examined for polysaccharide constituents, again affording different hypoglycemic glycans, panaxans M, N, O, and P from Chinese ginseng, and panaxans Q, R, S, T, and U from Japanese ginseng.[123,124] These findings demonstrated that the polysaccharide composition of ginseng varied notably depending on the source, growth conditions, climate, and/or nature of soil, although all the isolated glycans more or less elicited hypoglycemic actions.

I. ACTIVITY ON ADRENAL GLANDS

It was reported that injection of a ginseng extract to rats resulted in a tendency to hypertrophy of the adrenals, decreased eosinophils in peripheral blood, reduced the ascorbic acid and cholesterol contents in the adrenal cortex, and increased the corticosteroid content in the urine, indicating promotion of adrenal function. It was also found that extracts and saponins of ginseng stimulates secretion of corticosteroids or adrenocortical function. Ginseng saponins were later shown to increase plasma ACTH and corticosteroids, and adrenal cAMP, the second messenger for hormone action, without significant change in plasma glucose in intact rats but not to increase plasma corticosterone in dexamethasone-treated rats or adrenal cAMP in hypophysectomized rats, showing that ginseng saponins indirectly stimulated the adrenocortical function. Ability to increase plasma corticosterone was in the following order: ginsenosides $Rd>Rb_2>Rc>Re=Rb_1>Rg_1$. The potency of the activity of prosapogenins also varied depending on the structure.[125] Increase of plasma corticosterone caused by saponin treatment was inhibited by hypophysis or pretreatment with dexamethasone, indicating that the saponins act on the hypothalamus and/or hypophysis primarily, and stimulate ACTH secretion which results in increased synthesis of corticosterone in the adrenal cortex. It is considered that secretion of these hormones and related neurohormones may act in concert as defense or adaptation mechanisms to adverse or new circumstances. Acceleration of secretion of ACTH and corticosterone by oral and i.p. administration of ginseng saponins was also confirmed.[126]

Ginseng saponins were found to be antagonistic to the suppressive action of cortisone on compensatory adrenal hypertrophy after unilateral adrenalectomy and to the decrease of thymus weight and the serum K^+ concentration induced by cortisone in rats, while the ginseng saponins by themselves exhibited no such effects.[127] These results not only support clinical findings that in the treatment of chronic hepatitis, the dose of a conticosteroid can be reduced by ginseng saponins, but also indicate a possibility that ginseng saponins can inhibit adverse effects of corticosteroids. It was further revealed that a ginseng extract bound to progesterone, mineralocorticoid, and glucocorticoid receptors in rat uterus *in vitro*, suggested that ginseng may at least partly elicit its effects as a consequence of its ability to occupy steroid receptors.[128] Quite recently, the mechanisms of the elevation of the blood corticosterone level by the ginesenosides was examined to reveal that although the saponins exerted no effect on steroidogenesis of the cultured adrenal cells, ginsenoside Rb_1 exhibited a tendency to increase ACTH secretion of cultured pituitary cells, a fact which was claimed to suport the results previously obtaind *in vivo*.[129]

J. ACTIVITY ON TUMOR

Originally, ginseng was shown to inhibit the growth of Ehrlich's ascites, sarcoma 180, and adenocarcinoma 755, but did not influence the growth of Walker's ascites carcinosarcoma and leukemia 1210. Later, it was found that ginseng extracts inhibited the growth of murine leukemia L5178 and ascitic sarcoma 180 cells *in vitro* when DNA, RNA, and protein synthesis was inhibited in the latter cells. The cytotoxicity was considered to be due to the inhibitory effects on the biosynthesis of special macromolecules.[130,131] Because the known ginseng constituents, including the ginsenosides, exert no such drastic and specific cytotoxicity against tumor cells, reconfirmation of the specific antitumor activity and clarification of active constituents are quite necessary. Recently, a water extract and a polysaccharide fraction of ginseng were shown to elicit antitumor activity against sarcoma 180 and as a result of examination on the immune system, the polysaccharides were revealed to stimulate the overall immune system.[132]

The effects on tumor cells in which active cell division occurs as in bone marrow and testis were examined. Ginseng saponins displayed no influence on biosynthesis of DNA, protein, and lipid in rat ascites tumor but improved hyperlipidemia and accelerated lipogenesis in the liver and adipose tissue.

Since a water extract of ginseng reduced a number of adverse effects shown by the antitumor agents, 5-fluorouracil and mitomycin C, in rabbits, ginseng may possibly decrease their side effects.[133]

On the other hand, a ginseng extract diminished the incidence and growth of tumors evoked by a number of carcinogenic chemicals. For instance, it decreased the lung adenoma incidence by urethane and aflatoxin B_1, the hepatoma incidence by aflatoxin B_1, and the proliferation of the lung adenomas by 9,10-dimethyl-1,2-benzanthracene and urethane, and prolonged the average life span of rats bearing *N*-methyl-*N*-nitroso-*N'*-nitroguanidine-induced tumor.[134]

Because the mode of action of ginseng was variable in different experiments, it was thought that the antitumor effects of ginseng were attributed to unspecified increases of resistance in the host against external stimuli. A similar experiment was performed with 3-methylcholanthrene and it was found that the incidence and growth of tumor was decreased by a ginseng extract for tumor-bearing mice in which marked prolongation of the average life span was observed.[135] In relation to this, it was revealed that in animals a ginseng saponin fraction caused a selective induction of the inactivation enzyme epoxide hydratase and an elevation of the detoxifying enzyme glutathione transferase without a concurrent induction of benzopyrene monooxygenase, responsible for the formation of carcinogenic intermediates, indicating that ginseng has the potential to alter the metabolic pathway of carcinogenic aromatic hydrocarbons and thereby may participate in the prevention of tumors caused by carcinogens.[136] Further, in a study to examine the mechanism of the inhibition by ginseng of carcinogenesis induced by various chemical carcinogens, ginseng was found to increase natural killer cell activity and to lower incidences of the lung adenoma in urethane-injected mice.[137]

Ginsenosides have also been shown to cause reverse transformation of cultured Morris hepatoma cells *in vitro*. After subcultures with ginsenosides, the cells not only changed morphologically but also showed much less growth as compared to the control hepatoma cells. A marked increase in uptake of ornithine in an arginine-deficient medium, an increase of succinate-cytochrome *c* reductase activity, a decrease in 5'-nucleotidase activity, and a prominent increase of serum protein synthesis were also observed in the reversely transformed cells. Further, ginsenosides induced the development of subcellular organelles in cultures Morris hepatoma cells. Development of mitochondria and rough endoplasmic reticulum, formation of the stack of the endoplasmic reticulum, and decrease of aldolase activity were also found during the transformation. Among the saponin fractions, the ginsenoside Rh_1 fraction showed the most intense inhibitory effect on aldolase activity and also inhibited the efflux of 3-*O*-methylglucose and sulfate ion in the Morris hepatoma cells, indicating that the ginsenoside Rh_1 fraction has a direct effect on the cell membrane.[138] The mechanism of reverse transformation of tumor cells was examined and showed that ginsenoside Rh_2 inhibited the growth of cultured B16 melanoma cells, 20(*S*)- and 20(*R*)-ginsenoside Rh_1 and ginsenoside Rg_3 induced no inhibition of the growth but accelerated melanogenesis of B16 melanoma cells, indicating that inhibition of growth and induction of reverse transformation are closely related to structural features of the ginsenosides.[139,140] Further, it was revealed that inhibition of 20(*S*)-ginsenoside Rh_2 on growth of various tumor cells was not elicited through growth factors nor by inhibition of DNA synthesis.[141] Although the mechanism of the effects of ginsenosides on tumor cells has not been fully clarified, its elucidation may provide valuable information for understanding of cell differentiation and further for development of effective therapies for cancer.

Ginseng was found to be nonmutagenic by the Ames test in the presence or absence of the S-9 liver homogenate[142,143] and by the V79 Chinese hamster cell test,[143] and further revealed to contain antimutagenic constituents.[144]

K. ACTIVITY ON IMMUNE SYSTEM

The first paper on the immunological activity reported that although ginseng saponins induced no action on the activity of cytotoxic T-cells and natural killer cells in mice infected with the influenza virus, the saponins suppressed delayed-type hypersensitivity responses to the influenza virus and to sheep red cells. This seemed to indicate a prophylactic anti-inflammatory effect of the saponins which was suggested to be due to its steroid-like structures.[145] Next, polysaccharides of ginseng were found to markedly stimulate phagocytosis of the reticuloendothelial system and the production of antibodies and to cause an increase of serum complement content in guinea pigs, to raise the serum IgG level in mice, and to bring about an increase of the B-lymphocyte/T-lymphocyte cell ratio.[146]

It has also been noted that ginseng significantly enhanced the protective activity of 6-MFA (an interferon-inducing antiviral substance) against Semliki forest virus in mice.[147] Further, in *in vitro* studies on human peripheral blood lymphocytes, a ginseng extract, besides inducing interferon, augumented natural killer cells and antibody-dependent cytotoxic actions.[148] In agreement with the results of *in vitro* studies, ginseng extracts enhanced antibody formation against sheep red cells, increased the primary IgM and the secondary IgG and IgM responses, stimulated cell mediated immunity against Semliki forest virus antigen, potentiated natural killer cell activity, inhibited lymphocyte proliferation, and elevated interferon production in mice.[149,150] Because ginseng and its saponins inhibited PHA-P-induced transformation of peripheral blood lymphocytes,[151,152] which was potentiated by a combination with hydrocortisone, it was suggested that ginseng possessed a steroid-like effect and may have a potentiating action with hydrocortisone on T-cell-mediated immunity.[152] Recently, red ginseng was shown to promote the phagocytic activity on the reticuloendothelial system in mice and has a stimulatory effect on macrophage.[153]

Stimulation of the immune activity against invading foreign substances by ginseng must also contribute to its adaptogenic effects. The beneficial action of ginseng on chronic hepatitis as well as its antitumor activity is ascribed to stimulation of the immune system by polysaccharides as described above.

L. ACTIVITY ON RADIATION INJURY

In the 1960s, improvement of survival of X- and γ-irradiated mice and rats by ginseng was suggested. Since then several investigations pointed to antiradiomimetic activity of ginseng and its saponins and polysaccharides. Treatment of cultured V-79 cells with ginseng saponins prior to γ-irradiation increased radiation resistance, the mechanism being considered to be indirect due to changes in cell physiology than to a DNA repair process.[154] Further, the radioprotective mechanism was alleged to involve production of thrombocytes and consequently the recovery of the thrombocyte count which was an essential factor for survival of irradiated mice, rats, and guinea pigs.[155,156]

Although active principles have not been clarified, it was observed that ginseng saponins were not responsible for the radioprotective activity,[157] and heating of a ginseng extract gave a thermostable fraction which not only reduced the adverse effect of ginseng-produced splenic hyperplasia but also retained the radioprotective activity.[158] Further, an antiradiation effect was claimed to be due to ginseng proteins whose subunit molecular weights were estimated.[159] The recovering effect of ginseng was examined on the γ-radiation-induced bone marrow depression in cervical cancer patients to reveal that ginseng (p.o.) elicited no statistically significant changes in hematological parameters except the significant increase of the mean platelet counts.[160] Because there are no radioprotective compounds really effected by postirradiation administration, ginseng may be one of the promising drugs.

M. PRODUCTION AND PHYSIOLOGICAL ACTIVITY OF CALLUS

It is well known that secondary metabolites can be effectively produced in high yield and on a large scale by plant cell cultures. A number of investigations with ginseng have been performed on examination of culture conditions for better growth and more saponin production in cell lines selected from original callus. Although little or no saponin content was found in original callus just formed, in the selected cell line cultured under the best conditions thus far achieved, dried amount of cells was 20 g/L, growth rate was 0.6 g/L/day, the saponin content per dried weight was four times higher than that of the original roots and the saponin composition was almost the same as that of ginseng roots. Thus, mass production methods of ginseng callus have been developed and are now put into operation in a 20,000 L scale (Figures 6 and 7).[3,161,162] Comparison of pharmacological properties of ginseng callus with those of ginseng roots demonstrated that an antifatigue action of the former was weaker than that of the latter, and toxicity, CNS depressant and tranquilizing actions were approximately identical, although CNS stimulant and cholinergic actions were less in the former than in the latter.[163] Because these actions may largely be due to saponins, the observed potencies should depend on the quality and quantity of saponins in the materials employed.

Callus cultures of ginseng of Korean origin are also capable of producing polysaccharides. Different from saponins, these polysaccharides were already produced even in newly generated callus tissues, although the contents were at present less than those in the original roots. The polysaccharide constituents also lowered blood glucose concentration in normal and alloxan-induced hyperglycemic mice.[164]

Figure 6. Culture of ginseng callus (Japan). (Courtesy of Nitto Electric Industrial Co.)

Figure 7. Harvest of ginseng callus (Japan). (Courtesy of Nitto Electric Industrial Co.)

N. DISCUSSION

As was mentioned above, an enormous number of investigation have been performed on the pharmacological and biochemical properties of ginseng. As a result, data have been accumulated to support the clinical efficacy of ginseng in traditional usage and at the same time to indicate new pharmacological and biochemical effects which cannot be expected from the traditional usage. Nevertheless, it may be said that the pharmacological and biochemical actions of ginseng have been clarified only partly and complete understanding of their whole aspect waits for further investigations.

Ginseng has been utilized for a long time on the basis of its empirical therapeutic efficacy, but with little scientific evidence, and the world demand for this drug will not increase rapidly simply due to the recent confirmation of some of its pharmacological and biochemical effects. However, continuation of research as in the past, as well as evaluation of its effectiveness by novel methodologies, will possibly develop unique types of drugs.

One thing which must be emphasized here is that although ginseng is popularly called an elixir, it is not a cure-all. In contrast to its application in Oriental medicine, where the condition of a patient is carefully diagnosed prior to selection of the drugs (including ginseng and ginseng-containing multi-item herbal formulas) in order to obtain the best therapeutic efficacy and to avoid adverse actions, Westerners use ginseng simply as a tonic regardless of the condition of a patient. This is by no means the way to obtain the best therapeutic efficacy of ginseng, and often results in ineffectiveness, and rather detracts from its reputation because of side effects caused by obvious misuse. Correlation of the crude drug including ginseng (or the multi-item herbal preparations) with the condition of a patient for full display of their clinical effectiveness will be one of the most essential problems for understanding of traditional medicines employing these "mysterious" crude drugs.

III. CONCLUSION

Many efforts have recently been directed toward the clarification of the validity of traditional medicines in the light of modern science.

It is obvious that there are crude drugs found to be active by screening, well planned according to traditional usages, and found to be sources of active principles, thus confirming the previous experiences of traditional usage, while there are other crude drugs in which unexpected effects have now been observed by random screening. From the former examples, we can say that the ancient humans were wonderful because at periods when no scientific methods were available, they still discovered clinical efficacy which can be proven at present. From the latter examples, it is evident that traditional practice is not all-knowing and infallible.

It is sometimes considered that when one or more active constituents are found in a crude drug, the efficacy of the crude drug is fully elucidated. Even in the crude drug, ginseng, which has been most thoroughly investigated, new active principles have recently been found; it is therefore probable that when examined from a modern point of view, crude drugs which have not been thoroughly investigated may certainly reveal new biological effects and generate active constituents responsible for those effects. It often happens that known constituents which were not expected to have biological activity at the time of isolation, were later revealed to exhibit some pharmacological actions. Hence, crude drugs are actually like springs which cannot be exhausted.

It seems logical that as far as even a single crude drug is concerned, the therapeutic efficacy could be the combined effect of interations of its various constituents, and more so, in a composite preparation made of a number of crude drugs, where interactions among them come into existence, so that assessment of their action in such complexity cannot be easily accomplished.

In many cases when from crude drugs active principles have been isolated and by the displayed activity of these the traditional usages of the crude drugs can be indicated, modern doctors may accept their therapeutic value as far as the scientific evidence presented. However, traditional practitioners may not be satisfied by such evidence because they would consider that the effectiveness of a traditional medicine cannot be substantiated by simple pharmacological evaluation.

It seems a distant possibility that modern doctors who take a serious view of science and traditional doctors who attach importance to experience will understand each other on a common basis. However, accumulation of efforts toward assessment of the efficacy of crude drugs (elucidation of active constituents and of the mechanism of their actions) is believed to be the shortest way to bring on such a day.

In any event, once an active principle is isolated and its spectrum and mechanism of activity clarified, even if it is obtained from a local traditional crude drug like reserpine, it will serve the betterment of human health.

REFERENCES

1. **Tanaka, O. and Kasai, R.,** Saponins of ginseng and related plants, in *Fortschritte der Chemie Organischer Naturstoffe*, Herz, W., Grisebach, H., Kirby, G. W. and Tamm, Ch., Eds., Springer-Verlag, New York, 1984, 1.
2. **Tanaka, O.,** Application of ^{13}C-nuclear magnetic resonance spectrometry to structural studies on glycosides: saponins of *Panax* spp. and natural sweet glycosides, *Yakugaku Zasshi*, 105, 323, 1985.
3. **Shibata, S., Tanaka, O., Shoji, J., and Saito, H.,** Chemistry and pharmacology of Panax, in *Economic and Medicinal Plant Research*, Vol. 1, Wagner, H., Hikino, H., and Farnsworth, N. R., Eds., Academic Press, London, 1985, 217.
4. **Fulder, S. J.,** *The Root of Being: Ginseng and the Pharmacology of Harmony*, Hutchinson & Co., London, 1980.
5. Oura, H., Kumagai, A., Shibata, S., and Takagi, K., Eds., *Yakuyoninjin (Recent Studies on Ginseng)*, Kyoritsu Publishing, Tokyo, 1981.
6. **Seo, K.-L., Lee, Y.-Y., Chang, S.-H., Park, I., and Lee, S.-Y.,** Studies on the components of ginseng which stimulate the activity of adenylate cyclase. I., *Hanguk Saenghwa Hakhoe Chi*, 13, 81, 1980.
7. **Tsang, D., Yeung, H.-W., Tso, W.-W., and Peck, H.,** Ginseng saponins: Influence on neurotransmitter uptake in rat brain synaptosomes, *Planta Med.*, 221, 1985.
8. **Moon, J.-G., Park, I.-W., Lee, Y.-Y., Seo, K.-L., and Lee, K.-S.,** The properties of the blue fluorescent substance from ginseng and its effects on the several enzymes, *Han'huk Saenghwa Hakhoechi*, 17, 307, 1984.
9. **Kita, T., Hata, T., Kawashima, Y., Kaku, T. and Itoh, E.,** Pharmacological actions of ginseng saponin in stress mice, *J. Pharm. Dyn.*, 4, 381, 1981.
10. **Banerjee, U. and Izaquiedo, J. A.,** Antistress and antifatigue properties of Panax ginseng: comparison with piracetam, *Acta Physiol. Latinoam.*, 32, 277, 1982.
11. **Saito, H. and Bao, T.-T.,** Effect of red ginseng on mice exposed to various stresses, *Proc. 4th Int. Ginseng Symp.*, 97, 1984.
12. **Saito, J., Nishiyama, N., Fujimoto, H., Hinata, K., Kamegaya, T., Kato, Y., and Bao, T.-T.,** Effects of drugs on sex and learning behaviors, and tyrosine hydroxylase activities of adrenal gland and hypothalamic regions in chronic stressed mice, *Stress: Proc. 3rd. Int. Symp. Role Catecholamines Other Neurotransm.*, 1, 467, 1983 (Pub. 1984).
13. **Lewis, W., Zenger, V., and Lynch, R.,** No adaptogen response of mice to ginseng and *Eleutherococcus* infusion, *J. Ethnopharmacol.*, 8, 209, 1983.

14. **Shia, G. T. W., Ali, S., and Bittle, A. H.,** The effects of ginseng saponins on the growth and metabolism of himan diploid fibroblasts, *Gerontology*, 28, 121, 1982.
15. **Cho, Y.-D., Koo, B.-S., and Lee, S.-J.,** Studies of ginseng extract on age-related enzymes, *Proc. 4th Int. Ginseng Symp.*, 37, 1984.
16. **Fulder, S., Kataria, M., and Gethyn-Smith, B.,** A double-blind clinical trial of *Panax ginseng* in aged subjects, *Proc. 4th Int. Ginseng Symp.*, 215, 1984.
17. **Yokozawa, T., Oura, H., Ogita, Z., Shibata, T., Wakagi, K., Taniuchi, M., and Koizumi, T.,** On the effects of Oriental drugs on aged rats, *Wakan Iyaku Gakkaishi*, 2, 126, 1985.
18. **Sato, T., Kozima, S., Karasawa, H., Tonouchi, M., and Tsutsumi, S.,** Effects of medicated alcoholic beverage (Yomeishu) which includes crude drugs on the function of gastrointestinal secretion in the normal and the oscillation strss-loaded rats, *Oyo Yakuri*, 20, 425, 1980.
19. **Sato, T., Kojima, S., Toda, S., and Arichi, S.,** Effects of ginseng saponin on experimental gastric ulcer (1), *Oyo Yakuri*, 20, 715, 1980.
20. **Matsuda, H. and Kubo, M.,** Study on Panax ginseng C. A. Meyer II. Effets of red ginseng on the experimental gastric ulcer (1). *Yakugaku Zasshi*, 104, 449, 1984.
21. **Lee, D.-C., Lee, M.-O., Kim, C.-Y., and Clifford, D. H.,** Effect of ether, ethanol and aqueous extracts of ginseng on cardiovascular function in dogs, *Can. J. Comp. Med.*, 45, 182, 1981.
22. **Chen, X., Zhu, Q.-Y., Liu, L.-Y., and Tang, X.-L.,** Effects of ginsenosides on cardiac performance and hemodynamics in dogs, *Zhongguo Yaoli Zuebao*, 3, 236, 1982.
23. **Kim, N.-D., Kim, C.-K., Kim, B.-K., Han, B.-H., and Lee, S.-S.,** A study on the tonic effects of ginseng. Effects of ginseng saponins on the rat heart, *Yakhak Hoe Chi*, 24, 15, 1980.
24. **Kim, N.-D., Kim, B.-K., and Lee, H-.S.,** Study on the contractile force of isolated hearts from rats treated with ginseng components, *Yakhak Hoe Chi*, 26, 239, 1982.
25. **Yuan, W.-X., Gui, L.-H., Zhou, J.-Y., Li, Q.-Y., and Yu, Q.-H.,** Some pharmacological effects of ginseng saponins, *Zhongguo Yaoli Zuebao*, 4, 124, 1983.
26. **Qian, C.-G., Ye, Y.-W., and Yan, Y.-F.,** Counteraction of ginsenosides on myocardial ischemia and its relation to glucocorticoids and cyclic nucleotides in rabbits, *Hunan Yixueyuan Zuebao*, 9, 219, 1984.
27. **Sohn, E.-S., Huh, B.-Y., Park, S.-C., Park, C.-W., and Kim, H.-J.,** The effect of ginseng on blood pressure in spontaneously hypertensive rat and essential hypertension, *Proc. 3rd Int. Ginseng Symp.*, 1, 1980.
28. **Kishi, K., Watanabe, T., and Sokabe, H.,** Acute and chronic effects of Korean red ginseng powder on blood pressure of hypertensive rats, *Wakan Iyaku Gakkaishi*, 2, 7, 1985.
29. **Kishi, K., Watanabe, T. X., Kawashima, K., and Sokabe, H.,** Effects of steamed ginseng radix on blood pressure in hypertensive rats, *Wakan Iyaku Gakkaishi*, 2, 7, 1985.
30. **Lee, K.-S.,** Effect of ginseng saponin on the vascular smooth muscle, *Proc. 3rd. Int. Ginseng Symp.*, 71, 1980.
31. **Kaneko, H., Nakanishi, K., Murakami, H., and Kuwajima, K.,** On the vasolilative action of red ginseng saponins, *Wakan Iyaku Gakkaishi*, 2, 122, 1985.
32. **Kaneko, H., Nakanishi, K., Tahara, M., Murakami, H., and Kuwajima, K.,** Clinical examinations of Korean ginseng powder. II. Effect of acute administration of large dose on circulatory functions monitored by digital plethysmography, *Kiso to Rinsho*, 14, 839, 1980.
33. **Ueshima, T., Kitajima, S., Tanabe, T., Hayashi, M., Yamamoto, M., Uemura, T., Uemiya, M., Hara, H., Deguchi, H., Miki, Y., Yagyu, N., and Koyama, T.,** Effects of ginseng principles on the human cerebral circulation, *Proc. Symp. Wakan-Yaku*, 14, 65, 1981.
34. **Takemoto, Y., Ueyama, T., Saito, H., Horio, S., Sanada, S., Shoji, J., Yahara, S., Tanaka, O., and Shibata, S.,** Potentiation of nerve growth factor-mediated nerve fiber production in organ cultures of chicken embryonic ganglia by ginseng saponins: structure-activity relationship, *Chem. Pharm. Bull.*, 32, 3128, 1984.
35. **Saito, H., Lee, Y.-M., and Handa, S.,** The activity of tyrosine hydroxylase in superior cervical ganglion and sub-maxillary gland of senescent and irradiated mice, in *Catecholamines and Stress: Recent Advances*, Usdin, Kvetnansky, and Kopin, Eds., Elsevier-North Holland, 1980, 371.
36. **Otsuka, H., Komiya, T., Fujioka, S., Goto, M., Hiramatsu, Y., and Fujimura, H.,** Studies on anti-inflammatory agents. IV. Anti-inflammatory constituents from roots of Panax ginseng, *Yakugaku Zasshi*, 101, 1113, 1981.
37. **Rhee, D.-K., Lim, C.-J., Kim, D.-H., Hong, S.-K., Park, E.-H., and Han, Y.-N.,** Studies on the acute and subacute toxicity of ginseng saponin, *Yakhak Hoe Chi*, 26, 209, 1982.
38. **Hess, F. G., Jr., Parent, R. A., Stevens, K. R., Cox, G. E., and Becci, P. J.,** Effects of subchronic feeding of ginseng extract G115 in beagle dogs, *Food Chem. Toxicol.*, 21, 95, 1983.
39. **Soldati, F.,** Toxicological studies on ginseng, *Proc. 4th Int. Ginseng Symp.*, 119, 1984.
40. **Hong, S.-K., Lee, S.-J., Kim, Y.-S., Jeon, B.-S., and Lim, C.-H.,** Studies on the safety of Korean ginseng ingested as food substance, *Proc. 4th Int. Ginseng Symp.*, 199, 1984.
41. **Wang, J.-M., Hu, Y.-H., and Hu, Y.-J.,** A review on the ill effects occurred in the treatment with Chinese herbal drugs since the foundation of the People's Republic of China, *Zhongcaoyao*, 11, 322, 1980.
42. **Peng, Z.-R. and Niu, Y.-Z.,** Harmfulness by abuse of ginseng, *Yaoxue Tongbao*, 18, 464, 1983.
43. **Zhou, Q.-X. and Han, R.,** Effects of the total saponins in Panax ginseng on bone marrow hemopoietic stem cells in mice, *Zhongcaoyao*, 14, 27, 1983.
44. **Kim, K.-Y., Lee, H.-B., and Joo, C.-N.,** The effect of saponin fraction on the protein biosynthesis of rat reticulocytes, *Han'guk Saenghwa Hakhoechi*, 17, 312, 1984.
45. **Yamamoto, M.,** Panax ginseng — metabolic and endocrinological actions, *Chiryogaku*, 7, 685, 1981.
46. **Lee, S.-I. and Kang, B.-S.,** Effect of saponin on the Donnan ratio in human erythrocytes, *Yonse Uidae Nonmunjip*, 16, 618, 1984.
47. **Kang, B.-S. and Han, K.-H.,** Effect of Panax ginseng saponin on metabolism and ion transport in human erythrocytes, *Taehan Saengri Hakhoechi*, 17, 125, 1983.
48. **Yamamoto, M. and Uemura, T.,** Endocrinological and metabolic actions of ginseng principles, *Proc. 3rd Int. Ginseng Symp.*, 115, 1980.
49. **Yamamoto, M., Uemura, T., Nakama, S., Uemiya, M., and Kumagai, A.,** Effect of Panax ginseng on serum and hepatic lipids — anti-atherogenic actins, *Domyakukoka*, 9, 671, 1981.
50. **Yamamoto, M., Kumagai, A., and Yamamura, Y.,** Plasma lipid-lowering action of ginseng saponins and mechanism of the action, *Am. J. Clin. Med.*, 11, 84, 1983.
51. **Yamamoto, M., Kumagai, A., and Yamamura, Y.,** Plasma lipid-lowering and lipogenesis-stimulating actions of ginseng saponins in tumor-bearing rats, *Am. J. Clin. Med.*, 11, 88, 1983.
52. **Yamamoto, M., Uemura, T., Nakama, S., Uemiya, M., and Kumagai, A.,** Serum HDL-cholesterol-increasing and fatty liver-improving actions of Panax ginseng in high cholesterol diet-fed rats with clinical effect on hyperlipidemia in man, *Am. J. Clin. Med.*, 11, 96, 1983.
53. **Yamamoto, M., Uemura, T., Uemiya, M., Hara, H., Yagiu, N., Deguchi, H., Miki, Y., and Koyama, T.,** Effects of panax ginseng on various plasma hormones and lipids, *Proc. Symp. Wakan-Yaku*, 14, 117, 1981.
54. **Nakanishi, K., Tahara, M., Kaneko, H., Murakami, H., Kuwajima, K., Ikeda, K., Kondo, S., and Kaito, H.,** Clinical examinations of Korean ginseng powder. III. Effect of acute administration of large dose on platelet aggregation and lipid metabolism, *Rinsho to Kenkyu*, 57, 3089, 1980.
55. **Matsuda, H. and Kubo, M.,** Pharmacological study on *Panax ginseng* C. A. Meyer I. Effects of red ginseng on the experimental disseminated intravascular coagulation (1), *Yakugaku Zasshi*, 103, 1269, 1983.

56. **Nakanishi, K., Kaneko, H., Murakami, H., Kuwajima, K., Ikeda, K., Kaito, J., and Kondo, S.,** Clinical examinations of Korean ginseng powder. V. Platelet aggregation inhibitory action by Korean ginseng powder ingredients, *Rinsho to Kenkyu*, 59, 3671, 1982.
57. **Kaneko, H., Nakanishi, K., Murakami, H., Kuwashima, K., Kaido, H., Ikeda, K., and Okuda, H.,** Effect of Panax ginseng on platelet functions, *Thromb. Hemostas.*, 50, 199, 1983.
58. **Sekiya, K. and Okuda, H.,** Effect of medicinal plants components on arachidonic acid metabolism in platelet, *Proc. Symp. Wakan-Yaku*, 15, 58, 1982.
59. **Yamamoto, M., Uemura, T., Nakama, S., Uemiya, M., Tanaka, T., and Minamoto, S.,** Effect of saponins from ginseng and bupleurum, extracts from Xiao-chai-hu-tang and Gui-zhi-fu-ling-wan, on blood liproteins, apoproteins and prostanoids and on hepatic lipids in experimental hyperlipidemia, *Wakan Iyaku Gakkaishi*, 2, 377, 1985.
60. **Matsuda, H., Kubo, M., Tani, T., and Arichi, S.,** unpublished.
61. **Matsuda, H., Kubo, M., Tani, T., Arichi, S., and Kitagawa, I.,** Pharmacological study on *Panax ginseng* C. A. Meyer V. Effects of red ginseng on the experimental disseminated intravascular coagulation (4). On ginsenoside-Rg3, Rh1 and Rh2, *Shoyakugaku Zasshi*, 39, 123, 1985.
62. **Hase, J., Kobashi, K., Mitsui, K., Namba, T., Yoshizaki, M., and Tomimori, T.,** The structure-activity relationship of oleanolic acid derivatives and inhibition of the saponin-induced hemolysis with sapogenins, *J. Pharm. Dyn.*, 4, 833, 1981.
63. **Hikino, H., Kiso, Y., Kinouchi, J., Sanada, S., and Shoji, J.,** Antihepatotoxic actions of ginsenosides from *Panax ginseng* roots, *Planta Med.*, 62, 1985.
64. **Joo, C.-N.,** The effect of ginseng saponins on biochemical reactions, *Proc. 4th Int. Symp. Surfactants Solution*, 3, 2093, 1982 (Pub. 1984).
65. **Hu, M.-Q., Yang, L.-S., Liu, X.-M., Jiang, W., Ye, W.-F., An, F., Liu, F., Ma, L.-H., Wang, Q.-X., and Liang, C.-L.,** Study on clinical efficacy in chronic hepatitis and mechanism of immunological regulation of ginseng polysaccharides, *Zhongcaoyao*, 16, 171, 1985.
66. **Harper, N. and Bittles, A.-H.,** The effects of ginseng saponins on lactate dehydrogenase activity in the mouse, *Biochem. Soc. Trans.*, 11, 356, 1983.
67. **Harper, N., Osborne, A. J., Makov, U. E., and Bittles, A. H.,** The effects of oral ginseng administration on the activities and isoenzyme profiles of murine lactate dehydrogenase, *Biochem. Pharmacol.*, 33, 1571, 1984.
68. **Tai, G.-S. and Joo, C.-N.,** The effect of ginseng saponin fraction on rat hepatic microsomal ethanol oxidizing system, *Han'guk Saenghwa Hakhoechi*, 17, 424, 1984.
69. **Saito, H., Nagatome, Y., and Bao, T.-T.,** Effects of red ginseng, vitamins and their preparations (V). Effect on behaviors of alcohol-administered mice, *Yakuri to Chiryo*, 12, 1482, 1984.
70. **Choi, C.-W., Lee, S.-I., and Hu, K.,** Effect of ginseng on the hepatic alcohol-metabolizing enzyme system activity in chronic alcohol-treated mouse, *Taehan Yakrihak Chapchi*, 20, 13, 1984.
71. **Hu, K. and Choi, C.-W.,** Influence of Panax ginseng on lactic acid formation in ethanol-treated rat, *Yongnam Taehakkyo Nonmunjip, Chyon Kwahak Pyon*, 14, 217, 1980.
72. **Yu, H.-Y., Kim, C.-M., and Koo, K.-H.,** An experimental study on the effect of ginseng saponin upon alcoholic liver injury, *Hanyang Uidae Haksulchi*, 2, 287, 1982.
73. **Joo, C.-N.,** The preventive effect of the saponin fraction of *Panax ginseng* C. A. Meyer against ethanol intoxication of rat liver, *Proc. 4th Int. Ginseng Symp.*, 63, 1984.
74. **Han, B.-H., Park, M.-H., and Han, Y.-N.,** Studies on the antioxidant components of Korean ginseng. III. Identification of phenolic acids, *Seoul Taehakkyo Saengyak Yonguso Opjukjip*, 20, 14, 1981.
75. **Choi, J.-H. and Oh, S.-K.,** The inhibitory effects of diol and triol saponins on lipoperoxide formation, *Han'guk Saenghwa Hakhoechi*, 17, 445, 1984.
76. **Kim, C.-C.,** Effect of ginseng on visceral nucleic acid of rats and mice, *Inkan Kwahak*, 4, 336, 1980.
77. **Huh, K. and Choi, C.-W.,** Effect of *Panax ginseng* on xanthine oxidase activity in ethanol-induced hyperuricemia, *Proc. 3rd Int. Ginseng Symp.*, 131, 1980.
78. **Lu, Z.-Q. and Dice, J.-F.,** Ginseng extract inhibits protein degradation and stimulates protein synthesis in human fibroblasts, *Biochem. Biophys. Res. Commun.*, 126, 636, 1985.
79. **Joo, C.-N., Kim, D.-S., and Koo, J.-H.,** Biochemical studies on ginseng saponins. XVI. The effect of ginseng saponin on hypercholesterolemia induced by prolonged cholesterol feeding in rabbits, *Hanguk Saenghwa Hakhoe Chi*, 13, 51, 1980.
80. **Joo, C.-N., Koo, J.-H., and Balik, T.-H.,** Biochemical study of some pharmacological effects of Panax ginseng C. A. Meyer, *Hanguk Saenghwa Hakhoe Chi*, 13, 63, 1980.
81. **Joo, C.-N., Koo, J.-H., Lee, H.-B., Yoon, J.-B., and Byun, Y.-S.,** Biochemical studies on the absorption of ginseng saponin and its effect on metabolism in animal body, *Hanguk Saenghwa Hakhow Chi*, 15, 189, 1982.
82. **Koo, J.-H.,** The effect of ginseng saponin on the development of experimental atherosclerosis, *Hanyang Uidae Haksulchi*, 3, 273, 1983.
83. **Lim, C.-J., Park, E.-H., Rhee, D.-K., Lee, S.-J., and Hong, S.-K.,** Studies on the effects of ginseng total saponin on serum cholesterol and triglyceride levels in rats, *Hanguk Saenghwa Hakhoe Chi*, 14, 188, 1981.
84. **Yamamoto, M., Uemura, T., Nakama, S., Uemiya, M., Kishida, Y., Kasayama, S., Yamauchi, K., Komuta, K., and Kumagai, A.,** Comparative studies of ginseng saponins, saikosaponins and glycyrrhizin on hyperlipidemia in rats and in man, *Proc. Symp. Wakan-Yaku*, 15, 26, 1982.
85. **Yokozawa, T., Izawa, K., Kobayashi, T., Oura, H., and Yamamoto, M.,** Effect of ginseng saponin on lipid contents and glucose-6-phosphate activity, *Abstr. Pap., 102nd Annu. Meet. Pharm. Soc. Jpn.*, 193, 1982.
86. **Mitsuma, T., Izawa, K., Yokozawa, T., Oura, H., Yamamoto, M., and Kawashima, Y.,** Improving effects of ginseng saponin on cholesterol metabolism in rats, *Wakan Iyaku Gakkaishi*, 1, 15, 1984.
87. **Yokozawa, T., Kobayashi, T., Oura, H., and Kawashima, Y.,** Improving effects of ginsenoside-Rb_2 in streptozotocin diabetic rats with hyperglycemia and hyperlipemia, *Wakan Iyaku Gakkaishi*, 1, 22, 1984.
88. **Yamamoto, M. and Kumagai, A.,** Long term ginseng effects on hyperlipidemia in man with further study of its actions on atherogenesis and fatty liver in rats, *Proc. 4th Int. Ginseng Symp.*, 13, 1984.
89. **Yokozawa, T., Kobayashi, T., Kawai, A., Oura, H., and Kawashima, Y.,** Hyperlipemia-improving effects of ginsenoside-Rb_2 in cholesterol-fed rats, *Chem. Pharm. Bull.*, 33, 722, 1985.
90. **Joo, C.-N., Cho, Y.-D., Koo, J.-H., Kim, C.-W., and Lee, S.-J.,** Biochemical studies on ginseng saponins. XV. The effect of ginseng saponins on the absorption of fats and cholesterol in rats, *Hanguk Saenghwa Hakhoe Chi*, 13, 1, 1980.
91. **Yamamoto, M., Uemura, T., Nakama, S., Uemiya, M., Kumagai, A., Tamura, Y., Makino, H., and Saito, Y.,** Effect of long term administration of red ginseng powder in patients with hyperlipidemia, *Proc. Symp. Wakan-Yaku*, 16, 188, 1983.
92. **Lee, S.-O., Koo, J.-H., and Joo, C.-N.,** Biochemical study of ginseng saponin. XVII. The effect of ginseng saponin on the biosynthesis of phospholipids in rat liver, *Hanguk Saenghwa Hakhoe Chi*, 14, 161, 1981.
93. **Rhee, D.-K., Lim, C.-J., Park, E.-H., and Hong, S.-K.,** Effect of ginseng saponin on adrenaline-induced hyperglycemia and hyperlipidemia, *Yakhak Hoechi*, 25, 49, 1981.
94. **Okuda, H., Matsuno, H., and Lee, S.-J.,** Effect of rad ginseng powder on lipolytic and anorexigenic factor (toxohormone-L) from cancerous ascites fluid, *Proc. 4th Int. Ginseng Symp.*, 145, 1984.
95. **Park, H.-S., Kim, C.-M., and Koo, K.-H.,** Effect of ginseng saponin on hyperlipidemia and fatty liver induced by lipid administration in albino rabbits, *Hanyang Uidae Haksulchi*, 4, 445, 1984.
96. **Joo, C.-N.,** The preventative effect of Korean ginseng saponins on aortic atheroma formation in prolonged cholesterol fed rabbits, *Proc. 3rd Int. Ginseng Symp.*, 27, 1980.
97. **Lee, M.-W., Lee, T.-H., Ahn, B.-W., Park, B.-J., and Yang, S.-Y.,** Effects of ginseng components on rodenticide Vacor-induced diabetes mellitus in rats, *Proc. 4th Int. Ginseng Symp.*, 83, 1984.

98. **Martinez, B. and Staba, E. J.,** The physiological effects of Aralia, Panax and Eleutherococcus on exercised rats, *Jpn. J. Pharmacol.*, 35, 79, 1984.
99. **Bao, T.-T.,** Effect of total saponin of Panax ginseng on alloxan diabetes in mice, *Yaoxue Xuebao*, 16, 618, 1981.
100. **Yokozawa, T., Kobayashi, T., Oura, H., and Kawashima, Y.,** Stimulation of lipid and sugar metabolism in ginsenoside-Rb_2 treated rats, *Chem. Pharm. Bull.*, 32, 2766, 1984.
101. **Yokozawa, T., Kobayashi, T., Kawai, A., Oura, H., and Kawashima, Y.,** Stimulation of the lipogenic pathway in ginsenoside-Rb_2 treated rats, *Chem. Pharm. Bull.*, in press.
102. **Kimura, M., Waki, I., Tanaka, O., Nagai, Y., and Shibata, S.,** Pharmacological sequential trials for the fractionation of components with hypoglycemic activity in alloxan diabetes mice from ginseng radix, *J. Pharm. Dyn.*, 4, 402, 1981.
103. **Yokozawa, T., Kobayashi, T., Oura, H., and Kawashima, Y.,** Studies on the mechanism of the hypoglycemic activity of ginsenoside-Rb_2 in streptozoto-cin-diabetic rats, *Chem. Pharm. Bull.*, 33, 869, 1985.
104. **Yokozawa, T., Kobayashi, T., Oura, H., and Kawashima, Y.,** Hyperlipemia-improving effects of ginsenoside-Rb_2 in streptozotocin-diabeetic rats, *Chem. Pharm. Bull.*, 33, 3893, 1985.
105. **Yokozawa, T., Kiso, M., Oura, H., Yano, S., and Kawashima, Y.,** Improving effects of diabetic symptoms by ginsenoside-Rb_2 in streptozotocin diabetic rats, *Wakan Iyaku Gakkaishi*, 2, 372, 1985.
106. **Ando, T., Muraoka, T., Yamasaki, N., and Okuda, H.,** Preparation of anti-lypolytic substance from Panax ginseng, *Planta Med.*, 38, 18, 1980.
107. **Ohminami, H., Kimura, Y., Okuda, H., Tani, T., Arichi, S., and Hayashi, T.,** Effects of ginseng saponins on the actions of adrenaline, ACTH and insulin on lipolysis and lipogenesis in adipose tissue, *Planta Med.*, 41, 351, 1981.
108. **Kimura, M., Waki, I., Chujo, T., Kikuchi, T., Hiyama, C., Yamazaki, K., and Tanaka, O.,** Effects of hypoglycemic components in ginseng radix on blood insulin level in alloxan diabetes mice and on insulin release from perfused rat pancreas, *J. Pharm. Dyn.*, 4, 410, 1981.
109. **Kimura, M. and Suzuki, J.,** The pattern of action of blended Chinese traditional medicines to glucose tolerance curves in genetically diabetic KK-CA^y mice, *J. Pharm. Dyn.*, 4, 907, 1981.
110. **Waki, I., Kyo, H., Yasuda, M., and Kimura, M.,** Effects of a hypoglycemic component of ginseng radix on insulin biosynthesis in normal and diabetic animals, *J. Pharm. Dyn.*, 5, 547, 1982.
111. **Kimura, M., Suzuki, J., and Imamura, M.,** Mechanisms of insulin release stimulated by the hypoglycemic component (DPG 3-2) in ginseng radix, *Proc. Symp. Wakan-Yaku*, 14, 129, 1981.
112. **Kimura, M., Suzuki, J., and Arai, I.,** Calcium ion dependency of insulin release by hypoglycemic principle (DPG3-2) of ginseng, *Abrstr. Pap., 102nd Annu. Meet. Pharm. Soc. Jpn.*, 1982, 288.
113. **Kimura, M., Suzuki, J., Arai, I., Waki, I., and Tanaka, O.,** Non-physiological release of insulin by ginseng saponins, *Abstr. Pap., 102nd Annu. Meet. Pharm. Soc. Jpn.*, 1982, 288.
114. **Kasai, R., Hiyama, C., Yamamoto, S., Tanaka, O., Suzuki, J., Kimura, M., Matsuura, H., and Fuwa, T.,** Insulin release accelerating components (1): isolation and identification, *Abstr. Pap., 105th Annu. Meet. Pharm. Soc. Jpn.*, 1985, 473.
115. **Kimura, M., Suzuki, J., Onodera, S., Kasai, R., Tanaka, O., Matsuura, H., and Fuwa. T.,** Insulin release accelerating components (2): pharmacological verification of the activity by the pancreas perfusion method, *Abstr. Pap., 105th Annu. Meet. Pharm. Soc. Jpn.*, 1985, 473.
116. **Okuda, H., Sekiya, K., Ohminami, H., and Kimura, Y.,** Survey for physiologically active substances in crude drugs, *Abstr. Pap., Symp. Development of Natural Drugs*, Kyoto, 1981, 15.
117. **Sekiya, K., Masuno, H., and Okuda, H.,** Studies on insulin-like substances in Panax ginseng, *Proc. Symp. Wakan-Yaku*, 13, 6, 1980.
118. **Konno, C., Sugiyama, K., Kano, M., Takahashi, M., and Hikino, H.,** Isolation and hypoglycaemic activity of panaxans A, B, C, D and E, glycans of *Panax ginseng* roots, *Planta Med.*, 50, 434, 1984.
119. **Hikino, H., Oshima, Y., Suzuki, Y., and Konno, C.,** Isolation and hypoglycemic activity of panaxans F, G, and H, glycans of *Panax ginseng* roots, *Shoyakugaku Zasshi*, 39, 332, 1985.
120. **Tomoda, M., Shimada, K., Konno, C., Sugiyama, K., and Hikino, H.,** Partial structure of panaxan A, a hypoglycaemic glycan of *Panax ginseng* roots, *Planta Med.*, 50, 436, 1984.
121. **Tomoda, M., Shimada, K., Konno, C., Murakami, M., and Hikino, H.,** Structure of panaxan B, a hypoglycemic glycan of *Panax ginseng* roots, *Phytochemistry*, 24, 2431, 1985.
122. **Oshima, Y., Konno, C., and Hikino, H.,** Isolation and hypoglycemic activity of panaxans I, J, K and L, glycans of *Panax ginseng* roots, *J. Ethnopharmacol.*, in press.
123. **Konno, C. and Hikino, H.,** Isolation and hypoglycemic activity of panaxans M, N, O and P, glycans of *Panax ginseng* roots, *Int. J. Crude Drug Res.*, submitted.
124. **Konno, C., Murakami, M., Oshima, Y., and Hikino, H.,** Isolation and hypoglycemic activity of panaxans Q, R, S, T and U, glycans of *Panax ginseng* roots, *J. Ethnopharmacol.*, 14, 69, 1985.
125. **Hiai, S., Yokoyama, H., Oura, H., and Kawashima, Y.,** Evaluation of corticosterone secretion-induced activities of ginsenosides and their prosapogenins and saponins, *Chem. Pharm. Bull.*, 31, 168, 1983.
126. **Bombardelli, E., Cristoni, A., and Lietti, A.,** The effect of acute and chronic ginseng saponins treatment on adnerals function: biochemical and pharmacological aspects, *Proc. 3rd Int. Ginseng Symp.*, 9, 1980.
127. **Tanizawa, H., Numano, H., Odani, T., Takino, Y., Hayashi, T., and Arichi, S.,** Study of the saponin of *Panax ginseng* C. A. Meyer. I. Inhibitory effect on adrenal atrophy, thymus atrophy and the decrease of serum K^+ concentration induced by cortisone acetate in unilateral adrenalectomized rats, *Yakugaku Zasshi*, 101, 169, 1981.
128. **Pearce, P. T., Zois, I., Wynne, K. N., and Funder, J. W.,** Panax ginseng and Eleutherococcus senticosus extracts — *in vivo* studies on binding to steroid receptors, *Endocrinol. Jpn.*, 29, 567, 1982.
129. **Kotani, I., Arichi, S., and Odashima, S.,** Effects of ginsenosides on corticosterone generation of rat primary cultured adrenal cells, *Wakan Iyaku Gakkaishi*, 2, 154, 1985.
130. **Yun, Y.-S., Lee, S.-Y., Kim, B.-S., and Yun, T.-K.,** Studies on the mechanism of action of the cytotoxic fraction from Korean ginseng roots. I. Effects of petroleum ether fraction from Korean ginseng roots on the biosynthesis of macromolecules in mammalian neoplastic cells, *Hanguk Saenghwa Hakhoe Chi*, 13, 203, 1980.
131. **Yun, Y.-S., Lee, S.-Y., Kim, B.-S., and Yun, T.-K.,** Studies on the mechanism of action of the cytotoxic fractions from Korean ginseng roots. II. Effects of ethyl acetate fraction from Korean ginseng roots on the biosynthesis of macromolecules in mammalian neoplastic cells, *Hanguk Saenghwa Hakhoe Chi*, 13, 219, 1980.
132. **Moon, C.-K., Sim, K.-S., Lee, S.-H., Park, K.-S., Yun, Y.-P., Ha, B.-J., and Lee, C.-C.,** Antitumor activity of some phytobased polysaccharides and their effects on the immune function, *Arch. Pharmacol. Res.*, 6, 123, 1983.
133. **Kim, K.-J. and Kim, H.-S.,** Effects of Panax ginseng on the side effects of anticancer agent, *Chungnam Uidae Chapchi*, 9, 157, 1982.
134. **Yun, T.-K., Yun, Y.-S., and Han, I.-W.,** An experimental study on tumor inhibitory effect of red ginseng in mice and rats exposed to various chemical carcinogens, *Proc. 3rd Int. Ginseng Symp.*, 87, 1980.
135. **Ha, T.-Y. and Lee, J.-H.,** Effect of Panax ginseng on carcinogenesis induced by 3-methylcholanthrene in mice, *Taehan Uihak Hyophoe Chapchi*, 27, 541, 1984.
136. **Lee, F.-C., Park, J.-K., Kim, E.-K., Ko, J.-K., Lee, J.-S., and Kim, K.-Y.,** The role of *Panax ginseng* in detoxification of xenobiotics, *Proc. 4th Int. Ginseng Symp.*, 21, 1984.
137. **Yun, Y.-S., Jo, S.-K., Moon, H.-S., Kim, Y.-J., Oh, Y.-R., and Yun, T.-K.,** Effect of red ginseng on natural killer cell activity in mice with lung adenoma induced by urethan and benzo (a)pyrene, *Proc. 4th Int. Ginseng Symp.*, 27, 1984.
138. **Odashima, S., Nakayabu, Y., Ohtsuka, A., Abe, H., and Arichi, S.,** Effects of ginsenosides on cultured Morris hepatoma cells, *Proc. Symp. Wakan-Yaku*, 13, 1, 1980.

139. **Ohara, M., Kohno, H., Mune, S., Odashima, S., Kitagawa, I., Abe, H., Hayashi, T., and Arichi, S.,** Effects of ginsenosides on tumor cells. I. Inhibition of growth and induction of reverse transformation of melanoma cells, *Wakan Iyaku Gakkaishi*, 2, 170, 1985.
140. **Odashima, S., Ohta, T., Kohno, H., Matsuda, T., Kitagawa, I., Abe, H., and Arichi, S.,** Control of phenotypic expression of cultured B16 melanoma cells by plant glycosides, *Cancer Res.*, 45, 2781, 1985.
141. **Ohta, T., Matsuda, T., Ishii, E., Odashima, S., Nishikawa, K., Kitagawa, I., Abe, H., Arichi, S., and Hayashi, T.,** Effects of ginsenosides on tumor cells. II. On the effects on frowth of tumor cells, *Wakan Iyaku Gakkaishi*, 2, 172, 1985.
142. **Fegley, J., Fedak, L., and Turoczi, L. J.,** Studies in mutagenesis and anti-mutagenesis. I. Analysis of chromium, selenium, vitamin B-15 (pangamic acid), and ginseng (Panax), *Proc. Pa. Acad. Sci.*, 57, 181, 1983.
143. **Lee, I.-P., Yun, H.-C., Stammer, N., and Langenbach, R.,** Safety evaluation of *Panax ginseng* extracts: a lack of mutagenicity in the *Salmonella typhimurium* and Chinese hamster V79 cells, *Proc. 4th Int. Ginseng Symp.*, 75, 1984.
144. **Kimm, S.-W. and Park, S.-C.,** Evidences for the existence of antimutagenic factors in edible plants, *Kor. J. Biochem.*, 14, 47, 1982.
145. **Yeung, H.-W.,** Effect of ginseng on the immune responses to influenza virus infection in mice, *Proc. 3rd Int. Ginseng Symp.*, 245, 1980.
146. **Wang, B.-X., Cui, J.-C., and Liu, A.-J.,** The effect of polysaccharaides of roots of Panax ginseng on the immune funciton, *Yaoxue Xuebao*, 17, 66, 1982.
147. **Singh, V K., George, C. X., Singh, N., Agarwal, S. S., and Gupta, B. M.,** Combined treatment of mice with *Panax ginseng* extract and interferon inducer: amplification of host resistance to Semliki forest virus, *Planta Med.*, 47, 234, 1983.
148. **Gupta, S., Agarwal, S. S., Epstein, L. B., Fernandes, G., and Good, R. A.,** Panax — a new mitogen and interferon inducer, *Clinical Res.*, 28, 504A, 1980.
149. **Jie, Y.-H., Cammisuli, S., and Baggiolini, M.,** Immunomodulatory effects of Panax ginseng C. A. Meyer in the mouse, *Agents Actions*, 15, 386, 1984.
150. **Singh, V. K., Agarwal, S. S., and Gupta, B. M.,** Immunomodulatory activity of *Panax ginseng* extract, *Proc. 4th Int. Ginseng Symp.*, 255, 1984.
151. **Chong, S. K. F., Brown, H. A., Oberholzer, V., and Walker-Smith, J. A.,** Effect of ginseng saponins and hydrocortisone on phytohemagglutinin transformation of lymphocytes, *Lancet*, 2, 663, 1982.
152. **Chong, S. K. F., Brown, H. A., Rimmer, E., Oberholzer, V., Hindocha, P., and Walker-Smith, J. A.,** *In vitro* effect of Panax ginseng on phytohemagglutinin-induced lymphocyte transformation, *Int. Arch. Allergy Appl. Immunol.*, 73, 216, 1984.
153. **Matsuda, H., Hasegawa, T., and Kubo, M.,** Pharmacological study on *Panax ginseng* C. A. Meyer. VII. Protective effect of red ginseng on infection (1) On phagocytic activity of mouse reticuloendotherrial systems, *Yakugaku Zasshi*, 105, 948, 1985.
154. **Ben-Hur, E. and Fulder, S.,** Effect of Panax ginseng saponins and Eleutherococcus senticosus on survival of cultured mammalian cells after ionizing radiation, *Am. J. Clin. Med.*, 9, 48, 1981.
155. **Takeda, A., Yonezawa, M., and Katoh, N.,** Restoration of radiation injury by ginseng. I. Responses of X-irradiated mice to ginseng extract, *J. Radiat. Res.*, 22, 323, 1981.
156. **Yonezawa, M., Takeda, A., and Katoh, N.,** Restoration of radiation injury by ginseng extract II, *Proc. 4th Int. Ginseng Symp.*, 133, 1984.
157. **Yonezawa, M., Kato, N., and Takeda, A.,** Restoration of radiation injury by ginseng. II. Some properties of the radioprotective substances, *J. Radiat. Res.*, 22, 336, 1981.
158. **Takeda, A., Katoh, N., and Yonezawa, M.,** Restoration of radiation injury by ginseng. III. Radioprotective effect of thermostable fraction of ginseng extract on mice, rats and guinea pigs, *J. Radiat. Res.*, 23, 150, 1982.
159. **Kim, C.-M., Hwang, J.-J., and Kim, H.-L.,** Study on the radiation-protective ginseng proteins, *nonchong—Han" guk Saehwal Kwahak Yonguwon*, 34, 117, 1984.
160. **Chang, Y.-S., Park, C.-I., and Noh, H.-I.,** The effect of *Panax ginseng* of the postoperative radiation complication in cervical cancer patients, *Proc. 3rd Int. Ginseng Symp.*, 197, 1980.
161. **Furuya, T., Yoshikawa, T., Orihara, Y., and Oda, H.,** Saponin production in cell suspension cultures of Panax ginseng, *Planta Med.*, 48, 83, 1983.
162. **Ushiyama, K., Miyamoto, Y., Oda, H., and Ishida, Y.,** Production of *Panax ginseng* by tissue culture, *Nitto GIho*, 22, 1, 1984.
163. **Saito, H., Kohno, T., Tozuka, K., and Takagi, K.,** Biological evaluation of *Panax ginseng* callus, *Shoyakugaku Zasshi*, 34, 177, 1980.
164. Unpublished data.

Chapter 12

UTILIZATION OF MEDICINAL PLANTS: RECENT DEVELOPMENTS FROM THE CHINESE EXPERIENCE

Xiao Peigen

I. INTRODUCTION

China is one of the leading nations in the use of medicinal plants, and medicinal plants play an outstanding role within the framework of the official health services in China. The situation is well reflected in the 1990 edition of the *Chinese Pharmacopoeia*, in which the first volume is completely devoted to the crude drugs and their various preparations with a total of 784 items, the majority of which are of botanical origin. In comparison, the second volume is specific to chemical, antibiotic, and biological medicaments along with their preparations totaling 967 items. The importance of medicinal plants is also evidenced from the fact that traditional Chinese drugs, which consist of nearly exclusively medicinal plants, are estimated to account for 30 to 50% of total medicament consumption in recent years. Thus, it could be said that medicinal plants occupy an equally important position in comparison with synthetic drugs and antibiotics, etc., in China.[1-6]

In order to answer the continuing demand for drugs of plant origin, industrial production of the preparations derived from medicinal plants has been carried out in China both by modern as well as traditional methods.

II. GENERAL STATUS

Since Western medicine and traditional Chinese medicine are two main health care systems in China, the industrialization of medicinal plants is hence relevant to these systems, e.g., to the former digoxin from *Digitalis lanata* and steroidal hormones through semisynthesis from *Dioscorea nipponica*, while to the latter a traditional Chinese patent medicine 'Liu-wei-Di-huang Wan' is prepared from six crude drugs: *Rehmannia glutinosa, Cornus officinalis, Paeonia suffruticosa, Dioscorea opposita, Poria cocos,* and *Alisma orientalis* as a "honey pill" form.

In recent years, the Chinese government has emphasized systematizing traditional Chinese medicine by means of modern science and technology for the sake of raising it to a higher level. As a result, many pharmacologically active principles or fractions from the traditional Chinese drugs have been identified. At the same time, a series of new drugs and improved dosage forms have appeared (Figure 1).

III. CHINESE MEDICINAL PLANTS OF INDUSTRIAL IMPORTANCE

Nearly all the medicinal plants recorded in the 1985 edition of the *Chinese Pharmacopoiea* are industrialized either by traditional or modern methods. They are selectively listed in Table 1 and 2, with supplements in Table 2.

IV. SOME EXPERIENCES

A. INVESTIGATION AND SYSTEMATIZATION

Modern research on Chinese medicinal plants included as a necessary prerequisite, the comprehensive understanding of their origin, distribution, resources, and therapeutic usage. A nationwide survey has revealed that more than 5000 species of medicinal plants are being used in China today (Table 3).

This knowledge paved the way for the publication of several books:

1. *The Chinese Materia Medica:* It is regarded as a modern representative in this field. The old edition contained 994 items of crude drugs in 4 volumes, the new edition will be in 6 volumes, the first three (1979; 1982; 1985) have already been published. It was produced by the Institute of Materia Medica, and now the responsibility has been transferred to the Institute of Medicinal Plant Development (IMPLAD), Chinese Academy of Medical Sciences.
2. *The Atlas of Commonly Used Chinese Traditional Drugs* (1970): This booklet included 240 drug items, each of them illustrated by a colour painting. It serves as a tool for popularising the knowledge of traditional Chinese drugs, including identification, evaluation, collection, processing, and application.
3. *The Encyclopaedia of Chinese Materia Medica* (1977): It was edited by the Nanking College of Traditional Chinese Medicine, and included 5767 drug items of botanical, animal, and mineral origins, and is regarded as the most comprehensive compilation of traditional Chinese drugs and Chinese medicinal plants.
4. *The Colour Atlas of Chinese Herbal Drugs* (1982): Edited by the People's Medical Publishing House. It will eventually comprise 25 volumes, includes a total of 5000 drug items along with colour paintings, most of which are medicinal plants. At least six volumes have already been published.

As a result of this nationwide survey, a herbarium which included 50,000 specimens of medicinal plants has been set up in the Institute of Medicinal Plant Development; in the meantime, 35,000 items of ethnopharmacological data on Chinese medicinal plants have also been collected and subsequently systematized by electronic computer.

B. EXAMPLES OF THE UTILIZATION OF MEDICINAL PLANTS IN CHINA

1. *Scopolia tangutica*

This medicinal plant grows widely in the Qinghai-Tibet plateau. It has been commonly used in Tibetan traditional

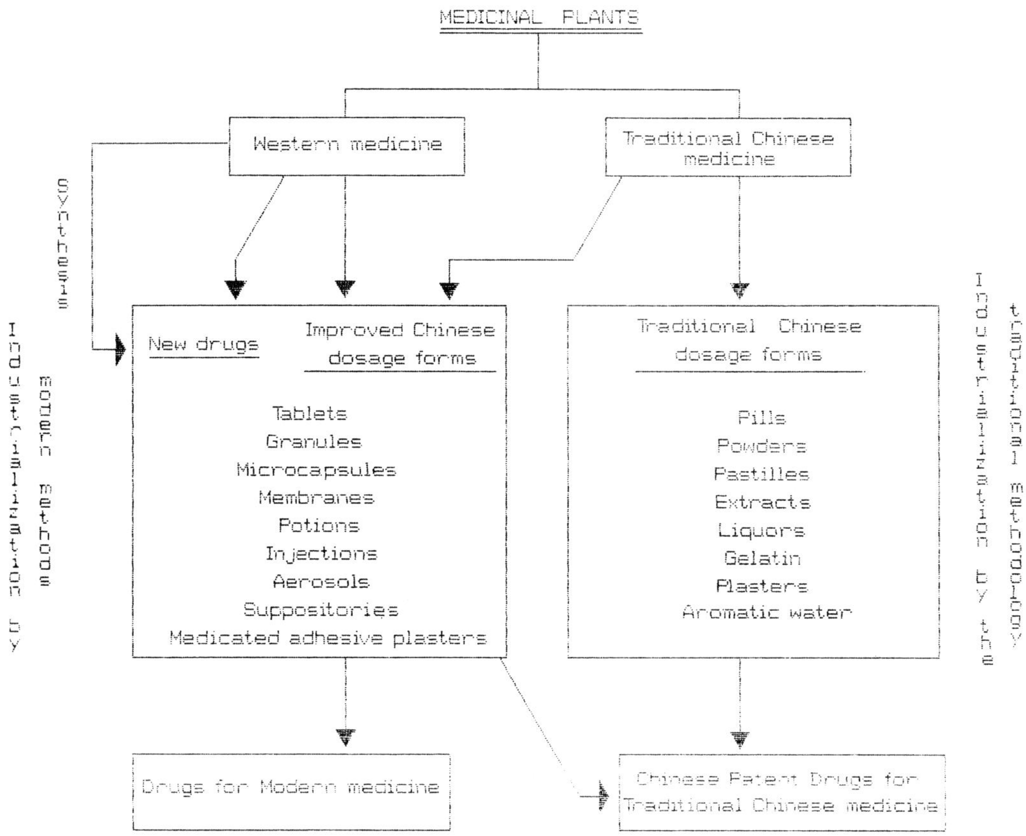

Figure 1. Diagram of the medicinal plant industry in China.

TABLE 1
Chinese Medicinal Plants with Greater Industrial Importance

Name ('Pin-Yin')	Part(s) used	Main constituent(s)	Product(s) or preparation(s)	Therapeutic use
Acanthopanax senticosus ('Ci-wu-jia')	Root Rhizome	Eleuthrosides, polysaccharide	Total extract, tablet, tincture	Used as a general tonic, for neurosis, neurasthenia, and weakness
Andrographis paniculata[a] ('Chuan-xin-lian')	Leaves	14-Deoxyandrographolide 1, andrographolide 2, neoandrographolide 3, 14-deoxy-11,12-didehydroandrographolide 4	Total extract, total lactones, tablet, injection (deoxyandrographolide)	For the treatment of infectious diseases of the gastrointestinal tract, respiratory organs and urinary system, and leprospirosis[3]

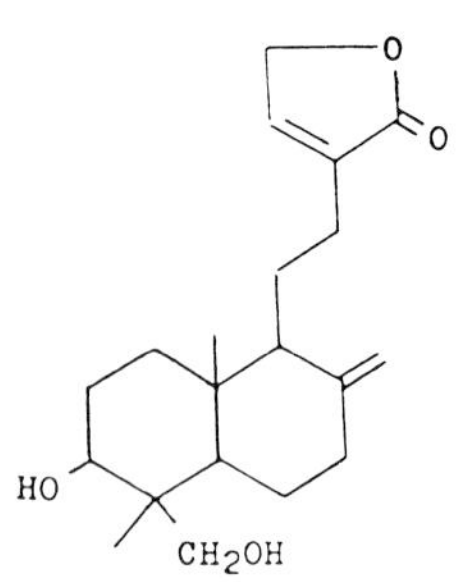

1

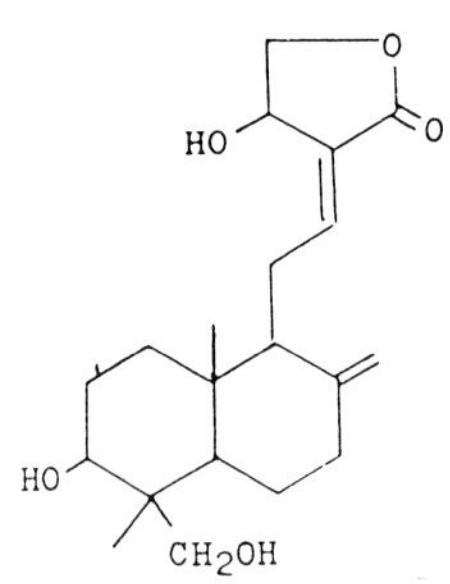

2

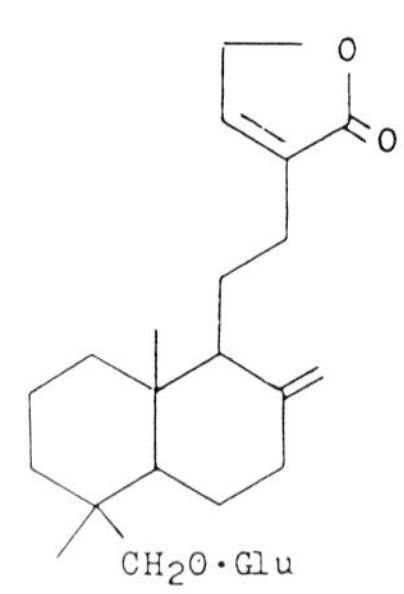

3

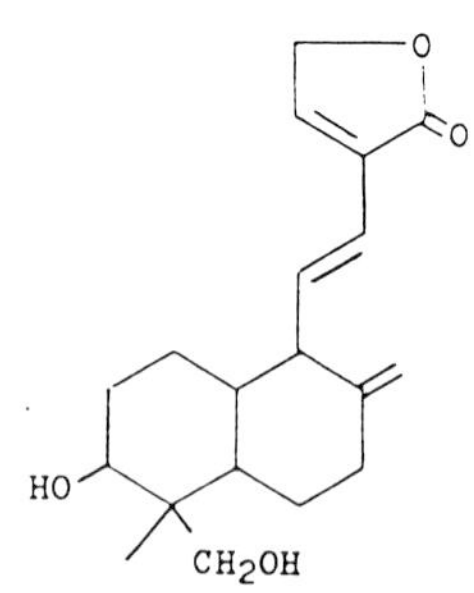

4

TABLE 1 (CONTINUED)
Chinese Medicinal Plants with Greater Industrial Importance

Name ('Pin-Yin')	Part(s) used	Main constituent(s)	Product(s) or preparation(s)	Therapeutic use
Angelica sinensis[a] ('Dang-gui')	Root	Ligustilide, *n*-butylidenepthalide, *n*-valerophenone-*o*-carboxylic polysaccharide, ferulic acid	Total extract, tablet, injection	For dysmenorrhea and abnormal menstruation, thromboangiitis obliterans
Artemisia argyi ('Ai-ye')	Leaves	Essential oil (mainly cineol, terpinen-4-ol, β-caryophyllene, artemisia alcohol, etc.)	Oil capsule aerosol	Antiasthmatic, for the treatment of chronic bronchitis
Atropa belladonna[a] ('Dian-qie')	Leaves, root	Hyoscyamine	Tincture	Anticholinergic
Bupleurum chinense ('Chai-hu')	Root	Saikosides Essential oil	Injection (distillation)	For the treatment of fever and hepatitis
Cartharanthus roseus[a] ('Chang-chun-hua')	Leaves Root	Vinblastine, vincristine	Injection	Antineoplastic
Chrysanthemum indicum ('Ye-ju-hua')	Inflorescens	Essential oil (camphor,α-pinene, carvone, etc.) Yejuhualactone	Suppository, injection	For the treatment of prostatitis
Cinnamomum camphora[a] ('Zhang')	Leaves, bark	Essential oil (α-camphor, cineole, etc.)	Essential oil	Antirhematic
Cinnamomum cassia[a] ('Rou-gui')	Bark	Essential oil mainly cinnamaldehyde)	Essential oil	Carminative, stomachic
Claviceps purpurea[a] ('Mai-jiao')	Sclerotium	Ergotamine, ergotoxine, ergometrine	Injection (ergometrine maleate)	Oxytocic antimigraine
Codonopsis pilosula[a] ('Dang-shen')	Root	Saponins, steriols polysaccharides	Total extract, potion (oral liquid)	General tonic, for weakness
Corydalis yanhusuo[a] ('Yan-hu-suo')	Rhizome	dl-Tetrahydropalmatine (THP) 5, dehydrocorydalin, d-corydaline,l-tetrahydrocoptisine, protopine	Total extract, THP (injection, tablet)	Used as an analgesic, hypnotic and sedative, also for the treatment of cardiovascular ailments

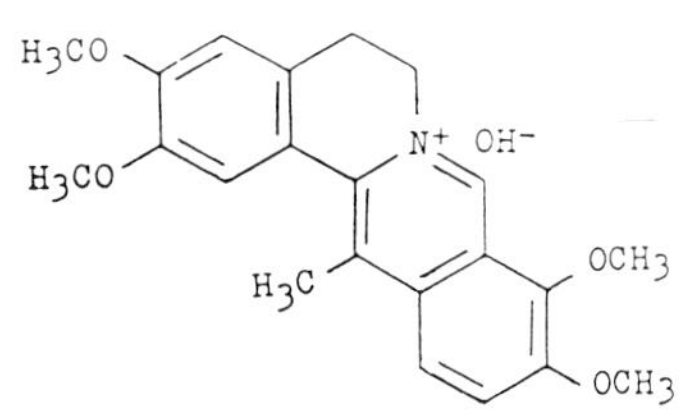

5

Name ('Pin-Yin')	Part(s) used	Main constituent(s)	Product(s) or preparation(s)	Therapeutic use
Coptis chinensis,[a] *C. deltoides* ('Huang-lian')	Rhizome	Berberine, coptisine, palmatine, etc.	Powder extract, berberine (tablet)	Various infections

TABLE 1 (CONTINUED)
Chinese Medicinal Plants with Greater Industrial Importance

Name ('Pin-Yin')	Part(s) used	Main constituent(s)	Product(s) or preparation(s)	Therapeutic use
Crataegus pinnatifidus[a] ('Shan-zha')	Fruit	Flavonoids, l-epicatechol	Extract, tablet, granules	Cardiovascular ailments, dyspepsia
Cymbopogon distans ('Yuan-xiang-cao')	Whole plant	Essential oil (piperitone, carene-4 geraniol, geranyl acetate, limonene)	Piperitone (aerosol)	Chronic bronchitis, asthmolytic
Datura metel[a] ('Yang-jin-hua')	Flower	Scopolamine, hyoscyamine	Scopolamine hydrobromide (tablet, injection)	Anticholinergic, also used in Chinese traditional anesthesia
Digitalis lanata[a] ('Yang-di-huang')	Leaves	Digoxin, lanatosides	Digoxin (tablet, injection)	Cardiotonic
Dioscorea nipponica ('Chuan-shan-long')	Rhizome	Diosgenin	Diosgenin	Manufacture of corticosteroids
Ephedra sinica ('Ma-huang')	Stem	l-Ephedrine	Ephedrine hydrochloride (tablet, injection)	Antiasthmatic, broncholytic, for hypotension
Eriobotrya japonica ('Bi-pa-ye')	Leaves	Essential oil (mainly *trans*-nerolidol), ursolic acid, oleanolic acid, etc.	Extract granule	Antitussive, expectorant
Eucalyptus globulus[a] ('An-ye')	Leaves	Essential oil (cineole, pinene, aromadendrene, etc.)	Oil of eucalyptus	For neuralgia
Fritillaria thunbergii[a] ('Zhe-bei-mu')	Bulb	Peimine, peiminine, etc.	Total extract (tablet, syrup)	Antitussive
Gastrodia elata[a] ('Tian-ma')	Rhizome	Gastrodin 6, vanillyl alcohol, vanillin	Total extract, gastrodin	Sedative, hypnotic, for vertigo[4]

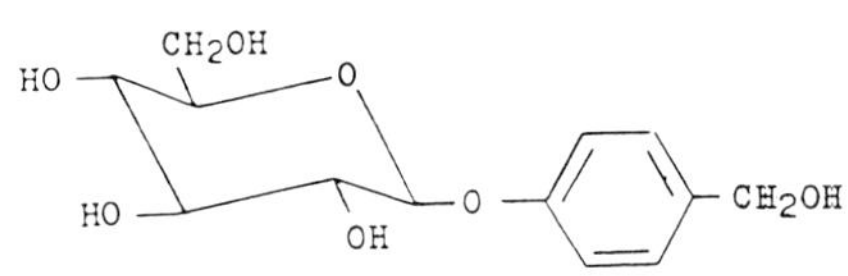

6

Name ('Pin-Yin')	Part(s) used	Main constituent(s)	Product(s) or preparation(s)	Therapeutic use
Glycyrrhiza uralensis ('Gan-cao')	Rhizome, root	Glycyrrhizin, 18-glycyrrhetic acid	Total extract	Emollient, expectorant, antiphlogistic for the treatment of ulcers
Houttuynia cordata ('Yu-xing-cao')	Fresh whole plant	Essential oil (decanoyl acetaldehyde 7, methyl-*n*-nonylketone, etc.)	Essential oil, decanoyl acetaldehyde (tablet, injection)	Respiratory infections

TABLE 1 (CONTINUED)
Chinese Medicinal Plants with Greater Industrial Importance

$$CH_3-(CH_2)_8-\overset{O}{\overset{\|}{C}}-CH_2-\overset{H}{\overset{|}{C}}=O$$

7

Name ('Pin-Yin')	Part(s) used	Main constituent(s)	Product(s) or preparation(s)	Therapeutic use
Iphigenia indica ('Li-jiang-shan-ci-gu')	Bulb	Colchicine, isocolchicine, etc.	Colchicine (tablet)	Acute gout, antineoplastic
Isatis indigotica[a] ('Da-qing-ye;Ban-lang-gen')	Leaves Root	Isatin B, indirubin, etc.	Total extract (granules)	Treatment of upper respiratory infections
Leonurus artemisia ('Yi-mu-cao')	Whole plant	Leonurine, stachydrine, leonuridine	Total extract, injection	Uterotonic, induces uterine contraction,[5] also for abnormal menstruation, menorrhagia
Ligusticum chuangxiong[a] ('Chuang-xiong')	Rhizome	Tetramethylpyrazine, 4-hydroxy-3-butyl phthalide	Total extract (tablet), tetramethylpyrazine (injection)	Coronary heart diseases, cerebral embolism[6]
Lonicera japonica[a] ('Jin-yin-hua')	Flower bud	Chlorogenic acid, isochlorogenic acid, essential oil (containing linalool, etc.)	Injection, tablet	Antimicrobial, for the treatment of various infections
Melia toosendan[a] ('Ku-lian-pi')	Root Bark	"Chuanliansu"	"Chuanliansu" (tablet)	Antiascaris
Menispermum dauricum ('Bei-dou-gen')	Rhizome	Menisperine, dauricine, sinomenine, etc.	Total alkaloids (tablet)	For tonsillitis, chronic bronchitis
Mentha haplocalyx[a] ('Bo-he')	Whole plant	Essential oil (containing menthol, menthone, isomenthone, etc.), d-8-acetoxycarvotanacetone (ACTC)	1. Mentha oil, menthol 2. ACTC to insects	1. Aromatic, carminative, symptomatic treatment of bronchitis 2. Repellent effect
Mucuna sempervirens ('Changchun-Youmateng')	Beans	L-Dopa	Tablet	Anti-Parkinson
Ophiopogon japonicus[a] ('Mai-dong')	Root	Steriodal saponins: ophiopogonins	Injection, potion (oral liquid)	Coronary heart disease, general tonic for weakness
Panax ginseng[a] ('Ren-shen')	Root	Ginsenosides polysaccharides	Extract, tincture granules	General tonic for weakness
Panax notoginseng[a] ('San-qi')	Root	Ginsenosides	Powder, tablet	For coronary heart diseases, hemoptysis, and hemorrhage
Physochlaina infundibularis ('Hua-shan-shen')	Root	Hyoscyamine, anisodamine	Total extract (tablet)	Antiasthmatic, antitussive, expectorant, for chronic and asthmatic bronchitis

TABLE 1 (CONTINUED)
Chinese Medicinal Plants with Greater Industrial Importance

Name ('Pin-Yin')	Part(s) used	Main constituent(s)	Product(s) or preparation(s)	Therapeutic use
Pueraria lobata ('Ge-gen')	Root	Daidzein 8, daidzin 9, puerarin 10, daidzein-4′,7-diglucoside 11	Total isoflavones extract (tablet)	Relief of hypertensive disease, angina pectoris, migraine, sudden deafness[7]

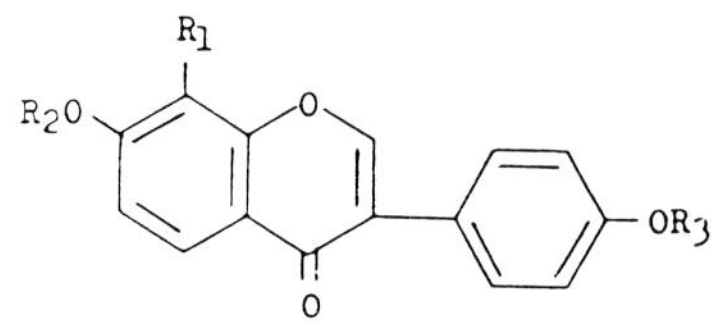

8 $R_1 = R_2 = R_3 = H$

9 $R_1 = R_3 = H$, R_2 = Glucopyranose

10 $R_2 = R_3 = H$, R_1 = Glucopyranose

11 $R_1 = H$, $R_2 = R_3$ = Glucose

Name ('Pin-Yin')	Part(s) used	Main constituent(s)	Product(s) or preparation(s)	Therapeutic use
Rauvolfia verticillata ('Lao-fu-mu')	Root	Reserpine, rescinnamine, ajmalicine, etc.	Total alkaloids, reserpine (tablet)	Hypertensive, sedative
Rhododendron dauricum ('Man-shan-hong')	Leaves	Essential oil	Oil capsule	Antitussive, expectorant, for bronchitis
Salvia miltiorrhiza ('Dan-shen')	Root	Cryptotanshinone 12, dihydrotanshinone I 13, dihydrotanshinone II-A 14, przewaquinone 15 tanshinone II-A, etc.	Total tanshinone (tablet, ointment) sodium tanshinone II_A sulfunate (injection) obliterans[9]	Antimicrobial[8] cardiovascular diseases, cerebral thromboembolism, thromboangiitis

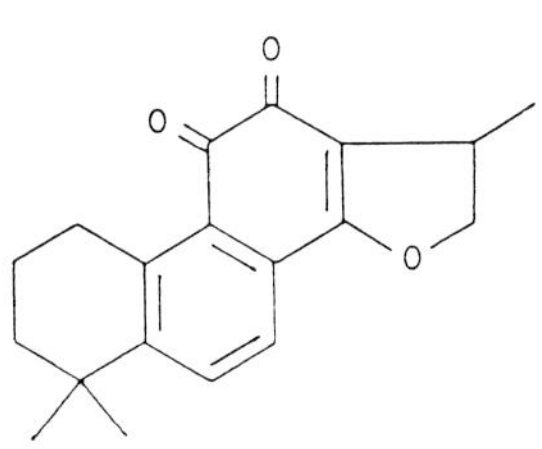

12

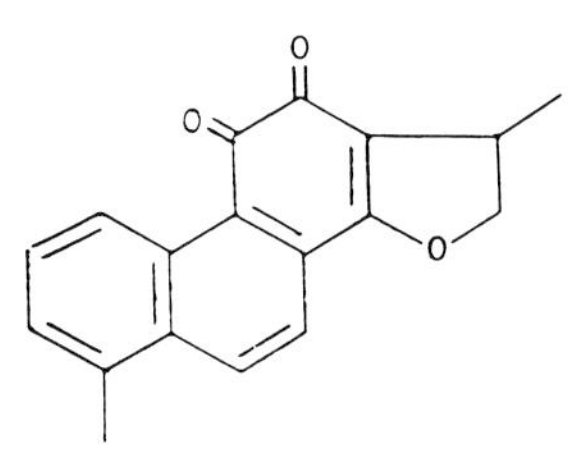

13

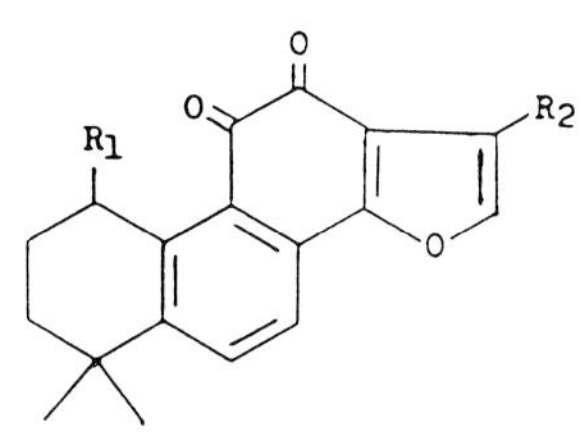

14 $R_1 = OH$, $R_2 = CH_3$

15 $R_1 = H$, $R_2 = CH_2OH$

Name ('Pin-Yin')	Part(s) used	Main constituent(s)	Product(s) or preparation(s)	Therapeutic use
Schisandra chinensis ('Wu-wei-zi')	Kernels	Schizandrin 16, γ-schizandrin 17, etc.	Ethanolic extract (tablet, capsule, syrup)	Lowering elevated serum glutamic pyruvic transaminase level[6]
Scopolia tangutica ('Shan-lang-dang')	Root Leaves	Hyoscyamine, scopolamine, anisodamine 18, anisodine 19, cuscohygrine	Anisodamine, anisodine (tablet, injection) atropine sulfate (tablet, injection)	Anticholinergic, septic shock from toxic bacillary dysentery, etc.[10]
Scutellaria baicalensis ('Huang-qin')	Root	Baicalein, baicalin, wogonin,wogonoside, etc.	Total extract (tablet), baicalin (injection)	Antipyretic, antimicrobial

TABLE 1 (CONTINUED)
Chinese Medicinal Plants with Greater Industrial Importance

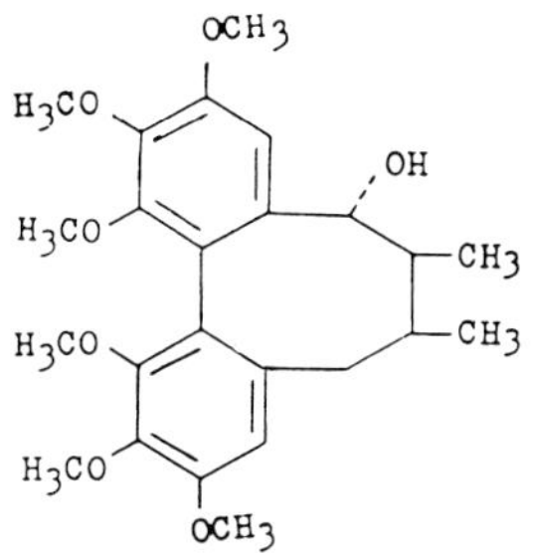

16

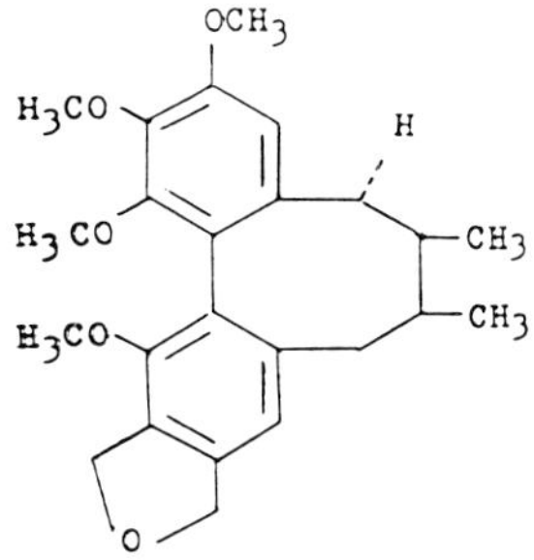

17

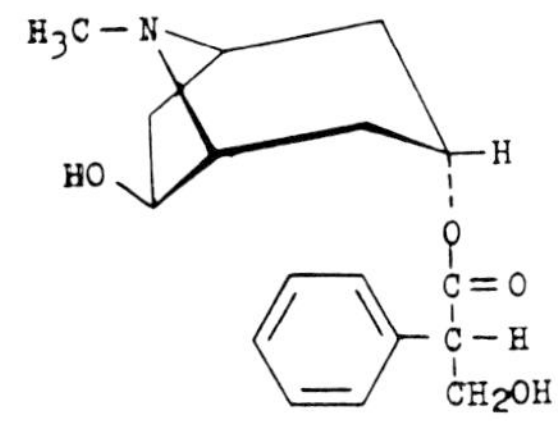

18

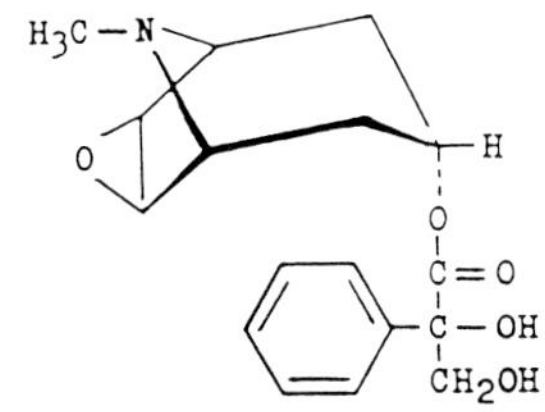

19

Name ('Pin-Yin')	Part(s) used	Main constituent(s)	Product(s) or preparation(s)	Therapeutic use
Sinomenium acutum ('Qing-teng')	Stem	Sinomenine	Sinomenine (injection)	Analgesic, sedative,[11] antirheumatic
Sophora japonica[a] ('Huai-mi')	Flower bud	Rutin, betulin	Manufacture of rutin	Decreasing capillary permeability
Stephania dielsiana ('Xuie-san-shu')	Tuber	Rotundine (l-tetra-hydropalmatine), stephanine, sinoacutine	Rotundine (tablet, injection)	Analgesic, hypnotic[12]
Vitex negundo var. *cannabifolia* ('Mu-jing')	Leaves	Essential oil (containing β-caryophyllene, caryophyllene oxide, β-elemene, cineole, α-pinene, etc.)	Oil capsule	Expectorant, antitussive, antiasthmatic, for chronic bronchitis

[a] Mainly cultivated.

TABLE 2
List of Chinese Medicinal Plants with Less Industrial Importance Including Those Recorded in the Local Standards and Those with Potential Industrial Importance

Name ('Pin-Yin')	Part(s) used	Main constituent(s)	Product(s)	Therapeutic use
Aconitum carmichaeli[a] ('Fu-zi')	Prepared lateral tuber	Higenamine 20 coryneine, etc.	Detoxic water soluble fraction (injection)	Cardiotonic[13]
Aconitum sinomontanum ('Ma-bu-qi')	Root	Lappaconitine 21	Lappaconitine	Analgesic[14]

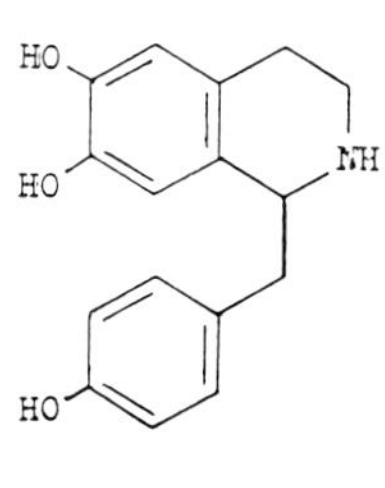

20

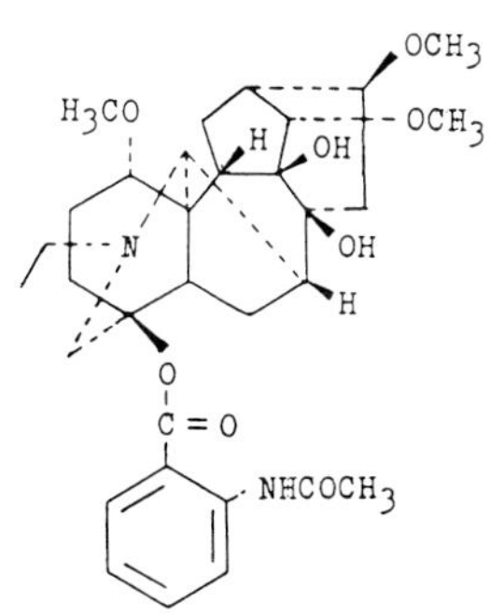

21

TABLE 2 (CONTINUED)
List of Chinese Medicinal Plants with Less Industrial Importance Including Those Recorded in the Local Standards and Those with Potential Industrial Importance

Name ('Pin-Yin')	Part (s) used	Main constituent(s)	Product(s)	Therapeutic use
Acorus gramineus ('Shi-chang-pu')	Rhizome	Essential oil (containing β-asarone, α-asarone, trans-4-propenyl- veratrole, etc.)	Essential oil (oil capsule)	CNS depressant, antiasthmatic[15]
Ajuga decumbens ('Jin-gu-cao')	Whole plant	Luteolin	Tablet	Antitussive[16]
Agrimonia pilosa ('He-cao-ya')	Winter bud	Agrimorphol 22	Agrimorphol (microcapsule)	Teanifuge[6]
Allium sativum[a] ('Da-suan')	Bulb	Diallyl thiosulfonate 23	Diallyl thiosulphonate microcapsule	Enhances the phagocytic action of the intraperitoneal macrophages mice, used for various infections[17,18]

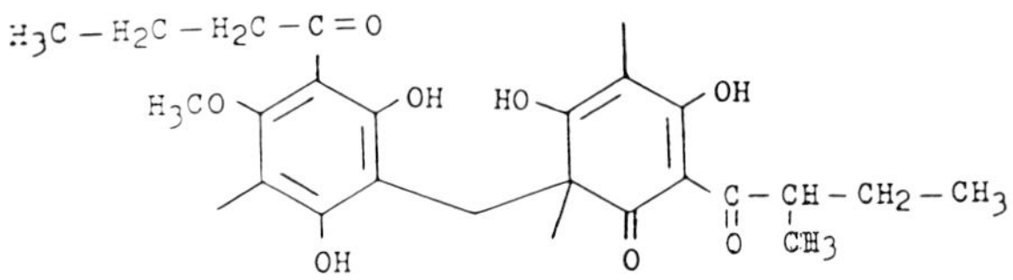

22

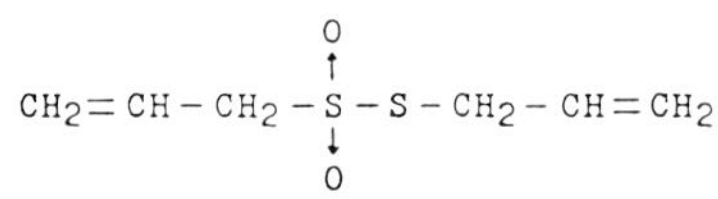

23

Name ('Pin-Yin')	Part (s) used	Main constituent(s)	Product(s)	Therapeutic use
Apocynum venetum ('Luo-bu-ma')	Leaves	Flavonoids	Total extract (tablet)	Hypotensive
Ardisia japonica ('Zi-jin-niu')	Whole plant	Ardisinal I 24, II 25 bergenin	Ardisinal I.II Bergenin	Antituberculosis[19] Antitussive

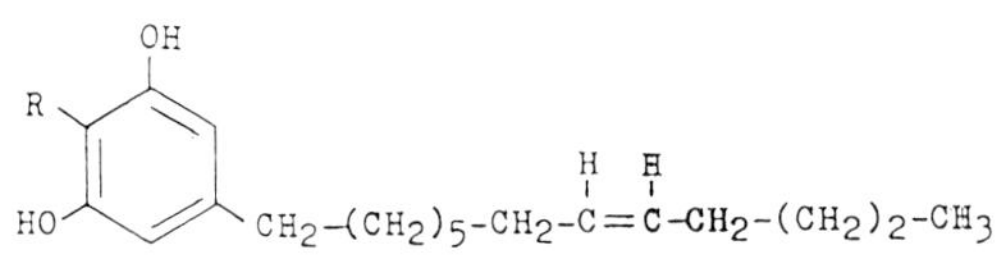

24 R = CH_3 25 R = H

Name ('Pin-Yin')	Part (s) used	Main constituent(s)	Product(s)	Therapeutic use
Artabotrys hexapetalus[a] ('Ying-chao')	Root	Yingzhaosu A 26	Yingzhaosu A	Antimalarial[20]
Artemisia annua ('Qing-hao')	Whole plant	Qinghaosu 27		Antimalarial[21]

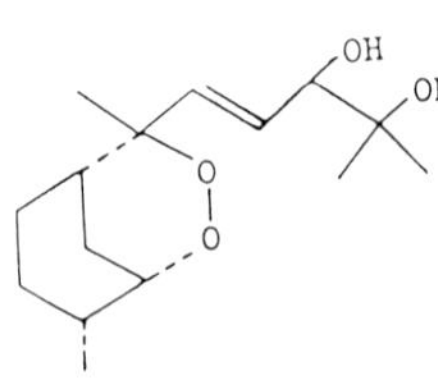

26

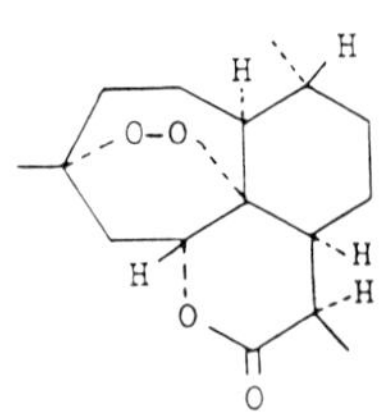

27

TABLE 2 (CONTINUED)
List of Chinese Medicinal Plants with Less Industrial Importance Including Those Recorded in the Local Standards and Those with Potential Industrial Importance

Name ('Pin-Yin')	Part (s) used	Main constituent(s)	Product(s)	Therapeutic use
Asarum forbesii ('Ma-xin')	Whole plant	Essential oil (methyleugenol, *trans*-isomethyl-eugenol, elemicin, etc.)	Essential oil	Strong sedative[22]
Astragalus mongholicus[a] ('Huang-qi')	Root	Astragalen (polysaccharides)	Astragalen	Immunostimulating
Atractylodes macrocephala[a] ('Bai-zhu')	Rhizome	Essential oil (containing atractylon, atractylol, butenolide A & B, etc.)	Essential oil	Active against esophageal cancer Ca 109 *in vitro*
Berberis poiretii ('Xiao-bo')	Root Root bark	Berberine, palmatine, berbamine 28, etc.	Berberine	Antibacterial, for infections of the gastro-intestinal tract
			Berbamine (tablet)	Leukocyte-increasing agent[23,24]

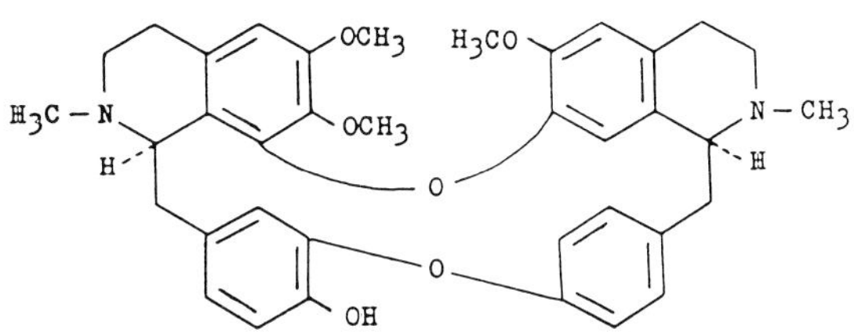

28

Name ('Pin-Yin')	Part (s) used	Main constituent(s)	Product(s)	Therapeutic use
Brucea javanica ('Ya-dan-zi')	Fruit	Brusatal 29, bruceine D 30, bruceine E 31	Relevant compounds	29 against sarcoma-180, 30.31 antimalarial[25]

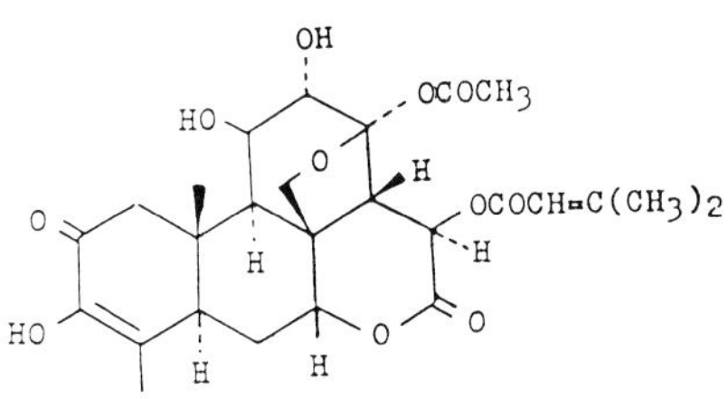

29

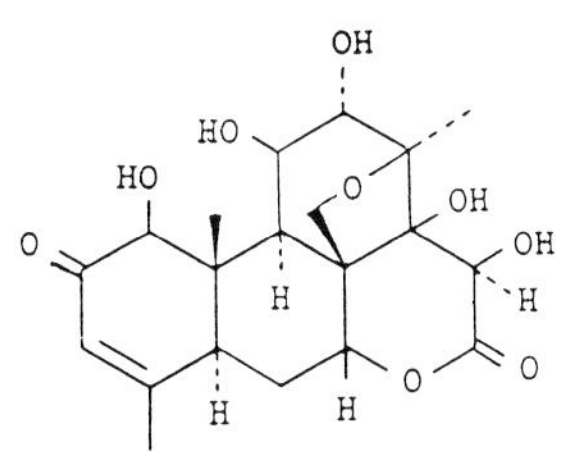

30

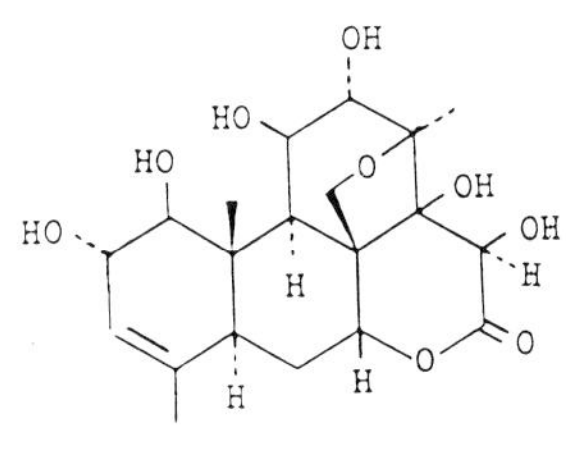

31

Name ('Pin-Yin')	Part (s) used	Main constituent(s)	Product(s)	Therapeutic use
Buxus microphylla var. *sinica* ('Huang-yang-mu')	Branch	Cycloviroxine D	Cycloviroxine D	Coronary heart diseases[26]
Camptotheca acuminata[a] ('Xi-shu')	Fruit	Camptothecine alkaloids camptothecine	Camptothecine 10-hydroxy-	For the treatment of stomach and intestinal cancer, as well as leukemia[27]
Cephalotaxus mannii (*C. hainanensis*) ('Hong-ke-song')	Bark	Cephalotaxine, harringtonine 32, homoharringtonine 33, isoharringtonine 34, deoxy-harringtonine 35, etc.	Harringtonine Homoharringtonine (injection)	Treatment of leukemia[28]

TABLE 2 (CONTINUED)
List of Chinese Medicinal Plants with Less Industrial Importance Including Those Recorded in the Local Standards and Those with Potential Industrial Importance

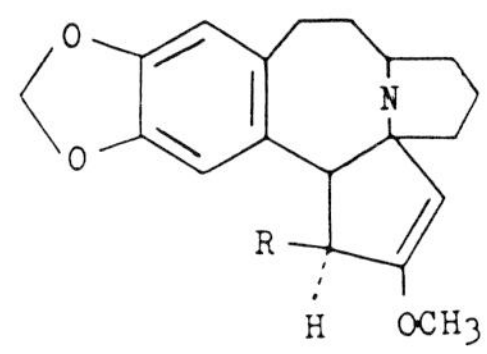

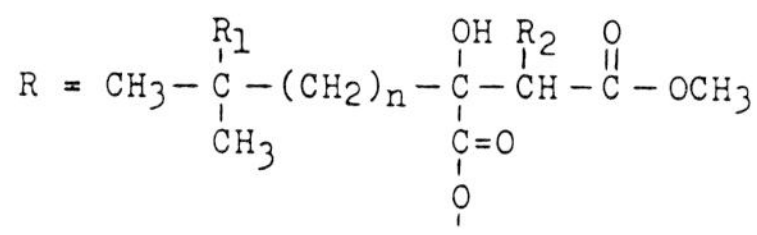

32 R_1 = OH , R_2 = H , n = 2

33 R_1 = OH , R_2 = H , n = 3

34 R_1 = H , R_2 = OH , n = 2

35 R_1 = H , R_2 = H , n = 2

Name ('Pin-Yin')	Part (s) used	Main constituent(s)	Product(s)	Therapeutic use
Cissampelos pareira var. *hirsuta* ('Ya-hu-nu')	Whole	Hayatine, hayatinine, hayatidine cissampreine, etc.	Hayatine methiodide	Muscle relaxant
Cucurbita moschata[a] ('Nan-qua-zi')	Seed	Cucurbitin 36	Cucurbitin	Inhibits growth of young worms of *Schistosoma japonica*[29]

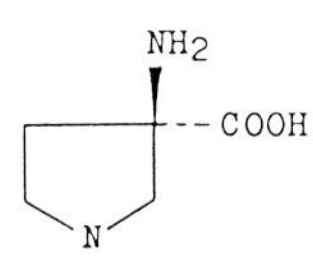

36

Name ('Pin-Yin')	Part (s) used	Main constituent(s)	Product(s)	Therapeutic use
Curcuma kwangsiensis[a] ('E-zhu')	Rhizome	Essential oil (curzerenone, curcumol 37, curdione 38, etc.)	Essential oil	For the treatment of the early stage of cervical cancer

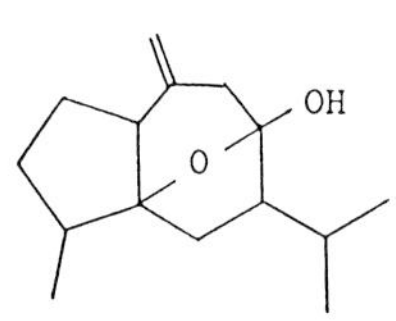

37

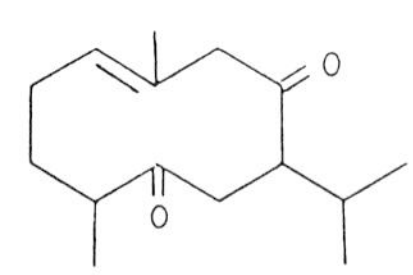

38

Name ('Pin-Yin')	Part (s) used	Main constituent(s)	Product(s)	Therapeutic use
Dactylicapnos scandens ('Zi-jin-long')	Root	Protopine d-isocorydine	Total extract, d-isocorydine (tablet)	Spasmolytic, analgesic
Daphne genkwa ('Yuan-hua')	Root	Yuanhuacine 39, yuanhuadine 40	39,40 intra-amniotic injection	Induces premature abortion[30,31]
Daphne giraldii ('Zu-si-ma')	Root Bark	Daphnetin, daphnin, etc.	Daphnetin (injection)	Analgesic[32]

TABLE 2 (CONTINUED)
List of Chinese Medicinal Plants with Less Industrial Importance Including Those Recorded in the Local Standards and Those with Potential Industrial Importance

Name ('Pin-Yin')	Part (s) used	Main constituent(s)	Product(s)	Therapeutic use

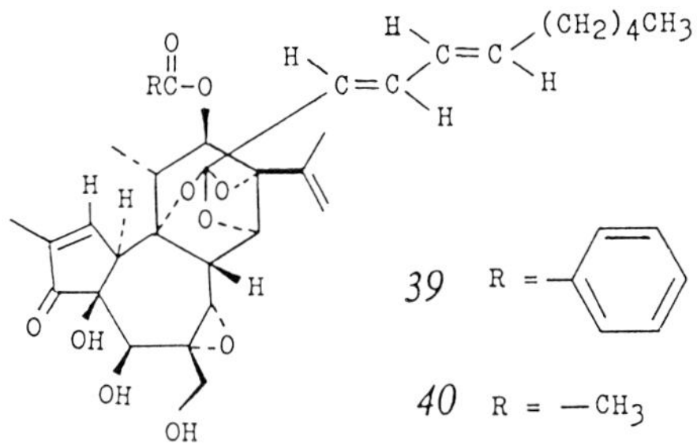

39 R = phenyl

40 R = —CH_3

Name ('Pin-Yin')	Part (s) used	Main constituent(s)	Product(s)	Therapeutic use
Dichroa febrifuga ('Chang-shan')	Root	Febrifugine 41 Isofebrifugine	Injection, tablet	Antimalarial Amebicidal[33]

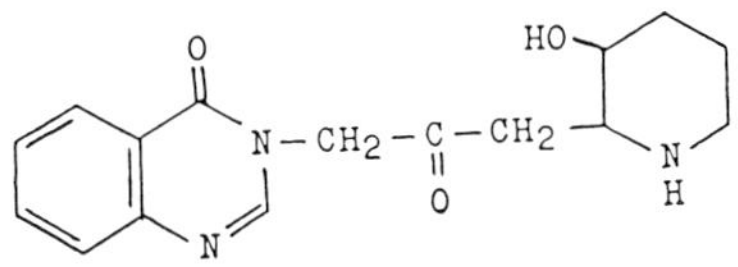

41

Name ('Pin-Yin')	Part (s) used	Main constituent(s)	Product(s)	Therapeutic use
Epimedium sagittatum ('Yin-yang-huo')	Up ground part	Icariine, des-*o*-methylicariine, β-anhydroicaritine	Total extract (tablet)	Coronary heart disease
Erigeron breviscapus ('Deng-zhan-xi-xin')	Whole plant	Pyromeconic acid, and its glucoside, scutellarin	Injection	Used as a hypotensive agent and for the treatment of sequelae of cerebral accident[34]
Erycibe obtusifolia ('Bao-gong-deng')	Stem	Bao-gong-teng A 44	Bao-gong-teng A	Treatment of glaucoma[35]

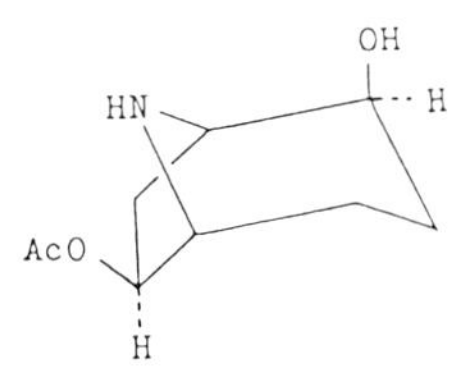

44

Name ('Pin-Yin')	Part (s) used	Main constituent(s)	Product(s)	Therapeutic use
Eucalyptus robusta[a] ('Da-ye-an')	Leaves	Robustanol A 45	Robustanol	Antimalarial[36]

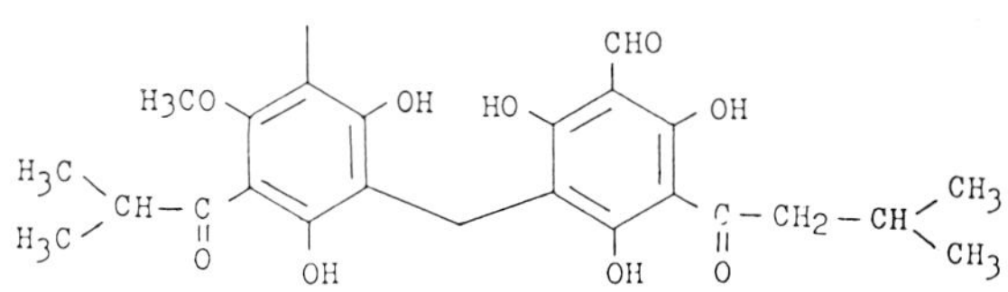

45

TABLE 2 (CONTINUED)
List of Chinese Medicinal Plants with Less Industrial Importance Including Those Recorded in the Local Standards and Those with Potential Industrial Importance

Name ('Pin-Yin')	Part (s) used	Main constituent(s)	Product(s)	Therapeutic use
Fagopyrum cymosum ('Jin-quio-mai')	Rhizome	5,7,3′,4′-tetra-Hydroxyflavane-3-nol C_4-C_8-dimer, hecogenin	Tablet	Treatment of pulmonary abscess, immunostimulating
Hemsleya amabilis ('Xue-dan')	Root	Cucurbitacin II_a 42, II_b 43 (tablet)	Total extract or crude saponins	Antimicrobial, for various infections[37]

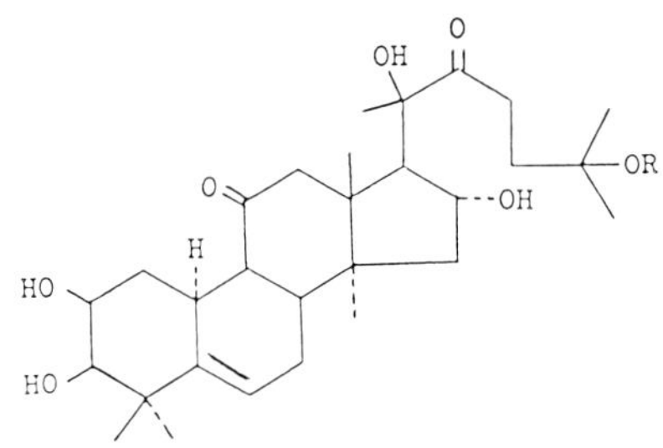

42 R = $COCH_3$

43 R = H

Name ('Pin-Yin')	Part (s) used	Main constituent(s)	Product(s)	Therapeutic use
Indigofera suffruticosa ('Jia-lan-dian')	Leaves	Indirubine 46	Indirubine	Treatment of leukemia[6]
Iris pallasii var. *chinensis* ('Ma-lin-zi')	Seeds	Irisquinone 47	Irisquinone	Marked efficacy in cervical cancer U_{14} in mice, also inhibited against lymphatic sarcoma, Ehrlich carcinoma[38]

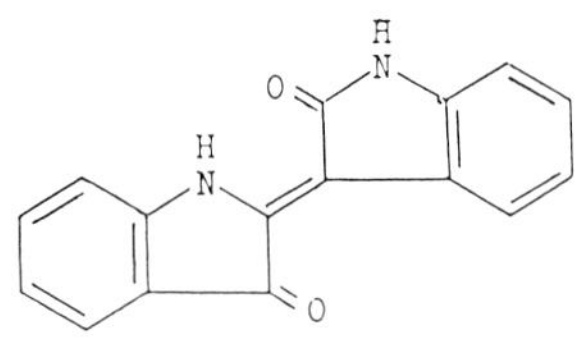

46

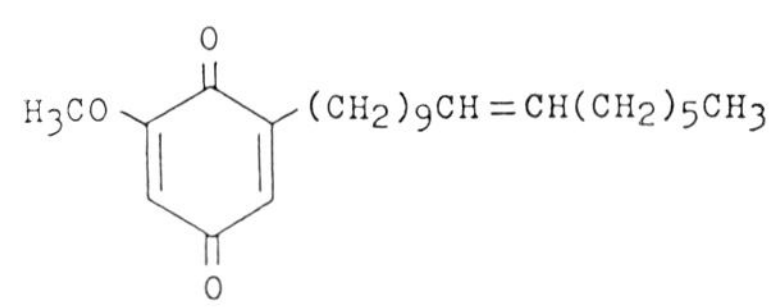

47

Name ('Pin-Yin')	Part (s) used	Main constituent(s)	Product(s)	Therapeutic use
Ganoderma lucidum ('Ling-zhi')	Spores	Uridine, uracil, etc.	Injection	Treatment of progressive muscular dystrophy, atropic myotonia[39]
Gentiana macrophylla ('Qin-jiao')	Root	Gentianine, gentia-nidine, gentianol	Total alkaloids (injection)	Antirheumatism[40]

TABLE 2 (CONTINUED)
List of Chinese Medicinal Plants with Less Industrial Importance Including Those Recorded in the Local Standards and Those with Potential Industrial Importance

Name ('Pin-Yin')	Part (s) used	Main constituent(s)	Product(s)	Therapeutic use
Gossypium hirsutum[a] ('Mian-zi')	Seed oil	Gossypol 48 6,6′-dimethoxy-gossypol, 6-methoxy-gossypol	Gossypol	Antifertility agent for males[6]

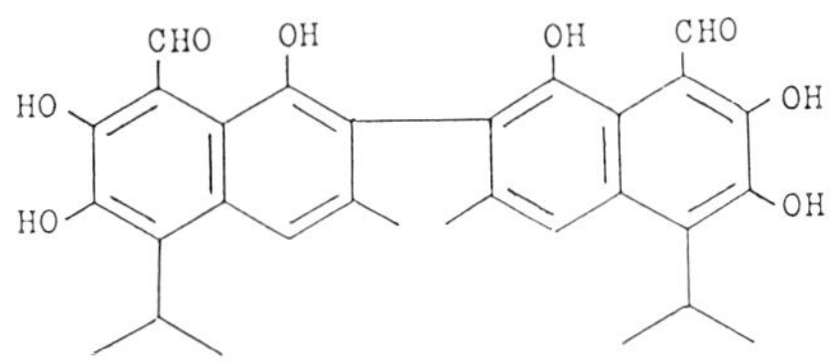

48

Name ('Pin-Yin')	Part (s) used	Main constituent(s)	Product(s)	Therapeutic use
Lysionotus pauciflorus ('Shi-diao-lan')	Whole plant	Nevadensin 49	Nevadesin (tablet)	Antituberculosis[41] expectorant, antitussive

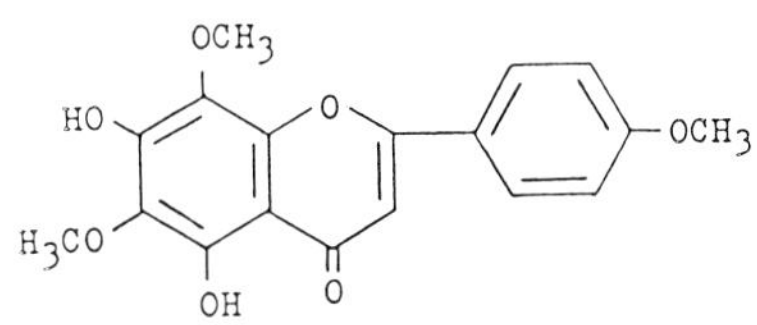

49

Name ('Pin-Yin')	Part (s) used	Main constituent(s)	Product(s)	Therapeutic use
Patrinia scabio-saefolia ('Bai-jiang-gen')	Rhizome, root	Scabiosides, essential oil (patrinene, isopatrinene, etc.)	Tincture, extract	Sedative, for insomnia and neurasthenia
Picrasma quassioides ('Ku-mu')	Stem	1-Carbomethoxy-β-carboline,4,5-dimethoxy-canthin-6-one 50, 4-methoxy-5-hydroxy-canthin-6-one 51, etc.	Total alkaloids	Antimicrobial[42]

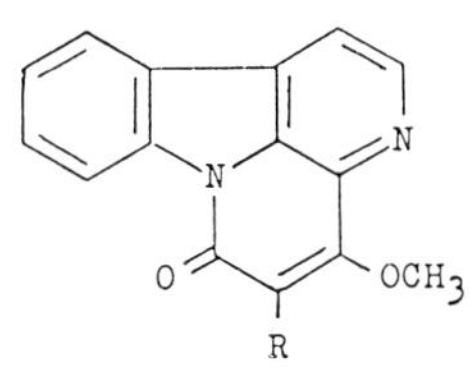

50 R = OCH_3

51 R = OH

Name ('Pin-Yin')	Part (s) used	Main constituent(s)	Product(s)	Therapeutic use
Piper nigrum[a] ('Hu-jiao')	Fruit	Piperine 52	Piperine and its derivatives	Antiepileptic[6]
Podophyllum emodii ('Tao-er-qi')	Rhizome Root	Podophyllotoxin, demethylpodophyllotoxin sikkimotoxin, etc.	Podophyllotoxin and its derivatives	Antineoplastic

TABLE 2 (CONTINUED)
List of Chinese Medicinal Plants with Less Industrial Importance Including Those Recorded in the Local Standards and Those with Potential Industrial Importance

Name ('Pin-Yin')	Part (s) used	Main constituent(s)	Product(s)	Therapeutic use
		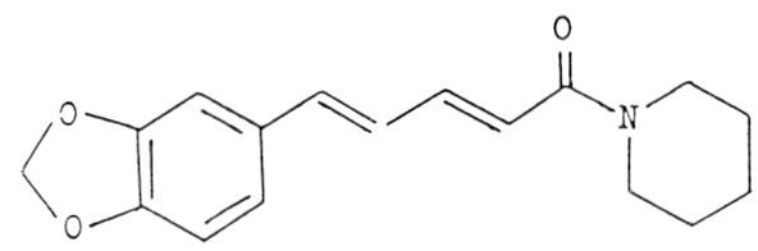 52		
Polygonum cuspidatum ('Hu-zheng')	Rhizome Root	Polydatin	Polydatin	Decreases blood lipids, particularly triglyceride
Polyporus umbellatus ('Zhu-ling')	Sclerotium	Ergosterol,α-hydroxy-tetracosanoic acid, zhuling polysaccharides	Extract	Against sarcoma 180, with an immunostimulating action
Przewalskia tangutica ('Ai-lang-dang')	Root and plant	Hyoscyamine, anisodamine, scopolamine	Hyoscyamine	Anticholinergic[43]
Pyrola rotundifolia ssp. *chinensis* ('Lu-xian-cao')	Whole plant	Methyl hydroquinone	Methyl hydroquinone	For infections of the respiratory, digestive, and urinary systems
Pyrrosia sheareri ('Shi-wei')	Leaves	Isomangiferin, saponin	Isomangiferin (tablet)	Antitussive
Rabdosia rubescens ('Dong-ling-cao')	Leaves	Rubescensines, oridonine, ponicidine	Extract (syrup, tablet)	For the treatment of esophageal cancer[44]
Rheum palmatum ('Da-huang')	Rhizome	Sennosides, rhein, emodin, tannin, etc.	Total sennosides Total extract	Purgative Hemostatic
Rhodea japonica[a] ('Wan-nian-qing')	Rhizome	Rhadexin A, B, C	Total extract (membrane)	Cardiotonic, in cases of arrhythmia and tachycardia
Securinega suffruticosa ('Yi-ye-qiu')	Leaves	Securinine allosecurinine, etc.	Securinine (injection)	Treatment of facial paralysis and neurasthenia
Sedum sarmentosum ('Chui-pen-cao')	Whole plant	Sarmentosine 53	Sarmentosine	Lowers SGPT level[45]
		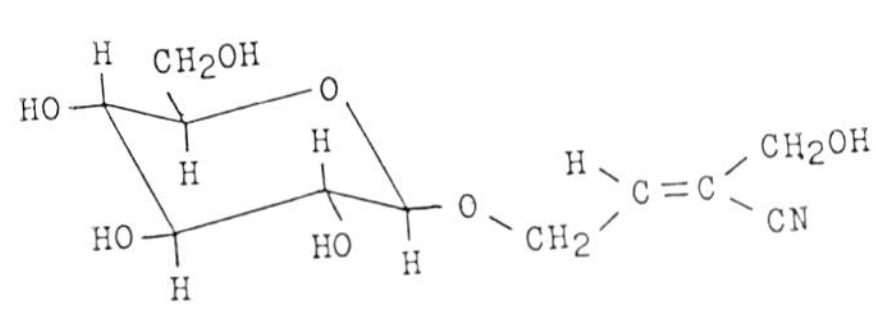 53		
Silybum marianum[a] ('Ru-ji-zi')	Seeds	Flavanolignans (silybin, silydianin, etc.)	Extract	Hepatitis, antihepatoxic
Sophora flavescens ('Ku-shen')	Root	Matrine, oxymatrine, sophoranol, etc.	Total alkaloids	Antineoplastic, antihepatotoxic

TABLE 2 (CONTINUED)
List of Chinese Medicinal Plants with Less Industrial Importance Including Those Recorded in the Local Standards and Those with Potential Industrial Importance

Name ('Pin-Yin')	Part (s) used	Main constituent(s)	Product(s)	Therapeutic use
Swertia milieensis ('Qing-ye-dan')	Whole plant	Swertiajaponin	Total extract	Lowers SGPT[46]
Thevetia neriifolia[a] ('Huang-hua-jia-zhu-tao')	Kernels	Thevetin A & B, peruvoside, neriifolin ruvoside, etc.	Mixture of secondary glucosides (tablet, injection)	Cardiotonic[47]
Trichosanthes kirilowii[a] ('Tian-hua-fen')	Root	Proteins of Trichosanthes	Proteins of Trichosanthes	Induces premature abortion
Tripterygium hypoglaucum ('Kunming-shanhaitang')	Root	Triptolide <u>54</u>, tripdiolide <u>55</u>	Alcoholic extract	Antileukemic, rheumatoid arthritis[48,49]
Typha angustifolia ('Pu-huang')	Pollens	Isorhamnetin	Total extract (tablet)	Coronary heart diseases
Uncaria rhychophylla ('Kou-teng')	Stem	Rhynchophylline	Total alkaloids	Hypotensive, sedative[50]
Valeriana wallichina ('Zhi-zhu-xiang')	Rhizome Root	Essential oil, (valtrate, etc.)	Total extract valepotriates	Tranquilizer, sedative
Wikstroemia indica ('Liao-ge-wang')	Root	Daphnoretin	Daphnoretin	Improves nutrient blood flow of the heart muscle in mice

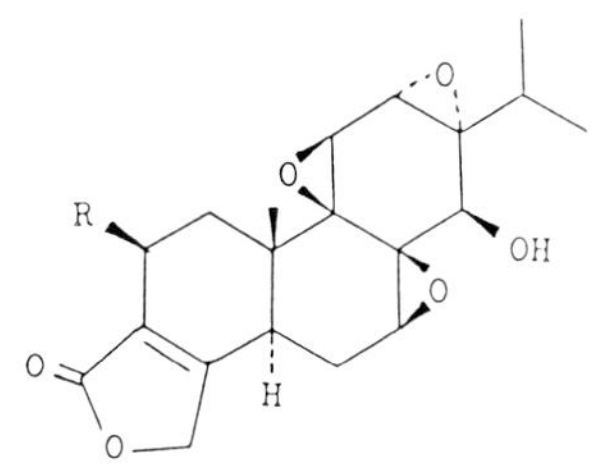

54 R = H

55 R = OH

TABLE 3
Up-To-Date Chinese Medicinal Plants

Origin	Medicinal plant included
Thallophytes	218
Bryophytes	39
Pteridophytes	395
Gymnosperms	55
Angiosperms	
Dicotyledones	3690
Monocotyledones	676
Total	5136

medicine as an analgesic, anesthethic, and antispasmodic agent. A phytochemical study revealed that besides hyoscyamine and scopolamine, it contained anisodamine, [(–)-6β-hydroxy-hyoscyamine], and anisodine. Pharmacologically[10] anisodamine has marked peripheral anticholinergic effects, and the inhibitory action on salivary secretion, the mydriatic action and influence on EEG, and other central actions of anisodamine were about 20 times, 10 times, and 6 to 20 times weaker, respectively, than that of atropine. Taking advantage of this weaker central action, anisodamine has been widely adopted as an emergency measure for septic shock from toxic bacillary dysentery, fulminant epidermic meningitis, and hemorrhagic enteritis, resulting in a markedly lower death rate. More recently, anisodamine was studied in the spinotrapezius muscle of the rat, and the result showed that this compound could improve the microvascular blood flow. Anisodine showed similar efficacy on central anticholinergic action in comparison with atropine, but on the peripheral anticholinergic effect it proved weaker than both atropine and scopolamine. Clinically, anisodine is being used for the treatment of migraine, diseases of the fundus occuli caused by vascular spasm, detoxication of organic phosphorus pesticides, and in some clinical cases for paralysis caused by various neurological diseases, it is also the main constituent in intravenous combination anesthesia.

Scopolia tangutica is now being used for manufacturing various preparations of anisodamine and anisodine.

2. *Schisandra chinensis*

This traditional Chinese drug has been commonly used as an astringent, but it was found clinically to exhibit therapeutic effects for certain forms of hepatitis, particularly in lowering the elevated serum glutamic pyruvic transaminase (SGPT) level. This immediately aroused the interest of physicians as well as pharmacists. Out of eight components isolated from this drug, all are derivatives of dibenzo (a,c) cyclooctene series, six of which could significantly impede the elevation of SGPT induced by CCl_4 in mice. Based on the pharmacological results, the ethanolic extract of the kernel of this plant is now being produced commercially for this purpose. Furthermore, a new synthetic drug, biphenyl dimethoxydicarboxylate (BDD) derived from an intermediate for the synthesis of schisandrin C analogs, has also been developed;[51,52] it has been shown to be effective and safe for the treatment of chronic viral hepatitis B. BDD is now being manufactured on a large scale by synthetic means.

3. *Salvia miltiorrhiza*[8,9]

The root of this plant called 'Dan-shen' is another commonly used traditional Chinese drug, from which a series of pharmacologically active principles have been isolated, e.g., cryptotanshinone, hydroxytanshinone II-A, dihydrotanshinone I, methyltanshinonate, and tanshinone II-B, exhibited stronger inhibition against *Staphylococcus aureus* and its drug-resistant strains. Clinically, the total tanshinones showed satisfactory results in the treatment of tonsilitis, furuncle of the external auditory canal, acne, carbuncles, burns, osteomyelitis, etc. Sodium tanshinone II-A sulfonate, prepared from tanshinone II-A, has been investigated clinically on coronary patients, and the results showed that the percentages of recovery for angina pectoris, chest opppression, and of electrocardiogram improvements are all satisfactory.

A series of preparations from *Salvia miltiorrhiza* were developed.

4. *Pueraria lobata*[7]

From an ancient medical publication "Sheng Han Lun" (Treatise on Febrile Diseases), it is recorded that 'Ge Gen Tang' (a decoction that has the root of *Pueraria lobata* as its chief ingredient) is recommended for the treatment of stiffness and soreness in the neck. This clue led Chinese phytochemists to isolate a series of isoflavones, i.e., daidzein, daidzin, puerarin, and daidzein-4′,7-diglucoside. At the same time, pharmacological studies revealed that the total isoflavone was capable of increasing the cerebral and coronary blood flow, decreasing the oxygen consumption of the myocardium, increasing the blood oxygen supply, and depressing the production of lactic acid by the blood oxygen deficient heart muscle; it has also shown a papaverine-like spasmolytic action. The single components daidzein, daidzin, and puerarin showed similar pharmacological actions to the total isoflavones. Clinical trials of the preparations from the Pueraria total isoflavones or the single daidzein have demonstrated their utility for relieving the symptoms caused by hypertensive diseases such as headache, dizziness, and a stiffness and soreness in the neck, as well as for angina pectoris, migraine, and sudden deafness. Thus, the total isoflavone preparation 'Yu-feng-ning-xin-pian' was developed. In the meantime, the total synthesis of daidzein is also being carried out for industrial production.

5. *'Gan-mao Tui-re Chong-ji'* (Antipyretic Granules for Cold)

The leaves and roots of *Isatis indigotica* fruits of *Forsythis suspensa* and rhizomes of *Polygonum bistorta* are all Chinese traditional and herbel drugs with supposed actions for eliminating toxic heat. After pharmacological and clinical studies, an improved dosage form "Antipyretic Granules for Cold" has been developed and recorded in the 1985 *Chinese Pharmacopoiea*. The preparation is described as follows:

> The mixture of *Folium isatidis* (200 g), *Radix isatidis* (200 g), *Fructus forsythiae* (100 g), and *Rhizoma bistortae* (100 g) is extracted twice with water, 1.5 h for each decoction. The decoctions are combined and filtered. The filtrate is concentrated at 90° to 95°C to sp-gr 1.08 and allowed to cool to room temperature before an equal amount of alcohol is added. After sedimentation, the clear supernatant liquid is separated and concentrated at 60° to 65°C to sp-gr 1.20. An equal amount of water is added and stirred well after standing for 8 h and the clear supernatant liquid is again concentrated at 60 to 65°C to a light extract with sp-gr 1.38 to 1.40. Granules are made by mixing 1 part this light extract, 3 parts sugar powder, 1.25 parts dextrin, and an appropriate amount of alcohol. The granules are dried and packed into bags of 18 g each. It is used for eliminating toxic heat, treatment of upper respiratory infections, acute tonsillitis, and laryngopharyngitis. The dose is 18 to 36 g one to three times daily, taken after dissolving the granules in boiling water.

C. TRAINING AND POPULARIZATION

In order to increase and promote the utilization of medicinal plants in various regions, it is necessary to train professionals. Official training is being carried out in the colleges of traditional Chinese medicine, the colleges of pharmacy, and a college of traditional Chinese drugs in Nanjing. In addition, postgraduate training is also being under taken in the above-mentioned colleges along with several research institutions, Institute of Medicinal Plant Development, Institute of Chinese Material Medica, etc., where a series of curricula including traditional medicine and Western medicine are provided. There are also several scientific periodicals that specialize in this field, such as the *Journal of Chinese Traditional and Herbal Drugs*, the *Bulletin of Chinese Traditional Drugs*, etc. Each province and municipality has its own books or booklets on medicinal plants for the popularization of knowledge including the identification, collection, processing, application, and cultivation of the local medicinal plants.

D. DRUG CONTROL

Medicinal plants and their preparations are controlled by the *Chinese Pharmacopoiea* and the *Local Standard*. When the active principles or active fractions of a medicinal plant or preparation are ascertained, they must be evaluated by specific methods qualitatively and quantitatively. In most cases, however, where the active principles of any medicinal plants are not yet clarified, quality is ensured by adherence to fixed standards which include factors such as origin, characteristics, identification, formulation, manufacturing process of the drug, and some other relevant parameters.

V. DISCUSSION

For the purpose of industrial production of medicaments, medicinal plants appear to serve as increasingly important sources. Since medicinal plants have been used for centuries among vast numbers of people, there should be ample opportunity for the development of favorable therapeutic agents with less toxicity and undesired side effects.

By means of multidisciplinary research including pharmacognostical, chemical, pharmacological, and clinical studies, a knowledge of the active principles as well as the mechanism of their therapeutic function could be acquired. The skeletons of new structures or an important pharmacological response of a known compound could thus be treated as prototypes of new drug design. Then, by means of total synthesis or modification of the chemical structures, satisfactory new medical agents could be obtained. This integration of modern science and technology with traditional knowledge is gradually being accepted as an important strategy in the field of drug research. Based on statistical information, there have been 104 new drugs developed in China since 1949, and 53 to 60 of these originated directly or indirectly from medicinal plants. This fact has shown that the increasing emphasis on the utilization of medicinal plants is undoubtedly a correct and valuable approach in the development of the pharmaceutical industry.

V. CONCLUSION

There are about 110 selected medicinal plants which have either been already industrially utilized or which have potential industrial importance in China. Thus, Chinese medicinal plants have already made and will continue to make a great contribution to the health care of the Chinese people. Furthermore, the experience with Chinese medicinal plants and their ethnopharmacological data offer a fundamental basis for further development of new drugs.

As a current trend, increasing emphasis has now been laid on natural products for medicinal use all over the world. We expect that a flourishing medicinal plant industry will appear in the near future.

ACKNOWLEDGMENT

My grateful thanks are due to Professor Sung Weilang of my Institute for his valuable help in preparing this manuscript.

REFERENCES

1. **Xiao Peigen,** Some experience on the utilization of medicinal plants in China, *Fitoterapia*, LII, 65, 1981.
2. **Xiao Peigen,** Recent developments on medicinal plants in China, *J. Ethnopharmacol.*, 7, 95, 1983.
3. **Deng Wenlong Nie Renji, and Liu Jiayu,** Pharmacological comparison of four kinds of andrographolides, *Chin. Pharm. Bull.*, 17, 195, 1982.
4. **Deng Shixian and Mo Yujiang,** Pharmacological Studies on *Gastrodia elata* I. The sedative and anticonvulsant effects of synthetic gastrodin and its genin, *Acta Bot. Yunnanica*, 1, 66, 1979.
5. **Yeung, H.W., Kong, Y.C., Lay, W.P., and Cheng, K.F.,** The structure and biological effect of leonurine, a uterotonic principle from the Chinese drug, I-mu Ts'ao, *Planta Med.*, 31, 51, 1977.
6. **Xiao Peigen,** Traditional experience of Chinese herb medicine, its application in drug research and new drug searching, in *Natural Products as Medicinal Agents*, Beal, J.L. and Reinhard, E. Eds., Hippokrates-Verlag, Stuttgart, 1981, 351.
7. **Tseng Kueiyun, Chou Yuanpeng, Chang Liying, and Fan Lili,** Pharmacologic studies on *Radix Puerariae* I. Its effects on dog arterial pressure, vascular reactivity, cerebral and peripheral circulation, *Chin. Med. J.*, 54, 265, 1974.
8. **Gao Yuqui, Song Yumei, Yang Youyi, Liu Wenfu, and Tang Jixue,** Pharmacology of tanshinone, *Acta Pharm. Sin.*, 14, 75, 1979.
9. **Chen Weizhou, Dong Yueli, Wang Changgen, and Ting Guangsheng,** Pharmacological studies of sodium tanshinone II-A sulfonate, *Acta Pharm. Sin.*, 14, 277, 1979.
10. Department of Paediatrics, Beijing Friendship Hospital, Department of Pharmacology, Institute of Materia Medica and No. 1 Laboratory, Hospital of Chinese Academy of Medical Sciences, Studies on the treatment of anisodamine on acute microcircular blocking diseases, *Natl. Med. J. China*, 53, 259, 1973.
11. **Feng Chingi and Chang Shihshan,** The pharmacological action of sinomenine on the central nervous system, *Acta Pharm. Sin.*, 12, 81, 1965.
12. **Kin Kuochang, Wang Tuee, and Hsu Bin,** Some neuropharmacological actions of rotundine, *Acta Pharm. Sin.*, 11, 754, 1964.
13. **Zhou Yuanpeng, Fan Lili, Zhang Liyin, and Zeng Guiyun,** The effect of higenamine on cardiovascular system, *Natl. Med. J. China*, 58, 664, 1978.

14. **Tang Xican, Zhu Meiying, Feng Jie, and Wang Yuee,** Studies on pharmacologic actions of lappaconitine hydrobromide, *Acta Pharm. Sin.*, 18, 679, 1983.
15. **Liu Guoqing, Sun Jianning, He Zhengzheng, and Jiang Yin,** Spasmolytic actions of active principles of volatile oil of Acorus gramineus, *Acta Pharmacol. Sin.*, 4, 95, 1983.
16. **Peng Huamin, Xiang Siyuan, and Bi Zhaoqing,** Pharmacological studies on the antitussive, expectorant and antiasthmatic effects of luteolin, an active principle of *Ajuga decumbens*, *Chin. Pharm. Bull.*, 16, 75, 1981.
17. **Lang Yijiang and Zhang Xingda,** Studies on the active principle of garlic, *Chin. Trad. Herb. Drugs*, 12, 4, 1981.
18. **Shen Lianci, Wang Yuanyu, and Feng Lihong,** Preparation of allicin microcapsule, *Chin. Trad. Herb. Drugs*, 14, 161, 1983.
19. **Huang Buhan, Chen Wensen, Hu Yan, Qiu Chenbo, Zhang Weijie, and Liang Bailing,** A study on the chemical constituents of Ardisia japonica, *Acta Pharm. Sin.*, 16, 27, 1981.
20. **Liang Xiaotian, Yu Dequan, Wu Weiliang, and Deng Huaichun,** The structure of Yingzhaosu A, *Acta Chim. Sin.*, 37, 231, 1979.
21. Qinghaosu Antimalaria Coordinating Research Group, Antimalaria studies on Qinghaosu, *Chin. Med. J.*, 92, 811, 1979.
22. **Jiang Ying, Liu Guoqing, Ma Junru, Xie Lin, and Wu Huiqiu,** The pharmacological studies on methyleugenol, *Acta Pharm. Sin.*, 17, 87, 1982.
23. **Xiao Peigen, Liu Guosheng, Chen Bizhu, and Song Wanzhi,** Study on the comprehensive utilization of San Ke Zhen (Berberis) plant resources, *Chin. Pharm. Bull.*, 14, 381, 1979.
24. **Yang Ke, Zhao Xiaolin, Xiao Peigen, and Liu Guosheng,** Clinical trial of Berbamine in 405 leucopenia patients, *Chin. Pharm. Sin.*, 17, 213, 1982.
25. **Lin Longze, Zhang Jinsheng, Chen Zhongliang, and Xu Rensheng,** Studies on chemical constituents of *Brucea javanica*. I. Isolation and identification of bruceaketolic acid and other four quassinoids, *Acta Chim. Sin.*, 40, 1982.
26. **Fang Taihui, Wu Jinshu, and Zhang Shiwei,** Experimental anitarrhythmic study of cyclovirobuxine-D, *Chin. Pharm. Bull.*, 16, 246, 1981.
27. Shaghai Institute of Materia Medica, Academia Sinica, Studies on the anticancerous action of 10-hydroxy-camptothecin, *Natl. Med. J. China,*, 58, 598, 1978.
28. Wang Yushen (chief ed.), *The Pharmacology and Application of Chinese Drugs*, People's Medical Publishing House, Beijing, 1983, 54.
29. **Shiao Shuhua, Shao Baorouh, Ho Yishun, and Yang Yuanching,** Prophylactico-therapeutic studies of cucurbitine in *Schistosomiasis japonica* in mice, *Acta Pharm. Sin.*, 9, 327, 1962.
30. **Ying Baiping, Wang Chengrui, Chou Pingnan, Pan Beichuan, and Liu Jiasen,** Studies on the active principles of the root of Yuan-Hua (*Daphne genkwa*). I. Isolation and structure of yuanhuacine, *Acta Chim. Sin.*, 35, 103, 1977.
31. **Wang Chengrui, Chen Zhengxiong, Ying Baiping, Zhou Bingnan, Liu Jiasen, and Pan Beichuan,** Studies on the active principles of the root of Yuan-Hua (*Daphne genkwa*). II. Isolation and structure of a new antifertile diterpene yuanhuacine, *Acta Chim. Sin.*, 39, 421, 1981.
32. **Jiang Ying, Zhao Zhicheng, and Li Fei,** The relationship between daphnetin analgesia and central neurotransmitters of the monoamines, *Acta Pharm. Sin.*, 19, 647, 1984.
33. **Chang Tanmu and Lue Fuhua,** Studies of β-dichroine, chloroquanide, cyclochloroquininide and baicalin on amebicides, *Acta Acad. Med. Wuhan*, 1, 11, 1958.
34. **Wang Sun, Yang Shengyuan, and Li Yin,** Treatment of postapoplectic paralysis with breviscapin injection (a report of 469 cases), *Chin. Trad. Herb. Drugs*, 14, 33, 1983.
35. **Yao Tianrong and Chen Zenai,** Chemical investigation of Chinese medicinal herb, Bao-Gong-Teng. I. The isolation and preliminary study on a new myotic constituent, Bao-Gong-Teng A, *Acta Pharm. Sin.*, 14, 731, 1979.
36. **Qin Guowei, Chen Zhengxiong, Wang Hongcheng, and Qian Mingkun,** The structure and synthesis of robustanol A, *Acta Chim. Sin.*, 39, 83, 1981.
37. **Chen Weihsin, Nieh Shuilin, Chen Yuchun, and Hsia Keming,** The structures of cucurbitacin II_a and II_b from *Hemsleya amabilis*, *Acta Chim. Sin.*, 33, 49, 1975.
38. **Li Dehua, Hao Xiaoge, Zhang Shukun, Wang Shixian, Liu Ruoyong, Ma Keshao, Yu Suping, Jiang Hui, and Guan Jingfang,** Antitumor action and toxicity of irisquinone, *Acta Pharmacol. Sin.*, 2, 131, 1981.
39. **Liu Gengtao, Wei Huaiting, Bao Tiangtong, and Song Zhenyu,** The effect of *Ganodermas* on elevated serum aldolase levels induced by 2,4-dichlorophenoxyacetic acid (2,4-D) in mice, *Acta Pharm. Sin.*, 15, 142, 1980.
40. **Sung Chenyu, Chi Hsiuchuan, and Liu Kengtao,** The pharmacology of gentianine. I. Antiphlogistic effect and action on pituitary-adrenal function of the rat, *Acta Physiologica Sinica*, 22, 201, 1958.
41. **Xu Yin, Hu Zibi, Feng Shengchu, and Fan Guangjin,** Studies on the antituberculosis principles from *Lysionotus pauciflora*. I. Isolation and identification of nevadensin, *Acta Pharm. Sin.*, 14, 447, 1979.
42. **Yang Junshan, Luo Shurong, Shen Xiulan, and Li Yuanxiang,** Chemical investigation of the alkaloids of Ku-mu (*Picrasma quassioides*), *Acta Pharm. Sin.*, 14, 167, 1979.
43. **Xiao Peigen and He Liyi,** *Przewalskia tangutica* — a tropane alkaloid-containing plant, *Planta Med.*, 45, 112, 1982.
44. Department of Pharmacology and Pharmaceutico-chemistry, Institute of Medical Science, Honan; Department of Chemistry, Honan Medical College; Laboratory of Phytochemistry, Institute of Botany, Yunnan and Chengchow Chemipharmaceutical Plant, A new antitumor substance-rubescensin, *Kexue Tongbao*, 23, 53, 1978.
45. **Fang Shengdin, Yan Xiuquan, Li Chingfang, Fan Zhiyun, Xu Xiaoyi, and Xu Rensheng,** The isolation and structure of active principle sarmentosin, *Kexue Tongbao*, 24, 431, 1979.
46. **Liang Qinxin and Gao Xiayun,** Studies on the antihepatitis flavanoids of *Swertia mileensis*, *Zhongcaoyao Tongxun*, 9, 385, 1979.
47. **Kao Shijin and Zeng Guiyun,** Cardiotonic and toxic effects of peruvoside and neriifolin, *Acta Pharm. Sin.*, 18, 572, 1983.
48. Laboratory of Phytochemistry, Institute of Botany, Yunnan and Pharmacologica Laboratory, Institute of Medical Science, Honan, Diterpenoid triepoxides, antitumor principles from *Tripterygium hypoglaucum* and *T. wilfordii*, *Kexue Tongbao*, 22, 458, 1977.
49. **Zheng Tanmu, Chen Zhengyu, and Li Chen,** Antineoplastic action of triptolide and its effect on the immunologic functions in mice, *Acta Pharmacol. Sin.*, 21, 128, 1981.
50. **Yuan Wenxue and Chang, I.,** The sedative and hypotensive action of *Uncaria rhynchophylla*, *Acta Physiol. Sin.*, 25, 161, 1962.
51. **Liu Gengtao, Wei Huiling, and Song Zhenyu,** The physiological dispositions of biphenyl dimethyl-dicarboxylate (BDD), *Acta Pharm. Sin.*, 18, 892, 1983.
52. **Peng, S.X. and Wang, E.H.,** A review and prospect of new drug research and development in China, *Chin. Pharm. Bull.*, 16, 27, 1981.

Chapter 13

RECENT DEVELOPMENTS FROM INDIAN MEDICINAL PLANTS

B.N. Dhawan and R.P. Rastogi

I. INTRODUCTION

India has been endowed with a very rich flora due to the extreme variations in geographical and climatic conditions prevalent in the country. It was natural, therefore, that plants have been used since ancient times for the treatment of human ailments. The traditional systems of medicine (Ayurveda, Siddha, and Unani-Tibb) together with folklore medicine still continue to serve, in spite of the advent of modern medicine, a large portion of the population, particularly in the rural areas. The pharmacopeias of these systems have mainly drawn upon the indigenous flora for the preparation of a wide variety of herbal medicaments and about 800 plants are mentioned in the relevant systems. Out of about 15,000 species of higher plants in India, medicinal uses are usually attributed to over 1500 species.[1] Chopra has included a large number of plants,[2] raising the total to over 2500, which includes many plants of not very definite pharmacological significance and sometimes species of extremely limited occurrence.

The modern system of medicine was introduced to India by the Portuguese in the 16th century and later established and consolidated by the French and the British. The 17th to 19th century period witnessed the appearance of a classified modern knowledge of the flora of India, establishment of herbaria, introduction of exotic plants, and the beginnings of the application of Western methodology to the investigation of indigenous drugs. Some valuable contributions of this period are *Dymock's Pharmacographica Indica, Watt's Dictionary of Economic Raw Materials,* and *Hooker's Flora of British India.* The systematized study of the botanical as well as chemical aspects as indeed the clinical investigations, based on Western methodology, evolved during the same period due to the contributions of Sir Buchanan Hamilton, Dr. Thomas Thomson, Dr. T. Anderson, Dr. C.B. Clarke, Sir George King, Sir David Prain, and Lt. Col. A.T. Cage.

The first attempt to bring the vast knowledge contained in the indigenous systems of medicine into the mainstream of modern medicine can be discerned in the pioneering efforts of Col. R. N. Chopra during 1935 to 1950 toward the objective appraisal of the indigenous pharmacopeias. His school carried out extensive investigations, both chemical and pharmacological, on the indigenous plants. Other pioneers in chemical studies on plants were Dr. S. Siddiqui, Prof. T.R. Seshadri, and Dr. K. Venkataraman, whose works on alkaloids of *Rauwolfia serpentina* and oxygen heterocycles, respectively, have drawn international recognition.

II. APPROACH TO RESEARCH ON MEDICINAL PLANTS

The methodology of the scientific evaluation of plant drugs as envisaged by Col. R. N. Chopra, is being followed to a large extent today. It involves, with minor variations at different research centres, the collection of fresh plants, their botanical identification, preparations of total extracts, and screening of the extracts in a broad spcetrum of biological test systems. Those plants showing activity are then subjected to more detailed assessment, which includes chemical fractionation monitored by biological evaluation, followed by toxicity study of active principles and clinical trials.

Such integrated programmes of coordinated chemical and biological studies have by and large replaced the earlier isolated working groups and have been going on over the last two decades at Central Drug Research Institute, Lucknow; Indian Institute of Chemical Biology, Calcutta; Regional Research Laboratory, Jammu; Ciba-Geigy Research Centre, Bombay; and also under the projects of the Indian Council of Medical Research, New Delhi. Mention may also be made of the notable chemical work done at the National Chemical Laboratory, Pune; Andhra University, Waltair; and the Chemistry Departments of Calcutta and Delhi Universities.

At the Ciba-Geigy Research Centre, over 1500 extracts from 500 species belonging to over 100 genera have been studied. At the Central Drug Research Institute, extracts from 1975 species belonging to about 1000 genera of the higher plants (angiosperms: 1928 and gymnosperms: 197) belonging to about 125 families in terms of the Engler and Prantle classification have been put through a very wide biological screening,[3-9] which includes anticancer testing under a collaborative programme with the National Cancer Institute, Bethesda, MD (U.S.). The presence of activity has been confirmed in 352 plants, i.e., 17.8% of the total plants screened. Of these, 218 plants showed promising activity while the rest were weakly active. These botanical, chemical, and pharmacological results have been summarized and the plants with confirmed biological activities are listed by activity in 12 tables.[10] Recently the results of biological screening of another 575 plant species have been published from Central Drug Research Institute, Lucknow.[11,12] The data on antiviral screening of 967 plant extracts against Ranikhet and Vaccinia viruses have been presented in a separate communication. The activity was observed in 17% of these plants and appears to be due to proteins biologically resembling interferon but different from that of chick interferon.[13] Similarly, the activity data have been exhaustively reviewed on the research done on Indian plants for fertility regulation in the female and male.[14,15] In another study the antifertility data on the extracts from different parts of 89 plant species are reported.[16]

III. PHARMACOGNOSTIC STUDIES

Simultaneously, the importance of correct identity of the

plants mentioned in the texts, knowledge of their substitutes, or adulterants available in commerce became apparent and a foremost concern both for the practitioners as well as research scientists. As a result great interest has been witnessed in the pharmacognostic studies and surveys of medicinal plants in various parts of the country which has led to an impressive increase in growth in this area in recent decades.

The pharmacognostic studies on Indian medicinal plants have been reviewed extensively, covering the published work of the period 1940 to 1968 and having a bibliography comprised of 1081 references.[17] In another report 312 plant parts have been listed on which anatomical studies had been carried out.[18] An account of the pharmacognostic studies of some indigenous drugs at National Botanical Research Institute, Lucknow, from 1955 to 1978 has also been published.[19]

Recent pharmacognostic studies on some important medicinal plants are mentioned here. The market samples of the tubers of *Aconitum heterophyllum* (ativisha) has revealed also the corms of *Cryptocoryne spiralis* or *C. retrospiralis* being sold under the same commercial name.[20] Pharmacognostic studies of *Acorus calamus* and its adulterants have been carried out, providing distinguishing features through which the genuine drug can be identified.[21] The distribution pattern of alkaloids of *Adhatoda vasica* with change of season has been elucidated.[22] The bark of *Alstonia scholaris* and its diagnostic features has been reported to distinguish it from apparently similar bark of *A. macrophylla*.[23] Pharmacognostic studies on *Anisomeles malabarica*,[24] detailed pharmacognosy of floral and vegetative parts of *Artemisia scoparia*,[25] morphology and anatomy of the roots of *Asparagus racemosus* and its five other species,[26] and diagnostic characteristics of the leaves, stem, and roots of *Barleria cristata*, *B. lupulina*, and *B. prionitis*[27] have been reported. The physicochemical characteristics of oleoresin of *Boswellia serrata* have been described.[28] The pharmacognostic characteristics of leaf, stem bark, fruits, and seeds of *Buchanania lanzan* have been reported.[29-31] A study on the senoside content with the maturation of the pods of *Cassia* sp. showed that it was maximum in 2-day-old pods and minimum in 20-day-old pods.[32] The pharmacognosy details have been reported on heartwood of *Caesalpinia sappan*[33] and flowers of *Calophyllum inophyllum*.[34] Pharmacognostic studies of *Coleus forskohlii* and seasonal variations in the active constituents of the plant have been carried out.[35] The pharmacognostic study of gum oleoresin of *Commiphora wightii* (*C. mukul*)[36] and macro- and microscopic features of the stem, leaf rhizome of *Costus speciosus*, which has assumed importance as an important source of disosgenin,[37-39] have been studied. The pharmacognosy of leaf of *Datura metel* has been reported along with the features which differentiate it from its adulterants, *D. stramonium*, *Chenopodium hybridum*, and *Xanthium orientale*.[40] The pharmacognostic evaluation of seven species of *Dioscorea* tubers detailing all the macro- and microscopic features and physicochemical analytical data to distinguish the tubers of highest diosgenin content and also to detect adulterants has been carried out.[41,42] The distinguishing pharmacognostic features of fruit of *Embelia ribes* have been reported to differentiate from the fruit of other species of *Embelia* sometimes admixed in market samples.[43] Studies describing chemicomorphological characteristics of stem bark of *Millingtonia hortensis* and *Nepta longibracteata*[44,45] and diagnostic macro- and microscopic characteristics of the roots and rhizomes of *Nymphoides macrospermum* and *Valeriana jatamansi* to distinguish these drugs from one another have been reported.[46] The diagnostic characteristics of leaf, fruits, and root of *Piper betle*, *P. cubeba*, *P. longum*, and *P. nigrum* have been reported which enable these species to be distinguished from one another.[47] The pharmacognosy and ethnobotanical studies of *Plumbago zeylanica*,[48] fluorescence analysis, quantitative microscopy, and microchemical tests of leaf powder of *Portulaca quadrifida*[49,50] and pharmacognostic details of *Psorelea corylifolia*[51] have been published. Scannning electron microscopy of the leaves of seven species of *Salvia* comprising mainly surface ornamentation of various epidermal cells and the appendages has provided parameters to distinguish one species from another.[52] The pharmacognosy of *Solanum nigrum* and *S. trilobatum*,[53,54] and pharmacognostical features of *Tephrosia purpurea* have been reported.[55] The comparative pharmacognostic characteristics of three parasitic herbs, *Cassytha filiformis*, *Cuscuta reflexa*, and *C. chinensis*, are reported to enable identification in the natural state as well as in the powder form.[56] The pollen morphology of 12 species of Leguminosae used as floral drugs in indigenous system has been studied.[57]

IV. CHEMICAL AND PHARMACOLOGICAL STUDIES

The results reported in the last two decades on the isolation and elucidation of structures of the chemical constituents are prodigious and due to constraints of space, cannot be presented in all-inclusive manner. The significant developments have been selected from the vast data where the studies have led to the discovery of new biologically active substances, revealing new structures possessing activity and thus providing new leads for synthetic chemists.

The approach is, therefore, selective and the presentation is based on classification according to chemical structure.

A. ALKALOIDS

The genus *Ancistrocladus* is represented only by one species in India, *Ancistrocladus heyneanus* Wall. A novel group of four isoquinoline alkaloids has been isolated from this species. The major alkaloid, ancistrocladine, has been assigned the structure and stereochemistry by chemical, spectroscopic as well as X-ray analysis as **1.** The X-ray analysis indicated that the naphthalene and isoquinoline rings were at an angle of 37° due to the restricted rotation about the bond joining the two nuclei, introducing an element of asymmetry at this bond. The absolute configurations at C-3 were found to be S in all the alkaloids, since perozonolysis yielded (+)L-β-aminobutyric acid. The chemistry of this group of alkaloids has been reviewed recently.[58,59] Of the four alkaloids, ancistrocladine and cedricladine showed spasmolytic activity on the isolated guinea pig ileum at a concentration of 5 μg/ml, comparable with that of papaverine.

Several interesting alkaloids isolated from *Alstonia scholaris* R.Br. and the isolation and structure elucidation (X-ray) of nareline **2** isolated from this plant is of great interest. This has the novel 2-azaadamantane skeleton.[60] One of its alkaloids, strictamine, showed antidepressant activity in addition to inhibiting the MAO enzyme.[61] Another alkaloid, picerinine, on the other hand, produced CNS depressant activity.[62] The different parts of *Alstonia venenata* are used for treatment of insanity and epilepsy in the Indian systems of medicine. The hypotensive activity of the extracts of the plant was shown to be due to the presence of reserpine. In a thorough study of all parts of *A. venenata* R.Br., Chatterjee et al. have reported the isolation and structures elucidation of 20 other alkaloids, in addition to the above. Broadly, the alkaloids belong to the yohimbine, aspidofractinine, and vincadifformine types. The occurrence of the monoterpenoid base, venoterpine, is of great biogenetic significance. The chemistry of the alkaloids of *A.venenata* has been reviewed by Chatterjee et al.[63] Venenatine, a 4-methoxyindole alkaloid, has been shown to produce reserpine-like effects in animals.[64] Another alkaloid, alstovenine, inhibits MAO enzyme. More recently three vincadifformine type alkaloids, echitoveniline **3**, (–)11-methoxyechitoveniline, and (–)11-methoxyechitovenedine from its fruits;[65] two indole alkaloids of kopsane group alstolenine, 19,20-dihydropolyneuridine **4** along with three minor (0.002%) known alkaloids from its leaves;[66] two stereomers (16-epivenenatine and 16-epialstovenine) and a carboline derivative from its root bark[67,68] have been isolated and their structures established. The isolation of this carboline base is significant as this unit has been found to be present in all yohimbine alkaloids isolated from this plant.

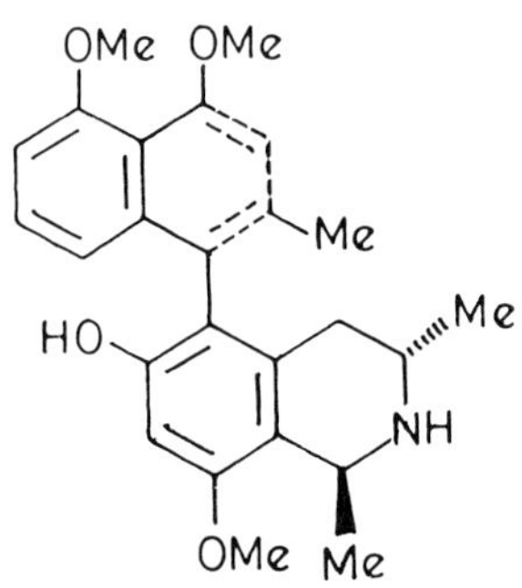

1 Anclstrocladine

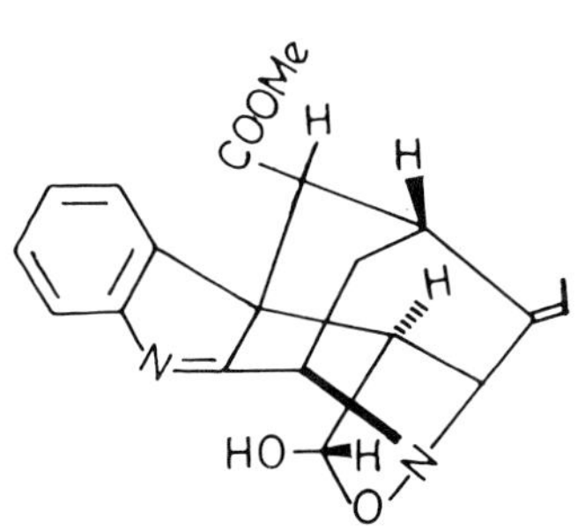

2 Nareline

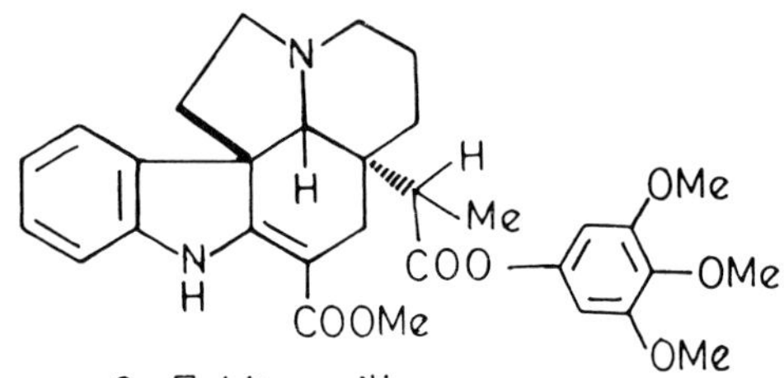

3 Echitoveniline

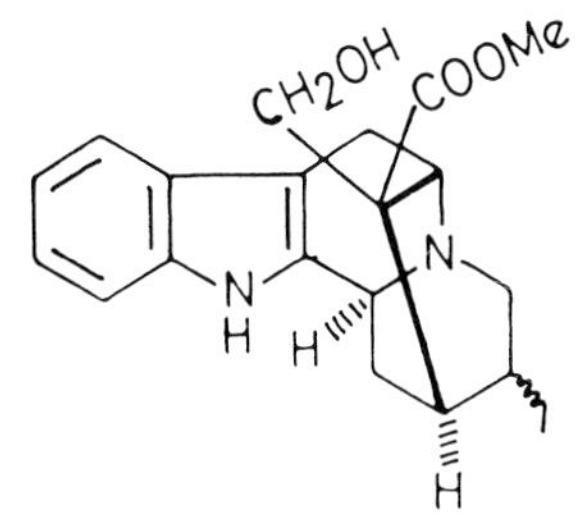

4 19,20-Dihydropoly-neuridine

Adhatoda vasica is a reputed drug in indigenous medicine. Its chemical constituents have been studied at different centres in the country (*vide infra*). New alkaloids, vasicol, vasicinolone, vasicinone, and a nor-harmal alkaloid, 9-acetamido-3,4-dihydropyrido-(3,4-b)-indole, have been isolated from the roots of the plant.[69] Harmal-type alkaloids have not so far been found in this plant. A unique protoberberine alkaloid, bharatamine **5**, unoxygenated in ring D and biogenetically derivable from the monoterpenoid precursor, loganin, a pathway hitherto unknown for this class of compounds, has been isolated from *Alangium lamarckii*.[70] The isolation of alangiside, a monoterpene alkaloidal glycoside containing a lactone group is considered of biogenetic importance.[71]

More than a dozen alkaloids belonging to normal and abnormal erythrina, bisbenzyl isoquinoline, aporphine, 1-benzyl isoquinoline, and dibenz(d,f)azonine groups have been isolated from *Cocculus laurifolius* DC. and their structures elucidated. The absolute configuration of sebiferine **6** and biosynthesis of isotetrandrine has been studied. Detailed studies on the biosynthesis of abnormal erythrina alkaloids, bisbenzyl isoquinoline alkaloids, tetrandrine and laurifinine have been reported.[72-74] The new erythrina alkaloid, isococculidine **7** is a neuromuscular blocking agent. It could not be developed further because in slightly higher doses it inhibited direct twitch responses and lowered the blood pressure.[75] Significant neuromuscular property was also exhibited by another quaternary alkaloid, isocorydine methochloride.[76]

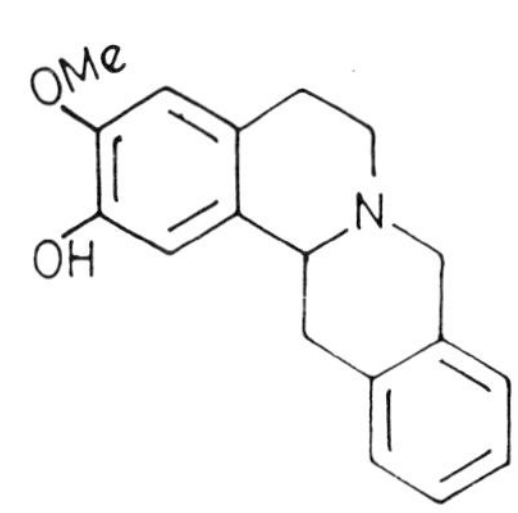

5 Bharatamine

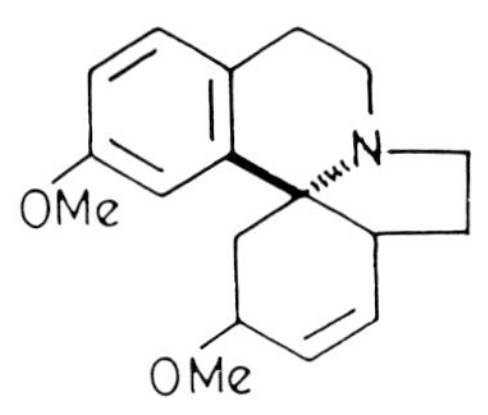

6 Isococculidine

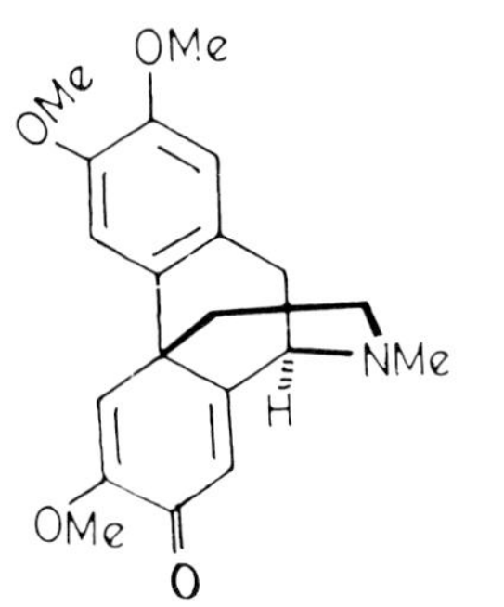

7 Sebiferine

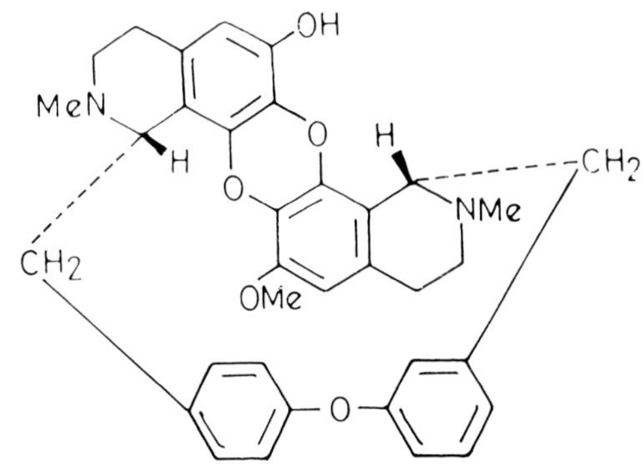

8 Cocsulinine

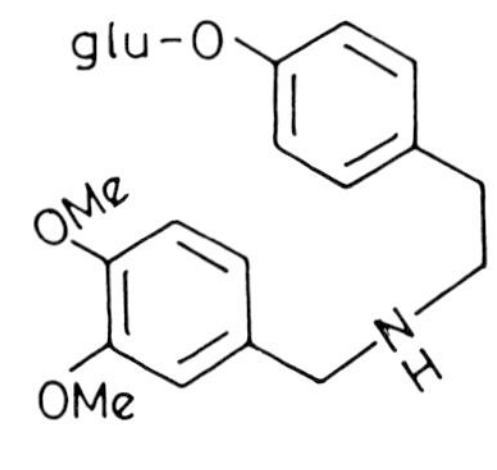

9 Latisoline

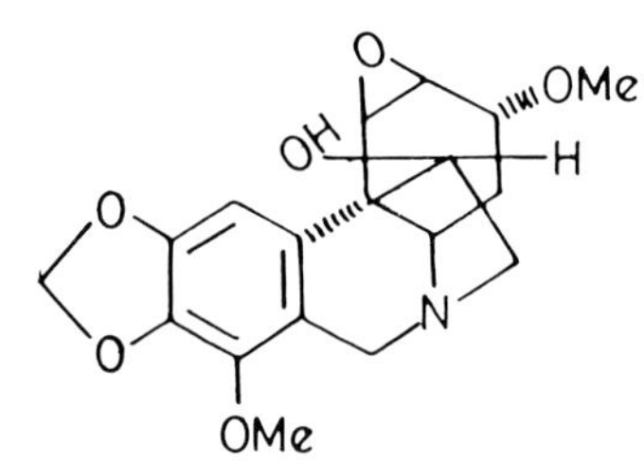

10 1,2-β-epoxyambelline

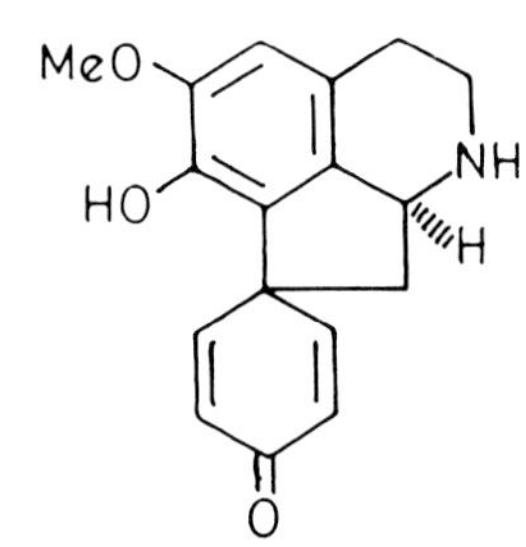

11 Crotsparine

Six bases of the bis-coclaurine group have been isolated from *Cocculus pendulus* Forsk. and characterized.[77] Of these bases, cocsulinin **8** showed activity against thehuman epidermoid carcinoma of nasopharynx in tissue culture (KB system) at a concentration of 2.7 μg/ml.

Two new pyrrolophenanthridone bases, pratorimine and pratosine, and a novel glucoalkaloid, latisoline **9** along with four known alkaloids have been isolated from *Crinum latifolium* and their structures established by chemical and physical methods.[78,79] Latisoline appears in the fluid of flower stems only during the time of blooming, covering a period of about 1 week, while the aglycone, latisodine, occurs in all parts of the plant. An immunostimulant alkaloid, 1,2-β-epoxyambelline **10** has also been isolated and identified from this plant.[80] A concentration of 5 μg/ml produced activation of mouse spleen lymphocytes (stimulation index 3.12, comparable to that of concanavalin 3.23).

Croton sparsiflorus Morong., a weed of common occurrence in the plains of India, has yielded three proaporphine bases (crotsparine **11**, *N*-methylcrotsparine, *O,N*-dimethylcrotsparine), two dihydroproaporphines (crotsparinine, *N*-methyl-crotsparinine), and an aporphine (sparsiflorine).[81] The absolute configuration of these alkaloids has been determined.[82] Biosynthetic studies showed that these alkaloids are derived from tyrosine and coclaurine.[83] *N*-Methylcrotsparine exhibited significant hypotensive activity in cats, at an i.v. dose of 5 to 15 mg, producing a lowering of blood pressure by 50 to 70% for 1 to 3 h. The hypotensive activity is mainly due to α-adrenergic blockade associated with a negative inotropic effect and direct suppression of supraspinal vasomotor loci.[84,85] This is perhaps the first report of hypotensive activity in aporphine and proaporphine bases.

From the roots of *Cissampelos pareira* L., a reputed drug in Ayurvedic medicine, several bisbenzylisoquinoline alkaloids have been isolated, many of them of known structure.[86] Hayatin is a major component of the alkaloidal fraction in roots collected from Kashmir. Extensive pharmacological studies have been carried out on hayatin (dl-berberine) methosalts and clinical studies have shown that it is about one third as potent as tubocurarine, the duration of activity being comparable at equipotent doses.[87] The stereostructure of a novel macrocyclic spermidine alkaloid, caesalpinine A **12** from *Caesalpinia digyna*, has been determined by X-ray analysis.[88] The genus *Crotalaria* is known to produce macrocylic ester alkaloids incorporating a pyrrolizidine system. Atal et al. have investigated a large number of species occurring in the Jammu region and isolated over a dozen new alkaloids; cromadurine **13** is a typical alkaloid of the group.[89] Hypotensive activity has been reported in usaramine and integerrimine isolated from *Crotolaria brevifolia*.[90,91] The hypotensive effect of *Crotolaria agatiflora* is also probably due to the presence of pyrrolizidine alkaloids such as madurensine.[92] Many semisynthetic analogues of these have also been prepared and tested. *N-p*-Hydroxybenzoylretronamine exhibits potent hypotensive activity.[93]

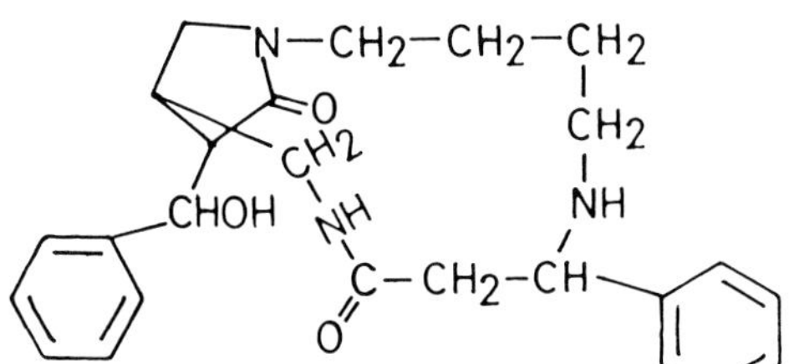

12 Caesalpinine A

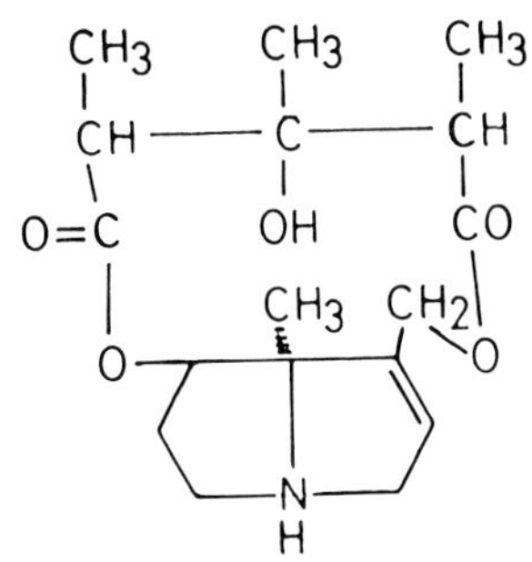

13 Cromadurine

Almost 30 alkaloids belonging to the carbazole group of the leaves of *Murraya koenigii* Spreng have been isolated. Chakraborty et al. were the first to report the isolation of the alkaloids girinimbine, mahanimbine, and murrayacine.[94] Subsequently, a number of other alkaloids were isolated in other laboratories and there is confusion due to the multiplicity of names given to the same alkaloid. Mahanimbine was correctly formulated as **14**,[95] girinimbine as **15**,[96] and structures of murrayacine,[97] koenimbine, koenimbidine,[98] and isomahanimbine (a position isomer of mahanimbine) and mahanine have been established.[99] The structures of two new alkaloids, mukoline and mukolidine,[100] isolated from the roots, and another base mukonicine, isolated from the leaves, have been proposed.[101] Among the more complex alkaloids, cyclomahanimbine has been examined in detail[102] and an X-ray study of murrayazoline (mahanimbidine, curryangin) has confirmed its structure as **16**.[103] Modified structures have been suggested for the related bases, bicyclomahanimbine **17** and bicyclomahanimbicine in place of those originally proposed.[104,105] The structure of a new hexacyclic alkaloid, isomurrayazoline, from its root bark has been established.[106] The chemistry of carbazole alkaloids from *M. koenigii* as well as other plants of the Rutaceae has been reviewed.[107]

Three oxindole alkaloids, mitraphylline **18**, isomitraphylline, and speciophylline N-oxide, have been isolated from the leaves of *Mitragyna parvifolia* (Roxb.) Korth., and a novel method has been developed for the conversion of mitraphylline to the corresponding heteroyohimbinoid base, ajmalicine.[108] The value of random screening of plants was well demonstrated when an extract of *Mappia foetida* Miers., for which no medicinal properties have been attributed, yielded over 0.25% of camptothecin.[109] Camptothecin has attracted wide attention as an anticancer drug,[110] available earlier only in low yield (0.005%) from *Camptotheca acuminata*. Besides camptothecin **19**, 9-methoxycamptothecin and mappicine were also isolated from the plant.[111,112] The structure of mappicine was established by a partial synthesis from camptothecin. 9-Methoxycamptothecin showed anticancer activity (145%) in lymphocytic leukaemia in mouse (PS system) at a dose of 0.5 mg/kg as compared to the activity of camptothecin (161%) at a dose of 1 mg/kg.

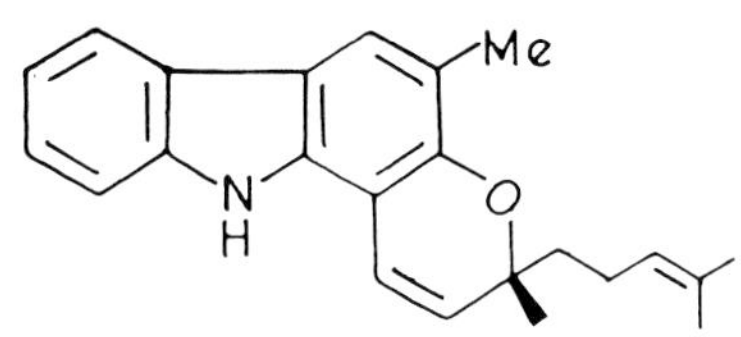

14 Mahanimbine

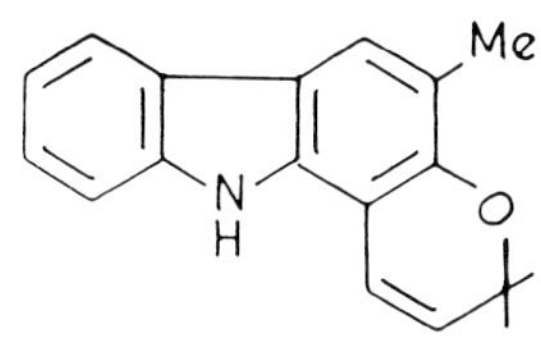

15 Girinimbine

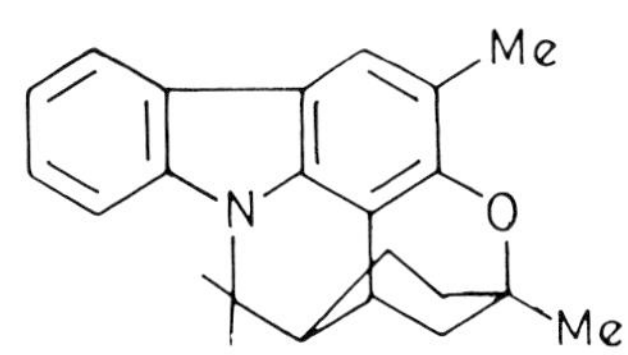

16 Murrayazoline

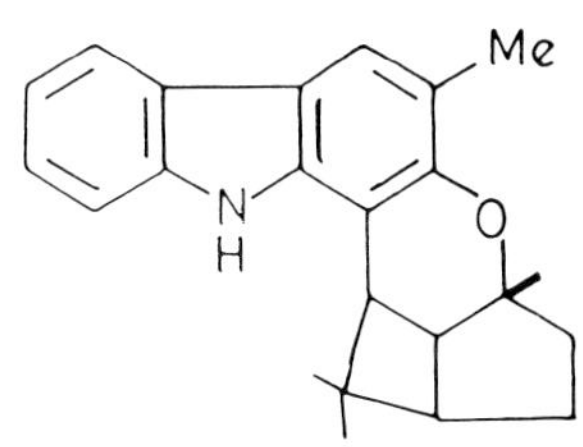

17 Bicyclomahanimbine

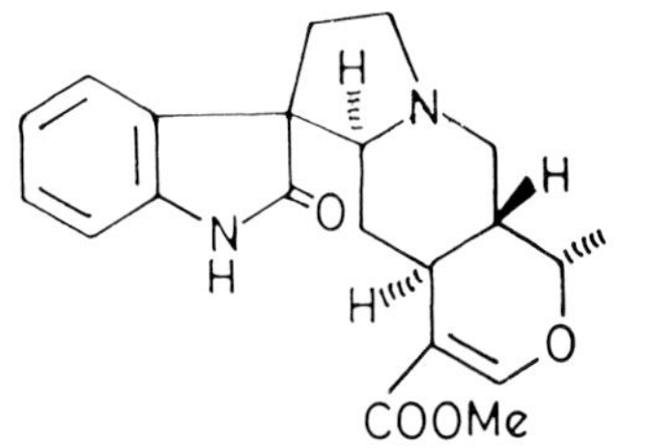

18 Mitraphylline

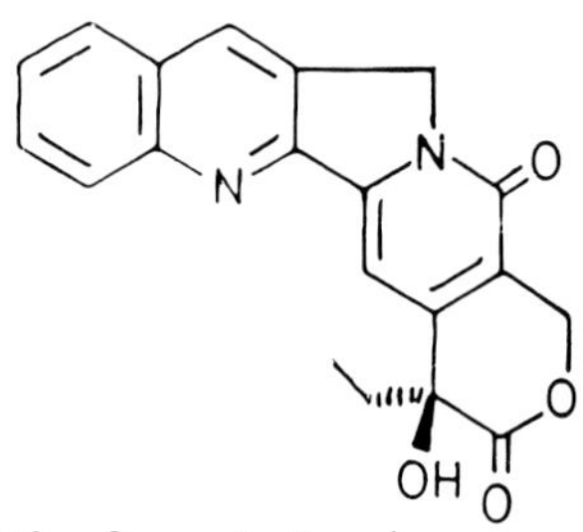

19 Camptothecin

The medicinal importance of the genus *Piper* has led to examination of several of its species.[113] From *Piper trichostachyon* C.DC. four new alkaloids, piperstachine **20**, cyclostachine A **21**, cyclostachine B, and cyclopiperstachine, have been isolated and their structures elucidated by a combination of chemical and spectroscopic methods and confirmed by syntheses.[114,115] The confirmation of the structure of cyclostachine A was obtained by X-ray analysis.[116] Cyclostachines A and B evidently arise by an internal Diels-Alder reaction of a triene-amide, and cyclopiperstachine from piperstachine. The cyclo-compounds are not artefacts but exist in the plant. Investigation of *Piper sylvaticum* has led to the isolation of two alkamides, PS-A and sylvamide, whose structures have been established;[117,118] and from *P. aurantiacum* another amide, aurantiamide **22**, has been isolated which belongs to a new class of modified dipeptides.[119] Several types of pharmacological activity have been demonstrated in the dried fruits of *Piper longum*. It showed antiallergic activity and has been used clinically to treat cases of paediatric asthma.[120,121] The interceptive activity of natural amides from *Piper* and some semisynthetic analogues have been reported.[122] It has been demonstrated that the mechanism of enhanced bioavailability by piperine is the inhibition of drug metabolising enzymes in the liver, and this probably also explains the results reported earlier with the crude material.[123,124]

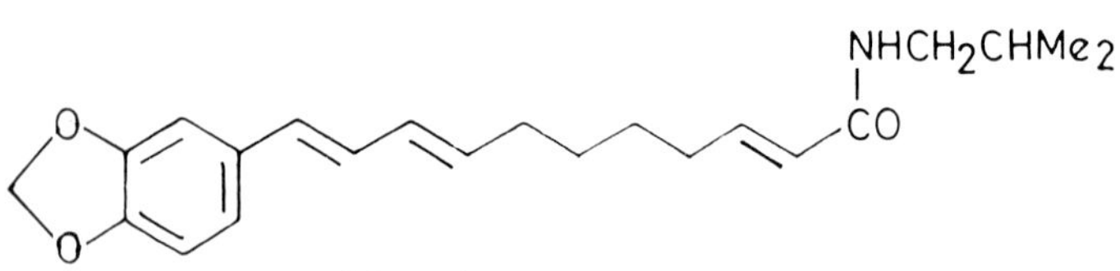

20 Piperstachine

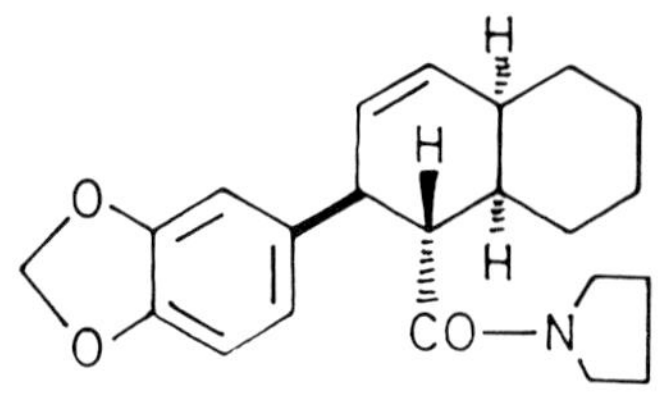

21 Cyclostachine A

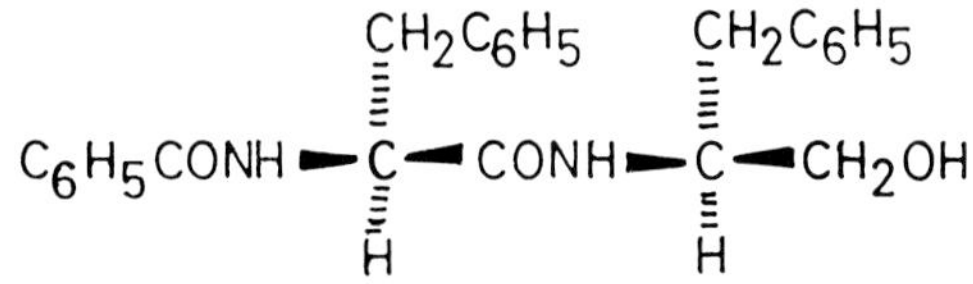

22 Aurantiamide

A new group of biogenetically interesting secotropane alkaloids, physoperuvine **23**, and two of its congeners have been characterised from *Physalis peruviana* L. roots.[125]

From *Tiliacora racemosa* Colebr. a number of bisbenzylisoquinoline alkaloids have been isolated, which are unique in having the isoquinoline moieties joined through a diphenyl unit and not through a diphenyl ether unit as is common. The essential features of the structures of tiliacorine and the accompanying alkaloids were established some years ago,[126] but it was only recently that the position of the phenolic hydroxyl group in tiliacorine could be established by a selective cleavage at one of the benzylic carbon atoms.[127] The absolute configurations of tiliacorine and the isomeric bases, nortiliacorinins A and B, have been established in an unique manner using biosynthesis as a tool.[128,129] The total alkaloid fraction from *T. racemosa* showed significant hypotensive activity due perhaps to some minor constituent, since none of the four pure alkaloids was active.

From the leaves of *Tylophora asthmatica* W.&A., used in folk medicine as a cure for asthma, several phenanthroindolizidine alkaloids, tylophorine **24**, tylophorinine, tylophorinidine, d-isotylocrebrine, and the *seco* alkaloid septicine **25** have been isolated. All aspects of the structures and stereochemistry of these alkaloids have been established. The absolute configuration of tylophorine has been established by exhaustive ozonolysis which gave S-pyrrolidine-2-acetic acid.[130] The absolute configuration of other alkaloids at C-12a has been assigned by comparison of their ORD curves with that of tylophorine. The structure and stereochemistry of the minor alkalid, tylophorinidine, has been confirmed by X-ray analysis of its diacetate methiodide.[131] Another minor alkaloid, tylophorinicine, has also been characterised by chemical and physical evidence.[132] The clinical trials conducted with leaves, total alkaloids, as well as pure tylophorine, support the common belief of the beneficial effects of the leaves in treating asthma and allergic rhinitis.[133] All the phenanthroindolizidine alkaloids show significant anticancer activity.[134]

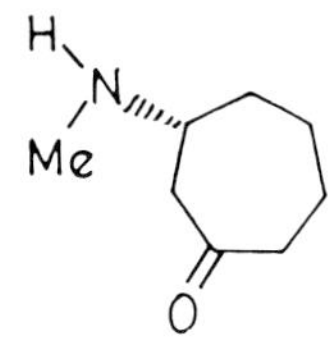

23 Physoperuvine

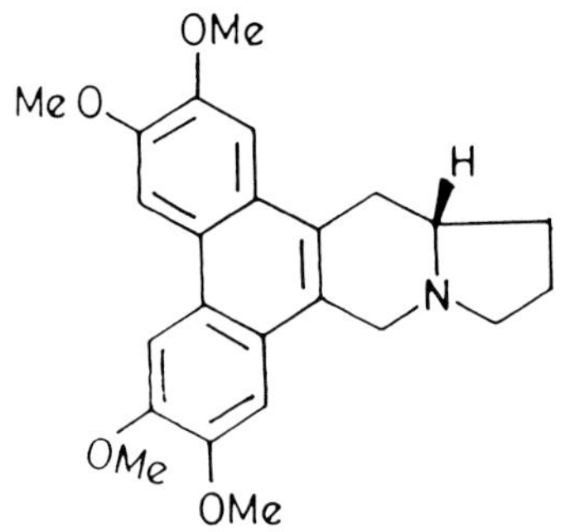

24 Tylophorine

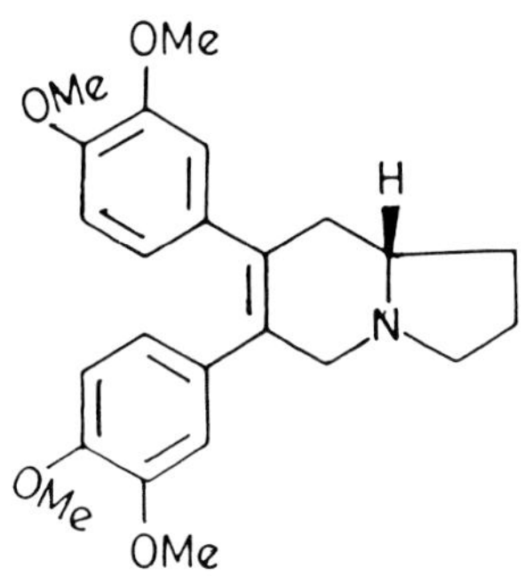

25 Septicine

A number of tertiary and quaternary alkaloids has been isolated from *Thalictrum foliolosum*. Of these, the quaternary alkaloid, magnoflorine, has shown marked hypotensive activity.[135] The structure of the very complex bisindoline alkaloid, amataine (grandifoline) isolated from *Voacanga grandifolia* Miq. was solved by collaborative chemical and spectrometric studies of several groups of workers.[136]

B. OXYGEN HETEROCYCLES

A systematic examination of plants belonging to the Umbelliferae occurring in the Himalayan region has led to the isolation of a large number of new coumarins. Typical compounds are heratomin **26** from *Heracleum thomsoni* and sesebrinol from *Seseli sibiricum*.[137]

Many plants belonging to the Rutaceae and Guttiferae have been examined, giving interesting new coumarins. Morellin, isolated from the seeds of *Garcinia morella* Desr., is a modified xanthone of a very complex structure **27** resolved by X-ray analysis.[138] Three new tetraoxygenated xanthones, garcinones A, B, and C, have been isolated from the fruit hulls of *G. mangostana* and their structures have been elucidated.[139] Mangostin and several of its derivatives had CNS depressant effects in rodents.[140,141] It also inhibited systemic anaphylaxis as well as adjuvant-induced arthritis in rats. A hypertensive effect was obtained with the 3,6-di-*O*-glucoside.[142]

The structure of the complex optically active polyisoprenylated benzophenones, xanthochymol **28** and isoxanthochymol from the fruits of *Garcinia xanthochymus* were confirmed by X-ray analysis.[143-145] Other substances belonging to the same group, garcinol, isogarcinol, and camboginol from *G. indica*, cambogin (enantiomer of isoxanthochymol), and camboginol from *G. cambogia* have been isolated and their structures established.[146-147]

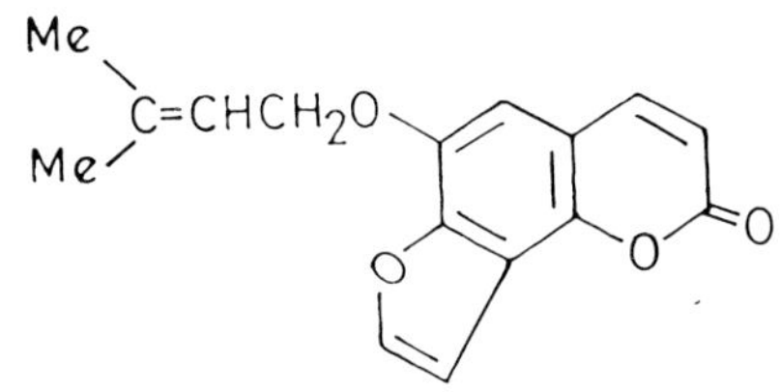

26 Heratomin

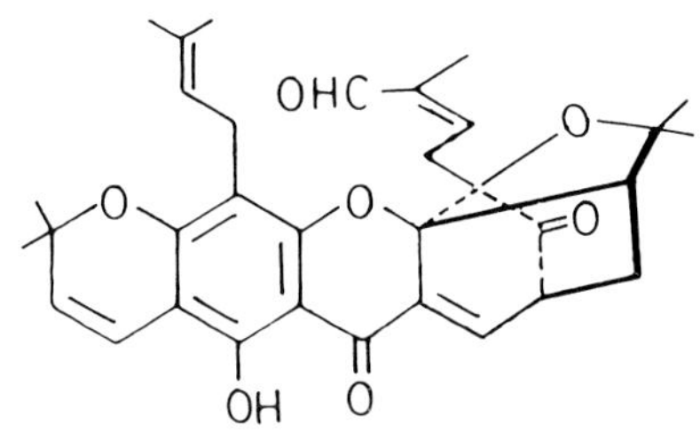

27 Morellin

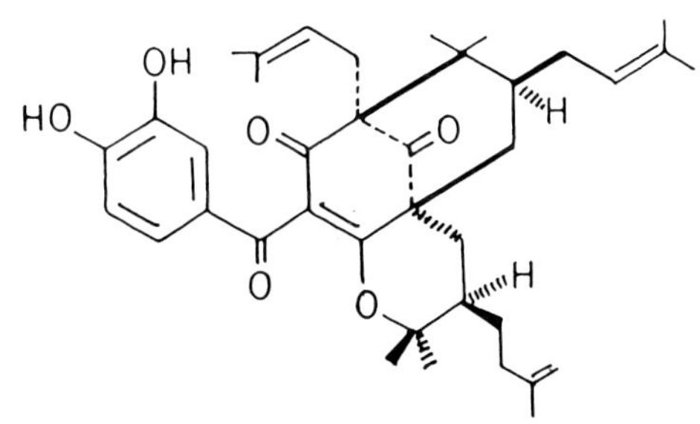

28 Isoxanthochymol

The known coumarins, bergapten and chalepin acetate, from *Boenninghausenia albiflora* have shown a good degree of spasmolytic activity.[148] From the roots of *Cryptocarya bourdilloni* Gamb., cryptocaryalactone **29** and cryptocaryone were isolated and their structures established.[149-151] Cryptocaryone undergoes a remarkable rearrangement on warming with 1 N NaOH, for which a mechanism has been proposed. Cryptocaryalactone showed moderate *in vitro* antifungal activity (*Candida albicans*) and produced a prolonged fall of blood pressure at 9 mg/kg in dogs. Coumarins of different species of *Clausena* have been investigated.[152] Clausmarins A **30** and B from *Clausena pentaphylla* (Roxb.) DC. were identified as the spasmolytic agents whose utility as a drug was found to be adversely affected due to poor oral absorption.[153-155]

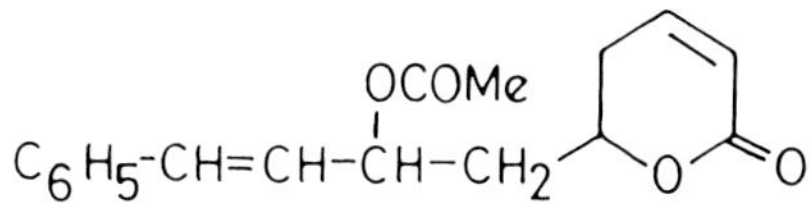

29 Cryptocaryalactone

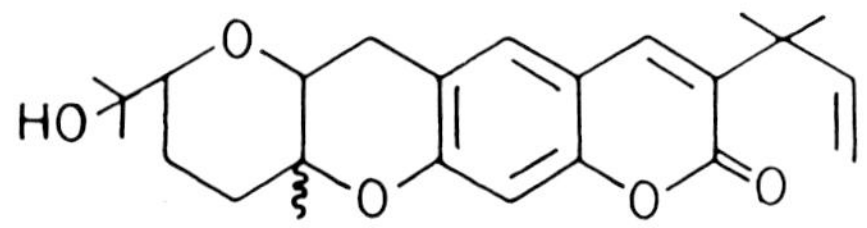

30 Clausmarin A

Coumarinolignoids are a new class of natural products which are represented by cleomiscosin A and B **31** isolated from *Cleome viscosa* L.,[156] and aquillochin from *Aquilaria agallocha* Roxb.[157] A novel type of furanobicoumarin rhamnoside, eriocephaloside **32**, has been isolated and characterised from *Lasiosiphon eriocephalus*.[158]

From the roots of *Mammea longifolia* (Wt.) Planch. two new coumarins, surangin A **33** and surangin B, have been obtained and their structures established entirely by spectroscopic methods.[159] Surangin A showed *in vitro* antistaphylococcal activity. Surangin B was active against mosquito larvae, the mustard beetle, and the housefly. A new antibiotic dimeric coumarin, mexolide **34** having an unusual cyclohexene system has been isolated and characterised from *Murraya exotica* L.[160]

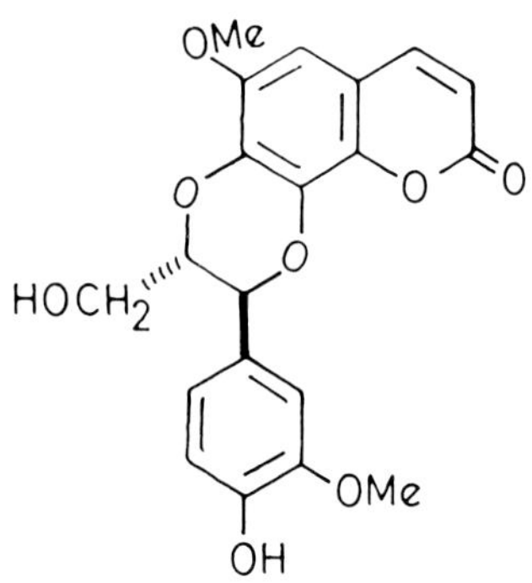

31 Cleomiscosin B

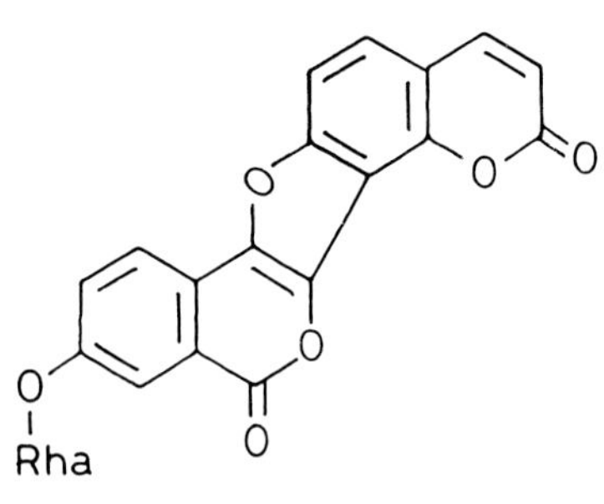

32 Eriocephalocide

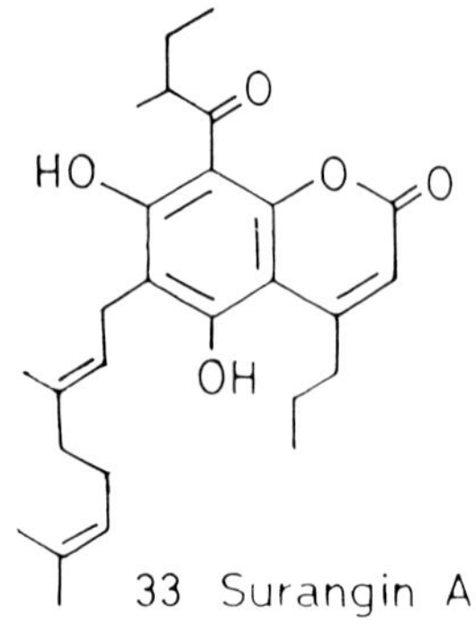

33 Surangin A

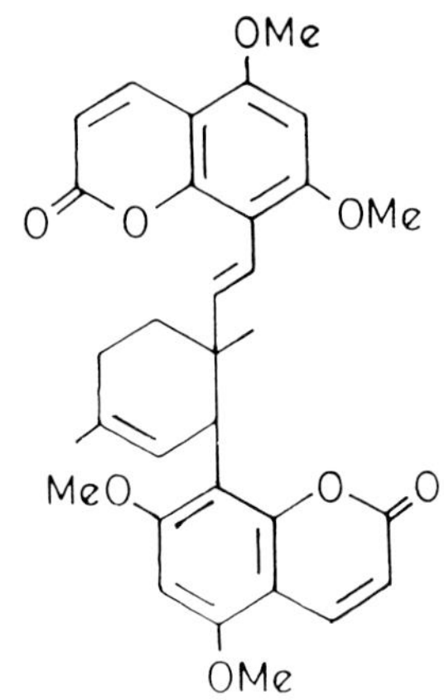

34 Mexolide

The extracts of *Pueraria tuberosa* DC. gave indications of antitubercular activity. It also showed antifertility activity in rodents.[161] Studies on the tuber extractive has led to the isolation of a new pterocarpan, tuberosin **35**, whose structure was established.[162] Tuberosin showed antistaphylococcal activity, antitubercular activity, and antifungal activity.

A dimeric coumarin, toddasin and the coumarins toddanol, toddanone, and toddalactone **36**, and four other known coumarins have been isolated from *Toddalia asiatica* Lamak.[163,164] All these compounds have 60 to 85% diuretic activity of hydrochlorthiazide at an equivalent dose level. Wikstrosin, the first example of tricoumarin, a new class of natural products, has been isolated from *Wikstroemia viridiflora* and its structure established as **37**.[165]

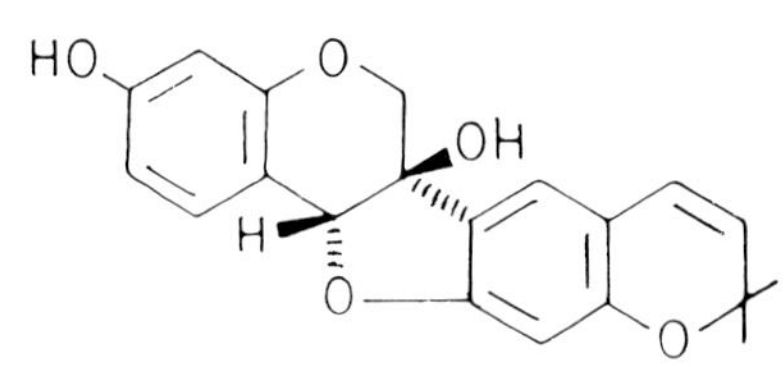

35 Tuberosin

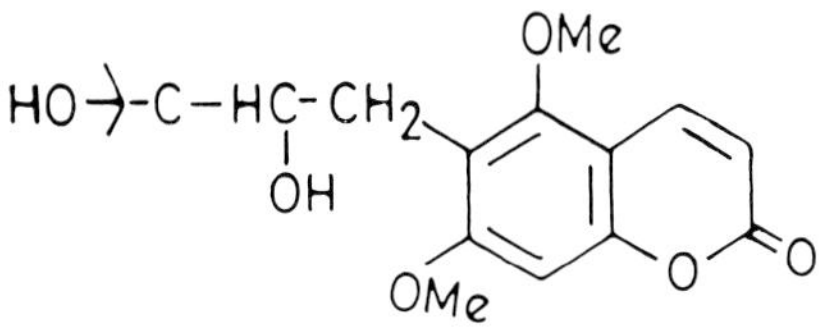

36 Toddalactone

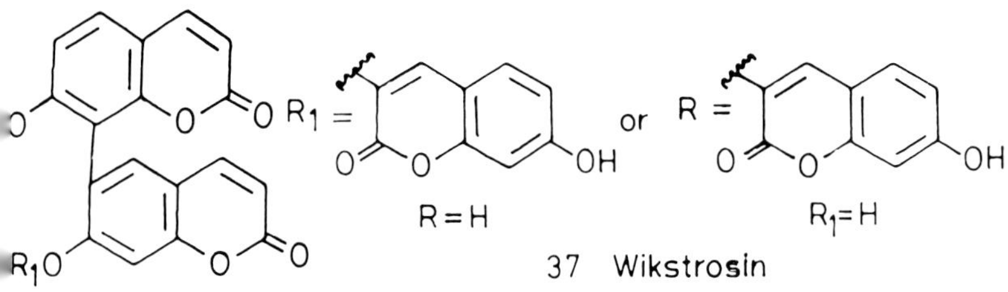

37 Wikstrosin

Among unusual flavonoids isolated are cycloheterophyllin **38** from *Artocarpus heterophyllus* Lamk. and chaplashin from *Artocarpus chaplasha* Roxb.[166,167] 6-Hydroxylated flavonoids are very rare; the first occurrence of 6-hydroxyluteolin-7-*O*-apioside from *Lepidagathis cristata* Willd. has been reported.[168] A novel isoflavone, milletin **39**, has been isolated from the seeds of *Millettia auriculata*.[169] Rhamnetin-dirhamnogalactoside has been isolated from *Rhamnus nepalensis* which accounted for the hypotensive activity of the plant.[170] A new flavone, arjunolone, has been obtained from stem bark of *Terminalia arjuna* (Roxb.) W.&A. showing oxytocic activity.[171] A new flavonol rhamnoside, tephdidoside **40**, isolated from *Tephrosia candida* has shown anticancer activity.[172]

The CNS-depressant activity was observed for the first time in a flavan glycoside, auriculoside **41**, isolated and characterised from *Acacia auriculiformis* A.Cunn.[173] A glucosyloxyflavan, diffutin, has been isolated and characterised from *Canscora diffusa* (Vahl.) R.Br., which was shown to be an adaptogenic principle.[174]

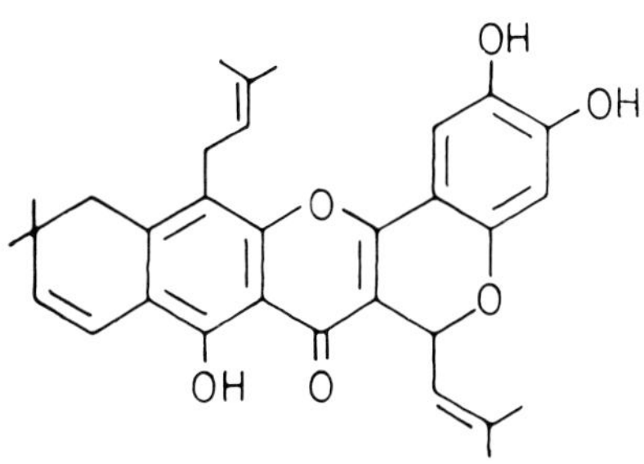

38 Cycloheterophyllin

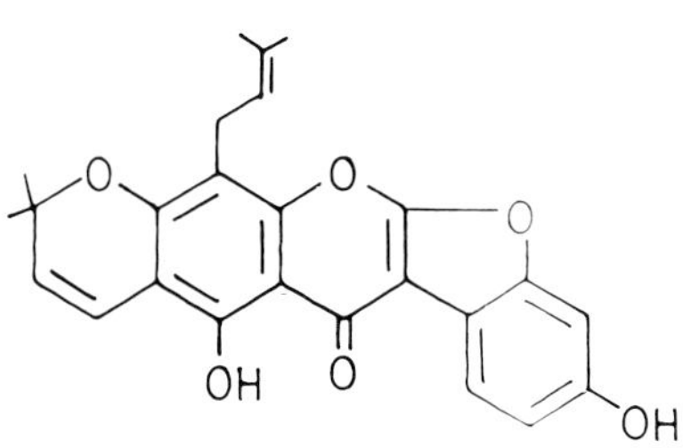

39 Milletin

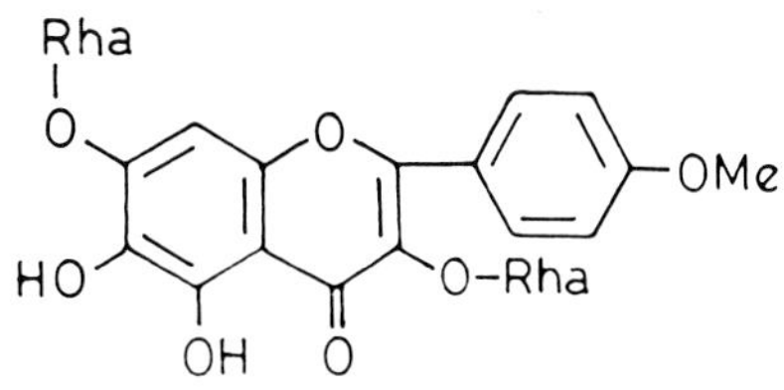

40 Tephdidoside

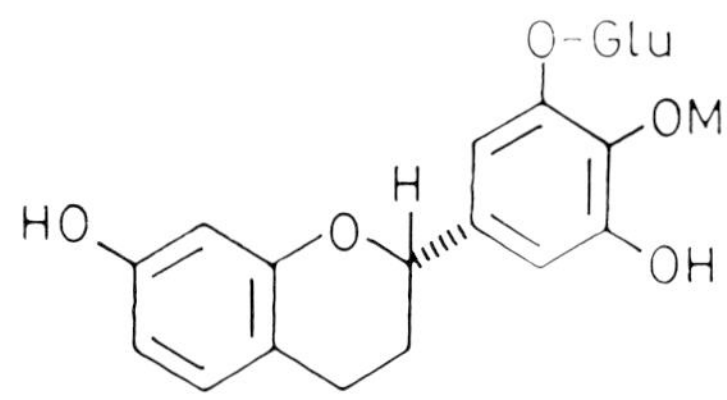

41 Auriculoside

C. TERPENOIDS

Important contributions have been made in the field of terpenoids mainly due to the work of the schools of research led by S. C. Bhattacharyya and Sukh Dev at the National Chemical Laboratory and R. P. Rastogi at the Central Drug Research Institute. Over 100 new compounds belonging to the sesqui-, di-, and triterpenoid groups have been isolated, many of them of novel types. From fungus-infected agarwood, *Aquilaria agallocha* Roxb., a spirocyclic sesquiterpene, agarospirol, furano-sesquiterpenes, α- and β-agarofurans, and dihydroagarofuran have been isolated and structurally characterised.[175] Two rare types of sesquiterpene ketones, agarol **42** and gmelofuran, have been reported from the stem wood.[176]

Five sesquiterpene alcohols have been reported from *Cedrus deodara* Loud.[177-182] Of these, himachalol **43**, allohimachalol, and centdarol exhibited pronounced concentration-dependent spasmolytic activity.[183,184] The ED_{50} (M) against $BaCl_2$-induced spasms in isolated guinea pig illeum was himacholol (1.1) and centdarol (14.13), compared to papavarine (4.85). The percentage antagonism of carbachol-induced spasms in anaesthetised cat was himacholol 60 and centdarol 72, compared to papavarine 80. These results are significant since such activity was not demonstrated in sesquiterpenes earlier.

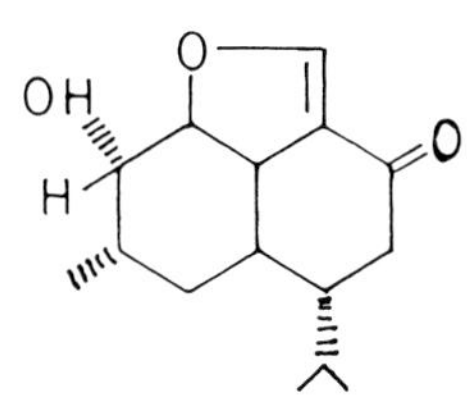

42 Agarol

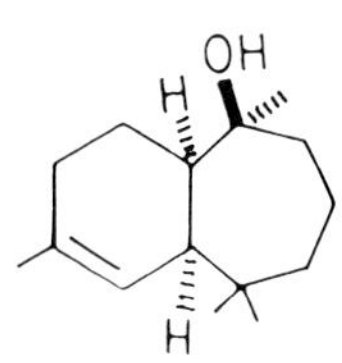

43 Himachalol

The sesquiterpene lactones, alantolactone and isoalantolactone, obtained earlier from *Inula racemosa* Hook.f., were identified as the antifungal constituents of the plant. They showed high *in vitro* activity but poor response against experimental ringworm infection in guinea pig.[185] A highly oxygenated germacranolide, enhydrin **44**, has been isolated from *Enhydra fluctuans* Lour. and its structure established.[186] This was also confirmed by X-ray analysis.[187]

Sassaurea lappa Clarke (costus) root oil has been intensively studied and several sesquiterpenes with the germacrane skeleton have been isolated and characterised. Typical studies are the structure elucidation of dehydrocostus lactone and the transfromation products of costunolide.[188,189] A number of germacranolides (tagitinins A-F) have been isolated from *Tithonia tagitiflora* Desf.,[190,191] of which tagitinin F **45** displayed significant anticancer activity at 12.5 mg, 161% in PS system. α-Humulene **46**, zerumbone, (–)-caryophyllene, and (+)-ar-curcumene are some of the constituents of *Zingiber zerumbet* Smith, and all aspects of the structure of the first two compounds have been elucidated.[192]

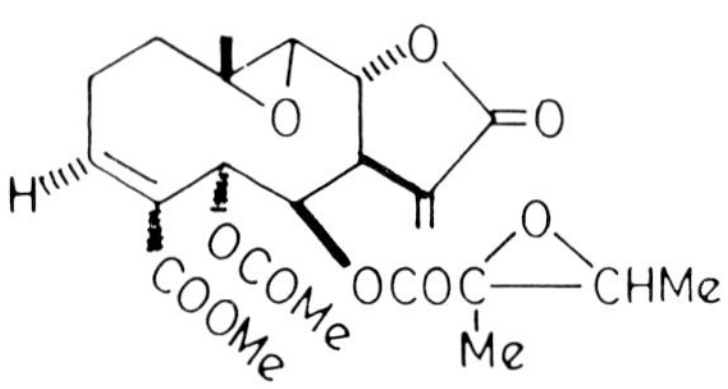

44 Enhydrin

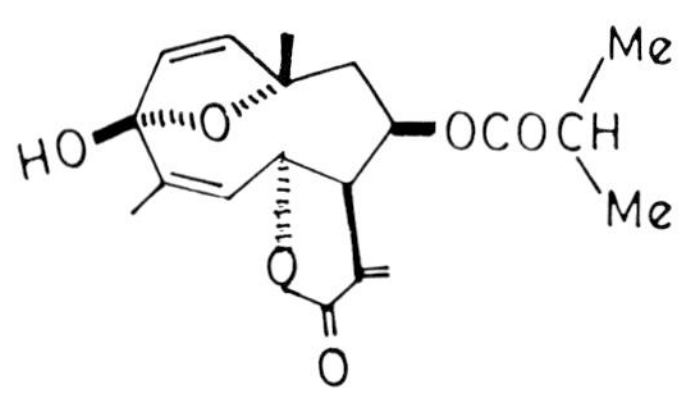

45 Tagitinin F

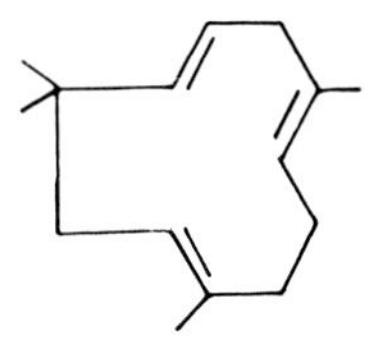

46 α-Humulene

The first tetracyclic sesquiterpene, longicyclene **47**, was also found to be of natural occurrence in *Pinus longifolia* Roxb.[193] Ishwarane and ishwarone from *Aristolochia indica* are the other rare tetracyclic sesquiterpens whose structures have been completely elucidated.[194,195]

Several diterpenes have been isolated from *Coleus forskohlii* Briq., of which the most interesting is coleonol, a manool oxide derivative **48**, and produces sustained hypotension (*vide infra*).[196-200] The structure of a novel diterpene lactone, cleomeolide **49** from *Cleome icosandra* L., which represents a new skeleton has been established by physical and X-ray analyses.[201]

The report that tubers of *Dioscorea bulbifera* L. have appetite-suppressing properties has led to their chemical examination which resulted in isolation of a norditerpene lactone, diosbulbine.[202] On the basis of spectral data and degradative evidence, diosbulbine was shown to have the structure **50** which was devoid of biological activity. A polyacylated jatrophone-type diterpene, euphornin **51** has been isolated from *Euphorbia maddeni* Boiss. and its structure elucidated by physicochemical methods.[203] A diterpene, precalyone **52** from *Roylea calycina* (Roxb.) Briq., has shown anticancer activity at 50 mg, 143% (PS system).[204]

47 Longicyclene

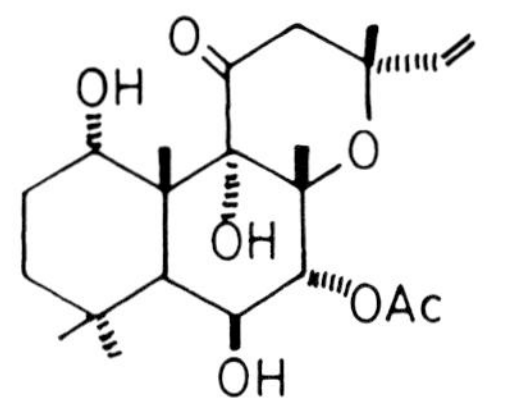

48 Coleonol

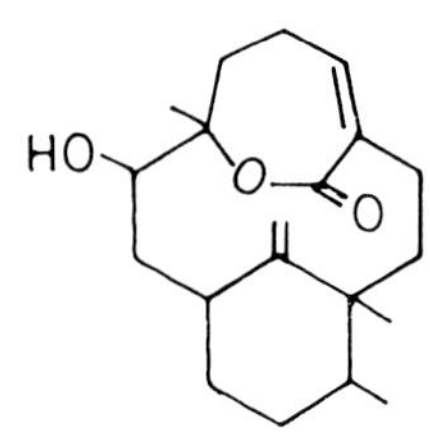

49 Cleomeolide

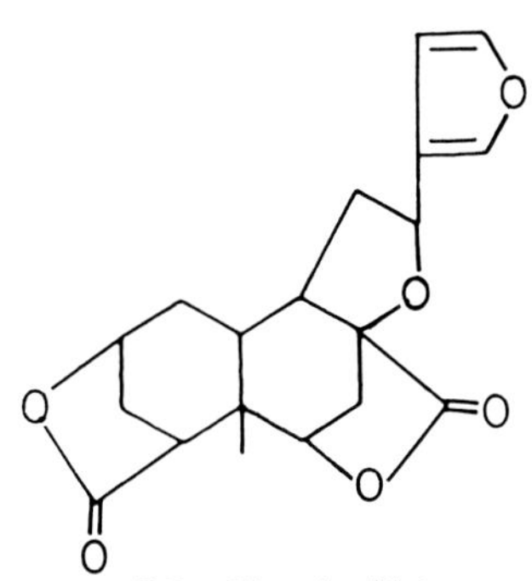

50 Diosbulbine

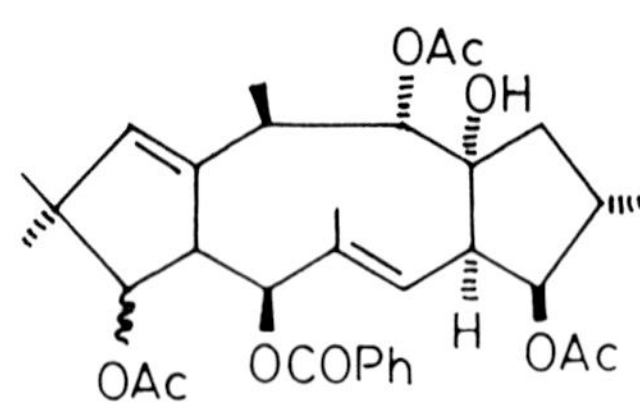

51 Euphornin

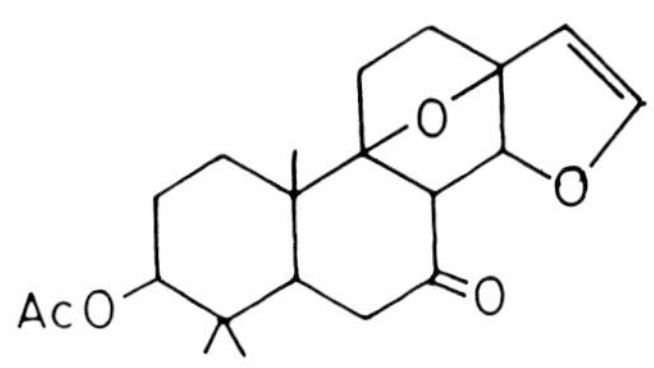

52. Precalyone

Cheilanthatriol **53** from the fern *Cheilanthes farinosa* Kaulf. belongs to the rare group of sesterterpenoids.[205]

Typical of the large number of triterpenoids which have been isolated and characterised are the 19-oxogenated oleanane arjunic acid (*Terminalia arjuna* W.&A.);[206] a 28-nor-oleanene, acacidiol **54**, from *Acacia concinna*;[207] two 28-nor-D:A-freidooleananes, elaeodendrol **55**, and elaeodendradiol isolated from *Elaeodendron glaucum*;[208] the hopane, mollugogenol-A from *Mullugo hirta* Roxb.;[209] the ursene, lantic acid **56**, from *Lantana camara* L.;[210] 24-nor-lupane lactones, caltholide **57**, and its 3α-epimer as well as a lupane lactone, palustrolide from *Caltha palustris* L.[211]

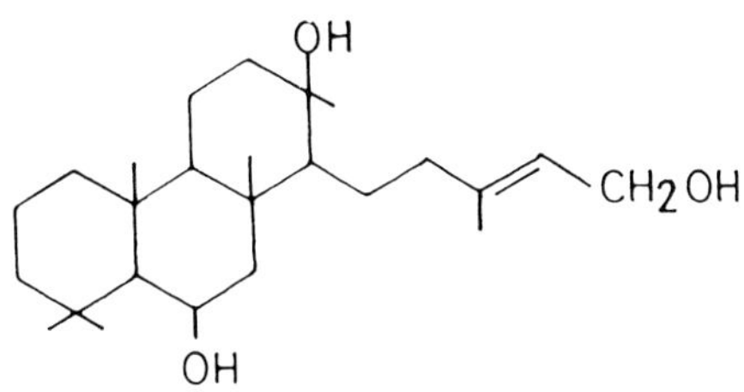

53. Cheilanthatriol

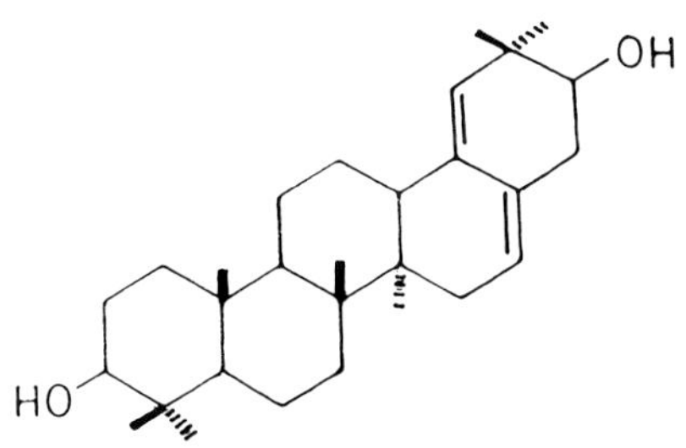

54. Acacidiol

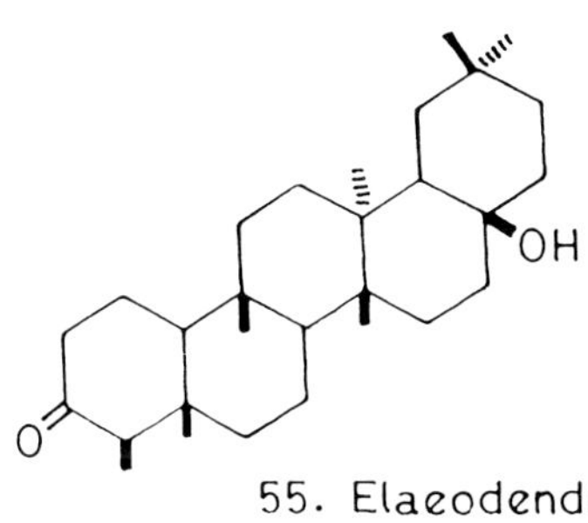

55. Elaeodendrol

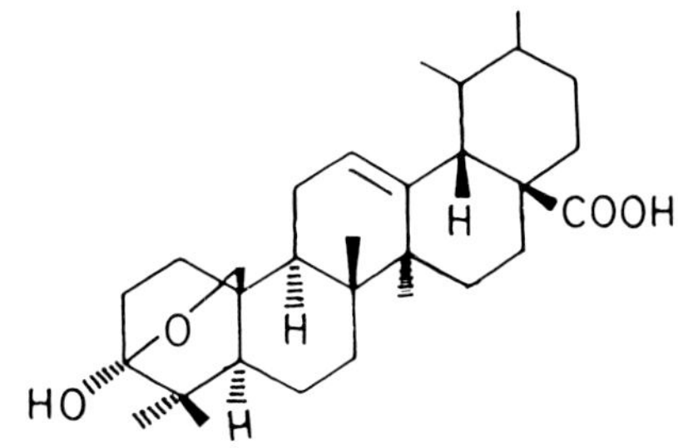

56 Lantic acid

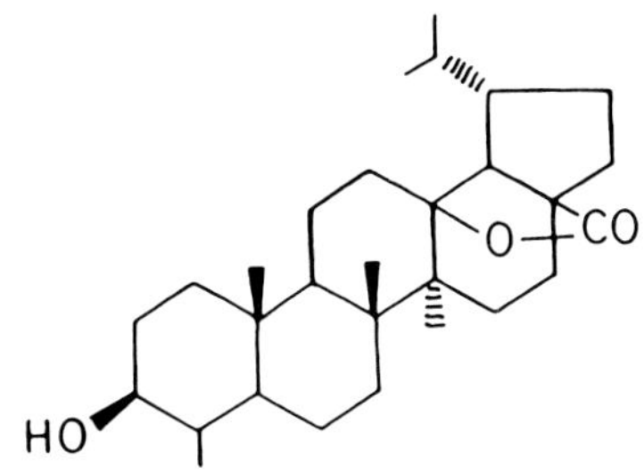

57. Caltholide

The roots and bark of *Salacia prinoides* DC. are used widely in indigenous medicine for the treatment of diabetes. Chemical studies have led to the isolation of six closely related triterpenes which have been shown to be 1,3-diketofriedelane derivatives.[212,213] The structures of substances Q, T, and U were revised by the aid of X-ray analysis having functionalities at C-25 instead of at C-24, as originally proposed. The position of the ether linkage in substances R **58** could be established only by X-ray analysis.[214]

Litsomentol isolated from *Litsea tomentosa* Heyne is the only number of the cucurbitacin group isolated from a Lauraceae plant. It represents the simplest member of the group and is unique in lacking an oxygen function at C-11. It was found to be exceedingly toxic.[215] Vilasinin **59**, from the leaves of *Azadirachta indica* A.Juss., belongs to the interesting group of tetranortriterpenoids.[216] Nimbidin isolated from the seed oil has exhibited good anti-inflammatory activity in several test models.[217] Undiluted neem oil possesses strong spermicidal action against the sperm of several species, including man,[218] but well-controlled safety evaluation must be done before clinical studies. A new cucurbitacin, trewinine **60**, from *Trewia nudiflora* has been characterised which exhibited plant growth inhibitory activity showing maximum inhibition at 100 ppm using rice second leaf.[219]

A new meliacin, dysobinin **61**, isolated from *Dyoxylum binectariferum* has exhibited significant CNS-depressant activity, as well as mild anti-inflammatory activity.[220] Malabaricol **62**, isolated from *Ailanthus malabarica* DC., represents a new fundamental type of triterpene not encountered thus far in nature.[221,222]

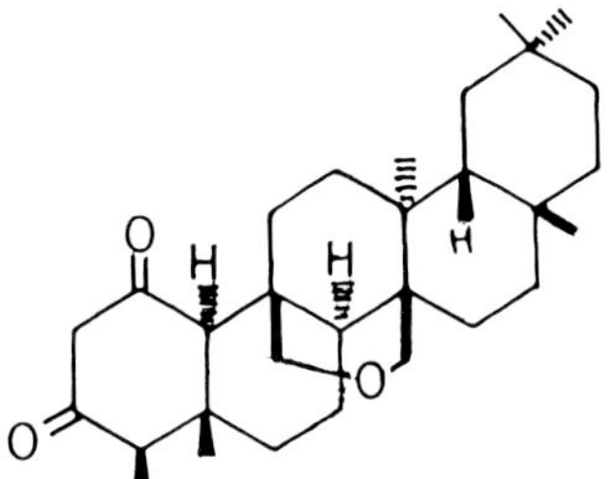

58 Substance R

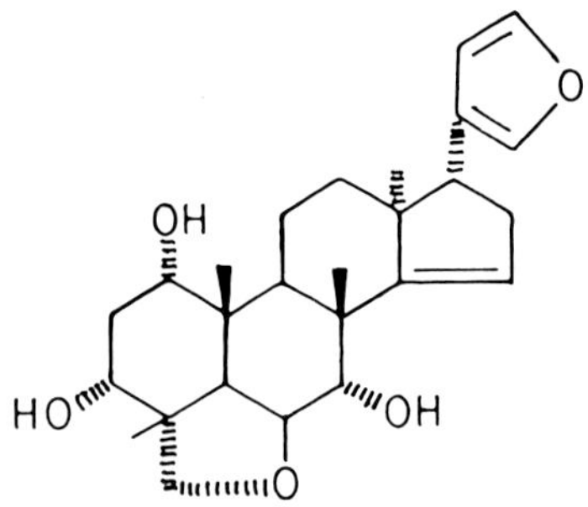

59 Vilasinin

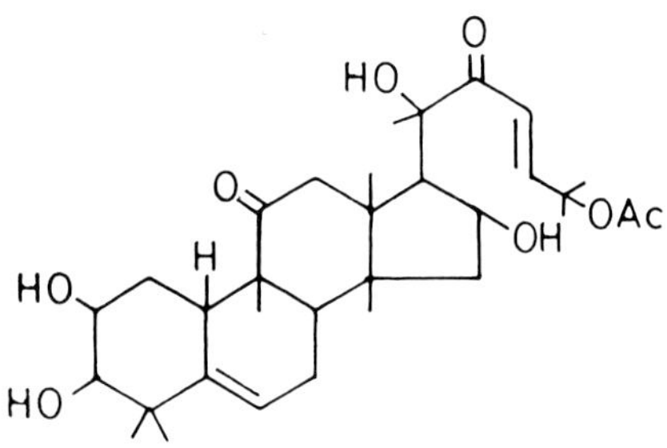

60 Trewinine

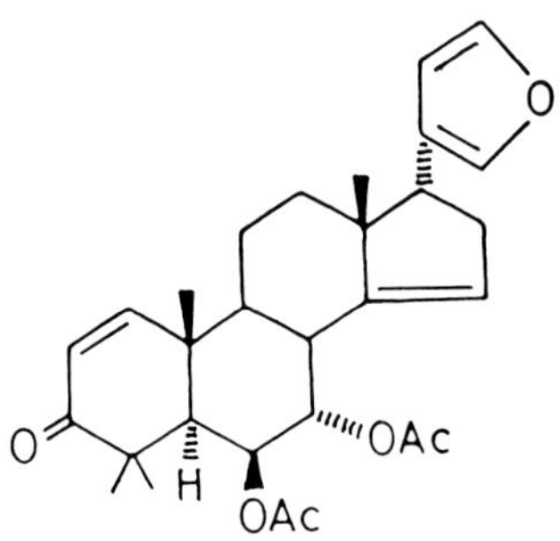

61 Dysobinin

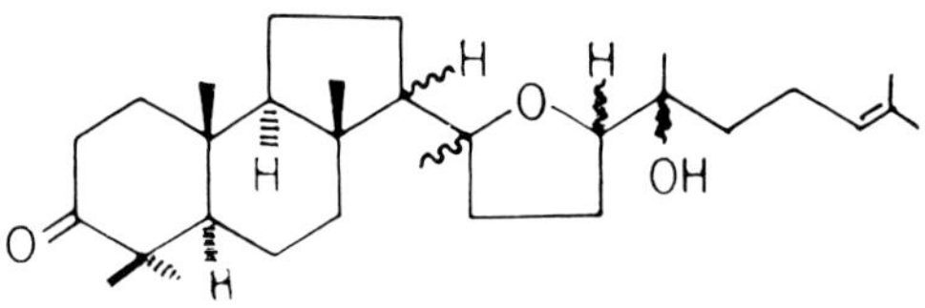

62 Malabaricol

Various types of saponins have shown a wide range of biological activities. Spermicidal saponins from *Schefflera capitata*,[223] *Pittosporum nilghrense*,[224] *Polemonium coeruleum* L.,[225] sigmastadienol-rhamnoside from *Cleome viscosa* as fish poison,[226] trichonin **63** and santholin showing hypoglycemic activity from *Trichosanthes palmata* Roxb.,[227] celsioside from *Celsia coromandeliana* Vahl. showing high anticancer activity against Walker carcinoma 256 in rats (WM system);[228] schiwallin, which crystallises as a stable adduct of five closely related saponins, from *Schima wallichii* Choisy showing antifungal activity, have been isolated.[229] Consap, a spermicidal cream containing saponins from *Sapindus mukorossi*, is in phase II of clinical testing.[230] The bacosides A and B from *Bacopa monniera* have been shown to improve the performance of rats in several learning tests as evidenced by better acquisition, improved retention, and delayed extinction.[231] The saponins of *Albizzia lebbeck* have potent antiallergic property shown to be due to several effects including chromoglycate-like action in mast cells and suppression of synthesis of reaginic antibodies, T-lymphocyte activity, and initial process of sensitization.[232,233] The aqueous decoction of the bark reduces the release of catecholamines in response to histamine.[234]

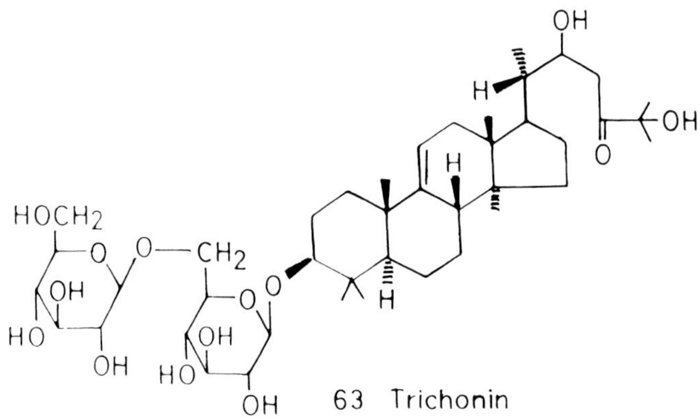

63 Trichonin

D. GLYCOSIDES

Examination of *Asclepias curassavica* L. collected from Dehradun yielded results differing from the plant of Brazilian origin.[235] The principle glycoside was asclepin (3-*O*-acetylcalotropin) **64**. Asclepin showed excellent cardiotonic activity. The oral absorption in cats was 22.5% in 5 h which is comparable to digoxin. A single dose of asclepin persisted up to 96 h (digoxin, 72 h) and the cumulative toxicity was lower than that of digoxin.[236,237] Its cardiotonic activity is comparable with digoxin and ouabain but with a better safety margin. Four new glycosides, shatavarins I-IV, were isolated from the medicinal plant *Asparagus racemosus* Willd. Shatavarine IV is a glycoside of sarsasapogenin having two molecules of rhamnose and one molecule of glucose (2:1). Shatavarine I has an additional glucose unit, whose position has not been determined. It exhibited significant antioxytocic activity.[238]

From the leaves of the highly poisonous *Cleistanthus collinus* (Roxb). Hook f. & Benth. ellagic acid, diphyllin and two new lignan lactones, collinusin and cleistanthin were isolated. The structure **65** has been assigned to cleistanthin.[239] Cleistanthin induces a significant increase in neutrophilic granulocyte count (NGC) in peripheral blood of rats, mice, cats, and monkeys when administered by oral, intraperitoneal, intravenous, intramuscular, and subcutaneous routes. A dose of 0.86 mg/kg/day maintained increased level of NGC for 1 month in rats. It is postulated that cleistanthin might more effectively protect the bone marrow from the damage caused by cytostatic agents.[240]

A 3,11-dihydroxy hexadecanoic acid glycoside (ipolearoside) having the structure **66** was isolated from *Ipomoea leari* Paxt.[241] It showed high anticancer activity in WM system (rats). A new lignan, justisolin, and a lignan-glucoside, simplexoside, exhibiting growth regulatory properties in plants, has been isolated from *Justicia simplex* D.Don.[242] Simplexoside produced a weak CNS depressant action in mice and rats, while the free lignans of the plant produced CNS stimulant action. Different proportions of mixtures of simplexoside and free lignans showed a complex psychopharmacological action as was observed before with mixtures of swertamarin and mangiferin.[243]

Scutelarein 5-glucuronide isolated from *Millingtonia hortensis* was half as active as diuretic as chlorthiazide.[244] The alcoholic extract of the stem bark had diuretic and antipyretic activity in rats. The natriuretic effect was more pronounced than the kaliuretic effect.[245] A glycoside, vicine **67**, of a nonprotein nitrogenous base which is a favism-inducing toxin, has been isolated and characterised from *Momordica charantia* L. seeds.[246]

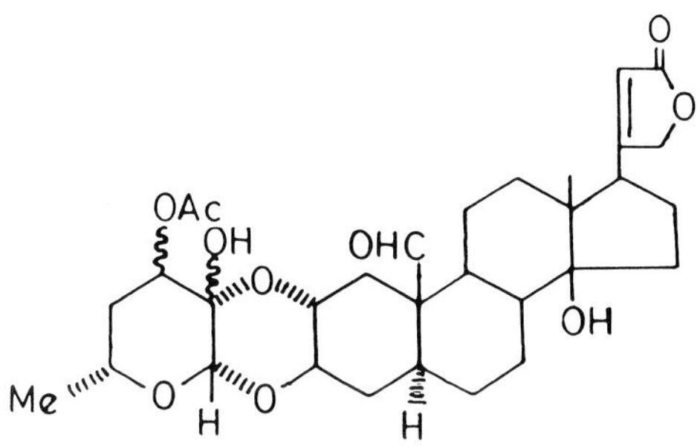

64 Asclepin

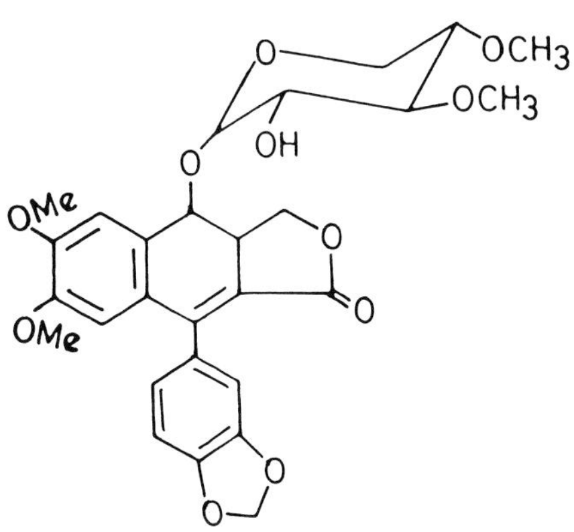

65 Cleistanthin

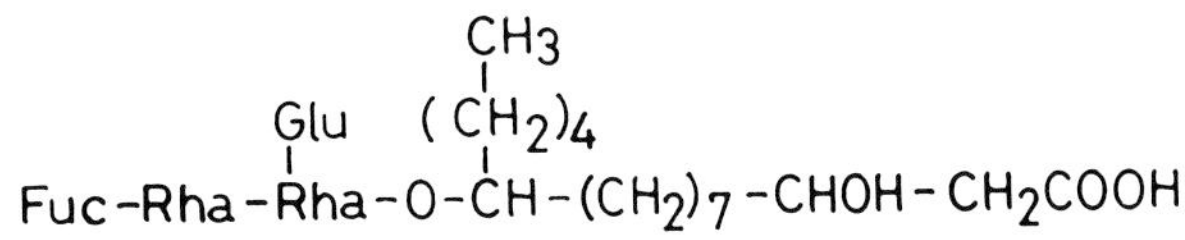

66 Ipolearoside

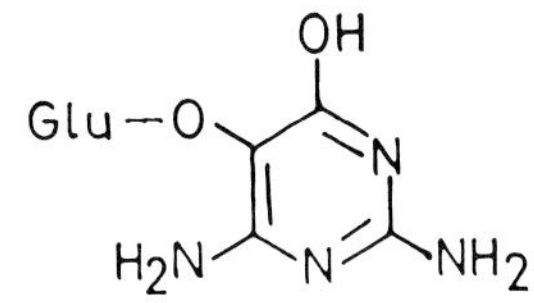

67 Vicine

In a reinvestigation of the reputed medicinal plant, *Picrorhiza kurrooa*, two iridoid glucosides, picroside A and kutkoside, have been isolated and their structures determined.[247] Kutkin, which is a stable mixed crystal of these two glucosides, showed weak diuretic activity, about a sixth of that of chlorthiazide. Kutkin has also been found to have antiallergic activity in the passive cutaneous anaphylaxis test and to afford protection against lanthanum and galactosamine-induced liver damage in rats. In both cases it has to be administered for at least 3 days. Clinical reports of usefulness of *P. kurrooa* in bronchial asthma and rheumatoid arthritis are also available.[248] Its usefulness in liver disorders is discussed in the next section.

Stigmast-7,2,-dien-3β-*O*-glucoside (vittadinoside) was isolated from *Vattadinia australis* A.Rich. It possessed diuretic activity comparable to chlorthiazide.[249] The glycoside from the leaves of *Vallaris solanacea* were reported to show digitalis-like activity. The most active glycoside, *O*-acetyl-solanoside, from this mixture has been isolated and shown to be very active with the added advantage of being completely absorbed from the intestine of cat.[250]

E. MISCELLANEOUS

From the roots of *Arnebia nobilis* seven naphthaquinones were isolated, six of them being derivatives of alkanin;[251-253] of these arnebin-1 **68** showed good anticancer activity (WM 6 mg; PS 3 mg).[254] This substance is also a potent inhibitor of reverse transcriptase (RNA-dependent DNA synthesis) and poliovirus replicase (RNA-dependent RNA synthesis), but is toxic because it also inhibits DNA-dependent RNA synthesis. Since antileukaemic activity has been highlighted in several quassinoids, investigation of *Ailanthus excelsa* was undertaken and a new quassinoid, 13,18-dehydroexcelsin **69**, and an ailanthol derivative, glaucarubol, were isolated.[255]

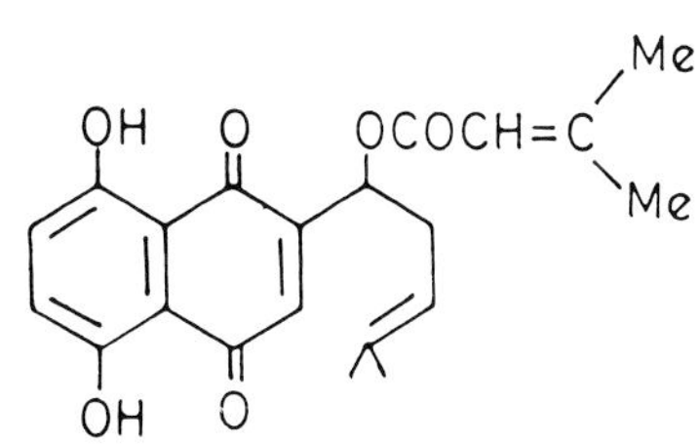

68 Arnebin – 1

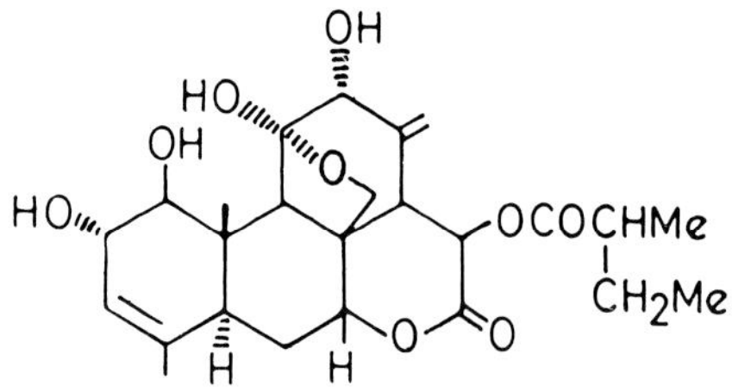

69 13,18-dehydro-excelsin

A novel dimeric lactone, angeolide **70**, has been isolated from *Angelica glauca* whose structure and stereochemistry has been established by X-ray diffraction studies.[256] *Aristolochia indica* is a good source of natural phenanthroids and has been reported to show anti-implantation activity (*vide infra*). New phenanthrene derivatives, aristolactam-*N*-β-D-glucoside, and a lactone, aristolide, have been isolated and characterised. This lactone is the first 10-oxygenated artistolic acid derivative encountered in nature.[257a] A sterol, amasterol **71**, has been isolated from roots of *Amaranthus viridis* L. It caused inhibition of lettuce seed germination and growth of seedlings and was also found to strongly inhibit the growth of a pathogenic fungus, *Helminthosporium oryzi*.[257b]

From the roots of *Boerhaavia diffusa*, an insect moulting hormone, β-ecdysone has been isolated as the first reported ecdysone from a plant source.[258] Antifibrinolytic activity was demonstrated in the crude extracts of the roots of the plant and in its phenolic glycoside, punarnavoside **72**. These are effective in controlling IUD-induced uterine bleeding in monkeys.[259]

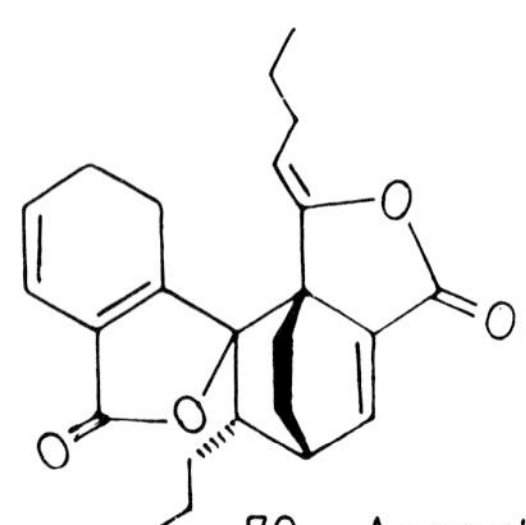

70 Angeolide

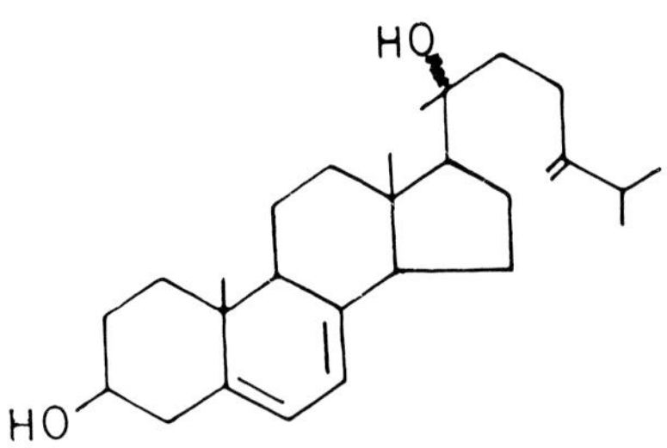

71 Amasterol

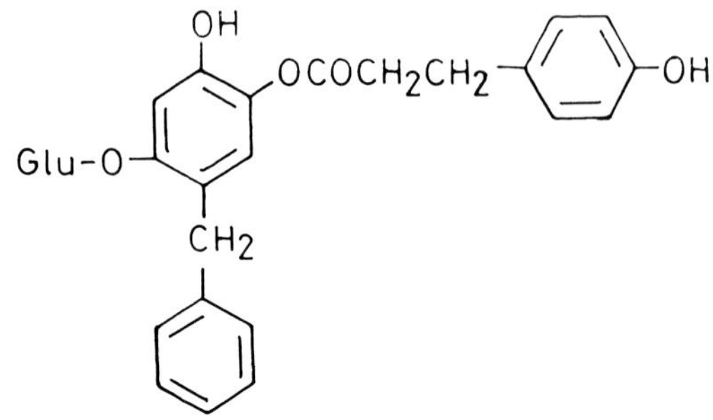

72 Punarnavoside

The gum resin exudate from *Commiphora mukul* (Hook ex. Stocks) Engl. is highly rated in Ayurvedic medicine for the treatment of rheumatoid arthritis, obesity, and several other disorders (*vide infra*). The active constituents are present in the steroid fraction which yielded a number of new steroids, Z-guggulsterone **73**, guggulsterol-I, guggulsterol-II, guggulsterone-III, and guggulsterone-IV.[260-263] A stamina-building, antistress, and antihyperlipidemic principle, pangamic acid, has been isolated from *Cicer arietinum*, and characterised as D-gluconodimethylaminoacetic acid.[264]

Extracts of the fresh fruits of *Diospyros mollis* Griff. are widely used in Thailand as an anthelmintic particularly against hookworms. The active principle was recognized to be diospyrol **74**, which has been evaluated for antihelminthic activity. The LD_{50} of diospyrol is over 3 g/kg in several species of experimental animals. In hamsters infected with human hookworm *Necator americanus*, a dose of 200 mg/kg for 3 days cured 80% of the infected hamsters (nonpatent infection) while a single dose of 500 mg/kg affected a 100% cure of the patent infection. Diospyrol was also effective against *Hymenolepis nana* and *Nematospiroides dubius* parasites in mice.[265] The disadvantage of the compound is its high susceptibility to aerial oxidation. Thus, there is sound basis for the use of the juice of fresh fruits against parasitic infections.

An unusual type of cardenolide, elaeodendroside-B **75** has been isolated from *Elaeodendron glaucum* Pers.[266] A novel spironaphthoquinone, microphyllone, was isolated from *Ehretia macrophylla* and its structure **76** established by X-ray analysis.[267]

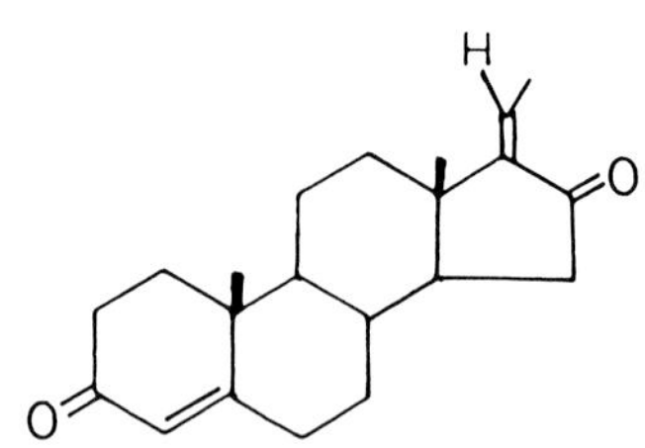

73 Z-Guggul sterone

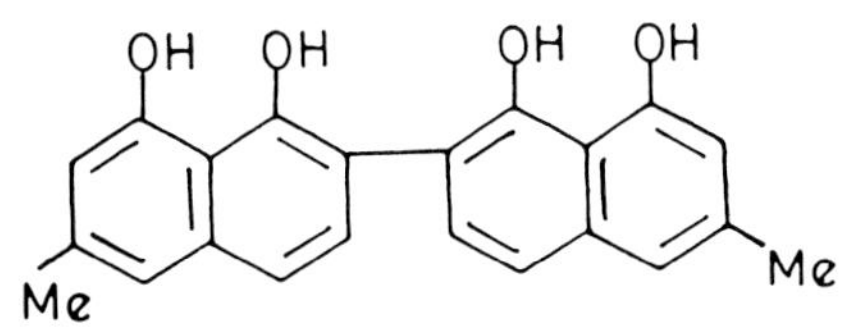

74. Diospyrol

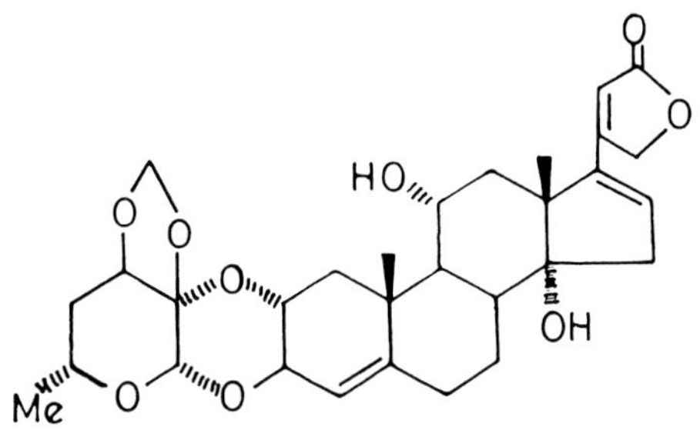

75. Elaeodendroside-B

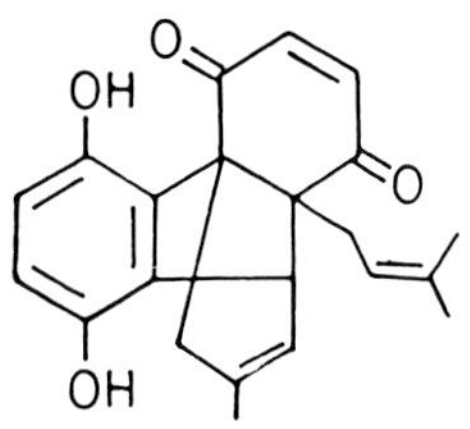

76. Microphyllone

A dihydrochalkone, lyonogenin **77** and its 2′-glucoside (lyonotin), have been characterized from *Lyonia formosa* which accounted for the hypotensive activity shown by the plant.[268] Bavachinin, a flavone isolated from seeds of *Psorelea corylifolia* exhibited analgesic and anti-inflammatory activities in animal tests.[269] A new coumestan, psoralidin oxide **78**, has been characterised from it.[270] The structure of hypophyllanthin **79** isolated from *Phyllanthus niruri* has been confirmed by X-ray analysis.[271]

A novel withanolide, daturalactone-4, from *Datura quercifolia* had anti-implantation activity.[272,273] A number of new ergostane-type steroids (withanolides) have been isolated from *Physalis peruviana* and *P. pubescence*; of these withaperuvin D **80** is a novel withanolide.[274-276] The structure of a spasmolytic principle isolated from *Rheum webbianum* Royle has been established as 3′,5′-dihydroxy-4-methoxystilbene.[277] It showed significant *in vitro* spasmolytic activity at a concentration of 50 μg/ml.

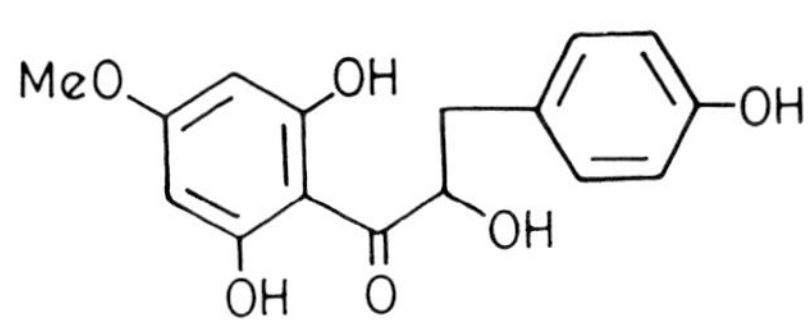

77. Lyonogenin

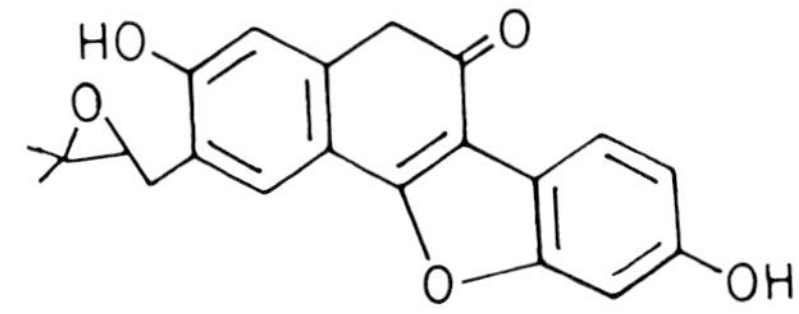

78. Psoralidin oxide

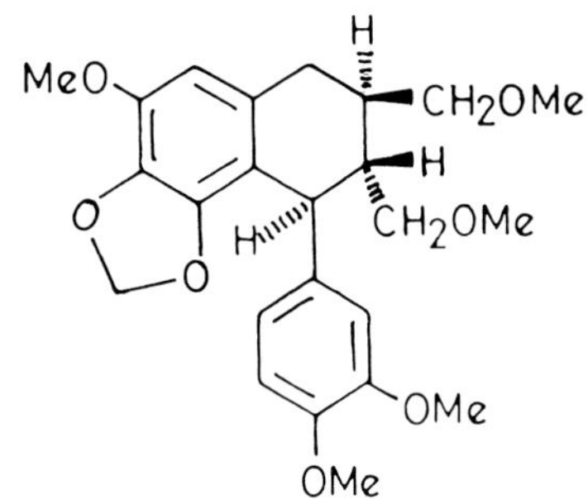

79. Hypophyllanthin

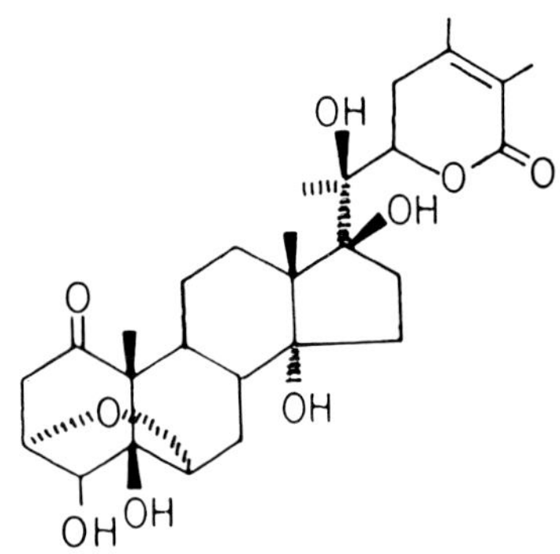

80. Withaperuvin - D

V. DETAILED PHARMACOLOGICAL AND CLINICAL STUDIES

Some significant in-depth studies are presented here; however, the voluminous data on *Cannabis*, *Rauwolfia*, and opium have been excluded because the respective well-known chemical constituents are not new and the numerous studies, numbering hundreds, cannot be included here due to constraints of space.

A. *ADHATODA VASICA*

Atal and his colleagues reported abortifacient activity in vasicine, carried out detailed experimental and clinical studies, and published a comprehensive review.[278] Several derivatives have also been tested but none had a better profile than vasicine, even though some were quite potent.[279] A sustained release formulation has also been developed.[280] Vasicine has been shown to potentiate oxytocin-induced responses in isolated mammary strips and prostaglandin-induced uterine contraction.[281,282] Absorption and distribution studies indicated maximal concentration in the uterus.[283]

The bronchodilator activity of vasicine and its several analogues, derivatives, and related compounds have also been investigated and reviewed.[278] A synthetic analogue, RLX, appears promising as a bronchodilator. Some 7-hydroxy-vasicinone analogues have mild CNS depressant effects.[284] Vasicine also produced an increase in platelet count associated with significant hyperplasia of megakaryocytes in bone marrow following repeated administration in several species and the results indicate another possible area of its clinical utility.[285]

B. *ARISTOLOCHIA INDICA*

The plant has been extensively studied for anti-implantation activity originally reported in the extracts of the root by Pakrashi et al.[286] Among the isolated compounds aristolic acid, its methyl ester, *p*-coumaric acid, and a sesquiterpene, *seco*-ishwaran-12-ol, have been found effective anti-implantation agents in mouse and rabbits.[287-290] Weaker activity was found in the aristolactum glucoside.[291] The methyl ester, however, produced toxic effects in liver and kidney.[292] Some of these compounds have antioestrogenic activity. Some of the extracts and *p*-coumaric acid have been shown to arrest spermatogenesis but there was also associated reduction in the weight of associated sex organs.[293,294]

C. *BOSWELLIA SERRATA*

The gum resin from the plant exhibited good anti-inflammatory activity in acute and chronic models of inflammation,[295,296] which has been shown due to the presence of boswellic acids.[297] Its efficacy has been demonstrated in limited clinical studies as well.[298,299] It also has hypolipidaemic activity.[296]

D. *COLEUS FORSKOHLII*

The hypotensive activity of crude extract of root was subsequently shown to be due to a diterpene, coleonol,[300] whose chemical structure was established.[196] Detailed pharmacological studies demonstrated antihypertensive, positive inotropic, and smooth muscle relaxant activities of coleonol.[301] Bhat et al. independently reported the isolation of a diterpene, termed by them forskolin,[302] and its hypotensive and inotropic properties were reported.[303,304] The latter report also included data on structural modifications. Data on structural variations have also been published by Tandon et al.[305] It has now been established that the two molecules are identical.[306] The compound has been shown to be an activator of adenylate cyclase and the available literature has been reviewed.[307] It is also an inhibitor of platelet aggregation.[308] It is a potential antimetastatic agent[309] capable of lowering raised intraocular tension.[310] Clinical studies with the compound are in progress at several centres.

E. *COMMIPHORA MUKUL*

Interest in the gum resin of this plant, which is widely used in the Ayurvedic system of medicine, has been aroused following demonstration of its hypolipidaemic activity.[311] Various fractions have since been investigated experimentally,[312,313] as well as clinically.[314,315] The active fraction has a number of sterols which have been isolated and characterised.[260] A standardized fraction containing the steroids has been extensively investigated at CDRI and termed "gugulipid". It exhibits hypolipidaemic activity experimentally[261,316] and clinically,[317] and a multicentric double blind comparison with clofibrate has been undertaken. Gugulipid produces a dose-dependent increase in catecholamines and dopamine β-hydroxylase enzyme following chronic feeding in monkeys.[318] It also inhibits platelet aggregation experimentally[319] as well as clinically.[320] The z-guggulosterone displays a thyroid-stimulating action.[321] Anti-inflammatory activity against adjuvant arthritis in rats has also been found in a petroleum ether extract of the resin.[322]

F. *CURCUMA LONGA*

Curcumin, the yellow pigment of the rhizomes of *Curcuma longa*, is diferuloyl methane and is the major constituent responsible for the anti-inflammatory activity of various extracts of the rhizome. The activity was first reported by Srimal et al. who subsequently showed it to be effective in various models of acute and chronic inflammation and to have a low ulcerogenic index.[323,324] These results have been confirmed by other groups of workers who have also shown the volatile oil to be active.[325-327] Its antiulcerogenic effect has also been reported.[328] Toxicity studies in rats and monkeys have shown the compound to be safe and show no evidence of GI toxicity.[329] Curcumin prevents inflammation-induced increase in the activity of lysosomal enzymes, acid phosphatase, and cathepsin D, and mildly inhibits prostaglandin synthesis.[330,331] Curcumin inhibits lipid peroxide formation in the liver *in vivo* as well as *in vitro*,[332,333] and this may be involved in its anti-inflammatory activity. Initial clinical studies demonstrated good responses in cases of rheumatoid arthritis, comparable to phenylbutazone, with no side effects.[334] Studies on its metabolic disposition showed 75% absorption 5 h after oral administration.[335] The elimination was completed in 48 h. Anti-inflammatory activity has also been demonstrated in certain naturally obtainable analogues of curcumin.[336]

G. *THEVETIA NERIIFOLIA*

Peruvoside, isolated from *Thevetia neriifolia* Juss. by Rangaswami and Rao, has been subjected to extensive experimental and clinical studies in India.[337-340] The earlier data have been reviewed.[341] It compares with ouabain in its rapidity of action by intravenous route. It is absorbed rapidly and effectively from the GI route. Its significant inotropic and chronotropic effect, low acute and cumulative toxicity, and efficacy by oral and parenteral route[342] has been commented upon by several investigators.[343-345] The drug is being marketed in several Western countries and is a valuable addition to available drugs for management of congestive heart failure.

H. HEPATOPROTECTIVE PLANTS

The nonavailability of effective drugs in the modern system of medicine led to several studies on possible hepatoprotective action of traditional drugs. The experimental evaluation has largely been performed against CCl_4-induced hepatic damage which is not a satisfactory model. Protection, however, has been reported to be obtained with extracts of *Tinospora cordifolia* Miers.,[346,347] *Picrorhiza kurrooa, Eclipta alba*,[348,353] *Ricinus communis*,[349] *Andrographis paniculata*, and *Luffa echinata* Roxb.[350] There have been few properly controlled clinical studies. Antarkar et al.[351] in a double blind clinical trial obtained significant improvement in clinical and laboratory parameters with the Ayurvedic formulation 'Arogyavardhini' in cases of viral hepatitis. Its major constituent is *Picrorhiza kurrooa*. Improvement has also been reported with *Berberis aristata*.[352] Vaidya et al. reported a marked reduction in serum bilirubin and SGPT along with clinical relief following nasal instillation of the juice of *Luffa echinata* which was followed by profuse rhinorrhea.[353]

VI. SUMMARY AND PROSPECTS

The literature reviewed above indicates clearly the continued interest of medicinal chemists, biologists, and clinicians in investigating medicinal plants in Indian laboratories. Such investigations also figure as high priority areas for research funding, etc., with national agencies in developing countries, including India, and also with international agencies such as WHO. Does the output justify continued input into this area? Can one hope to obtain breakthroughs such as reserpine in the future as well? These and other related questions are repeatedly being posed and discussed. It is against this background that future prospects must be assessed.

As mentioned earlier, Col. R. N. Chopra was the pioneer in evolving a strategy for developing new drugs from medicinal plants and indigenous remedies. The strategy developed by him in the 1930s was contingent upon the needs then and the available expertise and facilities. These have changed significantly during the last 50 years, whereas the strategy has undergone minimal changes. For example, very sophisticated experimental models have been developed, particularly for pure and synthesised compounds. Many of these are less suitable for use with crude indigenous drugs or semipurified plant materials. The problem was less acute during Col. Chopra's time since the test systems were also rather crude. Further, the regulatory requirements for introduction of new drugs were much less stringent. With the present regulatory procedures in most countries it is almost impossible for any academic institution to do the total developmental work on a new drug. Due to the difficulty in obtaining patent protection, procurement of adequate quantities of plant materials, regulatory problems, etc., the large drug houses in the developed countries have not taken adequate interest in natural products and have almost exclusively been involved in synthesising novel molecules as drugs. Since many traditional drugs have long been used, a different regulatory approach as well as modified test models are needed for their expeditious development and evaluation.

The research efforts on natural products and synthetic agents are complementary and both types of drug play an important role in the health care programme. The discovery of a novel biological activity in a natural product or of a novel structural lead is often the starting point of a synthetic programme wherein the activity as well as the safety may be optimised. There are areas of therapy where the synthetic programmes have yielded effective, safe, and cheap remedies, e.g., drugs for treatment of intestinal helminths. It appears unlikely that a natural product can effectively substitute them. There are several other areas, however, wherein no satisfactory drugs are available and these would be the most suitable areas for exploitation of medicinal plants; these should, for example, include development of new hepatoprotective agents, immunostimulants and imunomodulators, adaptogens, and remedies against drug-resistant malaria. There is a need, therefore, to redefine the strategy of work on medicinal plants so that the available resources can be optimally utilised on worthwhile areas.

The studies in India and elsewhere have clearly demonstrated the necessity for collaborative multidisciplinary efforts for the development of drugs from medicinal plants. Such teamwork is still not adequate in most countries. Many constituents isolated from the medicinal plants, therefore, are never biologically evaluated, and conversely plants in which biological activity has been observed may not be chemically investigated. Fortunately, the international agencies such as WHO have appreciated its significance and their support is increasingly given to laboratories where teamwork is possible.

Attention has been hitherto focused on the land flora in India, largely because these plants are used in the traditional systems of medicine. India has a vast marine flora which needs systematic study. A start in this area has been made,[354] and in the years to come the marine flora should receive much greater attention worldwide.

REFERENCES

1. Kirtikar, K.R. and Basu, B.D., *Indian Medicinal Plants*, Vol. I-IV, Lalit Mohan Basu, Allahabad, 1933.
2. Chopra, R.N., Chopra, I.C., Handa, K.L., and Kapur, L.D., *Indigenous Drugs of India*, Dhur, U.N. & Sons Pvt. Ltd., Calcutta, 1958.
3. Dhar, M.L., Dhar, M.M., Dhawan, B.N., Mehrotra, B.N., and Ray, C., Screening of Indian plants for biological activity. I., *Indian J. Exp. Biol.*, 6, 232, 1968.
4. Bhakuni, D.S., Dhar, M.L., Dhar, M.M., Dhawan, B.N., and Mehrotra, B.N., Screening of Indian plants for biological activity. II., *Indian J. Exp. Biol.*, 7, 250, 1969.
5. Bhakuni, D.S., Dhar, M.L., Dhar, M.M., Dhawan, B.N., Gupta, B., and Srimal, R.C., Screening of Indian plants for biological activity. III., *Indian J. Exp. Biol.*, 9, 91, 1971.
6. Dhar, M.L., Dhar, M.M., Dhawan, B.N., Gupta, B., Srimal, R.C., and Tandon, J.S., Screening of Indian plants for biological activity. IV., *Indian J. Exp. Biol.*, 11, 43, 1973.
7. Dhar, M.L., Dhawan, B.N., Prasad, C.R., Rastogi, R.P., Singh, K.K., and Tandon, J.S., Screening of Indian plants for biological activity. I., *Indian J. Exp. Biol.*, 12, 512, 1974.
8. Dhawan, B.N., Patnaik, G.K., Rastogi, R.P., Singh, K.K., and Tandon, J.S., Screening of Indian plants for biological activity. VI., *Indian J. Exp. Biol.*, 15, 208, 1977.
9. Dhawan, B.N., Dubey, M.P., Mehrotra, B.N., Rastogi, R.P., and Tandon, J.S., Screening of Indian Plants for biological activity. IX., *Indian J.Exp. Biol.*, 18, 594, 1980.
10. Rastogi, R.P. and Dhawan, B.N., Research on medicinal plants at Central Drug Research Institute Lucknow (India), *Indian J. Med. Res.*, 76 (Suppl.), 27, 1982.
11. Aswal, B.S., Bhakuni, D.S., Geol, A.K., Kar, K., Mehrotra, B.N., and Mukherjee, K.C., Screening of Indian plants for biological activity. X., *Indian J. Exp. Biol.*, 22, 312, 1984.

12. Aswal, B.S., Bhakuni, D.S., Goel, A.K., Kar, K., and Mehrotra, B.N., Screening of Indian plants for biological activity. XI., *Indian J. Exp. Biol.*, 22, 487, 1984.
13. Babbar, O.P., Joshi, M.N., and Madan, A.R., Evaluation of plants for antiviral activity, *Indian J. Med. Res.*, 76(Suppl.), 54, 1982.
14. Setty, B.S., Kamboj, V.P., and Khanna, N.M., Screening of Indian plants for biological activity. VII., *Indian J. Exp. Biol.*, 15, 231, 1977.
15. Kamboj, V.P. and Dhawan, B.N., Research on plants for fertility regulation in India,*J. Ethnopharmacol.*, 6, 191, 1982.
16. Choudhury, R.R., Haq, M., and Gupta, U., Review of plants screened for antifertility activity. I-IV., *Bull. Med. Ethnobot. Res.*, 1, 408, 419, 542, 546, 1980.
17. Mehra, P.N., Bhatnagar, J.K., and Handa, S.W., Researches in pharmacognosy in India. III & IV., *Res. Bull. (N.S.) Punjab Univ., India*, 20, 261, 1969.
18. Puri, H.S., Anatomical studies on Indian medicinal plants, *Herba Hung.*, 10, 47, 1971.
19. Mitra, R., Pharmacognostical studies on some indigenous drugs, in *Progress in Plant Research*, Vol. 1, Khoshoo, T.N. and Nair, P.K.K., Eds., Today and Tomorrow's Printers and Publishers, New Delhi, 1979, 239.
20. Anand Kumar, A., Rajendran, V., Tirugnanasabhan, P., Balasuramanium, M., and Murlidharan, R., Ativisha - its botanical identity, *J. Sci. Res. Plant Med.*, 2, 43, 1981.
21. Dey, D. and Das, H.N., Pharmacognostic studies of *Acorus calamus* and its adulterants, *Acta Bot. Indica*, 10, 28, 1982.
22. Kanwal, P., Bhatia, M.S. Thappa, R.K., Agarwal, S.G., Dhar, K.L., and Atal, C.K., Seasonal variation of alkaloids of *Adhatoda vasica* , *Planta Med.*, 48, 81, 1983.
23. Dutta, S.K. and Dutta, P.C., Pharmacognosy of *Alstonia* bark drug, *Int. J. Crude Drug Res.*, 22, 151, 1984.
24. Brindha, P., Rukmani, P., and Purshothaman, K.K., Pharmacognostical studies on *Anisomeles malabarica*, *Bull. Med. Ethnobot. Res.*, 2, 74, 1982.
25. Shome, U., Joshi, P., and Sharma, H.P., Pharmacognostic studies on *Artemisia scoparia*, *Proc. Indian Acad. Sci. (Plant Sci.)*, 93, 151, 1984.
26. Inamdar, P.C. and Mahabale, T.S., Comparison between 'Shatawar' and *Asparagus* species, *Bio. Vigyanam*, 6, 27, 1980.
27. Murty, Y.S. and Varma, S.K., Pharmacognostic studies on *Barleria*, *Bull. Med. Ethnobot. Res.*, 2, 346, 1982.
28. Dennis, T.J., Billore, K.V., and Mishra, K.P., Pharmacognostic study of gum-oleoresin of *Boswellia serrata*, *Bull. Med. Ethnobot. Res.*, 1, 353, 1980.
29. Mehrotra, S. and Mitra, R., Pharmacognostical studies on 'Pariyal' (*Buchanania lanzan*) leaf, *Bull. Med. Ethnobot. Res.*, 1, 338, 1980.
30. Mitra, R. and Mehrotra, S., Pharmacognostical studies on stem bark of 'Pariyal' (*Buchanania lanzan*), *Herba Hung.*, 20, 15, 1981.
31. Mitra, R. and Mehrotra, S., Pharmacognostical studies on 'Pariyal' fruits and seeds (*Buchanania lanzan*, Anacardiaceae), *Bull. Bot. Surv. India*, 22, 68, 1980.
32. Sharma, H.K., Daiya, K.S., and Chawan, D.D., Sennosides and carbohydrates in developing pods of *Cassia* species, *Pharm. Weekbl.*, 4, 83, 1982.
33. Mehrotra, S. and Sharma, H.P., Pharmacognostic studies on 'Sappan' (*Caesalpinia sappan*) and its market samples,*Proc. Indian Acad. Sci. (Plant Sci.)*, 93, 135, 1984.
34. Mehrotra, S., Shome, U., and Sharma, H.P., Pharmacognostic studies on the flowers of *Calophyllum inophyllum*, *Proc. Indian Acad. Sci. (Plant Sci.)*, 93, 643, 1984.
35. Tewari, A., Pharmacognostic Investigations of Some Biologically Active Plants, Ph.D. thesis, Doctor Harisingh Gour University, Sagar, 1984.
36. Dennis, T.J., Yadav, B.B.L., Joshi, J.G., and Mishra, K.P., Pharmacognostic study of gum oleoresin of *Commiphora wightii*, *Bull. Med.-Ethnobot. Res.*, 1, 72, 1980.
37. Singh, P. and Srivastava, G.N., Pharmacognostic study of leaf and stem of *Costus speciosus*, *Bull. Med. Ethnobot. Res.*, 1, 203, 1980.
38. Singh, P., Srivastava, G.N., and Tomy, P., Pharmacognostic study of *Costus speciosus* , *Q. J. Crude Drug Res.*, 19, 103, 1981.
39. Gupta, M.M., Farooqui, S.U., and Lal, R.N., Diosgenin content of *Costus speciosus* at different stages of growth, *Indian Drugs*, 18, 285, 1981.
40. Gupta, S.D. and Dutta, P.C., Pharmacognosy of *Datura metel* leaf, *Bull. Med. Ethnobot. Res.*, 2, 37, 1981.
41. Tomy, P., Srivastava, G.N., and Singh, P., Diagnostic characters of tropical *Dioscorea*: pharmacognostic study of some common Indian *Dioscorea*, *Q. J. Crude Drug Res.*, 18, 169, 1980.
42. Dutta, K. and Dutta, S.K., Pharmacognostic evaluation of Yams *(Dioscorea)*, *J. Econ. Tax. Bot.*, 5, 181, 1984.
43. Sanyal, M., Roy, S.K., and Dutta, P.U., Pharmacognostic evaluation of *Embelia ribes* fruit drug, *J. Econ. Tax. Bot.*, 5, 1235, 1984.
44. Agarwal, V.K., Wahi, S.P., and Wahi, A.K., Pharmacognostical studies on bark of *Millingtonia hortensis*, *Bull. Med. Ethnobot. Res.*, 2, 229, 1981.
45. Issar, R.K., Preliminary comparative chemico-morphological study of the indigenous drug 'Dyanku',*Bull. Med. Ethnobot. Res.*, 1, 289, 1980.
46. Mary, Z., Pattan Shetty, J.K., and Yoganarasimhan, S.N., Pharmacognostical studies on *Nymphoides macrospermum* — (Menyanthaceae) and comparison with *Valeriana jatamansi* (Valarianaceae), *Proc. Indian Acad. Sci. (Plant Sci.)*, 90, 323, 1981.
47. Das Gupta, A. and Dutta, P.C., Medicinal species of *Piper* — pharmacognostic delimitation, *Q. J. Crude Drug Res.*, 18, 17, 1980.
48. Karnic, C.R., Tiwari, K.C., and Mazumdar, R., Cultivation trials, pharmacognosy and ethnobotanical investigations of *Plumbago zeylanica* (Chitraka) of the Indian system of medicine, *Int. J. Crude Drug Res.*, 20, 193, 1982.
49. Lal, J., Folia pharmacognosy of *Portulaca quadrifida*, *Bull. Med. Ethnobot. Res.*, 1, 46, 1980.
50. Lal, J. and Khan, A.M., Pharmacognosy of the stem of *Portulaca quadrifida* and *P. oleracea*, *Proc. Indian Acad. Sci. (Plant Sci.)*, 91, 235, 1982.
51. Karnik, C.R., Ethnobotanical, pharmacognostical and cultivation trial studies on *Psorelea corylifolia*, *Herba Hung.*, 19, 27, 1980.
52. Sharma, H.P. and Shome, U., Leaf surface studies of some medicinal Salvias, *Proc. Indian Acad. Sci. (Plant Sci.)*, 91, 449, 1983.
53. Brindha, P., Sasikala, B., Rao, B.R., and Purshothaman, K.K., Pharmacognostic studies on *Solanum nigrum* — an Ayurvedic drug, *J. Econ. Tax. Bot.*, 3, 125, 1982.
54. Brindha, P., Sasikala, B., Rao, B.R., and Purshothaman, K.K., Pharmacognostic studies on *Solanum trilobatum*, *Bull. Med. Ethnobot. Res.*, 1, 25, 1980.
55. Karnik, C.R., Tewari, K.C., and Mazumdar, R., Ethnobotanical, pharmacognostical and cultivation trial studies on *Tephrosia purpurea* used in Indian system of medicine, *Herba Hung.*, 20, 65, 1981.
56. Issar, R.K., Pharmacognostic study of three parasitic herbs (*Cassytha filiformis*, *Cuscuta reflexa* and *C. chinensis* used differently in traditional systems of medicine, *Bull. Med. Ethnobot. Res.*, 1, 478, 1980.
57. Modal, M.S. and Mondol, M., Pollen morphological studies of some indigenous floral drugs of India (Leguminosae),*J. Econ. Tax. Bot.*, 4, 517, 1983.
58. Sharma, S.C., Shukla, Y.N., and Tandon, J.S., Alkaloids and terpenoids of *Ancistrocladus heyneanus*, *Sagittaria sagitifolia*, *Lyonia formosa* and *Hedychium spicatum*, *Phytochemistry*, 14, 578, 1975.
59. Govindachari, T.R. and Parthasarthy, P.C., Alkaloids of Ancistrocladaceae, *Hetrocycles*, 7, 661, 1977.
60. Morita, Y., Hesse, M., Schmid, H., Banerji, A., Banerji, J., Chatterjee, A., and Oberhanoli, W.E., *Alstonia scholaris*: structure of indole alkaloid narelin, *Helv. Chim. Acta*, 60, 1419, 1977.
61. Bhattacharya, S.K., Bose, R., Dutta, S.C., Ray, A.B., and Guha, S.R., Neuropharmacological studies on strictamine isolated from *Alstonia scholaris*, *Indian J. Exp. Biol.*, 17, 598, 1979.
62. Dutta, S.C., Bhattacharya, S.K., and Ray, A.B., Flower alkaloids of *Alstonia scholaris*, *Planta Med.*, 30, 86, 1976.
63. Chatterjee, A., Mukhopadhyay, S., and Ray, A.B., Alkaloids of *Alstonia* venenata, *J. Sci. Ind. Res.*, 37, 187, 1978.
64. Bhattacharya, S.K., Ray, A.B., and Dutta, S.C., Psychopharmacological investigations of the 4-methoxyindole alkaloids of *Alstonia venenta*, *Planta Med.*, 27, 164, 1975.

65. Mazumdar, P.L., Joardar, S., Chandra, T.K., Dinda, B.H., Banerji, M., Ray, A.B., Chatterjee, A., and Das, B.C., Structures and absolute stereochemistry of (–) echitoveniline, (–)-11-methoxyechitoveniline and (–)-11-methoxyechitovenedine — new indole alkaloids of *Alstonia venenata*, *Tetrahedron*, 35, 1151, 1979.
66. Mazumdar, P.L. and Basu, A., Alstolenine, 19,20-dihydropolyneuridine & other minor alkaloids of the leaves of *Alstonia venenata*, *Phytochemistry*, 21, 2389, 1982.
67. Chatterjee, A., Roy, D.J., and Mukhopadhyay, A., 16-Epivenanatine and 16-epialstovenine, new stereomers from *Alstonia venenata*, *Phytochemistry*, 20, p. 1981, 1981.
68. Banerji, J., Chatterjee, A., Roy, D.J., and Shoolery, J.N., 5-Methoxy-l-oxo-tetrahydro-β-carboline, an alkaloid from *Alstonia venenata*, *Phytochemistry*, 21, 2765, 1982.
69. Jain, M.P., Koul, S.K., Dhar, M.L., and Atal, C.K., Novel nor-harmal alkaloid from *Adhatoda vasica*, *Phytochemistry*, 19, 1880, 1980; Jain, M.P. and Sharma, V.K., *Planta Med.*, 46, 250, 1982; Dhar K.L., Jain, M.P., Koul, S.K., and Atal, C.K., *Phytochemistry*, 20, 319, 1981.
70. Pakrashi, S.C., Mukhopadhyay, R., Ghos, Dastidar, P.P., Bhattacharya, A., and Ali, E., Bharatamine — a unique protoberberine alkaloid from *Alangium lamarckii*, *Tetrahedron Lett.*, 24, 291, 1983.
71. Shoeb, A., Raj, K., Kapil, R.S., and Popli, S.P., Alangiside, the monoterpenoid alkaloidal glycoside from *Alangium lamarckii*, *J. Chem. Soc. Perkin Trans. I*, 1245, 1975.
72. Pandey, H. and Bhakuni, D.S., New dibenz-(d,f)-azonine alkaloids from *Cocculus laurifolius*, *J. Chem. Soc. Perkin Trans. I*, 2197, 1976.
73. Singh, A.N., Pandey, H., and Bhakuni, D.S., cocculitine, a new abnormal erythrina alkaloid from *Cocculus laurifolius*, *Lloydia*, 40, 322, 1977.
74. Bhakuni, D.S. and Jain, S., Alkaloids of *Cocculus laurifolius*, *Tetrahedron*, 36, 3107, 1980.
75. Kar, K., Mukherjee, K.C., and Dhawan, B.N., Characterisation of the neuromuscular blocking property of isococculidine, *Indian J. Exp. Biol.*, 15, 547,1977.
76. Mukherjee, K.C., Patnaik, G.K., Bhakuni, D.S., and Dhawan. B.N., Mechanism of neuromuscular blocking action of isocorydine methochloride, a new quaternary alkaloid of *Cocculus laurifolius* *Indian J. Exp. Biol.*, 22, 54, 1984.
77. Bhakuni, D.S. and Joshi, P.P., Alkaloids of *Cocculus pendulus*, *Tetrahedron*, 31, 2575, 1975.
78. Ghosal, S., Saini, K.S., and Arora, V.K., Latsoline, a novel glucoalkaloid from *Crinum latifolium*, *J. Chem. Res.*, (S), 238, 1983.
79. Ghosal, S., Saini, K.S., and Frahm, A.W., Alkaloids of *Crinum latifolium*, *Phytochemistry*, 22, 2305, 1983.
80. Ghosal, S., Saini, K.S., and Arora, V.K., 1,2-β-Epoxyambelline, an immuno-stimulant alkaloid from *Crinum latifolium*, *J. Chem. Res.*, (S), 232, 1984.
81. Bhakuni, D.S., Satish, S., and Dhar, M.L., the alkaloids of *Croton sparsiflorus*, *Phytochemistry*, 9, 2573, 1970.
82. Bhakuni, D.S., Satish, S., and Dhar, M.L., Absolute configuration of crotsparine, crotsparinine and sparsiflorine, *Tetrahedron*, 28, 4579, 1972.
83. Bhakuni, D.S., Satish, S., Upreti, H., and Kapil, R.S., Biosynthesis of crotsparine, crotsparinine and sparsiflorine, *Phytochemistry*, 13, 2767, 1974.
84. Dubey, M.P., Srimal, R.C., and Dhawan, B.N., Analysis of hypotensive activity of N-methylcrotsparine, an alkaloid from *Croton sparsiflorus*, *Indian J. Pharmacol.*, 3, 18, 1971.
85. Dubey, M.P., Srimal R.C., and Dhawan, B.N., Pharmacological studies on aporphine alkaloids from *Croton sparsiflorus*, *Indian J. Pharmacol.*, 9, 77, 1977.
86. Bhatanagar, A.K., Bhattacharji, S., Roy, A.C., Popli, S.P., and Dhar, M.L., Chemical examination of roots of *Cissampelos pareira*. IV Structure and stereochemistry of hayatin, *J. Org. Chem.*, 32, 819, 1967.
87. Patnaik, G.K., Pradhan, S.N. and Vohra, M.M., Effects of hayatin methochloride and (±)-Tubocurarine chloride on autonomic ganglia of cats, *Indian J. Exp. Biol.*, 11, 89, 1973.
88. Mahato, S.B., Sahu, N.P., and Luger, P., Structure of Caesalpinine A: a novel spermidine alkaloid of *Caesalpinia digyna*, *J. Am. Chem. Soc.*, 105, 4441, 1983.
89. Rao, P.G., Sawhney, R.S., and Atal, C.K., Cromaduline, a new pyrrolizidine alkaloid from *Crotalaria madurensis*, *Indian J. Chem.*, 13, 835, 870, 1975.
90. Singh, G.B., Sharma, M.L., and Ray Ghatak, B.J., Pharmacological investigations of usaramine, an alkaloid isolated from *Crotolaria brevifolia*, *Indian J. Exp. Biol.*, 7, 144, 1969.
91. Sharma, M.L. and Ray Ghatak, B.J., Pharmacological properties of integerrimine, an alkaloid isolated from *Crotolaria brevifolia*, *Indian J. Exp. Biol.*, 9, 177, 1971.
92. Sharma, M.L., Singh, G.B., and Ray Ghatak, B.J., Pharmacological investigations on *Crotolaria agatiflora*, *Indian J. Exp. Biol.*, 5, 149, 1967.
93. Gupta, O.P., Singh, G.B., Ray Ghatak, B.J., and Atal, C.K., Hypotensive activity of semisynthetic pyrrolizidine amides, *Indian J. Exp. Biol*, 14, 282, 1976.
94. Chakraborty, D.P. and Das, K.C., Structure of murrayacine, *Chem. Commun.*, 967, 1968.
95. Narasimhan, N.S., Paradkar, M.V., and Chitguppi, V.P., Structure of mahanimbin and koenimbin, *Tetrahedron Lett.*, 5501, 1968.
96. Dutta, N.L. and Quasim, C., Constituents of *Murraya koenigii;* structure of girinimbin, *Indian J. Chem.*, 7, 307, 1969.
97. Charkraborty, D.P., Das, K.C., and Chowdhury, B.K., Structure of murrayacine, *J. Org. Chem.*, 36, 725, 1971.
98. Joshi, B.S., Kamat, V.N., and Gawad, D.H., On the structures of girinimbine, mahanimbine, isomahanimbine, koenimbidine and murrayacine, *Tetrahedron*, 26, 1475, 1970.
99. Narasimha, N.S., Paradkar, M.N., and Kelkar, S.L., Structure of mahanine, koenine, koenignine and koenidine, *Indian J. Chem.*, 8, 473, 1970.
100. Roy, S., Bhattacharyya, L., and Chakraborty, D.P., Structure and synthesis of mukoline and mukolidine — two new carbazole alkaloids from *Murraya koenigii*, *J. Indian Chem. Soc.*, 59, 1369, 1982.
101. Mukherjee, M., Mukherjee, S., Shaw, A.K., and Ganguly, S.N., Muconicine, a carbazole alkaloid from leaves of *Murraya koenigii*, *Phytochemistry*, 22, 2328, 1983.
102. Bandaranayake, W.M., Begley, M.J., Brown, B.O., Clarke, D.G., Crombie, L., and Whiting, D.A., Crystal and molecular structure of dibromocannibicyclol and its bearing on the 'cyclol' alkaloids, *J. Chem. Soc. Perkin Trans. I*, 998, 1974.
103. Bordner, J., Chakraborty, D.P., Chowdhury, B.K., Ganguly, S.N., Das, K.C., and Weinstein, B., The crystal structure of murrayazoline, *Experientia*, 28, 1406, 1972.
104. Kureel, S.P., Kapil, R.S., and Popli, S.P., Terpenoid alkaloids of *Murraya koenigii*. II. Constitution of cyclomahanimbine, bicyclomahanimbine and mahanimbidine, *Tetrahedron Lett.*, 3857, 1969.
105. Begley, M.J., Clarke, D.G., Crombi, L., and Whiting, A., The X-ray structure of dibromocannabicyclol: structure of bicyclomahanimbine, *Chem. Commun.*, 1547, 1970.
106. Bhattacharya, L., Roy, S.K., and Chakraborty, D.P., Structure of the carbazole alkaloid isomurrayazoline from *Murraya koenigii*, *Phytochemistry*, 21, 2432, 1982.
107. Chakraborty, D.P., Carbazole alkaloids, in *Progress in the Chemistry of Organic Natural Products*, Vol. 34, Zechmeister, L., Ed., Springer-Verlag, Berlin, 1977, 299.
108. Chatterjee, A., Dhara, K.P., and Banerji, J., Alkaloids of *Mitragyana* parvifolia and their transformations, *J. Indian Chem. Soc.*, 59, 1360, 1982.
109. Agarwal, J.S. and Rastogi, R.P., Chemical constituents of *Mappia foetida*, *Indian J. Chem.*, 11, 969, 1973.
110. Danieli, B. and Palmisano, G., Total synthesis of camptothecin, *Fitoterapia*, 45, 87, 1974.
111. Govindachari, T.R., Ravindranath, K.R., and Viswanathan, N., Mappicine, a minor alkaloid from *Mappia foetida*, *J. Chem. Soc. Perkin Trans. I*, 1215, 1974.
112. Govindachari, T.R. and Visawanathan, N., 9-Methoxycamptothecin — a minor alkaloid from *Mappia foetida*, 10, 453, 1972.
113. Atal, C.K., Dhar, K.L., and Singh, J., The chemistry of Indian *Piper* species, *Lloydia*, 38, 256, 1975.
114. Joshi, B.S., Viswanathan, N., and Gawad, D.H., Piperaceae alkaloids. I. Structure of piperstachine, *Helv. Chim. Acta*, 58, 1551, 1975.

115. Viswanathan, N., Balakirshnan, V., Joshi, B.S., and Von Philipsborn, W., Synthesis of Piperstachine, *Helv. Chim. Acta,* 58, 2026, 1975.
116. Joshi, B.S., Viswanathan, N.,Gawad, D.H., Balakirshnan, V., Von Philipsborn, W., and Quick, A., Structure and synthesis of cyclostachine A, a novel alkaloid from *Piper trichostachyon, Experientia,* 31, 880, 1975.
117. Banerji, A., Banerji, J., Chatterjee, A., and Shoolery, J.N., Structure and stereochemistry of an alkamide from *Piper sylvaticum, Indian J. Chem.,* 19B, 346, 1980.
118. Banerje, A. and Pal, S.C., A new alkamide from *Piper sylvaticum, Phytochemistry,* 21, 1321, 1982.
119. Banerji, A. and Ray, R., Aurantiamides: a new class of modified dipeptides from *Piper aurantiacum, Phytochemistry,* 20, 2217, 1981.
120. Dahanukar, S.A. and Karandikar, S.M., Evaluation of antiallergic activity of *Piper longum, Indian Drugs,* 21, 377, 1984.
121. Dahanukar, S.A., Karandikar, S.M., and Desai, M.P., Clinical evaluation of *Piper longum* in childhood asthma, in Abstr. II World Conf. Clin. Pharmacol. Ther., Washington, 1983, 59.
122. Chandoke, N., Gupta, S., and Dhar, S., Interceptive activity of various species of *Piper*, their natural amides and semisynthetic analogs, *Indian J. Pharm. Sci.,* 48, 113, 1978.
123. Atal, C.K., Dubey, R.K., and Singh, J., Biochemical basis of enhanced drug bioavailability by piperine: evidence that piperine is a potent inhibitor of drug metabolism, *J. Pharmacol. Exp. Ther.,* 232, 258, 1985.
124. Atal, C.K., Zutshi, U., and Rao, P.G., Scientific evidence on the role of Ayurvedic herbals on bioavailability of drugs, *J. Ethnopharmacol.,* 4, 229, 1981.
125. Sahai, M. and Ray, A.B., Secotropane alkaloids of *Physalis peruviana, J. Org. Chem.,* 45, 3265, 1980.
126. Anjaneyulu, B., Govindachari, T.R., Sathe, S.S., Viswanthan, N., Gopinath, K.W., and Pai, B.R., Alkaloids of *Tiliacora racemosa, Tetrahedron,* 25, 3091, 1969.
127. Shamma, M., Foy, J.E., Govindachari, T.R., and Viswanathan, N., The position of phenolic function in tiliacorine and related alkaloids, *J. Org. Chem.,* 41, 1293, 1976.
128. Bhakuni, D.S., Singh, A.N., Jain, S., and Kapil, R.S., Absolute configuration and biosynthesis of tiliacorine and tiliacorinine, *Chem. Commun.,* 226, 1978.
129. Bhakuni, D.S., Singh, A.N., and Jain, S., The structure, absolute configuration and biosynthesis of nortiliacorinine A, *Tetrahedron,* 37, 2651, 1981.
130. Govindachari, T.R., Rajagopalan, T.G., and Viswanathan, N., Absolute configuration of tylophorine, *J. Chem. Soc., Perkin Trans. I,* 1161, 1974.
131. Govindachari, T.R. and Viswanathan, N., Stereochemistry of tylophorinidine and tylophorinine, *Indian J. Chem.,* 12, 886, 1974; Wadhawan, V.K., Sinha, S.K. and Mulchandani, N.B., Minor alkaloids of *Tylophora asthmatica*, X-ray analysis of tylophorinidine, *Tetrahedron Lett.,* 5091, 1973.
132. Mulchandani, N.B. and Venkatachalam, S.R., Tylophorinicine — a phenanthroindolizidine alkaloid from *Tylophora asthmatica* and *Pergularia pallida, Phytochemistry,* 23, 1206, 1984.
133. Shivpuri, D.N., Studies on a new anti-asthmatic *Tylophora indica* on bronchial asthma, *Aspects Allergy Appl. Immunol.,* 8, 61, 1975; Gore, K.V., Rao, K.A., and Guruswamy, M.M., Physiological studies with *Tylophora asthmatica (T. indica)* in bronchial asthma, *Indian J. Med. Res.,* 71, 144, 1980.
134. Hartwell, J.L. and Abbot, B.J., Antineoplastic principles in plants: recent developments in the field, in *Advances in Pharmacology and Chemotherapy,* Vol. 7, Academic Press, New York, 1969, 117.
135. Bhakuni, D.S. and Singh, R.S., The alkaloids of *Thalictrum foliolosum, J. Nat. Prod.,* 45, 252, 1982.
136. Agwada, V.C., Naranjo, J., Hesse, M., Schmid, H., Rolland, Y., Kunescth, N., Poisson, J., and Chatterjee, A., The structure of bisindole alkaloid amatain, *Helv. Chim. Acta,* 60, 2830, 1977.
137. Gupta, B.D., Banerjee, S.K., Handa, K.L., and Atal, C.K., Heratomin and heratomol, new coumarins from *Heracleum thomsoni, Phytochemistry,* 15, 1319, 1976; Kumar, R., Gupta, B.D., Banerjee, S.K., and Atal, C.K., New coumarins from *Seseli sibiricum, Phytochemistry,* 17, 2111, 1978.
138. Kartha, G., Ramachandran, G.N., Bhat, N.B., Nair, M.P., Raghavan, V.K.V., and Venkataraman, K., The constitution of morellin, *Tetrahedron Lett.,* 459, 1963.
139. Sen, A.K., Sarkar, K.K., Mazumdor, P.C., Banerji, N., Uusvuori, R., and Hase, T.A., The structure of garcinones A, B & C — three new xanthones from *Garcinia mangostana, Phytochemistry,* 21, 1747, 1982.
140. Pai, B.R., Natarajan, S., Sugira, H., Kameswaran, L., and Shankaranarayan, D., Synthesis and pharmacology of monogostin — 3,6-di-O-glucoside, *J. Nat. Prod.,* 42, 361, 1979.
141. Shankarnarayan, D., Gopalkrishnan, C., and Kameswaran, L., Pharmacology of mangiferin, *Indian J. Pharm. Sci.* 41, 78, 1979.
142. Gopalkrishnan, C., Shankarnarayanan, D., Kameshwaran, L., and Nazimuddin, S.K., Effect of mangostin, A xanthone from *Garcinia mangostana*, in immunopharmacological and inflammatory reactions, *Indian J. Exp. Biol.,* 18, 843, 1980.
143. Karanjgaokar, C.G., Rama Rao, A.V., Venkataraman, K., and Palmer, K.J., The constitution of xanthochymol and isoxanthochymol, *Tetrahedron Lett.,* 4977, 1973.
144. Venkatswamy, G., Yemul, S.S., Rama Rao, A.V., and Palmar, K.J., Absolute configuration of xanthochymol & isoxanthochymol, *Indian J. Chem.,* 13, 1355, 1975.
145. Rama Rao, A.V., Venkatswamy, G., and Yemul, S.S., Xanthochymol & isoxanthochymol, two novel polyisoprenylated benzophenones from *Garcinia xanthochymus, Indian J. Chem.,* 18B, 627, 1980.
146. Krishnamurthy, N., Lewis, Y.S., and Ravindranath, B., On structures of garcinol, isogarcinol & camboginol, *Tetrahedron Lett.,* 22, 793, 1981.
147. Rama Rao, A.V., Venkatswamy, G., and Pendse, A.D., Camboginol and cambogin, *Tetrahedron Lett.,* 21, 1975, 1980.
148. Rizvi, S.H., Shoeb, A., Kapil, R.S., and Popli, S.P., Spasmolytic coumarins from *Boenninghausenia albiflora, Indian J. Pharm. Sci.,* 41, 205, 1979.
149. Govindachari, T.R. and Parthasarathy, P.C., Cryptocaryalactone, a novel 5,6-dihydro-2H-pyran-2-one from *Cryptocarya bourdilloni, Tetrahedron Lett.,* 3401, 1971.
150. Govindachari, T.R., Parthasarathy, P.C., and Modi, J.D., Cryptocaryalactone, a new kawa-type lactone, *Indian J. Chem.,* 10, 149, 1972.
151. Govindachari, T.R., Parthasarathy, P.C. Desai, H.K., and Shanbhag, M.N., Structure of cryptocaryone, a constituent of *Cryptocarya bourdilloni, Tetrahedron,* 29, 3091, 1973.
152. Subba Rao, G.S.R., Raj, K., and Kumar, V.P.S., Chemical Examination of *Clausena willdenovii*, isolation of 3-(1,1-dimethylallyl) xanthyletin from root and bark, *Indian J.Chem.,* 20B, 88, 1981.
153. Anwar, F., Shoeb, A., Kapil, R.S., and Popli, S.P., Clausrin, a novel coumarin from *Clausena pentaphylla, Experientia,* 33, 412, 1977.
154. Patnaik, G.K., Shoeb, A., Manandhar, M.D., Kapil, R.S., Popli, S.P., and Dhawan, B.N., Spasmolytic activity of clausmarin, a new coumarin from *Clausena pentaphylla, Indian J. Pharmacol.,* 10, 82, 1978.
155. Shoeb, A., Manandhar, M.D., Kapil, R.S., and Popli, S.P., Clausmarins A and B: two novel spasmolytic terpenoid coumarins from *Clausena pentaphylla, Chem. Commun.,* 281, 1978.
156. Ray, A.B., Chattopadhyay, S.K., Konno, C., and Hikino, H., Structure of cleomiscosin B, a coumarino-lignoid of *Cleome viscosa* seeds, *Heterocycles,* 19, 19, 1982 .
157. Bhandari, P., Pant, P., and Rastogi, R.P., Aquillochin, a coumarinolignan from *Aquilaria agallocha, Phytochemistry,* 21, 2147, 1982.
158. Bhandari, P. and Rastogi, R.P., A novel type of bicoumarin rhamnoside from *Lasiosiphon eriocephalus, Phytochemistry,* 20, 2044, 1981.
159. Joshi, B.S., Kamat, V.N., Govindachari, T.R., and Ganguly, A.K., Surangin A and B, two new coumarins from *Mammea longifolia, Tetrahedron,* 25, 1453, 1969.
160. Chakraborty, D.P., Roy, S., Chakraborty, A., Mandal, A.K., and Choudhury, B.K., Structure and synthesis of mexolide, a new antibiotic dicoumarin from *Murraya exotica, Tetrahedron,* 36, 3563, 1980.
161. Chandhoke, N., Gupta, S. Daftari, P., Dhar, S.K., and Atal, C.K., Effect of *Pueraria tuberosa* on the reproductive performance in rats, *Indian J. Pharmacol.,* 12, 57, 1980.
162. Joshi, B.S. and Kamat, V.N., Tuberosin, a new Pterocarpan from *Pueraria tuberosa, J. Chem. Soc. Perkin Trans. I,* 907, 1973.

163. Sharma, P.N., Shoeb, A., Kapil, R.S., and Popli, S.P., Toddasin, a new dimeric coumarin from *Toddalia asiatica*, *Phytochemistry*, 20, 335, 1981.
165. Tandon, S. and Rastogi, R.P., Wikstrosin, a tricoumarin from *Wikstroemia viridiflora*, *Phytochemistry*, 16, 1991, 1977.
166. Rama Rao, A.V., Varadan, M., and Venkataraman, K., Cyclohetero-phyllin, a flavone linked to three isoprenoid groups from the wood of *Artocarpus heterophyllus*, *Indian J. Chem.*, 9, 7, 1971.
167. Rama Rao, A.V., Rathi, S.S., and Venkataraman, K., Chaplashin, a flavone containing an oxepine ring from *Artocarpus chaplasha*, *Indian J. Chem.*, 10, 905, 1972.
168. Ranganathan, R.M., Nagarajan, S., Mabry, T.J., Yong Long, L., and Newman, P., 6-Hydroxyluteolin 7-O-apioside from *Lepidagathis cristata*, *Phytochemistry*, 19, 2505, 1980.
169. Subba Raju, K.V., Srimannarayana, G., Ternai, B., Stanley, R., and Markham, K.R., Structure of milletin — a novel isoflavone isolated from *Millettia auriculata*, *Tetrahedron*, 37, 957, 1981.
170. Tripathi, V.D., Agarwal, S.K., and Rastogi, R.P., Chemical constituents of *Rhamnus nipalensis*, *Indian J. Chem.*, 17B, 89, 1979.
171. Sharma, P.N., Shoeb, A., Kapil, R.S., and Popli, S.P., Arjunolone, a new flavone from stem bark of *Terminalia arjuna*, *Indian J. Chem.*, 21B, 263, 1982.
172. Sarin, J.P.S., Singh, S., Khanna, N.M. Garg, H.S., and Dhar, M.M., Tephdidoside, a new anticancer flavonol glycoside from *Tephrosia candida*, *Phytochemistry*, 15, 232, 1976.
173. Sahai, R., Agarwal, S.K., and Rastogi, R.P., Auriculoside, a new flavan glycoside from *Acacia auriculiformis*, *Phytochemistry*, 19, 1560, 1980.
174. Ghosal, S., Saini, K.S., and Sinha, B.N., Diffutin, a new adaptogenic glycosyloxyflavan from *Canscora diffusa*, *J. Chem. Res.*, (S), 330, 1983.
175. Maheshwari, M.L., Jain, T.C., Bates, R.B., and Bhattacharyya, S.C., Structure and absolute configuration of alpha and beta-agarofurans and dihydroagarofuran, *Tetrahedron*, 19, 1079, 1963.
176. Pant, P. and Rastogi, R.P., Agarol, a new sesquiterpene from *Aquilaria agallocha*, *Phytochemistry*, 19, 1869, 1980.
177. Bisarya, S.C. and Dev. S., Studies in sesquiterpenes. XXXIII. Himachalol, *Tetrahedron*, 24, 3861, 1968.
178. Bisarya, S.C. and Dev., S., Studies in sesquiterpenes. XXXIV. Structure of allohimachalol, *Tetrahedron*, 24, 3869, 1968.
179. Bajaj, A.G., Dev, S., Tagle, B. Tesler, J., and Clardy, J., The stereochemistry of allohimachalol, *Tetrahedron Lett.*, 325, 1980.
180. Puri, V.N., Kar, K., Patnaik, G.K., Dhawan, B.N., Kulshreshtha, D.K., and Rastogi, R.P., Spasmolytic constituents of *Cedrus deodara*. II. Isolation and evaluation of allhimachalol and three new sesquiterpenoid alcohols, *Indian J. Exp. Biol.*, 13, 369, 1975.
181. Kulshreshtha, D.K. and Rastogi, R.P., Constitution of the sesquiterpenoid, centdarol, *Phytochemistry*, 14, 2237, 1975; Idem., The constitution of isocentdarol, a sesquiterpene from *Cedrus deodara*, *Phytochemistry*, 15, 557, 1976.
182. Agarwal, P.K. and Rastogi, R.P., Terpenoids from *Cedrus deodara*, *Phytochemistry*, 20, 1319, 1981.
183. Kar, K., Puri, V.N., Patnaik, G.K., Suri, R.N., Dhawan, B.N., Kulshreshtha, D.K., and Rastogi, R.P., Spasmolytic constituents of *Cedrus deodara*, pharmacological evaluation of himachalol, *J. Pharmacol. Sci.*, 64, 258, 1975.
184. Patnaik, G.K., Puri, V.N., Kar, K., and Dhawan, B.N., Spasmolytic activity of sesquiterpenes from *Cedrus deodara*, *IDMA Bull.*, 18, 238, 1976.
185. Tripathi, V.D., Agarwal, S.K., Srivastava, O.P., and Rastogi, R.P., Antidermatophytic constituents from *Inula racemosa*, *Indian J. Pharm. Sci.*, 40, 129, 1978.
186. Joshi, B.S. and Kamat, V.N., Structure of enhydrin a germacranolide from *Enhydra fluctuans*, *Indian J. Chem.*, 10, 771, 1972.
187. Kartha, G., Go, K.T., and Joshi, B.S., X-ray structure and configuration of enhydrin bromohydrin, *Chem. Commun.*, 1327, 1972.
188. Mathur, S.B., Hiremath, S.B., Kulkarni, G.H., Kelkar, G.R., and Bhattacharyya, S.C., Structure of dehydrocostus lactone, *Tetrahedron*, 21, 3575, 1965.
189. Joshi, R.S., Kulkarni, G.H., Kelkar, G.R., and Bhattafcharyya, S.C., Transformation products of costunolide, *Tetrahedron*, 22, 2331, 1966; Hiremath, S.V., Kulkarni, G.H., Kelkar, G.R., and Bhattacharyya, S.C., Transformation products of costunolide: thermal rearrangement of cyclodeca-1, 6-dienes to a guanine system, *Indian J. Chem.*, 6, 243, 1968.
190. Pal, R., Kulshreshtha, D.K., and Rastogi, R.P., Antileukemic and other constituents of *Tithonia tagitiflora*, *J. Pharm. Sci.*, 65, 918, 1976.
191. Pal, R., Kulshreshtha, D.K., and Rastogi, R.P., Chemical constituents of *Tithonia tagitiflora*, Tagitinins C,D,E, & F, *Indian J. Chem.*, 15B, 208, 533, 1977; Chowdhury, P.K., Barua, N.C., Sharma, R.P., Thyagarajan, G., and Herz, W., Structure of deacetyl viguiestenin (tagitinin E), an addendum, *J. Org. Chem.*, 45, 535, 1980.
192. Damodaran, N.P. and Dev, S., Sesquiterpenoids from essential oil of *Zingiber zerumbet*, *Tetrahedron*, 24, 4113, 1968.
193. Nayak, U.R. and Dev, S., Longicyclene, the first tetracyclic sesquiterpene, *Tetrahedron*, 24, 4099, 1968.
194. Govindachari, T.R., Mohamad, P.A., and Parthasarthy, P.C., Ishwarane and aristolochene, two new sesquiterpene hydrocarbons from *Aristolochia indica*, *Tetrahedron*, 26, 615, 1970.
195. Fuhrer, H., Ganguly, A.K., Gopinath, K.W., Govindachari, T.R., Nagarajan, K., Pai, B.R., and Parthasarathy, P.C., Ishwarone, *Tetrahedron*, 26, 2371, 1970.
196. Tandon, J.S., Dhar, M.M., Ramakumar, S., and Venkatesan, K., Structure of coleonol, a biologically active diterpene from *Coleus forskohlii*, *Indian J. Chem.*, 15B, 880, 1977.
197. Jauhairi, P.K., Katti, S.B., Tandon, T.S., and Dhar, M.M., Coleosol, a new diterpene from *Coleus forskohlii*, *Indian J. Chem.*, 16B, 1055, 1978.
198. Tandon, J.S., Jauhara, P.K., Singh, R.S., and Dhar, M.M., Structure of three new diterpenes, Coleonol B., Coleonol C and deoxycoleonol, isolated from *Coleus forskohlii*, *Indian J. Chem.*, 16B, 341, 1978.
199. Katti, S.B., Jauhari, P.K., and Tandon, J.S., New diterpenes from *Coleus forskohlii*: structure of coleonol-D, Coleol and coleonone, *Indian J. Chem.*, 17B, 321, 1979.
200. Painuly, P., Katti, S.B., and Tandon, J.S., Diterpenes from *Coleus forskohlii*: structures of coleonol-E and coleonol-F, *Indian J. Chem.*, 18B, 241, 1979.
201. Mahato, S.B., Pal, B.C., Kawasaki, T., Miyahara, K., Tanaka, O., and Yamasaki, K.C., Structure of cleomeolide, a novel diterpene lactone from *Cleome icosandra*, *J. Am. Chem. Soc.*, 101, 4720, 1979.
202. Joshi, B.S. and Rane, D.F., The structure of diosbulbine, *Indian J. Chem.*, 7, 452, 1969.
203. Sahai, R., Rastogi, R.P., Jakupovic, J., and Bohlmann, F., A diterpene from *Euphorbia maddeni*, *Phytochemistry*, 20, 1665, 1981.
204. Prakash, O., Bhakuni, D.S., Kapil, R.S., Subba Rao, G.S.R., and Ravindranath, B., Diterpenoids of *Roylea calycina*, *J. Chem. Soc. Perkin Trans. I*, 1305, 1979.
205. Khan, H., Zaman, A., Chetty, G.L., Gupta, A.S., and Dev, S., Cheilanthatriol — a new fundamental type in sesterpenes, *Tetrahedron Lett.*, 4443, 1971.
206. Row, L.R., Murty, P.S., Subba Rao, G.S.R., Sastry, C.S.P., and Rao, K.V.J., Isolation and structure determination of arjunic acid, a new trihydroxy triterpene carboxylic acid from *Terminalia arjuna* bark, *Indian J. Chem.*, 8, 716, 1970.
207. Anjaneyulu, A.S.R., Row, L.R., and Sree, A., Acacidiol, a new nortriterpene from sapogenins of *Acacia concinna*, *Phytochemistry*, 18, 1199, 1979 .
208. Anjaneyulu, A.S.R. and Rao, M.N., Elaeodendrol and elaeodendradiol, new nortriterpenes from *Elaeodendron glaucum*, *Phytochemistry*, 19, 1163, 1980.
209. Chakrabarti, P., Structure and sterochemistry of mollugogenol A, a new sapogenine from *Mollugo hirta*, *Tetrahedron*, 25, 2082, 1984.
210. Barua, A.K., Chakrabarti, P., Sanyal, P.K., Basu, K., and Nag, K., The structure of Lantic acid from *Lantana camara*, *J. Indian Chem. Soc.*, 49, 1063, 1972 .
211. Bhandari, P. and Rastogi, R.P., Two nortriterpene lactones from *Caltha palustris*, *Phytochemistry*, 23, 1699, 1984; Idem., Triterpene constituents of *Caltha palustris*, *ibid.*, 23, 2082, 1984.
212. Joshi, B.S., Kamat, V.N., and Viswanathan, N., Triterpenes of *Salacia prinoides*, *Tetrahedron*, 29, 1365, 1973.

213. Tewari, N.C., Ayengar, K.N.N., and Rangaswami, S., Triterpenoids of the root bark of *Salacia prinoides*, *J. Chem. Soc. Perkin Trans. I*, 146, 1974.
214. Rogers, D., Williams, D.J., Joshi, B.S., Kamat, V.N., and Viswanathan, N., Structure of a new triterpene ether from *Salacia prinoides*, X-ray investigation of the dibromo derivative, *Tetrahedron Lett.*, 63, 1974.
215. Govindachari, T.R., Viswanathan, N., and Mohamed, P.A., Structure of Litsomentol, a new tetracyclic triterpene, *Tetrahedron*, 27, 4991, 1971.
216. Pachapurkar, R.V., Kornule, P.M., and Narayanan, C.R., *Chem. Lett.*, 357, 1974.
217. Pillai, N.R. and Santha Kumari, G., Antiarthritic and antiinflammatory actions of nimbidin, *Planta Med.*, 43, 59, 1981.
218. Sinha, K.C., Riar, S.S., Tiwari, R.S., Dhawan, A.K., Bardhhan, J., Thomas, P., Jain, A.K., and Jain, R.K., Neem oil as a vaginal contraceptive, *Indian J. Med. Res.*, 79, 131, 1984.
219. Saha, P.K., Mukherjee, S., Shaw, A.K., Ganguly, T., and Ganguly, S.N., Trewinine, a new cucurbitacin from *Trewia nudiflora*, *Fitoterpia*, 52, 231, 1981.
220. Singh, S., Garg, H.S., and Khana, N.M., Dyosobinin, a new tetranortriterpene from *Dysoxylum binectariferum*, *Phytochemistry*, 15, 2001, 1976.
221. Chawla, A. and Dev, S., A new class of triterpenes from *Ailanthus malabarica*, derivatives of malabaricane, *Tetrahedron Lett.*, 4837, 1967.
222. Sobti, R.R. and Dev. S., A direct correlation of (+)-malabaricol with (+)-ambreinolide, *Tetrahedron Lett.*, 2215, 1968.
223. Jain, G.K., Sarin, J.P.S., and Khanna, N.M., Constitution of schefferoside — a spermicidal saponin from *Schefflera capitata*, *Indian J. Chem.*, 15B, 1139, 1977.
224. Jain, G.K., Pal, R., and Khanna, N.M., Spermicidal saponins from *Pittosporum nilghrense*, *Indian J. Pharm. Sci.*, 42, 12, 1980.
225. Tandon, R., Jain, G.K. Pal, R., and Khanna, N.M., Chemical examination of *Polemonium coeruleum*, *Indian J. Chem.*, 20B, 46, 1981.
226. Srivastava, S.K., Stigma-5,24(28)-diene-3Beta-O-Alpha-L-rhamnoside from *Cleome viscosa*, *Phytochemistry*, 19, 2510, 1980.
227. Bhandari, P. and Rastogi, R.P., Chemical constituents of *Trichosanthes palmata*, *Indian J. Chem.*, 22B, 252, 1983; idem., Santholin, a new saponin from *T. palmata*, *ibid.*, 22B, 624, 1983.
228. Agarwal, S.K. and Rastogi, R.P., Chemical constituents of *Celsia coromandeliana*, *Indian J. Chem.*, 12, 304, 907, 1974. Gupta H.P., Mathur, I.S., and Gupta, S.K., *Indian Vet. Med. J.*, 3, 69, 1979.
229. Chandel, R.S. and Rastogi, R.P., Saponins of *Schima wallichii*, *Indian J. Chem.*, 19B, 283, 1980.
230. Setty, B.S., Kamboj, V.P., Garg, H.S., and Khanna, N.M., Spermicidal potential of saponins isolated from Indian medicinal plants, *Contraception*, 14, 571, 1976.
231. Singh, H.K. and Dhawan, B.N., Effects of *Bacopa monniera* extract on avoidance responses in rat, *J. Ethnopharmacol.*, 5, 205, 1982.
232. Tripathi, R.M., Sen, P.C., and Das, P.K., Studies on the mechanism of action of *Albizzia lebbeck*, an Indian indigenous drug used in the treatment of atopic allergy, *J. Ethnopharmacol.*, 1, 385, 1979.
233. Tripathi, R.M., Sen. P.C., and Das, P.K., Further studies on the mechanism of the anti-anaphyllactic action of *Albizzia lebbeck*, an Indian indigenous drug, *J. Ethnopharmacol.*, 1, 397, 1979.
234. Tripathi, P., Tripathi, Y.B., Dey, P.K., and Tripathi, S.N., Release of catecholamines from the adrenal medulla by histamine and its prevention by *Albizzia lebbeck* in guinea pigs, *Indian J. Physiol. Pharmacol.*, 27, 176, 1983.
235. Singh, B. and Rastogi, R.P., Chemical examination of *Asclepias curassavica*, *Indian J. Chem.*, 7, 1105, 1969.
236. Patnaik, G.K. and Dhawan, B.N., Pharmacological investigations of asclepin, a new cardenolide from *Asclepias curassavica*. I. Cardiotonic activity and acute toxicity, *Arzeim. Forsch.*, 28, 1095, 1978.
237. Patnaik, G.K. and Kohler, E., Pharmacological investigations of asclepin, a new cardenolide from *Asclepias curassavica*. II. Comparative studies on the inotropic and toxic effects of asclepin, g-strophanthin, digoxin and digitoxin, *Arzeim. Forsch.*, 28, 1368, 1978.
238. Dev, S., personal communication, 1980.
239. Govindachari, T.R., Sathe, S.S., Viswanathan, N., Pai, B.R., and Srinivasan, M., Chemical constituents of *Cleistanthus collinus*, *Tetrahedron*, 25, 2815, 1969.
240. Rao, H.R. and Nair, T.N., Effect of cleistanthin on experimentally induced leucopenia in rats and mice, *Arnzeim. Forsch.*, 21, 828, 1971.
241. Sarin, J.P.S., Garg, H.S., Khanna, N.M., and Dhar, M.M., Ipolearoside, a new glycoside from *Ipomoea leari* with anticancer activity, *Phytochemistry*, 12, 2461, 1973.
242. Ghosal, S., Banerjee, S., and Jaiswal, D.K., New furofurano lignans from *Justicia simplex*, *Phytochemistry*, 19, 332, 1980.
243. Bhattacharya, S.K., Reddy, P.K.S.P., Ghosal, S., Singh, A.K., and Sharma, P.V., Chemical constituents of Gentianaceae CNS-depressant effects of swertiamarin, *J. Pharm. Sci.*, 65, 1547, 1976.
244. Kar, K., Singh, S., and Khanna, N.M., A diuretic flavone glycoside from *Millingtonia hortensis*, *Indian J. Pharm.*, 38, 26, 1976.
245. Agarwal, V.K., Some pharmacological studies on the stem bark of *Millingtonia hortensis*, *J. Res. Ayur Siddha*, 2, 50, 1982.
246. Dutta, P.K., Chakrabarty, A.K., Choudhury, U.S., and Pakrashi, S.C., Vincine, a favism-inducing toxin from *Momordica chrantia*, *Indian J. Chem.*, 20B, 669, 1981.
247. Singh, B. and Rastogi, R.P., Chemical examination of *Picrorhiza kurrooa*. VI. A reinvestigation of Kutkin, *Indian J. Chem.*, 10, 29, 1972.
248. Longer, J.G., Gupta, O.P., and Atal, C.K., Clinical trials on *Picrorhiza kurrooa* as an immunoregulator, *Indian J. Pharmacol.*, 13, 98, 1981.
249. Sarin, J.P.S., Garg, H.S., Khanna, N.M., and Dhar, M.M., Sterol elycoside and triterpenes from *Vittadinia australis*, isolation of a diuretic constituent, *Indian J. Chem.*, 13, 199, 1975; Kar, K., Sarin, J.P.S., and Khanna, N.M., Diuretic activity of vittadinoside, a sterol glucoside from *Vittadinia australis*, *Indian J.Pharm.*, 39, 17, 1977.
250. Vohra, M.M., Patnaik, G.K., Kapil, R.S., and Anand, N., Chemistry and pharmacology of a glycoside of *Vallaris solanacea*, *J. Pharm. Sci.*, 55, 1425, 1966.
251. Shukla, Y.N., Tandon, J.S., Bhakuni, D.S., and Dhar, M.M., Chemical constituents of the antibiotic fraction of *Arnebia nobilis*, *Experientia*, 25, 357, 1969.
252. Shukla, Y.N., Tandon, J.S., Bhakuni, D.S., and Dhar, M.M., Napthaquinones of *Arnebia nobilis*, *Phytochemistry*, 10, 1909, 1971.
253. Shukla, Y.N., Tandon, J.S., and Dhar M.M., Arbebin-7, a new napthaquinone from *Arnebia nobilis*, *Indian J. Chem.*, 11, 528, 1973.
254. Gupta, S.K. and Mathur, I.S., The effect of *Arnebia nobilis* & its naphthaquinones in rat Walker carcinosarcoma 256, *Indian J. Cancer*, 9, 50, 1972.
255. Khan, S.A. and Shamsuddin, K.M., 13,18-Dehydroexacelsin, a quassinoid and glaucarubol from *Ailanthus excelsa*, *Phytochemistry*, 19, 2484, 1980.
256. Banerjee, S.K., Gupta, B.D., Sheldrick, W.S., and Hofle, G., Angeolide, a novel lactone from *Angelica glauca*, *Liebigs Ann. Chem.*, 44, 699, 1982.
257a. Achari, B., Chakrabarty, S., and Pakrashi, S.C., A N-glycoside and steroids from *Aristolochia indica*, *Phytochemistry*, 20, 1444, 1981; Achari, B., Bangyopadhyay, S., Saha, C.R., and Pakrashi, S.C., A phenanthroid lactone, steroid and lignans from *Aristolochia indica*, *Heterocycles*, 20, 771, 1983.
257b. Roy, S., Dutta, A.K., and Chakraborty, D.P., Amasterol, an ecdysone precurser and a growth inhibitor from *Amaranthus virdis*, *Phytochemistry*, 21, 2417, 1982.
258. Suri, O.P., Ravi Kant, R.S., Jamwal, R.S., Suri, K.S., and Atal, C.K., *Boerhaavia diffusa*, a new source of phytoecdysones, *Planta Med.*, 44, 180, 1982.
259. Srivastava, K., Jain, G.K., Sarin, J.P.S., Khanna, N.M., Liedholm, P., Astedt, B., and Bladh, B., Effect of *Boerhaavia diffusa* and ε-aminocaproic acid on fibrinolytic activity, in *IUD Pathology and Management: Progress in Contraceptive Delivery Systems*, Vol. 3, Hafez, E.S.E. and Van Os, W.A.A., Eds., MTP Press, Lancaster, England, 1980, 121.
260. Patil, V.D., Nayak, U.R., and Dev, S., Chemistry Ayurvedic crude drugs I. Gaggulu (resin from *Commiphora mukul*) steroidal constituents, *Tetrahedron*, 28, 2341, 1972.

261. Nityanand, S. and Kapoor, N.K., Cholesterol lowering activity of various fractions of guggul, *Indian J. Exp. Biol.*, 11, 395, 1973.
262. Bajaj, A.G., Dev, S., Arnold, E., Tagle, B., and Clardy, J., The stereochemistry of guggulsterol-1, *Tetrahedron Lett.*, 22, 4623, 1981.
263. Bajaj, A.G. and Dev, S., Guggulu (resin from *Commiphora mukul*). V. Some new steroidal components and stereochemistry of guggulsterol-1, *Tetrahedron*, 38, 2949, 1982.
264. Singh, J., Handa, G., Rao, P.R., and Atal, C.K., Pangamic acid, a stamina building, antistress and antihyperlipidemic principle from *Cicer arietinum*, *J. Ethnopharmacol.*, 7, 239, 1983.
265. Sen, H.G., Joshi, B.S., Parthasarathy, P.C., and Kamat, V.N., Anthelminthic efficacy of diospyrol and its derivatives, *Arzeim. Forsch.*, 24, 2000, 1974.
266. Anjaneyulu, A.S.R. and Rao, M.N., Elaeodendroside-B, a novel cardiac glycoside from *Elaeodendron glaucum*, *Indian J. Chem.*, 19B, 944, 1980.
267. Agarwal, S.K., Rastogi, R.P., Van Koningsveld, H., and Olthof, G.J., The molecular structure of 4a,5,8,8a-tetrahydro-11,14-dimethoxy-7-methyl 4a(3-methyl-2-butenyl)-5,8a,O-benzo-l,4-naphthoquinone, *Tetrahedron*, 36, 1435, 1980.
268. Shukla, Y.N., Tandon, J.S., and Dhar, M.M., Lyonogenin, a new dihydrochalkone from *Lyonia formosa*, *Indian J. Chem.*, 11B, 720, 1973.
269. Anand, K.K., Sharma, M.L., Singh, G.B., and Ray Ghatak, B.J., Anti-inflammatory, antipyretic and analgesic properties of bavachinin — a flavone isolated from seeds of *Psorelea corylifolia* (Babchi), *Indian J. Exp. Biol.*, 10, 1216, 1978.
270. Gupta, B.K., Gupta, G.K., Dhar, K.L., and Atal, C.K., Psoralidin oxide, a coumestan from seeds of *Psorelea corylifolia*, *Phytochemistry*, 19, 2232, 1980.
271. Bhadbhade, M.M., Subba Rao, G.S.R., and Venkatesan, K., Concerning hypophyllanthin, *Tetrahedron Lett.*, 21, 3097, 1980.
272. Querishi, M.A., Kalla, A.K., and Dhar, K.L., A novel withanolide from *Datura quercifolia*, *Phytochemistry*, 18, 1756, 1979.
273. Chandhoke, N., Daturalactone isolated from *Datura quercifolia*, a new interceptive agent, *Indian J. Exp. Biol.*, 16, 419, 1978.
274. Sahai, M., Gottlieb, H.E., Ray, A.B., Ali, A., Glotter, E., and Kirson, I., Perulactone-B, a new natural ergostane-type steroid with an alpha-oriented sidechain, *J. Chem. Res.*, (S), 346, 1982.
275. Sahai, M., Ali, A., Ray, A.B., Slatkin, D.J., and Kirson, I., Withaperuvin D, a novel withanolide from *Physalis peruviana*, *J. Chem. Res.*, (S), 152, 1983.
276. Row, L.R., Reddy, K.S., Dhaveji, K., and Matsuura, T., Pubescenol, a withanolide from *Physalis pubescence*, *Phytochemistry*, 23, 427, 1984.
277. Chaudhari, M., Jain, G.K., Sarin, J.P., and Khanna, N.M., Spasmolytic principle from *Rheum webbianum*, *Indian J. Chem.*, 22B, 1163, 1983.
278. Atal, C.K., Chemistry and Pharmacology of Vasicine, a New Oxytocic and Abortifacient, Regional Research Laboratory, Jammu Tawi, 1980.
279. Gupta, V.N., Jain, M.P., Atal, C.K., and Bhardawaj, M., Synthesis of vasicine analogue, *Indian J. Chem.*, 21B, 74, 1982.
280. Gupta, V.N., Gupta, V.K., Atal, C.K., Shah, H., and Naqavi, B.G., Stability studies on vasicine stearate, *Indian J. Pharm. Sci.*, 46, 89, 1984.
281. Gautam, C.S. and Sharma, P.L., Potentiation of oxytocin evoked responses by (+)sotalol, deoxysotalol and vasicine hydrochloride on isolated rat and rabbit uterus, *Indian J. Med. Res.*, 76(Suppl.), 107, 1982.
282. Lal, R. and Sharma, P.L., Potentiation of prostaglandin evoked contractions of isolated rat uterus by vasicine hydrochloride, *Indian J. Med. Res.*, 73, 641, 1981.
283. Zutshi, V., Rao, P.G., Soni, A., Gupta, O.P., and Atal, C.K., Absorption and distribution of vasicine — a novel uterotonic, *Planta Med.*, 40, 373, 1980.
284. Devi, G., Kapil, R.S., and Popli, S.P., Potential CNS and CVS agents — synthesis based on vasicine, *Indian J. Chem.*, 14B, 354, 1976.
285. Atal, C.K., Sharma, M.L., Khajuria, A., and Kaul, A., Thrombopoietic activity of vasicine hydrochloride, *Indian J. Exp. Biol.* 20, 704, 1982.
286. Pakrashi, A., Chakrabarti, B., and Das Gupta, A., Effect of the extracts from *Aristolochia indica* on interception in female mice, *Experientia*, 32, 394, 1976 .
287. Pakrashi, A. and Chakrabarti, B., Antifertility effect of aristolic acid from *Aristilochia indica* in female albino rabbits, *Experientia*, 34, 1377, 1978 .
288. Pakrashi, A. and Shaha, C., Effect of methylester of aristolic acid from *Aristolochia indica* on fertility of female mice, *Experientia*, 34, 1192, 1978 .
289. Pakrashi, A. and Pakrashi, P., Biological profile of p-coumaric acid isolated from *Aristolochia indica*, *Indian J. Exp. Biol.*, 16, 1285, 1978.
290. Pakrashi, A. and Shaha, C., Antiimplantation and anti-estrogenic activity of a sesquiterpene from *Aristolochia indica*, *Indian J. Exp. Biol.*, 15, 1191, 1977.
291. Pakrashi, A. and Shaha, C., Short-term toxicity study with methyl ester of aristolic acid from *Aristolochia indica* in mice, *Indian J. Exp. Biol.*, 17, 437, 1979.
292. Pakrashi, A. and Pakrashi, P., Antifertility effect of *Aristolochia* indica on mouse, *Contraception*, 20, 49, 1979.
293. Pakrashi, A. and Pakrashi, P., Antispermatogenic effect of the extract of *Aristolochia indica* on male mice, *Indian J. Exp. Biol.*, 15, 256, 1977.
294. Pakrashi, A., Pakrashi, P., and Kabir, S.N., Effect of para-coumaric acid (PCA) on fertility of male mice, *IRCS Med. Sci. Endocr. Syst.*, 7, 8, 1979.
295. Atal, C.K., Singh, G.B., Batra, S., Sharma, S., and Gupta, O.P., Salai guggal ex-*Boswellia serrata* — promising antihyperlipidaemic and antiarthritic agent, *Indian J. Pharmacol.*, 12, 59, 1980.
296. Atal, C.K., Gupta, O.P., and Singh, G.B., Salai guggal: a promising antiarthritic and antihyperlipidaemic agent, *Br. J. Pharmacol.*, 74, 203, 1981.
297. Singh, G.B., Singh, B., and Atal, C.K., Assessment of total boswellic acids for antiinflammatory activity, *Indian J. Pharmacol.*, 16, 51, 1984.
298. Pachnanda, V.K., Kant, S., Singh, D., Singh, G.B., Gupta, O.P. and Atal, C.K., Clinical evaluation of salai guggal in patients of arthritis, *Indian J. Pharmacol.*, 13, 63, 1981.
299. Pachnanda, V.K., Gupta, M.N., Singh, D., Sashi Kant, Singh, G.B., and Atal, C.K., Clinical evaluation of active fraction of resin of *Boswellia serrata* (Salai guggal) for the treatment of rheumatoid arthritis, in Abstr. Symp. Recent Advances in Mediators of Inflammation and Anti-Inflammatory Agents, Jammu, 1984, 37.
300. Dubey, M.P., Srimal, R.C., Patnaik, G.K., and Dhawan, B.N., Hypotensive and spasmolytic activities of coleonol, active principle of *Coleus forskohlii*, *Indian J. Pharmacol.*, 6, 15, 1974.
301. Dubey, M.P., Srimal, R.C., Nityanand, S., and Dhawan, B.N., Pharmacological studies on coleonol, a hypotensive diterpene from *Coleus forskohlii*, *J. Ethnopharmacol.*, 3, 1, 1981.
302. Bhat, S.V., Bajwa, B.S., Dornauer, H., de Souza, N.J., and Fehlhaber, H.W., Structure and sterochemistry of new labdane diterpenoids from *Coleus forskohlii*, *Tetrahedron Lett.*, 19, 1669, 1977.
303. Lindner, E., Dohadwala, A.N., and Bhattacharya, B.K., Positive inotropic and blood pressure lowering activity of a diterpene derivative isolated from *Coleus forskohlii*, *Arzeim. Forsch.*, 28, 284, 1978.
304. Bhat, S.V., Dohadwala, A.N., Bajwa, B.S., Dadkar, N.K., Dornauer, H., and de Souza, N.J., The anti-hypertensive and positive inotropic diterpene forskolin: effects of structural modifications on its activity, *J. Med. Chem.*, 26, 486, 1983.
305. Tandon, J.S., Painuly, P., Katti, S.B., and Singh, S., Chemistry and anti-hypertensive activity of coleonol derivatives, *Indian J. Chem.*, 23B, 67, 1984.
306. Saksena, A.K., Green, M.J., Shue, H.J., Wong, K.J., and McPhail, A.T., Identity of coleonol with forskolin: structure revision of a base-catalysed rearrangement product, *Tetrahedron Lett.*, 26, 551, 1985.
307. Seamon, K.B., Forskolin and adenylate cyclase — new opportunities in drug design, *Annu. Rev. Med. Chem.*, 19, 293, 1984.

308. Seigl, A.M., Daly, J.W., and Smith, J.B., Inhibition of aggregation and stimulation of cyclic AMP generation in intact human platelets by the diterpene forskolin, *Mol. Pharmacol.*, 21, 680, 1982.
309. Agarwal, K.C. and Parks, R.E., Jr., Forskolin - a potential antimetastatic agent, *Int J. Cancer*, 32, 801, 1983.
310. Badian, M., Dabrowski, J., Grigoleit, H.G., and Liete, W., Decrease of intraoccular pressure (1OP) by forskolin (F) in normal subjects, in Abstr. 9th Int. Congr. Pharmacol., London, 1984, 1314.
311. Satyavati, G.V., Dwarkanath, C., and Tripathi, S.N., Experimental studies on the hypocholesterolaemic effect of *Commiphora mukul* (Guggul), *Indian J. Med. Res.*, 51, 1950, 1969.
312. Ahuja, M.M.S., Malhotra, S.C., and Tandon, H.D., Effect of fraction 'A' of *Commiphora mukul* (Guggulu) on Mongolian gerbils (*Meriones unguiculatus*), *Indian J. Exp. Biol.*, 15, 143, 1977.
313. Malhotra, C.L., Agarwal, Y.K., Mehta, V.L., and Prasad, S., The effect of various fractions of gum guggal on experimentally-produced hypercholesterolaemia in chicks, *Indian J. Med. Res.*, 58, 394, 1970.
314. Malhotra, S.C., Ahuja, M.M.S., and Sunderan, K.R., Long term clinical studies on the hypolipidaemic effect of *Commiphora mukul* (Guggulu) and clofibrate, *Indian J. Med. Res.*, 65, 390, 1977.
315. Kotiyal, J.P., Bisht, D.B., and Singh, D.S., Double blind cross over trial of Gum guggulu (*Commiphora mukul*) fraction 'A' in hypercholesterolaemia, *J. Res. Indian Med. Yoga Homeo.*, 14, 11, 1979.
316. Nityanand, S. and Kapoor, N.K., Hypolipidaemic effect of ethylacetate fraction of *Commiphora mukul* (guggul) in rats, *Indian J. Pharmacol.*, 7, 106, 1975.
317. Nityanand, S., Asthana, O.P., Gupta, P.P., Kapoor, N.K., and Dhawan, B.N., Clinical studies with guggulipid, a new hypolipidaemic agent, *Indian J. Pharmacol.*, 13, 59, 1981.
318. Srivastava, M., Nityanand, S., and Kapoor, N.K., Effect of guggulipid on non-epinephrine biosynthesis in monkey tissues, *Curr. Sci.*, 53, 1131, 1984.
319. Mester, L., Mester, M., and Nityanand, S., Inhibition of platelet aggregation by guggul steroids, *Planta Med.*, 37, 367, 1979.
320. Bordia, A. and Chuttani, S.K., Effect of gum guggulu on fibrinolysis and platelet adhesiveness in coronary heart disease, *Indian J. Med. Res.*, 70, 992, 1979.
321. Tripathi, Y.B., Malhotra, O.P., and Tripathi, S.N., Thyroid stimulating action of Z-guggulosterone obtained from *Commiphora mukul*, *Planta Med.*, 50, 78, 1984.
322. Sharma, J.N. and Sharma, J.N., Comparison of the anti-inflammatory activity of *Commiphora mukul* (an indigenous drug) with those of phenylbutazone and ibuprofen in experimental arthritis induced by mycobacterial adjuvant, *Arzeim. Forsch.*, 27, 1455, 1977.
323. Srimal, R.C., Khanna, N.M., and Dhawan, B.N., A preliminary report on anti-inflammatory activity of curcumin, *Indian J. Pharmacol.*, 3, 10, 1971.
324. Srimal, R.C. and Dhawan, B.N., Pharmacology of diferuloyl methane (curcumin), a new steroidal anti-inflammatory agent, *J. Pharm. Pharmacol.*, 25, 447, 1973.
325. Ghatak, N. and Basu, N., Sodium curcuminate as an effective anti-inflammatory agent, *Indian J. Exp. Biol.*, 10, 235, 1972.
326. Chandra, D. and Gupta, S.S., Anti-inflammatory and anti-arthritic activity of volatile oil of *Curcuma longa* (Haldi), *Indian J. Med. Res.*, 60, 138, 1972.
327. Yeganarayan, R., Sarraf, A.P., and Balwani, J.H., Comparison of anti-inflammatory activity of various extracts of *Curcuma longa*, *Indian J. Med. Res.*, 64, 601, 1976.
328. Sinha, M., Mukherjee, B.P., Mukherjee, B., Sikdar, S., and Dasgupta, S.R., Study of the mechanism of action of curcumin: an antiulcer agent, *Indian J. Pharmacol.*, 7, 98, 1975.
329. Sethi, N., Mukherjee, S.K., Srivastava, G.N., and Roy, A.K., Toxicity studies of curcumin, a nonsteroidal anti-inflammatory agent in rat and monkey, *U.P. Vet. J.*, 4, 153, 1976.
330. Srivastava, R. and Srimal, R.C., Biochemical mechanism of action of anti-inflammatory drugs, *Biol. Mem.*, 10, 66, 1985.
331. Srivastava, R. and Srimal, R.C., Modification of certain inflammation induced biochemical changes by curcumin, *Indian J. Med. Res.*, 81, 215, 1985.
332. Sharma, S.C., Mukhtar, H., Sharma, S.K., and Krishnamurti, C.R., Lipid peroxide formation in experimental inflammation, *Biochem. Pharmacol.*, 21, 1210, 1972.
333. Sharma, O.P., Anti-oxidant activity of curcumin and related compounds, *Biochem. Pharmacol.*, 25, 1811, 1976.
334. Deodhar, S.D., Sethi, R., and Srimal, R.C., Preliminary studies on antirheumatic activity of curcumin (diferuloyl methane), *Indian J. Med. Res.*, 71, 632, 1980.
335. Khanna, N.M., Singh, S., and Sarin, J.P.S., The metabolic disposition of curcumin in rats, *Indian Drugs*, 19, 31, 1981.
336. Rao, T.S., Basu, N., and Siddique, H.H., Anti-inflammatory activity of curcumin analogues, *Indian J. Med. Res.*, 75, 574, 1982.
337. Rangaswami, S. and Rao, E.V., New crystalline glycosides from the seeds of *Thevetia nerifolia*, *J. Sci. Ind. Res.*, 17B, 331, 1958.
338. De, N.N., Bohra, M.M., and Kohli, J.D., A comparative pharmacological study of new cardiac glycosides from *Thevetia nerifolia*, in *Proc. 1st Int. Pharmacol. Meeting*, Vol. 3, Wilbrant, W., Ed., Pergamon Press, Oxford, 1963, 227.
339. Araro, R.B., Sharma, J.N., and Bhatia, M.L., Pharmacological evaluation of peruvoside, a new cardiac glycoside from *Thevetia nerifolia*, with a note on its clinical trials in patients of congestive cardiac failure, *Indian J. Exp. Biol.*, 5, 31, 1967.
340. Bhatia, M.L., Manchanda, S.C., and Roy, S.B., Haemodynamic studies with peruvoside in human congestive heart failure, *Br. Med. J.*, 3, 740, 1970.
341. Arora, R.B. and Rangaswami, S., *Peruvoside and Other Cardiotonic Glycosides from Thevetia nerifolia*, Thompson Press (India) Ltd., New Delhi, 1972.
342. Arora, R.B. and Sundersan, P.R., Studies on peruvoside I, early manifestations and comparative toxicity in the guinea pig and the dog, *Indian J. Exp. Biol.*, 11, 295, 1973.
343. Dalal, N.J., Billimoria, A.R., Vakil, B.J., and Shah, S.J., A preliminary report on clinical trial of Peruvoside — a new cardiac glycoside, *Indian Heart J.*, 24, 129, 1972.
344. Schorscher, E., Sommer, S., and Wild, A.J.N., Internal efficacy of peruvoside, *Arzeim. Forsch.*, 18, 1590, 1968.
345. Schwind, K., Report of clinical testing of peruvoside, *Med. Welt.*, 20, 132, 1969.
346. Rege, N., Dahanukar, S., and Karandikar, S.M., Hepatoprotective effects of *Tinospora cordifolia* against carbon tetrachloride-induced liver damage, *Indian Drugs*, 21, 544, 1984.
347. Singh, B., Sharma, M.L., Gupta, D.K., Atal, C.K., and Arya, R.K., Protective effect of *Tinospora cordifolia* on carbon tetrachloride-induced hepatotoxicity, *Indian J. Pharmacol.*, 16, 139, 1984.
348. Mogre, K., Vora, K.K., and Seth, V.K., Effect of *Picrorhiza kurrooa* and *Eclipta alba* on Na^+ K^+ ATPase in hepatic injury by hepatotoxic agents, *Indian J. Pharmacol.*, 13, 253, 1981.
349. Natu, M.V., Agarwal, S., Agarwal, S.L., and Agarwal, S., Protective effect of *Ricinus communis* leaves in experimental liver injury, *Indian J. Pharmacol.*, 9, 265, 1977.
350. Choudhuri, S.K., Influence of *Andrographis paniculata* (Kalmegh) on bile flow and hexabarbitone sleeping in experimental animals, *Indian J. Exp. Biol.*, 16, 830, 1978.
351. Antarkar, D.S., Vaidya, A.B., Doshi, J.C., Athavale, A.V., Vinchoo, K.S., Natekar, M.C., Tathed, P.S., Ramesh, V., and Kale, N., A double blind clinical trial of Arogya wardhani — an Ayurvedic drug in acute viral hepatitis, *Indian J. Med. Res.*, 72, 588, 1980.
352. Das Adhikary, B.M. and Debnath, P.K., Preliminary report of Daruharidra (*Berberis aristata*) on clinical jaundice, *Indian J. Pharmacol.*, 13, 61, 1981.
353. Vaidya, A.B., Bhatia, C.K., Mehta, J.M., and Sheth, U.K., Therapeutic potential of *Luffa echinata* in viral hepatitis, *Indian J. Pharmacol.*, 8, 245, 1976.
354. Naqvi, S.W.A. et al., Screening of marine organisms from the Indian coasts for biological activity. I., *Bot. Mar.*, 24, 51, 1980.

Chapter 14

RECENT RESEARCH ON MEDICINAL PLANTS IN THE AFRICAN REGION

G. H. Mahran

I. INTRODUCTION

Medicinal plants and the drugs derived from them constitute great economic and strategic value for the African continent.[1-3] Most of the life forms of plants are represented within the African flora. The different regions in Africa support thousands of species.[4] Some of these species are considered to be constant, i.e., found in most of the African regions, while other species are considered to be indifferent, i.e., found in any region but not characteristic of it. Absolutely exclusive species are those confined to a singe region only.[1]

The flora includes tropical, subtropical, and temperate species in several regions.[5-8] The vegetation is not restricted only to wild plants, since Africa produces other plant species which generate cereals, vegetables, fruits, oilseeds, as well as fibres, colouring matters, tanning materials, drug products, condiments, gums, and essential oils.[1] Many of the cultivated as well as the wild plants yield products which are already used in modern medicine as drugs.[9]

Since antiquity, the primitive African found in nature his food, clothing, and health. By trial and error, he began to observe which plants were useful, i.e., those having therapeutic action, and which ones were toxic. This vast experience, which frequently demonstrated conspicuous and unique effects on certain diseases has accumulated over the centuries.

Most of the truly effective medicinal plants and drugs were kept secret by the practitioners and knowledge was handed down only to their descendants and favorite apprentices. Early practices of treating disease were even mixed with magic and religion during certain periods. This is why traditional African medicine is an amalgamation of dynamic know-how and ancestral experience.[9] Traditional African medicine might also be considered to be the sum total of practice, measures, ingredients, and procedures of all kinds, whether material or not, which from time immemorial enabled the African to guard against disease, to alleviate his sufferings, and cure himself. The African herbalists, chiefs, and "Attarins" made a significant contribution by using these medicinal plants and drugs for the health of the African people.[9]

African scientists started rather late to evaluate their traditional pharmacopoeias, by exploring the potentialities of the African medicinal plants used through the centuries.[10-13] At first they were content to study the active ingredients, isolating them, elucidating their chemical structures, and observing their therapeutic and pharmacological effects. However, by these they succeeded in developing the basis of modern African research in the field of new drugs from African medicinal plants. Priority has now also been given to operational work oriented toward the validation or invalidation of popular knowledge with the aid of scientific research. The latter approach has further permitted close collaboration with the traditional healers and their systems of practice, with special attention to medicinal plants already in use in folk and traditional medicine. This has, in turn, generated welcome feedback to the traditional healers themselves, and to the community in general. Research in traditional medicine should have as its goal the transfer of results to the traditional healers and the health professionals so as to improve the efficiency of their services and to eliminate harmful practices. Scientific research has indeed revealed many harmful side effects which could be eliminated from practice or corrected, as the research findings suggest.[14-17]

II. INDUSTRIAL USAGE

Having obtained experience in exploring the potentialities of African medicinal plants and oriented contemporary research in this field, African scientists have begun to examine the potential for the industrialisation of their indigenous medicinal plants and to formulate them into modern pharmaceutical preparations to be used in primary health care. They see in this the possibility to save hard currency spent on importing modern medicinal and pharmaceutical preparations. It could be that their own native African medicinal plants, exported from their homeland at relatively low prices, are returned to Africa in the form of modern pharmaceutical preparations at relatively high prices.[9]

Industrialisation has taken different forms and stages in the various African countries, depending on their state of development and independence, cultural patterns, and beliefs. It has also depended to some extent on how much people accepted the integration of Western medicine and the African traditional systems. Thus, industrialisation of African medicinal plants took the following steps and forms.

A. PREPARATION OF MEDICINAL "TEAS" FROM NATIVE AFRICAN MEDICINAL PLANTS

On the nearly 5000-year-old pylon of Karnak Temple in Egypt is written the everlasting wisdom of Imhotep, the God of medicine in ancient Egypt: "Cure every patient with the medicinal plants and drugs of his own native homeland, as this is the best for his recovery". It was a primitive African belief that prevails today that treatment with medicinal plants is more effective treatment with other means such as with synthetic chemicals.

The concept exists that the active ingredients in the medicinal plants help synergise the effect of each other, and that minerals and salts within the medicinal plants play a great role in the therapeutic effect. Since antiquity, Africans employed several native medicinal plants and drugs as boiled water extracts, "teas", taken orally for healing several diseases. Today, several African countries produce these medicinal teas in modern pharmaceutical forms, either as crude drugs mixed together in a powdered form or in tea bags. They learned how to combine the several ingredients, using very dense as well as lighter materials, to add the dense ones in smaller amounts, and then mix these with equal parts of the lighter, more voluminous constituents so as to obtain the most even mixtures. Modern mixing and packing machines are now available in several African countries. Several preparations of medicinal teas are now available in the various African markets. The most commonly used are

1. *Diabetic tea*, for the treatment of diabetes mellitus; containing *Jambosa* spp., *Catharanthus* spp., *Lupinus* spp., *Matricaria* spp., etc.
2. *Renal tea*, for the treatment of renal colic, dilatation of the ureters, and to aid in the passage of renal calculi; containing *Ammi visnaga, Ambrosia maritima, Maize stigmas, Cympobogon proximus, Carum copticum*, etc.
3. *Cough tea*, mostly as expectorant and usually containing linseed, guava leaves, liquorice, tilia, etc.
4. *Laxative tea*, for the treatment of habitual constipation; containing senna, aloe, tamarindus, etc.
5. *Slimming tea*, to reduce weight and help in slimming; consisting mainly of marine plants such as hypnea, fucus, sargassum, etc.
6. *Sedative tea*, to relief flatulence and colic, mainly in small children suffering from indigestion and usually containing chamomile, caraway, anise, fennel, etc.

B. PREPARATION OF EXTRACTS AND TINCTURES FOR VARIOUS PHARMACEUTICAL PREPARATIONS[9]

In some African countries, the pharmaceutical industry commenced with the art of preparing extracts and tinctures from medicinal plants. A variety of methods are available for extraction and the choice depends on the nature of the plant material and the compounds to be isolated. It is recognised that all plant materials used should be properly authenticated; much time and money can be wasted on the examination of drugs of doubtful origin. Dried materials are usually powdered before extraction, whereas fresh plants (leaves, etc.) can be homogenised or macerated with a solvent such as alcohol. Alcohol is a general solvent for many plant constituents and as such may cause problems in the subsequent elimination of pigments, resins, etc. Water-immiscible solvents are widely used in industrial preparation, e.g., light petroleum for the extraction of essential and fixed oils, steroids and ether, and chloroform for alkaloids and quinones. The extraction of organic bases, e.g., alkaloids usually necessitates basification of the plant material if a water-immiscible solvent is to be used; for aromatic acids and phenols, acidification may be required.

Extraction itself may be performed by repeated maceration with agitation, percolation, or by continuous extraction, e.g., in a Soxhlet-type extraction apparatus. Examples of such extracts prepared in the above fashion are stramonium, belladonna, hyoscyamus, senna, liquorice, cynara, kola, auranti, etc.

C. PREPARATION OF VOLATILE OILS FROM AROMATIC MEDICINAL PLANTS[9]

Volatile oils or essential oils are defined as complex mixtures of odorous and steam-volatile compounds which are excreted by glandular hairs or deposited in the plant body cell organelles, idioblasts (oil cells, specialised cells), or schizogenous or lysigenous ducts or cavities. Moreover, some woody plants deposit essential oils in their woods. As the name implies, they are volatile in steam. They are frequently associated with other substances such as gums and resins and themselves tend to resinify on exposure to air. Volatile oils are used for their therapeutic action[1] (e.g., oil of eucalyptus), for flavouring[1] (e.g., oil of lemon), or in perfumery[1] (oil of rose).

Volatile oils are extracted by distillation, water, or steam, as has long been practised in many African countries.

1. Medicinal and Commercial Uses[9]

Many crude drugs are used medicinally because of their volatile oil content, however, the volatile oils separated from the drugs, in numerous cases, are used as drugs themselves. The most common use is for flavouring. They all possess carminative action, but a few possess additional therapeutic properties. Many volatile oils possess antiseptic properties; the antibacterial, antimicrobial, and antifungal activities of certain oils are established.[19] Thus, because of their preservative qualities as well as their fragrance, spices were mixed with foodstuffs in early times. Undoubtedly, the presence of the antiseptic oils in the spices prevent excessive growth of bacteria which are responsible for food spoilage.

The most common African medicinal plants producing volatile oils are[3] *Coriandrum sativum, Pelargonium* sp., *Cymbopogon* spp., *Lavendula officinalis, Rosmarinus officinalis, Mentha spicata, Carum carvi, Anethum graveolens, Thymus vulgaris, Pimpinella anisum, Illiceum verum, Foeniculum vulgare, Eucalyptus globulus, Cinnamomum camphora, Petroselinum sativum, Myristica fragrans, Chenopodium ambrosioides, Brassica* sp., *Rosa* spp., *Citrus* spp., *Lavendula officinalis, Origanum* spp., *Melissa* spp., *Salvia* spp., *Eucalyptus* spp., *Mentha piperata, Cinnamomum cassia*, and *Eucalyptus citrodora*.

Oil of rose is obtained by distillation from the fresh flowers of *Rosa damascena, R. gallica, R. centifolia*, and other species growing all over Africa. It is of great importance in perfumery.

Oil of rosemary is obtained by distillation from the flowers of *Rosmarinus officinalis* (Fam. Labiatae).

Oil of lavender is distilled from the fresh flowering tops of *Lavendula officinalis* (Fam. Labiatae).

Oil of peppermint is prepared by distillation from the fresh flowering plants of *Mentha piperita* (Fam. Labiatae) and rectified, but no part of its menthol should be removed.

Oil of fennel is prepared from dried ripe fruits of *Foeniculum vulgare* (Fam. Umbelliferae) by steam distillation. It has aromatic and carminative properties.

Oil of caraway is prepared from the dried ripe fruits of *Carum carvi* (Fam. Umbelliferae) by steam distillation.

Oil of coriander is similarly prepared from *Coriandrum sativum* (Fam. Umbelliferae). It is used as a flavouring agent and carminative.

Oil of anise is similarly prepared from *Pimpinella anisum* (Fam. Umbelliferae).

Oil of cumin is prepared from *Cuminum cyminum* (Fam. Umbelliferae).

Oil of bitter orange is prepared from *Auranti amara* peel, and may be extracted by means other than distillation. It is used as a flavouring agent and as a bitter tonic.

Oil of lemon peel is prepared from *Citrus limon* peel by sponge process, *ecuelle a piquer* or machine processes. It is used in perfumery and as a flavouring agent.

Cinnamon oil is prepared from *Cinnamomum zeylanicum* bark (Fam. Lauraceae). The oil is a powerful germicide.[19]

Oil of clove is largely produced in Zanzibar and the Pemba Islands, from *Eugenia caryophyllus* (Fam. Myrtaceae) by steam distillation. The oil is used as a flavouring agent, stimulant, aromatic, and antiseptic.

Eucalyptus oil is distilled from the fresh leaves of various species of *Eucalyptus* (Fam. Myrtaceae) and rectified. The oil is classed as a flavour, but is frequently used as an antiseptic, diaphoretic and expectorant.

Oil of mustard is the oil obtained by maceration and subsequent distillation of the dried ripe seeds (freed from fixed oil) of *Brassica nigra* (Fam. Cruciferae). It is usually prepared by hydrolysis from the ground cake from which the fixed oil has been removed by expression. The oil is used externally as a rubifacient or vessicant.[18] It is a severe gastrointestinal irritant.[19]

Chamomile oil is the oil prepared by distillation from the fresh flower heads of *Matricaria chamomilla* (Fam. Compositae). The oil is blue when freshly distilled owing to the presence of azulene. Considerable quantities of the oil are used in domestic medicine, for dyspepsia,[1] and in shampoos.[1]

Citronella oil is prepared from *Cymbopogon nardus* (Fam. Graminae); the cut grass is usually dried for 2 to 3 days on the ground and bundled before being hauled to distillers. It is mainly used as an insecticide.[19]

Chenopodium oil is obtained by steam distillation of the whole aerial parts of *Chenopodium ambrosioides* (Fam. Chenopodiaceae). The highest yield of oil is obtained from plants during the pollination period when the ascaridole content is then low. It increases from 6 to 10% during flowering and rises to 50 to 56% during ripening of the fruits. The fruits are richest in essential oil (0.6 to 1.3%). Ascaridole seems to form largely during the last period of flowering. The oil is largely used as an anthelmintic.[20]

Sage oil is obtained by distillation of the herb of *Salvia officinalis* (Fam. Labiatae). The plant gives the most and finest oil before flowering. It is interesting to note that sage leaves contain estrogens in concentrations high enough to explain their action as an emenagogue, as prescribed in old herbals; however, the oil is free from estrogens.

D. UTILISATION OF MEDICINAL PLANT RESIDUES[9]

Certain residues are used in industry. The fresh pericarp of bitter and sweet orange, citron, and lime are extracted and used as flavorings for pharmaceuticals, liqueurs, drinks, etc. They also contain bitter substances and flavonoids, which could be extracted and produced on an industrial scale. Dried sweet orange peel, bitter orange peel, and lemon peel are used in a similar way. Coffee bean residue after extracting caffeine is used as a fuel. Sugarcane marc is used in the alcohol industry, in yeast production, and in the manufacture of wood boards.

E. PREPARATION OF FIXED OILS, FATS, AND WAXES[9]

Many drugs contain fixed oil and fats as their principal constituent. The fixed oils and fats are important products used pharmaceutically, industrially, and as foods. Some waxes are also of plant origin. Most commercial oils are derived from either seeds or fruits and the initial treatment before extraction depends on the botanical structure, e.g., cottonseed requires delinting and castor seeds and ground nuts require decorticating. Special machines are available for this purpose. Removal of the oil may take the form of cold or hot expression, centrifuging, or solvent extraction, again, depending on the commodity. With seeds, the remaining cake usually forms a valuable cattle food and for this reason complete removal of the oil is not always necessary. The crude oil requires refining. Cold pressed oils usually require nothing further than filtration, castor oil requires steaming to inactive lipase, the addition of a determined amount of alkali may be required to remove free acids; washing and decolourization may be performed.

Sometimes organic solvents are used for the extraction of oils. Oils may be further clarified by filtration and bleached with ozone. Stearins are often removed by chilling and filtration.

Vegetable oils and fats may occur in various parts of the plant but as a general rule seeds contain larger quantities of fats and oils than do other plant parts. Seeds are the usual source of fixed oils and, as a few examples, the following are mentioned: cottonseed, castor, almond, linseed, sesame, and hemp seed. In certain instances, other parts may yield considerable quantities of fixed oil, e.g., coconut and pericarp of olive. The most common fixed oils in Africa are as follows.

Almond oil: It is obtained by expression from seeds of *Prunus amygdalis* var. dulcis (sweet almond) or *P. amygdalis* var. amara (bitter almond). The oil may be produced extensively from countries bordering the Mediterranean. The fixed oil is obtained by grinding the seeds and expressing them in canvas bags between slightly heated iron plates. They are sometimes blanched before grinding, although not to any particular advantage. The oil is clarified by subsidence and filtration. It is a pale yellow liquid with a slight odour and a bland, nutty taste. It contains a considerable amount of olein with smaller quantities of the glycerides of linoleic and other acids. Almond oil is used in the preparation of many toilet articles and as a vehicle for oil injection. Taken internally, it is a mild laxative; mainly administered to children.

Arachis oil (earthnut, groundnut, peanut): obtained by expression from the seeds of *Arachis hypogaea* (Fam. Fabaceae); a small annual plant growing throughout tropical Africa. Various genotypes exist which show differences in the relative amounts of fatty acids in the oil. Groundnuts are the world's third largest source of fixed oil. During ripening the fruits bury themselves in the sandy soil in which the plant grows. Each fruit contains from one to three reddish-brown seeds. The fruits are shelled by a machine. The kernel contains from 40 to 50% oil. Owing to the high oil content, the seeds when crushed are somewhat difficult to express. After the initial cooking, part of the oil is removed in a low pressure expeller and the cake is solvent extracted. The two oil fractions are then mixed before purification. The pressed cake forms an excellent cattle food. Arachis oil is an ingredient of camphorated oil, but is used mainly in cooking and margarine production.

Olive oil: It is the fixed oil which is expressed from the ripe fruits of *Olea europea* (Fam. Oleaceae). The oil is mainly expressed in African countries bordering the Mediterranean. It is usually used in the preparation of soaps, plasters, etc.

Sesame oil: (Gingelly oil, Teel oil) is obtained by expression from the seeds of *Sesamum indicum* (Fam. Pedaliaceae); a herb widely cultivated in Africa. The seeds contain about 50% oil.

Corn oil: It is the refined oil obtained from the embryo of *Zea mays* (Fam. Graminae), and recommended for general use in place of other edible oils by patients suffering from arteriosclerosis.

Castor oil: It is the fixed oil obtained from the seeds of *Ricinus communis* (Fam. Euphorbiaceae); distributed throughout Africa. The seeds are deprived of their testa and kernels and cold-expressed in suitable hydraulic presses. The oil is refined by steaming, filtration, and bleaching. Cold expression yields about 33% medicinal oil and further quantities of oil of lower quality may be obtained by other methods. The oil is widely used as a purgative. It is a constituent of flexible collodion. Undecylenic acid, which is prepared from castor oil, is used in fungistatic preparations, as is its zinc salt.

Fractionated coconut oil: It is the fractionated and purified oil of the coconut; *Cocos nucifera* (Fam. Palmae) or thin vegetable oil. It is unique in containing triglycerides having only the short and medium chain-length fatty acids (e.g., octanoic, decanoic). It maintains its low viscosity until near the solidification point (about 0°C) and is a useful nonaqueous medium for the oral administration of some medicaments.

Cottonseed oil: *Oleum gossyppii* is expressed from the seeds of various species of *Gossypium* (Fam. Malvaceae); produced in most tropical and subtropical African countries and mainly in Egypt and Sudan. The oil is prepared by hot expression at a pressure of about 10,000 KPa (1500 lb/in.^{2}), then refined.

Linseed oil : It is the oil prepared from the dried ripe seeds of *Linum usitatissimum* (Fam. Linaceae). It is prepared by hot expression. The oil is used in liniments.

Palm kernel oil is the oil obtained from the seed of palm trees common all over Africa.

F. PREPARATION OF INSECTICIDES AND PESTICIDES[9]

Insect and pest control in recent years has become a major problem in homes and in agricultural programmes. Rodents are responsible for millions of pounds of damage annually and the destructive insect populations of the world has increased to an alarming extent.[21]

The pharmacist in suburban and country areas may act as distributor for extensive supplies of insecticides. In city communities, he is concerned mainly with insecticides and rodenticides. Several insecticides and rodenticides can be prepared from medicinal plants growing in Africa on an industrial scale. The following plants are of great value in this respect.

1. Tobacco

Tobacco consists of the cured and dried leaves of *Nicotiana tabacum* (Fam. Solanaceae). Finely powdered leaves had become well established as an insecticide before nicotine was isolated in 1882. In the form of a "dust", tobacco is widely used for vegetable crops.

Nicotiana glauca R.C. Grah, is a glabrous, blue-green shrub or small tree, reaching up to 6 m with long petiolated leaves.

Nicotiana tabacum L.is an annual plant with woody base; the stem is robust; attains 2 m in height. The most important constituents[22] of *Nicotiana* species are the steam volatile liquid alkaloids (2%) 1-nicotine, about 64 to 99% of the total alkaloids anabasine, nicotyrine, 1-nornicotine, d-nornicotine *N*-methyl pyrollidine, 3,2-dipyrridyl, *N*-methyl anabasine, anatabine, *N*-methyl anatabine, nicotelline, and nicoteine.[3]

This species contains the glycoside, tobaien,[23] which produces nicotine by hydrolysis. Drying and fermentation of the leaves causes the formation of several organic acids in the leaves: malic, citric, fumaric, oxalic, succinic, acetic, butyric, lactic, gallic, caffeic, quinic, chlorogenic, and nicotinic.[4] The leaves also contain aspargine, betaine,[24] isoamyl amine, pectin, tannins, laevulose, gum, carotene resin, mixture of paraffins, hentriacontane, heptacosane, and oil characterized by the isobutyl acetic and isovalerianic acids.[4] A volatile oil could be extracted from the leaves (0.00425%), this oil has a mint odour and contains ketone, borneol, and linalol; gives upon saponification acetic acid. The dry leaves contain the flavonoid rutin.

The flowers contain the pigment anthocyanin and a flavonoid glycoside. The concrete oil of the flower is a thick liquid having a coffee and clove odour, it contains eugenol, and the free and combined acetic and formic acids. The seeds contain 3.6% water, 4.2% proteins, and 30 to 42% fixed oil formed principally from glycerides of palmitic (32%), oleic (24.5%), and linoleic (15%) acids, and unsaponifiable matter (3%).[4]

The leaves are toxic;[24] certain creatures are particularly sensitive, especially those that are cold-blooded such as insects and frogs, and warm-blooded animals such as cats, dogs, and pigs. Tobacco is relatively distasteful but may be consumed in certain circumstances: by mistake when boiling leaves as greens, the poisoning of smugglers by carrying tobacco next to their skin, and human intoxication after treatment with parasiticide lotions containing leaf decoction or juice of tobacco. The acute poisoning begins with paleness, sensation of coldness, trembling, vertigo, irregular pulse, dyspnea, cholinergic syndrome, paralysis, and finally terminates by fatal syncope. The lethal dose is 320 to 640 mg of powdered leaves.[5]

Nicotine is extremely poisonous;[25] LD = 40 to 60 mg. The leaves are used for intestinal stimulation, 0.15 g dose, and as lotions in hernia, in cigarettes, in veterinary medicine as parasiticide, and as an insecticide.

2. Insect Flower

Pyrethrum is the dried flower heads of *Chrysanthemum cinerarifolium* (Fam. Compositae) and yields not less than 0.5% of total pyrethrins and is extensively cultivated in East Africa, mainly in Kenya and Tanzania.

Pyrethrum flowers and pyrethrins are superior to DDT in that insects do not show a tolerance to them and they lack toxicology to warm-blooded animals. Pyrethrum extracts and pyrethrins are noted for their quick "knockdown" to flying insects, particularly flies and mosquitos.

Chrysanthemum cinerarifolium:[1] Visiani (Synonyms: *Chrysanthemum pyrethrum-pyrethrum cinerariaefolium* Treviranus) is one of the species of the genus *Chrysanthemum* which belongs to the subfamily tubuliflorae, Fam. Compositae.

The Pyrethrum plant is a perennial herb, about 1 m high, very widely cultivated, being grown in various districts of Africa; mainly Kenya, Tanzania, the Congo, and Uganda.

The plant blooms in spring and early summer and continues to bear flowers in the same season for about 6 years, and sometimes for 15 years, especially when growing on well-drained hills.

Until 1923, the insecticidal power of pyrethrum flowers was wrongly attributed to previously separated resinous material.[26-34]

Staudinger et al.[35-39] reported that the insecticidal activity of pyrethrum flowers was mainly due to a mixture of pyrethrins I and II. They stated that these pyrethrins are esters of a ketoalcohol called pyrethrolone with two unsaturated acids, chrysanthemum monocarboxylic acid (synonyms: chrysanthemum acid I, chrysanthemic acid) and chrysanthemum dicarboxylic acid (synonyms: chrysanthemum acid II, pyrethric acid).

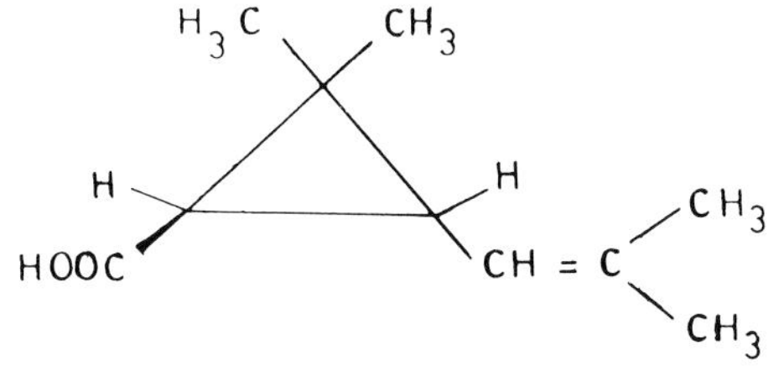

Chrysanthemic acid

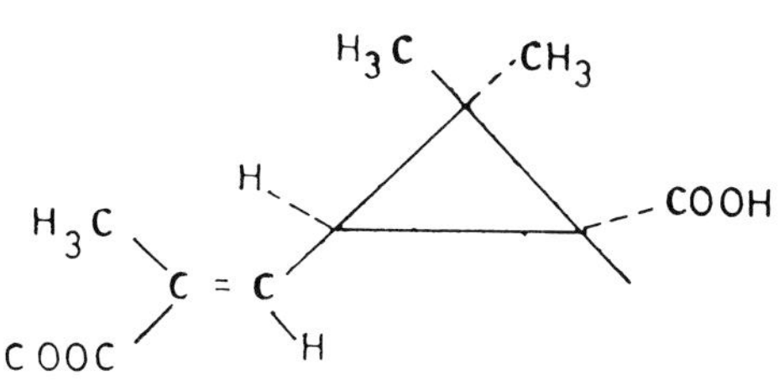

Pyrethric acid

La Forge and Barthel[36,37] showed that pyrethrolone was not a single substance but a mixture of two closely related cyclopentenolones which they named pyrethrolone ($C_{11}H_{14}O_2$) and cineralone ($C_{10}H_{14}O_2$).

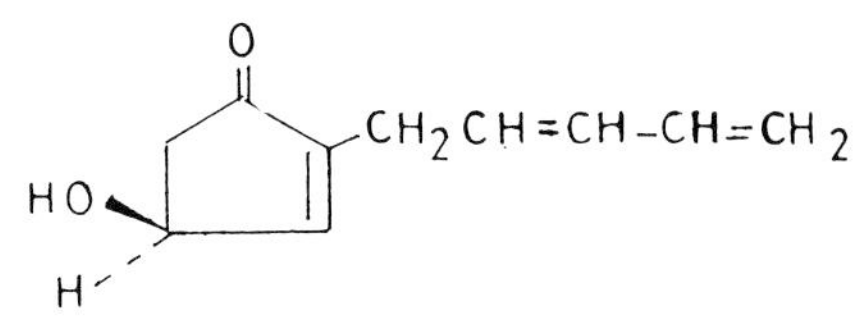

Pyrethrolone

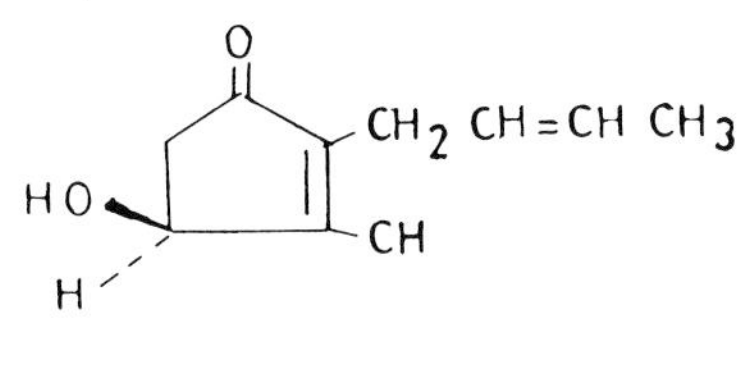

Cineralone

Thus, four insecticidal principles were recognised: the chrysanthemum monocarboxylic acid ester of pyrethrolone, which was known as pyrethrin I ($C_{21}H_{28}O_3$), and of cineralone which was known as cinerin I ($C_{20}H_{28}O_3$). The dicarboxylic acid similarly gives pyrethrin II ($C_{22}H_{28}O_5$) and cinerin II ($C_{21}H_{28}O_5$). Further work for elucidation of the structure of pyrethrins was done by Harper,[38] La Forge et al.,[39] and Crombie et al.[40]

In 1964 a new constituent of pyrethrum extract which was termed jasmolin II ($C_{22}H_{30}O_5$) was described by Golin et al.[41] It was found that this constituent differed from pyrethrin II in that the terminal double bond of the alcoholic side chain was saturated.

Beevor et al.[40,41] detected the presence of a new insecticidal active compound in pyrethrum oleoresin which they named jasmolin I and found it to be 4,5-dihydropyrethrin I ($C_{21}H_{30}O_3$).

The percentage of pyrethrins in the flower varies from 0.5 to 1. Investigations carried out by Gnadinger and Corl[42] showed that the active constituents were more or less completely confined to the flower heads and were present in small amounts (about 1/10 of that of the flower head) in the stems of the plant.

Pyrethrum and its preparations are widely used because of the great advantages they have: their safety, their lack of toxicity to warm-blooded animals, they are nonflammable, and they leave no oily residue. Ziegler,[43] Chevalier et al.,[44] and Chevalier[45] stated that pyrethrum was highly effective against flies, fleas, body lice, yellow fever, and malarial mosquitoes.

Pyrethrins were tried successfully against different worms by Anglade et al.,[46] who stated that specially prepared pills which released the pyrethrins gradually in the intestine caused rapid elimination of Ascaris, Ankylostoma, Taenia, and Oxyuris. Lemaire and Gaudin[47] mentioned that pyrethrins, when applied locally, gave good results in the treatment of scabies. Sweitzer and Teddar[48] reported that pyrethrum preparations penetrated the burrows and killed both mites and eggs. Tobias,[49] however, stated that pyrethrum ointments became unpopular in the treatment of scabies because of the high

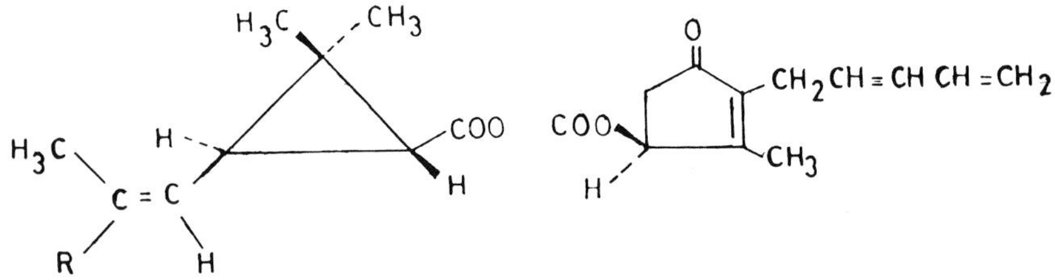

Pyrethrins

Pyerethrin I ; R = CH_3
Pyrethrin II ; R = $COOCH_3$

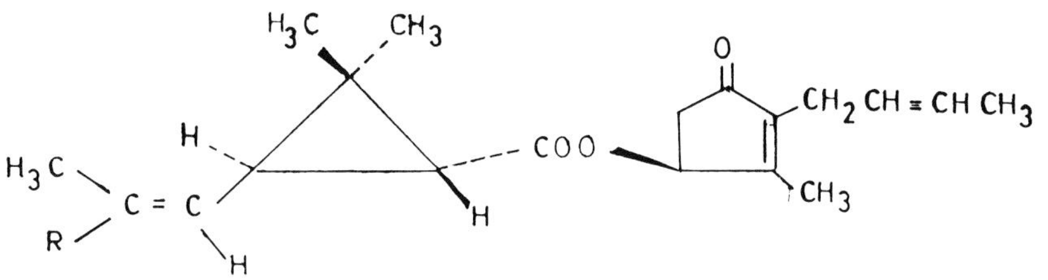

Cineririns

Cinerin I ; R = CH_3
Cinerin II ; R = $COOCH_3$

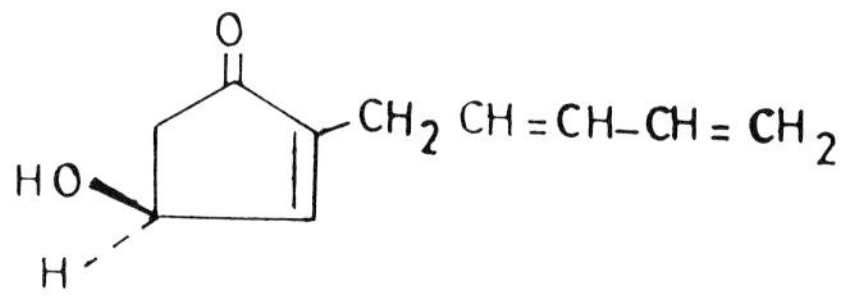

Pyrethrolone

relapse rate and frequency of sensitizations. El-Defrawi and El-Bahrawi[50] stated that by topical application, pyrethrins were very toxic to the larvae of the Egyptian cotton leafworm (*Prodenia litura*).

Sawicki and Elliott[48] examined the insecticidal activity of pyrethrum extract and its four insecticidal constituents against four strains of houseflies, and checked the relative toxicity of pyrethrins I and II. Their work confirmed the earlier results showing that pyrethrin II was 1.3 to 1.6 times more toxic than pyrethrin I, but the relative toxicity of pyrethrins I and II against the four strains of flies differed a little.

Jasmolin II was equal in its activity to cinerin II showing much less toxicity to insects than the pyrethrins.[50]

3. Derris

The roots of *Derris elliptica* L. (Fam. Compositae) produce an important insecticide due to their rotenone content. It is an insecticide widely employed in agriculture for controlling both chewing and sucking insects. It leaves no harmful residue. Rotenone may be an ingredient in sprays and aerosols containing pyrethrins to aid in knockdown effect.

4. Red Squill

Red squill consists of the bulbs of *Urginea maritima* (red var.) The bulbs are dried, powdered, and are largely used as rodenticide.

G. MANUFACTURE OF PLANT-BASED CONTRACEPTIVES AND STEROIDAL HORMONES[3]

Most oral contraceptives in use today are a combination of estrogen and progesterone in pill form. One pill contains an average of 0.5 to 1 mg of active steroid; the estrogen dosage being usually as small as 50 µg. Certain medicinal African plants represent rich sources of steroids. They can be industrially utilised for the isolation of steroidal hormones, to be eventually developed as contraceptives. Among these plants are the following.

1. *Dioscoria* or Yam

There are several species in Africa, mainly in South Africa: *D. sylvatica* Ecklon, containing botogenin and diosgenin, is a steroid sapogenin. Diosgenin has been the most versatile precursor for steroid drugs. Until recently it accounted for over two thirds of total raw material sources for steroid production.

2. *Solanum* sp.

Several Solanum species grow wild in Africa: *S. laciniatum, S. nigrum, S. dubium,* and *S. aviculare,* containing steroidal glycoalkaloids in the green shoot, leaves, and fruits, and mainly solasodine which is the N-analogue of diosgenin and is equally accepted as a raw material as diosgenin. Following ingestion, the glycoalkaloids are not readily absorbed, but first undergo hydrolysis to release the free alkamines.

As diosgenin accounts for such a substantial part of the total raw material sources for steroid production, care must be taken to concentrate on production of steroidal hormones from diosgenin. Diosgenin occurs as glycoside (saponin) in the tubers. The average content is 1% fresh weight and 4 to 5% dry weight. Two methods are generally used for sapogenin isolation:

1. Extraction of the saponin from plant tissue with a water-miscible organic solvent, e.g., ethanol, followed by purification of their glycosides before the acid hydrolysis to liberate the sapogenins.

2. Acid hydrolysis of the saponins within the plant tissue and extraction of the sapogenin from the neutralised acid-insoluble material with a hydrocarbon solvent.

3. *Agave sisalana* L.

Agave sisalana L.is a good source of steroidal sapogenin hecogenin.

H. MANUFACTURE OF GUMS AND RESINS FOR THE PHARMACEUTICAL INDUSTRY[9]

Gums, in the form of naturally occurring polysaccharides or modified polysaccharides, are consumed in large quantities by the pharmaceutical industry. Surprisingly, Africa is compelled to import gums, one of their own economic products. Importing organisations will give serious consideration to the physical and chemical properties of the gum, cost, constancy of cost, and constancy of supply, and exporting countries will have to be conscious of market requirements.

Gums are either hydrophobic or hydrophilic, high molecular weight molecules, usually with colloidal properties, which in an appropriate solvent or swelling agent produce gels, highly viscous suspensions, or solutions at low dry substance content. Most commonly, however, the term "gum" as technically employed in industry refers to plant polysaccharides or their derivatives which are dispersible in either cold or hot water to produce viscous mixtures or solutions.

Gum arabic is the dried gummy exudation obtained from the trees belonging to the various species of the genus *Acacia*, subfamily Mimosoideae, family Fabaceae. There are about 500 species of *Acacia* distributed over the tropical and subtropical areas of Africa. Sudan is the main producing area, followed by Senegal, Mauritania, Nigeria, Tanzania, Morocco, Ethiopia, and Somaliland. In Sudan, the great majority of gum is collected from *Acacia senegal*. They are mainly in the province of Kordofan, forming a large, uninterrupted belt stretching westward through the province of Darfur and eastward to the Nile River. Gum arabic is extensively used in the pharmaceutical industry. Its inherent emulsifying and stabilising properties and its demulcent and emolient characteristics led to a number of uses, from the stabilisation of emulsions to the formation of tablets and pills. The range of possible applications is further enhanced by the fact that gum arabic retains its viscosity and stabilising properties over a wide pH range. In comparison to various gums, gum arabic is the best emulsifying and suspending agent for many substances, and is also used as an adhesive or binder for tablets and as an excipient in the manufacture of pills and plasters. Many types of coatings for pills employ gum arabic in their manufacture. Because of its emollient and demulcent properties, gum arabic is used in the production of cough drops and syrups. It produces a smooth, viscous syrup and also prevents crystallisation of sugars in these products. In the cosmetic industry, gum arabic finds certain applications in a wide variety of products. In lotions and protective creams, it stabilises the emulsion, increases the viscosity, assists in imparting spreading properties, adds a smooth feel to the skin, and forms a protective coating. It is also used as a binding agent in the formulation of compact cakes and rouges, and as an adhesive in the preparation of facial masks. Gum arabic has the advantage in cosmetics of being nontoxic, with comparative freedom from dermatological and allergic reactions. It is used also in hair creams and fixatives.

Gum tragacanth is one of the oldest natural emulsifiers in the pharmaceutical industry. It is the exudation of several varieties of small, shrub-like plants of the *Astragalus* species. Chemically, tragacanth is a complex mixture of polysaccharides containing D-galacturonic acid. Other sugars produced on hydrolysis are D-galactose, L-fucose, D-xylose, and L-arabinose. Industrially, it is used as an emulsifier in oil emulsions, toothpastes, lubricating jellies, hair lotions, hand creams and lotions, facial clays, depilatory creams, and to form stable emulsions of insect repellants against mosquitoes, mites, chiggers, ants, and certain fleas.

I. PREPARATION OF COLOURING MATTERS FROM PLANT ORIGIN[9]

Certain colouring matters are prepared from medicinal plants. The following explores some examples of plants found in Africa from which colouring matters are prepared on an industrial scale:

1. *Curcumin* is prepared from the rhizomes of *Curcuma xanthorrhiza* (Fam. Zingiberaceae). Curcumin (diferuloylmethan colouring matter) and desmethoxycurcumin (hydroxycinnamoyl-ferulomethan colouring matter) are used as colouring matters.
2. *Hibiscin* is prepared from the flowers of *Hibiscus sabdariffa* L. (Fam. Malvaceae); also known as karkade or Sudanese tea. Hibiscin is delphinidin-3-xylosylglycoside, and is used to colour wine and certain pharmaceutical preparations.

J. INDUSTRIALISATION OF MARINE AND AQUATIC PLANTS[9]

As land resources throughout the world become strained to supply the needs of man, a further study of ocean potential is evident, particularly as the seas occupy some three quarters of the earth's surface.

The main emphasis of the study is concerned with food production, but such investigations become multidisciplinary and an interest in the sea as a source of new biomedicinals and other pharmacologically active compounds has been stimulated.

Early civilisations utilised marine life as a source of drugs. Marine organisms present a further problem in collecting and ensuring a constant supply of material. Some of the important marine products widely used in pharmaceutical industry follow.

1. Algin

It is found in all species of brown algae, in which it is the main structural constituent of the cell wall in Phaeophycae. Several species of *Macrocystis*, *Laminaria*, and *Fucus* are numerous in African waters and algins can be prepared from them. Algin finds extensive use in pharmaceuticals and cosmetics. It is used as a suspending agent and a protective colloid in many preparations. Smooth-flowing, nontacky surgical lubricants and medical jellies are produced from algin having good lubrication and emollient properties. It is used in ointments, emulsions, as a tablet binder and disintegrating agent, in toothpastes, shaving creams, hand creams and lotions, liquid shampoos, dental impression materials, as well as other industrial applications.

2. Agar

Agar is a hydrophilic colloid extract from certain red marine algae. It is insoluble in cold water, but soluble in boiling water. A 1.5% solution is clear and when cooled forms a firm gel which melts below 80°C. Several *Gelidium*, *Gracilaria*, *Sargassum pterocladia,* and *Suhria* species are common in African waters. Agar is classified as a suspending agent and is used as a bulk laxative, but mainly as a gel in bacteriological culture media. Industrially, it is employed for making certain emulsions and adhesives.

3. *Sargassum* Species (Fam. Sargassaceae)

Approximately 250 species of *Sargassum* are recorded; found mostly on rocky coasts of warm seas in Africa and oceans are potential sources of new drugs and foods.[51] The thalli of *Sargassum* species are known as Beerentag, Gulf weed, *Sargassum* weed.[52]

They are of extensive use in treatment of goiter,[53] renal colic, and lithiasis, and are anthelmintics, lipolytic, hypoglycemic, anticoagulant, and hypoglycemic, besides their use as fodder, fertiliser, and in the production of alginate. *Sargassum vulgare* thalli are found on protected rocky areas of the coast over an area reaching about 10 m in diameter.[54] Plants show large vegetation in flourishing seasons, i.e., spring and autumn, and shed most of their vegetative branches during the dormant season.[55]

Thalli are herbaceous, showing vertical branching. The terminals of the phyloclades rarely show dichotomous branching near their tips. The main thallus bears two types of branching: long, cylindrical laterals which reach up to 40 cm long and 2 mm thick, and numerous short, lateral branches arranged in alternate manner on the long laterals. They are cylindrical in shape and reach up to 6 cm long. They frequently carry globular stalked air vesicles, i.e., air bladders. The air vesicles vary in number, reaching up to five on each branch. They are occasionally cylindrical and have a stalk which is up to 7 mm long and 1 mm thick. The globular head is up to 0.5 cm in diameter. Occasionally, the short, lateral branches carry only flattened phylloclades, i.e., leaf-like thalli of varying number. These are found near the lower end of the branch. The phylloclades are mostly lanceolate, reaching up to 5 cm in length and 1 cm in breadth. They are brown in colour and their outline is slightly irregular. They occasionally show dichotomous branching near the tips. They have a smooth surface, showing numerous scattered dark brown spots.

A raised central ridge extends along each phylloclade. The thalli are attached to the short lateral branches with a short stalk. They show both symmetric and asymmetric bases. The short lateral branches carry numerous receptacles, which are male and female sexual organs. The receptacle shows dichotomous branching. The thallus has a sea-like odour and salty mucilagenous taste.[56]

Klenk et al.[57] examined the fatty acids in brown algae and found that C 16 and C 18 saturated fatty acids were dominant.

Moussa[58] observed the presence of arachidonic, oleic, palmitic, myristic, and lauric acids in brown algae.[58] He also reported the presence of fucosterol as the major sterol and cholesterol as the minor sterol in brown algae. Sarganin was isolated from several *Sargassum* sp., and is a complex, broad spectrum antimicrobial agent of low toxicity.

Analysis of the carbohydrates in *Sargassum* is helpful in the identification of the species. The most common are fructose, galactose, and mannitol.

Analysis of the ash showed it to be rich in potassium, sodium, calcium, magnesium, phosphorous, iron, and zinc. They also contain chloride, bromide, and iodine ions.[59-61]

Sargassum species show antitumor activity both *in vitro* and *in vivo*.[60] They show no antimicrobial, antiwormal, or molluscicidal activity.

III. ISOLATION OF ACTIVE INGREDIENTS ON AN INDUSTRIAL SCALE AND FORMULATION IN MODERN PHARMACEUTICAL PREPARATIONS[18]

This is the most important step and the backbone of industrialisation of medicinal plants. The rich flora of Africa of medicinal plants offers good raw material for the pharmaceutical industry, particularly as certain plants grow only in Africa. As the quantities of wild-growing plants are consistently declining, there must be an immediate response to increase cultivation and plantations of medicinal plants, using modern agricultural techniques to increase the yield of active ingredients and to have always a regular and constant supply for the pharmaceutical industry. Processing units should be established in areas producing medicinal plants to serve African factories.

There is evidence that on a semicommercial and, in certain examples, on a pilot scale, the isolation of active constituents from certain African medicinal plants has been successful. Among the most important active constituents which could be successfully isolated on an industrial scale for the pharmaceutical industry are the following examples.

A. ISOLATION OF ONCOLECTIC ALKALOIDS FROM *CATHARANTHUS ROSEUS* L.

Catharanthus roseus L. (Fam. Apocynaceae) is a common shrubby plant with pink or white flowers, found in the tropical and subtropical countries of Africa, such as in Madagascar, Mozambique, Nigeria, Tanzania, Kenya, Sudan, and Egypt. It is used in Madagascar as a hypoglycemic and the alkaloids have antitumor activity.

Madagascar perwinkle[62-67] is widely distributed in various parts of tropical and subtropical Africa. It is cultivated in several African countries as an ornamental plant, in gardens where it is multiplicated spontaneously. It contains about 70 alkaloids.[68-72] William[10] reported that the important biological properties were noted in four compounds, including the oncolytic alkaloids: vincaleukoblastine (vinblastine) leukocristine (vincrystine), diacetyl vinblastine, and vicristine-4-chloroacetate. The active alkaloidal content is exceptionally low; approximately 2 tons of crushed leaves yield 1 g of active alkaloid, equivalent to the supply needed for 1 month's treatment of one patient. This is why the alkaloids are extremely expensive. Vinblastine is used in treatment of monocytic leukemia, lymphosarcoma, Hodgkin's disease and other lymphomas, carcinomas, and histocytosis, while vincristine (vincaleukocristine) is used to treat acute leukemia, lymphosarcoma, carcinoma, and tumors.

In the first approach devised by Svoboda,[73] resolution of the complex alkaloid mixture was effected by successive extractions with an organic solvent at increasing pH increments. The first extraction at pH 2.5 to 3 contained the weak bases, while those at pH 8.7 contained the stronger alkaloids. Final purification of the alkaloids was accomplished by column chromatography using silicic acid or alumina partially deactivated with 10% acetic acid. Subsequent advances[73-76] in the isolation of catharanthus alkaloids permitted the separation of the monomeric form, the antitumor dimeric alkaloids using HPLC, and/or gel exclusion chromatography. Elution of the alkaloids vincristine and vincrosidine was affected by a citrate buffer at pH 3.0. The remaining polar bases were washed out with methanol and rechromatographed over alumina under 15 psi.

B. ISOLATION OF KHELLIN, VISNAGIN, AND KHELLININ FROM *AMMI VISNAGA* FRUITS

This genus comprises several species;[7] two (*Ammi visnaga* L. Khella and *Ammi majus* L. Khella shytani[77]) are common in North Africa, viz., Egypt, Morocco, and Algeria.

The plants are annual herbs attaining up to 2 m in height. They have solid, erect, slender, glabrous stems carrying alternate, exstipulate leaves. Flowers are grouped in compound umbels, and are small with a greenish to white colour. The primary and secondary rays gradually contract firmly in the case of *A. visnaga* at the fruiting stage, but spread in *A. majus*.

Fruit is cremocarp, splitting into two mericarps, reddish-brown in colour in *A. majus*, greenish-brown with violet tinge in *A. visnaga*. It has a slight odour, but has a pungent taste. *A. visnaga* L. contains cuptalline and bitter furochromone derivatives, viz., khellin visnagin and khellol glycoside.[78]

Ammi visnaga (Fam. Umbelliferae) fruits (powdered "Khella") are extracted with chloroform. The residue is kept for isolation of khellinin. The chloroform extract is concentrated and treated with hot water to separate fixed oil from khellin and visnagin. From the aqueous phase, both furanochromones are extracted with chloroform and separated through column chromatography. The residue left after isolation of khellin and visnagin is extracted with hot ethanol. The concentrated alcohol extract is left to crystallise to obtain pure khellinin (khellol glycoside).

Khellin is a vasodilator and spasmolytic, used to treat angina pectoris, chronic bronchial asthma, spastic conditions of stomach, intestines, gall and urinary ducts, in dysmenorrhoea, as well as for symptomatic treatment of whooping cough. Khellin is used to expel renal calculi, easing their passage. The plant grows abundantly in Egypt, Tunis, and Morocco.

C. ISOLATION OF AMMOIDIN FROM *AMMI MAJUS* FRUITS: WILD KHELLA

Ammi majus (Fam. Umbelliferae) fruits contain the furanocoumarins bergaptene, xanthotoxin, and imperatorin. Ammoidin is prepared in Egypt on an industrial scale for the world and is largely used in the treatment of leukodermia.

D. ISOLATION OF BABALOIN FROM CAPE ALOES

Powdered Cape Aloes, *Aloe ferox* Miller, Fam. Liliaceae and hybrids are extracted with hot methanol. Barbaloin is separated from the concentrated methanol through crystallisation or column chromatography. Several Aloe species grow abundantly in tropical and subtropical Africa. It is a safe to the colon, effective laxative.

E. ISOLATION OF CAPSAICINE FROM *CAPSICUM* FRUITS

Powdered *Capsicum annum* L. (Fam. Solanaceae) fruits are macerated with petroleum ether (defatting). The concentrated extract is saponified to free capsaicine, extracted with ether, and separated through column chromatography. Capsaicine is a counterirritant, promoting capillary dilatations, leading to strong blood flow. It is used to treat arthritis and related conditions.

F. ISOLATION OF CUBEBIN FROM CUBEBS

The volatile oil of *Piper cubeba* (Fam. Piperaceae) is prepared from powdered cubebs by water-steam distillation. The dried residue is extracted with chloroform and the concentrated chloroform passed over a column of kieselgel and crystallised. It is used as a urinary disinfectant and as an expectorant in bronchitis.

G. ISOLATION OF EPHEDRINE FROM *EPHEDRA* HERBS

Ephedra alata Decne (Figure 1) grows on sandy dunes, accumulating the sand into small hills. It occurs as a whitish, erect, sparingly leafy, diocious shrub, which attains up to 80 cm in height. It has a long, usually spirally twisted and branched tap root, with a knotty crown. The main root measures from 30 to 100 cm in length, and from 0.5 to 6 cm in diameter at its top portion. *Ephedra alata* flowers from February to May or even in June in some localities. The green branches and twigs contain the alkaloid ephedrine and pseudoephedrine[79] (isoephedrine) in varying amounts depending on the plant, altitude, time of year, and weather.

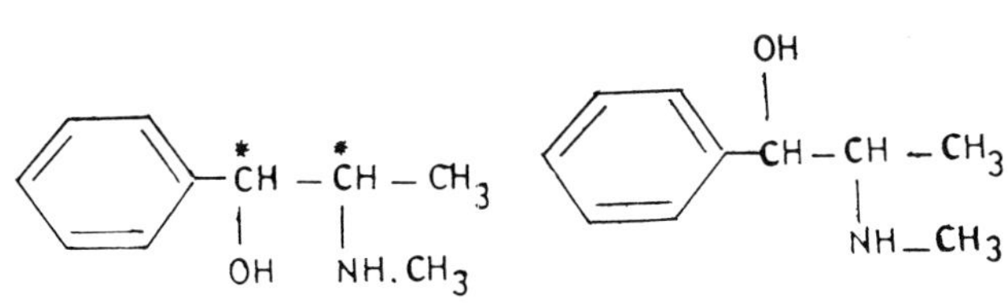

l- Ephedrine d-ψ Ephedrine

A

B

C

Figure 1. *Ephedra alata* Decne.

Ephedrine closely resembles epinephrine.[80] Pseudoephedrine stimulates both the inhibitory and accelerator mechanisms of the heart and has a stimulating effect on the myocardium. Ephedrine can be given orally and by injection. It is a mydriatic drug, stimulating the cardiac muscles, causing a marked rise in blood and pulmonary pressure, it is an effective dilator of the broncheoles, a uterine contractor, and is a diuretic. Pseudoephedrine is weaker in action on the heart, a more potent diuretic, and has no effect on the uterus. In larger doses, ephedrine induces nervousness, headache, insomnia, dizziness, palpitations, flushing of the skin, tingling and numbing of the extremities, nausea, vomiting, and possible pain in the cardiac region.

Leukocyanidins from *Ephedra* have vitamin P activity and inhibit the growth of different tumors.

The powdered *Ephedra* herb, when treated with sodium carbonate, releases the bases, which are then extracted by percolation with dichloromethane. Purification proceeds through multiple transfer between base and salt forms. The basic alkaloid is precipitated from the organic phase with chloroform. Ephedrine hydrochloride is obtained through fractional crystallisation. It is used in treatment of asthma.

The herb is common through subtropical Africa.

H. ISOLATION OF EUGENOL FROM CLOVES

From the petroleum ether extract used to obtain oleanolic acid from cloves (*Caryophyllum aromaticum* L. Fam. Myrtaceae), eugenol is separated as phenolate with sodium hydroxide. The aqueous alkaline phase is acidified and shaken with petroleum ether which is distilled, and the residue is distilled under vacuum to obtain eugenol.

The plant grows extensively in Zanzibar and the Pemba Islands.

I. ISOLATION OF 19 B-GLYCYRRHETIC ACID FROM LIQUORICE

Powdered peeled liquorice (*Glycyrrhiza glabra* L., Fam. Fabaceae) is extracted with chloroform to remove lipophilic substances. The residue is boiled with dilute sulfuric acid to hydrolyse glycyrrhizin to glycyrrhetic acid. From the hydrolysate, it is extracted with chloroform, concentrated, and passed over a kieselgel column. Glycyrrhetic acid is used to treat Addison's disease.

Liquorice is common in African countries bordering the Mediterranean.

J. ISOLATION OF L-HYOSCYAMINE FROM *HYOACYAMUS MUTICUE* HERB

Egyptian Henbane is a herbaceous plant with green, hairy, thick stems, about 30 to 90 cm high, with fleshy oval leaves; each leaf is sharply angled or lobed. Flowers are on one side of the stem, yellowish-white with violet colored spots and tube-shaped. The fruit is a small, globular light brown capsule enclosed in a prominent urn-shaped structure. The fruit contains numerous brown seeds. The plant flowers from March through April.

The coarse powdered *Hyoscyamus muticus* L. herb (Fam. Solanaceae) is extracted with petroleum ether to remove cellulose and wax. The alkaloids are extracted with dichloromethane. For further purification, they are transferred as salt in the aqueous phase, then made alkaline and alkaloids extracted, concentrated, and crystallised.

It is used as a spasmolytic and a mydriatic. The plant is abundant in Egypt.

K. ISOLATION OF HYOSCINE FROM *DATURA STRAMONIUM*

The thorn apple herb is an annual herb, stem green, erect, about 50 to 150 cm high, with spreading branches above. The plant flowers from July to September. It has a slight but characteristic narcotic odour and an unpleasant, bitter taste.

Hyoscine can be separated in a similar way as hyoscyamine from *Datura stramonium* (Figure 2) (Fam. Solanaceae), and hybrids grow in subtropical and tropical Africa.

Hyoscine hydrobromide is used as a sedative in acute mania and delirium. It helps the induction of general anaesthesia and is used in travel sickness preparations. It is also used for its mydriatic and cycloplegic action on the eyes.

Figure 2. *Datura stramonium* var. *inermis safford.*

L. ISOLATION OF HESPIRIDIN FROM ORANGE PEEL

Powdered, dried orange peel (*Citrus sinensis* L. Fam. Rutaceae) is extracted continuously with dichloromethane to be free from volatile oil. From the residue, hespiridin is extracted with methanol and recrystallised.

The plant is common in countries bordering the Mediterranean.

M. ISOLATION OF OUABAIN FROM *STROPHANTHUS GRATUS* WALL ET HOOK SEEDS

Powdered seeds of *Strophanthus gratus* Wall et Hook (Fam. Apocynaceae) are defatted with dichloromethane and then extracted with chloroform. The chloroform extract is concentrated and treated with lead acetate to precipitate resin and proteins. On concentration of the filtrate, ouabin is separated. *Strophanthus* species are common in tropical Africa. Ouabin is used as a heart tonic.

N. ISOLATION OF PIPERINE FROM BLACK PEPPER

Powdered black pepper (*Piper nigrum* L. Fam. Piperaceae) is extracted with chloroform. The concentrated extract is treated with alcoholic potassium hydroxide to remove the fixed oil. From the filtrate, piperine is crystallised.

Pepper is common in tropical Africa.

Piperine is a vasoconstrictor.

O. ISOLATION OF RUTIN FROM RUTA HERB

Powdered Ruta herb (*Ruta graveolens* L. Fam. Rutaceae) is extracted continuously with dichloromethane to extract lipophilic substances. The residue is extracted continuously with methanol extract rutin. The methanol extract is concentrated and recrystallised with water.

Rutin is used to treat varicose veins and in retinopathy of the eyes.

P. ISOLATION OF STRYCHNINE AND BRUCINE FROM NUX-VOMICA

Powdered Nux-vomica seeds (*Strychnos nux vomica* Fam. Loganiaceae) are defatted with petroleum ether, made alkaline, and bases extracted with chloroform. From the chloroform extract, alkaloids are separated with dilute acid as salt. After rendering alkaline, the water-insoluble alkaloids are separated as crystals.

The separation of strychnine from brucine is based on solubility in 50% ethanol, where brucine is better soluble than strychnine.

Q. ISOLATION OF *RAUWOLFIA* ALKALOIDS

Among 86 *Rauwolfia* species growing in many tropical and subtropical Africa, mainly Zaire, *R. serpentina* (Figure 3) and *R. vomitoria* are commercially the most important. They contain over 50 alkaloids, the major ones being reserpine, resinamine, ajmalcine, ajmaline, and serpentine. Reserpine is the most important and is used as an antihypertensive. Ajmaline is used in the treatment of cardiac arrythmia.

Rauwolfia vomitoria Afzel (Family Apocynaceae) *R. senegambiae* A. DC. — *Hylacinus owariense* P. Bearw. is a tree of 3.4 m long, widely growing in western Africa,[81] Senegal, Nigeria, Ivory Coast,[1] and Ghana.

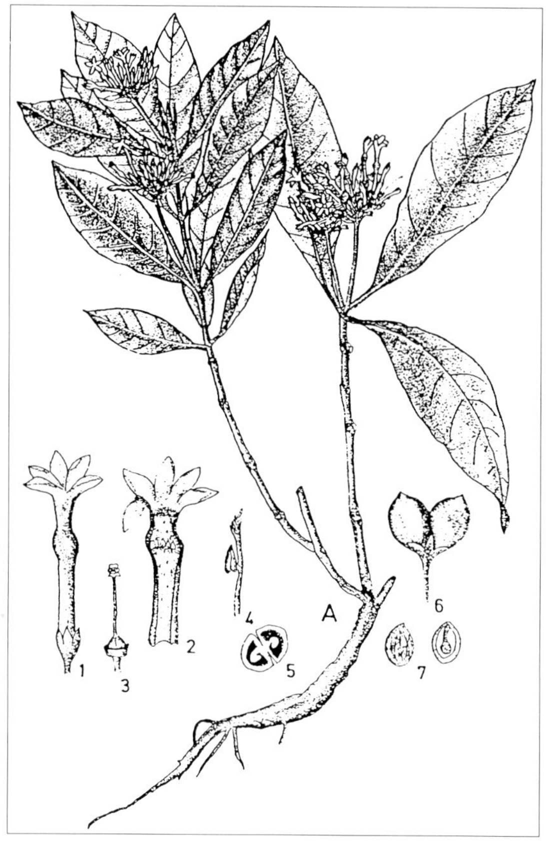

Figure 3. *Rauwolfia serpentina* (L.) Benth.

The leaves of African *Rauwolfia* contain about 0.8% total alkaloids,[81-86] but the roots contain a higher percentage.

Rauwolfia vomitoria contains the alkaloids ajmaline, reserpine, isoajmaline, ajmalinine, serpentinine, alstonine, as well as other constituents,[85] viz., 2,6-dimethoxybenzoquinone, 0.002%; the flavonoid glycosides nicotiflorin and astragalin; ursolic acid; and terpene alcohol vomifoliol. It is known as a remedy for neuropsychotic disorders and hypertension. Its preparations are dangerous because they are not only emetic but are also strong diuretics and purgatives.[86]

Reserpine is used as an antihypertensive drug and as a sedative.[83].

Ajmaline is a sympatholytic drug and is used as an antiarythmic drug in cases of tachycardia.[86]

REFERENCES

1. **Mahran, G.H.,** *Medicinal Plants*, Anglo-Egyptian Bookshop. Cairo (1967).
2. **Hilal, S. and Youngken, H.,** Certain poisonous plants of Egypt. National Information and Documentation Centre, Egypt (1983).
3. African Pharmacopoea 1985, OAU/STRC, Vol. 1, 1st ed. (1985).
4. **Mahran, G.H.,** *Principles of General Pharmacognosy*, Cairo University Press (1978).
5. **Raynal, J., Troupin, G., and Sita, P.,** *Flore et Medecine Tradionelle, Mission Detude 1978 au Rwanda.* Rapport a LA. C.C.T. (1978).
6. **Ake Assi, L., Abeye J., Giguet, R., and Bangavou Y.,** Contribution a Lidentification et au Recensement des Plantes Utilisees dans la

Medecine Tradionne le et la Pharmacopee en Empire Centrafrican, rapport presenta a L A.C.C.T (1977).

7. **Adjanohoun, E.J. et al.,** *Medecine Tradionnelle et Pharmacopee Contribution aux etude Ethnobotanique et Floristiques au Niger,* rapport presente a L A.C.C.T. (1978).
8. **Adjanohoun, E.J., Ake Assi, L., Floret, J.J., Guinko, S., Koumare, M., Ahyi, A.M.R., and Raynal J.,** *Medecine Tradionelle et Pharmacopee. Contribution aux etude Ethnobotaniques et Floristiques au Mali,* rapport presente a L A.C.C.T (1978).
9. **Mahran, G.H.,** *J. Afr. Med. Plants, 5,* 267 (1982).
10. **Boulus, L.,** *Medicinal Plants of North Africa,* Reference Publications, Michigan 48105 (1983).
11. **Sofowora, A.,** *African Medicinal Plants,* University of Ife Press, ILE-IFE, Nigeria (1979).
12. **Watt, J.M. and Breyer, B.W.,** *The Medicinal and Poisonous Plants of Southern Africa.,* Livingstone, London (1962).
13. **Atal, C.K. and Kapur, B.M.,** *Cultivation and Utilisation of Medicinal Plants,* Council of Scientific and Industrial Research, Jammu, Tawi-India (1982).
14. **Mercier, L.,** *Les Nouv. Remedes,* 31, 1 (1914).
15. **Derasari, R.,** *Indian J. Pharm.,* 27, 278 (1965).
16. **Flypiw, A.,** *Lloydia,* 28, 764 (1965).
17. **Bors, F.,** *Pharmazie,* 26, 764 (1971).
18. **Stahl, E. and Schild, W.,** *Pharmazeutische Biologie,* Gustav Fischer Verlag, Stuttgart (1981).
19. **Atal, C.K. and Kapur, B.M.,** *Cultivation and Utilisation of Aromatic Plants* Council of Scientific and Industrial Research, Jammu, Tawi-India (1982).
20. **Balbba, S.I.,** *Medicinal Plant Constituents.* General Organization for University and School Books, Cairo (1981).
21. **Fahmy, I.R.,** *Medicinal Plants and Their Vegetable Drugs.* Paul Barbeys Printing Office, Cairo (1932).
22. **Von Euler, U.S.,** *Tobacco Alkaloids and Related Compounds,* Vol. 1, Pergamon Press, London, 346 (1965).
23. **Garnier, G., Bezanger-Beauquesne, L., and Debraux, M.,** *Resources Medicinales de la Flore Francaise,* Vigot, Paris (1961).
24. **Planchon L. and Bretin P.,** *Precis de Matiere Medicale,* Maloine, Paris. 1790 (1946).
25. **Polson C.J. and Tattersall R.N.,** *Clinical Toxicology* (1959).
26. **Textor, O.,** *Am. J Pharm.,* 53, 491 (1881).
27. **Marino Zuco, F.,** *J. Chem. Soc.,* 60(1), 333 (1891).
28. **Ibid.,** 62, 84 (1892).
29. **Ibid.,** 59, 212 (1890).
30. **Thoms, V.,** *J. Chem. Soc.,* 62, 349 (1892).
31. **Fujitani,** *J. Arch. Exp. Pathol. Pharm.* 61, 47-75 (1909).
32. **Chevalier, J. and Fermand,** *Compt. Rend.,* 176, 1847 (1923).
33. **Yamamato, R.,** *J. Chem. Soc. Jpn.,* 44, 311, 1070 (1923).
34. **Juillet, A.,** *Comp. Rend.* 177, 294 (1923).
35. **La Forge, F.B.,** *J. Am. Chem. Soc.,* 71, 3165 (1949).
36. **La Forge, F.B. and Barthel, W.F.,** *J. Org. Chem.,* 9, 242 (1944).
37. **Ibid.,** 10, 106-114 (1945).
38. **Harper, S.H.,** *J. Chem. Soc.,* 32, 892 (1946).
39. **La Forge, G.B. and Soloway, S.B.,** *J. Am. Chem.* Soc., 69, 186, 2932 (1947).
40. **Crombie, L., Elliott. M., Harper, S.H., and Reed, H.W.B.,** *Nature,* 162, 222 (1948).
41. **Beevor, P.S., Golin, P.J., and Snarey, M.,** *Chem. Ind.,* 30, 1342 (1965).
42. **Gnadinger, C.B. and Corl., C.B.,** *J. Am. Chem. Soc.,* 51, 3054-3064 (1930).
43. **Ziegler, W.H.,** *J. Am. Pharm. Assoc.,* 12, 19, 26 (1923).
44. **Chevalier, J. and Ripert, J.,** *Comp. Rend,.* 184, 776 (1927).
45. **Chevalier, J.,** *Pharmacology,* 37, 154, 651 (1930).
46. **Anglad, G. O.,** *Bull. Sci. Pharmacol.,* 39, 23, 6 (1932).
47. **Lemaire, A. and Gaudin, O.,** *Bull. Sci. Pharmacol.,* 38, 692-4 (1951).
48. **Sweitzer, A. and Teddar,** *Minn. Med.,* 18, 793 (1935).
49. **Tobias, H.** *G.P.,* 4, 44 (1951).
50. **El Defrawi, M. and El-Bahrawi, J.,** *Econ. Entomol.* 58(1), 93 (1965).
51. **Rayakhorski, K.,** *Tr. Inst. Friziol. Akad. Nauk. Kaz. S.S.R.,* 16, 28 (1971).
52. **Levering, T.,** *Marine Algae,* De Gruyter, Hamburg (1969).
53. **Prescott, G.W.,** *The Algae,* Butler & Tanner, London (1966).
54. **Levering, T.,** *Marine Algae,* Gram, Hamburg (1969).
55. **Rond, F.F.,** *The Biology of Algae,* Arnold, London (1966).
56. **Smith, C.M.,** *Cryptogamic Botany,* Macmillan, London (1977).
57. **Klenk, E.,** *Hoppe Seyler's Z. Physiol. Chem.,* 344, 44 (1963).
58. **Chapman, V.J.,** *The Algae,* Arnold, London (1955).
59. **Youngken, H.W.,** *Apothekary,* 87, 10 (1975).
60. **Solimabi, D.,** *Indian J. Mar. Sci.,* 6, 180 (1977).
61. **Weinheimer, A.J.,** *Tetrahedron Lett.,* 923 (1977).
62. **Kato et al.,** *J. Am. Chem. Soc.,* 96, 2245 (1974).
63. **Mangoldzhier, N.,** *Apoteka.,* 22, 46 (1972).
64. **Stolle, Vonk. and Groger, D.,** *Pharmaz. Zentral.,* 106(5), 285 (1967).
65. **Svoboda G.H.,** *Excerpta Medica Int. Congr. Ser.,* Pays-Bas, 106 (1966).
66. **Paris R. and Moyse, H.,** *J. Agric. Trop. Bot. Appl.,* 4, 481 (1957).
67. **Roy S. and Chatterjee, S.,** *J. Indian. Chem. Soc.,* 34, 340 (1957).
68. **Forsyth, W.G.G. and Simmonds, W.G.,** *Nature,* 180, 247 (1957).
69. **Hertz, R. et al.,** *Cancer Res.,* 30, 1050 (1960).
70. **Andriamaialison, A.,** *Tetrahedron. Lett.,* 3, 163 (1976).
71. **Potier, G.,** *Appl.,* 73, 330 (1973).
72. **Flypiw, F.,** *Lloydia.,* 28, 354 (1965).
73. **Svoboda, G.H.,** Lloydia, 26, 243 (1962).
74. **Svoboda, G.H.,** Lloydia, 27, 299 (1964).
75. **Svoboda, G.H.,** J. *Am. Pharm. Ass.,* Sci. Ed., 48, 659 (1959).
76. **Svoboda, G.H.,** *J. Pharm. Sci.,* 51, 518 (1962).
77. **Wallis, T.E.,** *Text Book of Pharmacognosy,* 4th ed., J. & A. Churchill, London (1960).
78. **El-Keiy, M. et al.,** *Rep. Pharm. Soc. Eg.,* 3, 36 (1931).
79. **Bencze, B.,** *Helv. Chem. Acta,* 39, 923 (1956).
80. **Wahab, S.M.,** Ph.D. thesis, Cairo University (1960).
81. **Taraskina, T.,** *Khim. Prir. Soedin. Teknol.,* 1, 136 (1970).
82. **Heal, R.F. and Rogers, E.G.,** *Lloydia,* 13, 89 (1950).
83. **Paris, R.,** *Ann. Pharm. Fr.,* I, 138 (1943).
84. **Schlitter, E. et al.,** *Helv. Chem. Acta.* 35, 271 (1952).
85. **Kupchan, S.M. and Obasi, M.E.,** *J. Am. Pharm Assoc.,* 49, 257 (1960).
86. **Paris, R. and Etchepare, S.,** *Ann. Pharm. Fr.* 25, 783 (1967).

Chapter 15

SCREENING AND PROCESSING OF PLANT MATERIALS FOR POTENTIAL PHARMACEUTICAL NEEDS: EXPERIENCE AND APPLICATIONS IN THREE CONTINENTS

Jan G. Bruhn and Finn Sandberg

I. INTRODUCTION

The spectacular resurgence of use of medicinal plants by Western society during the past 20 years has been parallelled by a fast growing scientific interest. This interest has been rekindled in part because of the worldwide importance of medicinal plants and crude drugs in traditional medicine, in part because of the possibility of doing more precise studies, and in relatively small part because research funds became available.

The rate of increase in scientific inquiry into the use, effects, and constituents of medicinal plants has risen steeply over the last decade. It may soon be impossible to keep current with the rapidly growing literature.

The following discussion of the screening and processing of medicinal plants and/or crude drugs can therefore only serve the reader as an introduction to a very broad field of research.

II. GENERAL CONSIDERATIONS

A. EXISTING USE OF NATURAL PRODUCTS AS DRUGS

Although modern organic chemistry has provided us with an increasing number of synthetic compounds for medicinal use, many important drugs of modern medicine are still extracted from medicinal plants. Some of these plants were originally employed as traditional remedies, and some of them are still used in traditional therapy.

A survey of all registered Swedish drugs has shown that natural products account for 51% of all preparations,[1] and this percentage would seem to hold true for many other countries as well. Sticher estimated that only 50 to 55% of all drugs in the industrialized countries are synthetic.[2]

More than 33% of Russian drugs are derived from plants.[3] In France, plant drugs have a market share of 60 to 65%.[4] Digitoxin, quinine, atropine, reserpine, physostigmine, emetine, vincristine, and pilocarpine are a few examples of what traditional medicine has given us in the past.[5]

Indeed, it has been remarked that "one only needs to open the pages of any standard textbook of pharmacology to be impressed by the fact that virtually every pharmacological class of drug includes a natural product prototype that exhibits the classical effects of the pharmacological category in question; most of them are plant-derived".[6]

While modern pharmacologists sometimes appear ignorant of these facts, the majority of people in the world rely predominantly on medicinal plants as drugs. Because of this, WHO has emphasized the importance of scientific investigations into indigenous herbal medicines.[7-9]

In many countries of the world native medicinal plants are looked upon as possible additions to the WHO list of "essential drugs", once their value has been clinically proven.[10,11]

B. EXISTING KNOWLEDGE

Medicinal plant researchers today often forget to evaluate *the traditional knowledge* that existed about the plant or drug in question before they came along. Much of the work of the past to be sure was based on descriptions of effects caused by uncertain doses in uncertain medical conditions.

Still, much of the newer work with medicinal plants and/or traditional drugs involves the rediscovery of effects known for a long time. Modern pharmacological research may thus have contributed less that is really new than we like to believe. Careful documentation of traditional knowledge is necessary to achieve a true validation (or invalidation) of a medicinal plant through pharmacological experimentation.[12-19]

On the other hand, it must also be remembered that the advanced techniques of modern chemistry have made the isolation and structure determination of new compounds much easier during recent years.[20,21]

All plants suggested for further study should consequently be carefully searched out in *the modern scientific literature* to avoid unnecessary and uneconomic duplication of basic research.

It is clearly necessary for any research group in this area to have access to the leading scientific journals dealing with the subject, for example, *Planta Medica*, *Journal of Natural Products (Lloydia)*, *Journal of Ethnopharmacology*, *International Journal of Crude Drug Research*, as well as the indispensable *Medicinal & Aromatic Plants Abstracts*.

C. PROBLEMS OF PLANT-DERIVED DRUGS

There are several problems associated with the development of plant-based drugs:

- Patent protection is weak for plant-derived drugs.[6]
- The long-term supply of botanically well-defined raw materials can be difficult to achieve, especially for imported items.[22]
- The biological variation inherent in the raw material can result in quantitative and qualitative deviations.[23]
- Plant extracts are often difficult to dissolve in water and consequently difficult to test in various bioassay systems.[6,24,25]
- Active compounds are often present in minute concen-

trations in very complex mixtures which requires an advanced knowledge of modern separation methods and highly sensitive screening methods.[6,26]

- The biological activity can be lost during the work-up procedure, due to instability of the active compound(s) or the fact that the activity is a synergistic or additive effect between several structurally different compounds.[27]

D. THE PROSPECTS OF SCREENING

It is sometimes suggested that the field of medicinal plants and/or crude drugs has already been exploited to the extent that further research would yield only "academic" results and no novel drugs.

In contrast to this point of view, we believe that modern research has only scratched the surface of this field. Consider these two points:

- Whereas Carl von Linné believed that there were only about 10,000 species comprising the plant kingdom, modern estimates vary between 280,000 and 700,000 species.
- A vast knowledge of and experience with medicinal plants has accumulated during generations. This information has been annotated in herbals, herbarium notes, ethnobotanical studies, and pharmacopoeias, and is still part of the oral tradition of many ethnic groups. Penso has listed about 21,000 names of medicinal plants from 91 countries in a recent index.[28]

In the search for plant drugs for medicinal/pharmaceutical needs, it would obviously be foolish not to make use of this ethno-pharmacological experience. Reports that plants are used as insecticides, fish or arrow poisons, or are known to be toxic, are of course of equal interest.

E. INDUSTRIAL DEVELOPMENT

The processing of plant materials and formulation of a drug product requires a special approach which must include two fundamental points: (1) standardization of the manufacturing process know-how and quality assessment of active plant principles, and (2) biological testing to ensure safety and efficacy. These aspects are discussed later. One more important point, i.e., the genetic development to improve crop yield and the yield of active phytochemicals, is not addressed in this chapter.

III. SCREENING METHODS

Many different approaches have been employed in the search for natural products with biological activity.[19] The main purpose of the various screening methods is of course the *selection* of plant material for further study. Basically, there are two ways of looking for biologically active compounds: we can either search for the compound or for the effect it produces.

A *phytochemical screening* involves the testing of several different plant extracts for their content of different classes of compounds, such as alkaloids, flavonoids, saponins, and/or terpenes.

A *pharmacological/biological screening* starts by testing various extracts for possible effect(s) in one or more bioassay systems. The screening system can then be utilized to monitor the isolation of a biologically active compound (or group of chemicals).

An *ethnopharmacological screening* utilizes traditional knowledge in various cultures as a selection tool. This is a highly interdisciplinary approach which is attracting more and more interest.

Some examples and comments on the various methods are given in the following sections. A special section discusses some aspects of *field work*.

A. PHARMACOLOGICAL SCREENING

Biological screening can be performed on plants collected at random or on a selected group of plants, e.g., medicinal plants used in a special area or in a certain cultural complex.

In order to test the biological effects of plant material simple screening methods have been worked out using mice and rats.[25,27] The original method, elaborated by Malone, has been modified by one of us[19] and used in many different projects.[24,29,30]

In an inventory of medicinal and toxic plants from Equatorial Africa, one of us described the vernacular names and local uses of 196 medicinal plants collected in the southwest region of the Central African Republic and the northern region of the Congo. *Regardless of their local uses* these medicinal plants were all run through our pharmacological screening in rats.[24]

In the standardized protocol for this screening 53 parameters were monitored and annotated (Figure 1). On the basis of this Hippocratic screening the total material of 196 species could be divided into the following three groups: (a) 119 species with a strongly pronounced effect, (c) 70 species with a moderate effect, and (c) 7 species with no effect.

Of course, it is not necessary to observe all 53 parameters at all times. In Panama this procedure has been used in a reduced form for the screening of local medicinal plants, coupled with an LD_{50} determination of the various plant extracts.[29]

In contrast to these selected groups of plants, an extensive Indian screening project has tested more than 1000 Indian plants for various biological activities.[31] The test battery included antibacterial, antifungal, antihelminthic, antiviral, antifertility, hypoglycaemic, antilipaemic, effect on respiration, cardiovascular effects, effect on isolated ileum, effect on central nervous system and gross behavior, anticancer, diuretic, and antiinflammatory tests. Recent reviews[32,33] of generalized *in vitro* screening tests are also available.

These generalized biological screening procedures are giving way to more specialized tests, using fewer biological tests and critically selected plants. A useful organ that has proven its value in various medicinal plant projects is the guinea pig ileum (nonstimulated and stimulated). This *in vitro* procedure is especially useful in monitoring the isolation of active principles.[20,30]

In the future, biological/pharmacological screening will

Date: ________ Notebook No. ________ pp. ________ Project Code: ________ ;Reader Signature: ________

Hippocratic Qualitative/Semiquantitative Screen and Toxicity Report of: ________ Test Animal: ________ ; Fasted/Nonfasted: ________ ; Sex: ________ ;Weight: ________ g

________ Lot/Identification No. ________ Mark: ________ ;Dye Color: ________ ;Cage No. ________ ; Special Treatment: ________

Dosage: ________ mg/kg; Dosage Vehicle: ________ ;Conc.: ________ mg/ml Injection Volume: ________ ml; Injection Route: ________ ; Clock Time: ________

PARAMETERS	Response Rating (0 - +4) or Measurement													
	Con-trol	+5 min	+10 min	+15 min	+30 min	+60 min	+2 hrs	+ hrs	+ hrs	+24 hrs	+48 hrs	+ days	+7 days	*
Decr. Motor Activity														
Decr. Resp. Rate														
Decr. Resp. Depth														
Dyspnea														
Cheyne-Stokes Resp.														
Analgesia														
Anesthesia														
Loss, Corneal Reflex														
Loss, Pinna Reflex														
Back Plasticity														
Ataxia														
Hind Leg Grip Loss														
Foreleg Grip Loss														
Hind Leg Paralysis														
Foreleg Paralysis														
Neck Paralysis														
Incr. Motor Activity														
Incr. Resp. Rate														
Incr. Resp. Depth														
Fine Body Tremors														
Coarse Body Tremors														
Back Tonus														
Tonic Convulsions														
Clonic Convulsions														
Tail Erection														
Tail Grasping														
Tail Lashing														
Enophthalmos														
Exophthalmos														
Palpebral Ptosis														
Pupil Size, mm														
Pupil Size, (light)														
Nystagmus														

PARAMETERS	Response Rating (0 - +4) or Measurement													
	Con-trol	+5 min.	+10 min.	+15 min.	+30 min.	+60 min.	+2 hrs.	+ hrs.	+ hrs.	+24 hrs.	+48 hrs.	+ days	+7 days	*
Chromodacryorrhea														
Ear Blanching														
Ear Hyperemia														
Ear Cyanosis														
Ear Metachrosis														
Fasciculation														
Thick Salivation														
Watery Salivation														
Pilomotor Erection														
Robichaud Positive														
Writhing Movements														
Râles														
Micturition														
Colored Micturition														
Diarrhea														
Stereotypy														
Circling Motions														
Disorientation														
Statue Positions														
Startle Sensitivity														
Head Tap: Aggressive														
Head Tap: Fearful														
Head Tap: Passive														
Body Grasp: Aggressive														
Body Grasp: Fearful														
Body Grasp: Passive														
Persistent Grooming														
Priapism/Colpectasia														
Rectal Temperature, °C		x	x	x										
Body Weight, g		x	x	x										

Figure 1. Standardized protocol for Hippocratic screening. (From Reference 25 with permission.)

dominate chemical screening and more sophisticated activities (e.g., enzymatic) shall have to be tested. The evaluation of medicinal plants will have to find new ways, since the pharmacology of pure compounds is not always applicable to the testing of plant extracts. Furthermore, more attention should be paid to using determined amounts of plant material and to comparing the results with standards to avoid subjective evaluation.

B. PHYTOCHEMICAL SCREENING

The methods used to detect various phytochemicals must be simple and rapid, but they should also be selective and easy to use.

The utility of the preliminary chemical field screening lies in the identification of major chemical groups, thus providing guidance for further research.

Various field tests for alkaloids, etc. can be found in phytochemical and pharmacognostical textbooks, and can be very useful. However, false-positive reactions do occur, and will probably have to be accepted in field screening. The application of several different tests can reduce their number.

This preliminary research should not be aimed at the isolation of pure, biologically active compounds, but simply to detect certain major chemical groups for which general tests are available.

The methods used in these field tests should ideally have the following characteristics:[8,34]

1. They should only require the simplest equipment and a few reagents that can be easily transported.
2. They should be standardized and well adapted to the routine detection of alkaloids, saponins, glycosides, etc.
3. They should be easy to handle by technical staff after adequate training.

The classical example of alkaloid screening in the development of a folk medicine into a drug of therapeutic importance is the search for new sources of reserpine in the genus *Rauvolfia*. The root of *Rauvolfia serpentina* has been used for centuries in Indian medicine as a cure for insanity, epilepsy, and high blood pressure. In the 1930s Indian workers started an investigation of the plant and isolated several alkaloids. However, these alkaloids could not account for all of the interesting effects found in the plant extracts. The search for active principles continued, but not until 1952 was the alkaloid reserpine isolated and shown to possess all the main pharmacological properties of the root.

This discovery initiated an intensive screening for this therapeutically valuable alkaloid in the genus *Rauvolfia*, and subsequently also in other, closely related genera of the Apocynaceae. *Rauvolfia* is a genus of about 100 species, which are widespread in the tropics, and alkaloid screening revealed the presence of reserpine in several of these. On the basis of this screening, *R. vomitoria* of Africa is now used for the commerical extraction of reserpine.[34]

A recent example concerns the genus *Murraya* (Rutaceae) which has been widely used in traditional medicine in east Asia.[35] After the isolation of the anti-implantation alkaloid yuehchukene from *Murraya paniculata* a search was made for other natural sources of this alkaloid within the genus.

The yields of yuehchukene obtained from *M. paniculata* are exceedingly low, in the region of 5 to 20 ppm. Although yuehchukene has been synthesized, it was considered worthwhile to explore other natural resources so that if it proves to be acceptable for human use, then the most appropriate sources of the natural alkaloid will be available, e.g., when cultural or economic reasons speak for the preference of using natural plant materials.

It appears that a number of species of *Murraya* are potential sources of yuehchukene, but these species are rich in coumarins. It is obvious that the toxicity of these compounds must be thoroughly assessed before the use of a crude *Murraya* extract could be contemplated.[35]

Similarly, the phytochemical screening of Sri Lankan medicinal plants showed the occurrence of pyrrolizidine alkaloids in several crude drugs. The hepatotoxic properties of these alkaloids are well known and a limiting factor for the use of some traditional medicinal herbs.[47]

A screening procedure is not necessarily limited to the finding of and the preliminary characterization of given compounds. Actually, a chemical screening can be misleading in that it focuses attention on compounds that may not be active.

In traditional medicine remedies are almost exclusively prepared by extraction with water, and this fact has been long neglected by chemists, who have preferred to work with organic solvents. Aqueous extracts have been avoided because of their complexity and difficulties in work-up procedures. Logically, pharmacologically active compounds are extractable by water, and the use of other solvents may lead to a loss of activity.

Samuelsson et al.[20] have recently described a test system for the preliminary chemical characterization of pharmacologically active compounds in aqueous extracts. The system is based on the use of ion exchangers, gel filtration on Sephadex, and extraction with organic solvents, and gives information about the stability, molecular size, charge, and polarity of pharmacologically active compounds in aqueous plant extracts. Knowledge of these properties permits development of a suitable isolation method for the compounds.

The authors have used the guinea pig ileum for the detection of pharmacological activity, but the method can be applied regardless of the test procedure used for monitoring the activity. Figure 2 summarizes this promising test system, which can also be considered a pilot scale extraction.

The phytochemical screening should be carried out with determined amounts of plant material to make possible objective comparisons. On account of the *biological variation* met with in nature the screening results will have to be repeated on different plant individuals (and parts) of the same species from different localities.

C. ETHNOPHARMACOLOGICAL SCREENING

Ethnopharmacology can be defined as the interdisciplinary scientific exploration of biologically active agents traditionally employed or observed by man.[16]

Field observations and descriptions of the use and effects of native remedies by anthropologists and ethnobotanists, botanical identification, and phytochemical and pharmacological studies, are all within the scope of ethnopharmacology. The ultimate aim is the validation (or invalidation) of these preparations, either through the isolation of active substances or through pharmacological findings.

The basic disciplines involved are botany, chemistry, and pharmacology. The role of other disciplines should, however, not be overlooked or minimized.

There are several good examples of the profitability of ethnopharmacological screening. The Regional Research Laboratory at Jammu in India has a well-established pharmacology division which has been in existence for the last 2 decades. During the mid-1960s, a program for screening of the Indian flora with a view to discovering new herbal drugs possessing promising physiological and pharmacological activity was undertaken. The approach in this program was based on the then-existing pattern of evaluation of plant materials, the main feature of which was random selection of mostly 50% alcoholic extracts, and, in case of any promising lead based on animal experiments, further fractionation of these extracts for test purposes.

The pharmacological work included 15 initial screening tests for major types of activities. Out of 400 plants screened and a number of them studied in depth, the researchers did not discover any promising new drug. A decision of far-reaching significance was made in the early 1970s by introducing more meticulous selectiveness in the choice of plants for pharmacological investigations. Another change introduced in the programme was to carry out in-depth studies on a few selected plants rather than undertaking superficial screening of a large number of herbs. In the choice of plant drugs for investigations, the researchers therefore started relying heavily on information sources about the existing usage of certain herbs for relief and cure of specific diseases, based on ethnobotanical information. The work on these suggestions has been quite rewarding and three such examples are *Pueraria tuberosa*, an antifertility drug; *Allium cepa*, an antiasthmatic, and *Adhatoda vasica* as an oxytoxic and abortifacient (C.K. Atal, personal communication).

Based on the extensive screening results obtained by the National Cancer Institute, Spjut and Perdue have also dis-

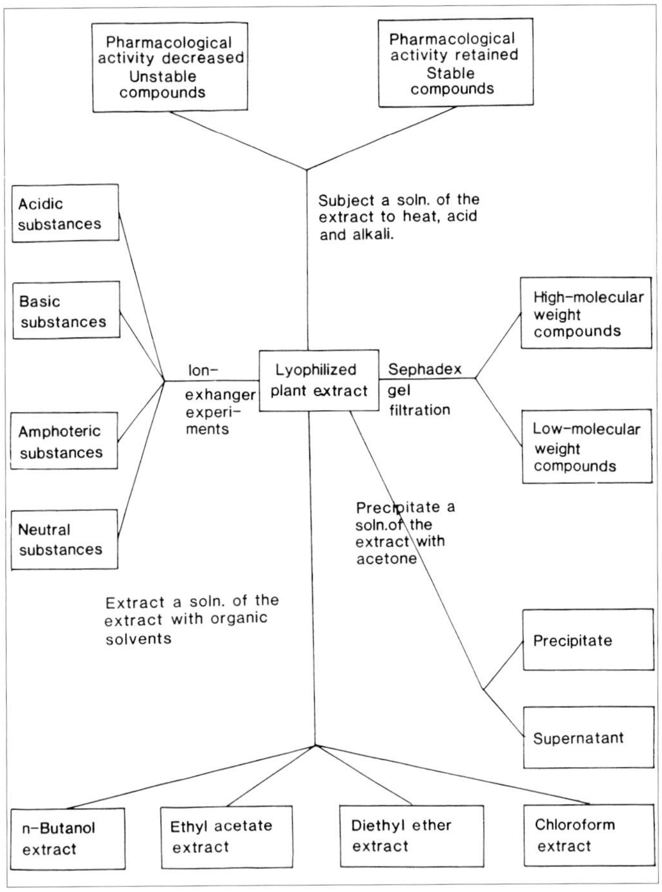

Figure 2. Fractionation of an aqueous plant extract for chemical characterization of pharmacologically active compounds. (From Reference 20 with permission.)

cussed the role of plant folklore in finding new sources of antitumor activity.[36]

The plant material studied at the National Cancer Institute (Bethesda, MD) has been collected at random, but the analysis performed by Spjut and Perdue (summarized in Table 1) shows that if antitumor screening had been guided by the knowledge of medicinal folklore and poisonous plants, the yield of active species would have been greatly increased.

It is impossible, however, to establish a dose-effect relationship unless the original drug preparations (water infusions, etc.) are analysed and evaluated chemically and pharmacologically. As a result we must, in ethnopharmacological research, have proper sampling and analysis methods.

Aqueous decoctions of *Oldenlandia affinis* are traditionally used to accelerate childbirth by the Lulua population in Zaire. In his capacity as a medical missionary Gran observed increased uterine activity in women receiving these preparations. He also isolated serotonin from the plant. However, the usual decoction from about 100 g of dried plant would not give a higher total intake than about 2 mg of serotonin, and since serotonin is rapidly destroyed when taken orally it could have no action. In *in vitro* experiments with isolated organs, Gran showed that the increased uterine contractions produced by the water extract also persisted after the treatment with methysergide. Therefore, another (or several other) compound(s) must be the active principle(s). In his research, which is a fine example of modern ethnopharmacology, Gran went on to isolate the active principles and found oxytocic peptides.[37]

TABLE 1
Plant Folklore: A Tool for Predicting Sources of Antitumor Activity?

Plant types	Active (%)[a]
Plants collected at random	10.4
Plants used against cancer	19.9
Anthelmintics	29.3
Fish poisons	38.6
Plants poisonous to man	50.0
Arrow poisons	52.2

[a] Extracts show a significant inhibitory effect in experimental tumor systems (National Cancer Institute, Bethesda, MD). (Table arranged from data presented in Reference 36.)

Most traditional drugs are administered as mixtures of many components, and with today's knowledge of the many possible interactions between drugs, and between food and drugs, ethnopharmacological research must also deal with this aspect. Additive, synergistic, or antagonistic effects are all possible. Various admixtures have also been shown to affect the bioavailability of pharmacologically active principles.

Pharmacological studies of traditional medicinal agents

should be initiated prior to or in parallel with chemical research and should guide the isolation of active principles.

In this way one of us has conducted an investigation to validate the usefulness of *Dorstenia drakena* roots for controlling diarrhoea. Extracts were first screened pharmacologically by the Hippocratic screening procedure.[24] Subsequent evaluations focused on the effects of various fractions on cholera toxin-induced intestinal secretion in mice, the coaxially stimulated guinea pig ileum, and inhibition of *Escherichia coli* and *Staphylococcus aureus* growths.[30]

Field observations of traditional therapies and the pharmacological effects in humans should be carried out by trained pharmacologists, and when interesting activity is found, controlled experiments should be initiated.[38-40]

Research into traditional medicine should follow these guidelines. Experimental pharmacological research should be designed to test water extracts, infusions from plants, etc., as used in aboriginal medicine, and to establish the effectiveness and toxicity (if any) of these remedies.

Interdisciplinary research teams should ideally include physicians, sociologists, botanists, anthropologists, chemists, pharmacognosists, pharmacologists and traditional healers, all contributing to make this approach a most effective one. Malone has suggested a research approach in ethnopharmacology which shows how this interdisciplinary team should work (see Figure 3).[25]

Various other data and criteria may be of additional value in the selection of plants for in-depth study, as shown by Farnsworth et al.[38,41,42]

The validation of traditional drugs must also take into account the disease or symptoms treated. This is an often overlooked aspect, which should receive more attention.[14,15,18,30,43]

We believe that an interdisciplinary bioscientific approach is essential to the attempts to verify the therapeutic potential of an indigenous drug. The goals and methods of ethnopharmacological research and the implications for drug development are summarized in Figure 4.

D. FIELD WORK

The accumulated experiences, positive and negative, of pharmacognostic field work in various parts of the world have made certain points clear.

The importance of the correct systematic authentication of the plant material cannot be overemphasized. For this reason appropriate herbarium material must be collected. A voucher specimen for reference should be kept in the department where the investigation is made and should be mentioned in the publication. The plant material is labelled with the collection number of the voucher specimen, used by the pharmacognosist or the accompanying botanist, but not both.[22,24,44]

The amount required for phytochemical and/or pharmacological screening is 1 to 3 kg. When the screening has shown the presence of substances worthy of further investigation, larger quantities are required, about 10 to 20 kg. Quite a lot of work is involved in collecting this amount of, e.g., an African *Strychnos* liana: the supporting tree has to be cut down, the lians carefully "dissected", and the bark (which contains the active substances) removed from the stem.

The drying of plant material in the tropics often creates great problems; therefore the dry season has to be chosen for collecting to allow the material to be dried in the sun or in a sheltered place, where possible. Transport to the home laboratory should be as quick as possible in order to avoid attack by mold and insects.

The investigation of folklore medicine is not easy for two main reasons: nowadays it is almost exclusively elderly people who have any knowledge of the local medicinal plants and their uses and they are often suspicious of foreigners and therefore liable to give faulty information or none at all. Our experience has taught us that nurses or physicians at missionary stations, who have the confidence of the population, are well suited to gathering reliable information. However, the information thus elicited must be cross-checked (cf. References 13, 38, and 44).

Appendixes 1 and 2 are presented at the end of this chapter as checklists for the inventory of plants used in traditional medicine and medicinal plant research in general. Appendix 3 shows how information from different disciplines can be utilized as criteria in the selection of the most promising plants for study as potential sedatives. Points are assigned to the various criteria according to their estimated value for the selection process, and a ranking list of plants can be constructed.[41]

IV. CLINICAL PHARMACOLOGY

Pharmacological screening is only a first step in the search for and development of plant-derived pharmaceuticals. Which step should be the next is a matter of discussion. Up to now the usual procedure has been to start a phytochemical investigation, elucidate the structure of some ingredients and, if enough material is isolated, let the pharmacologist do what he considers useful and meaningful.

However, when the final aim of the research is to develop a useful remedy, then the effects must be the most essential. Therefore, if the Hippocratic screening has shown an acceptable toxicity (during 7 days of observation) and the plant has been used in traditional (folklore) medicine, the next step will be a clinical pharmacology evaluation.[25] This involves testing single and multiple doses on volunteers, who either are treated and observed in the laboratory or, when necessary, in the hospital.[11,17,45]

This is a fairly recent concept in the development of plant-derived drugs. It is accepted, among other countries, in India. Thus, at the Central Drug Research Institute in Lucknow a lipid lowering drug from *Commiphora mukul* has been tested in this way. The Division of Pharmaceuticals and Natural Products of the Thailand Institute of Scientific and Technological Research has also accepted this concept: a good ex-

Figure 4. (p. 229) Goals and methods of ethnopharmacological research, and its implications for drug development in modern and traditional medicine.

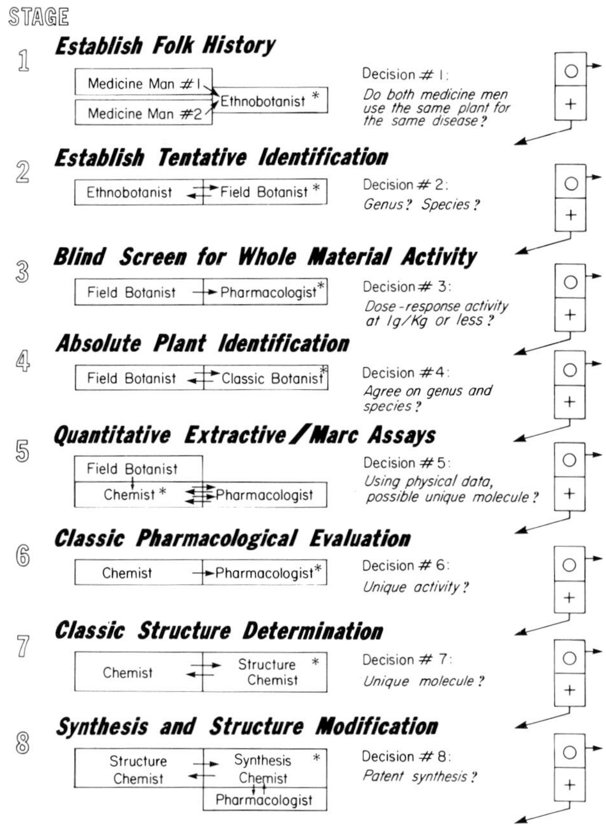

Figure 3. A functional scheme (developed by Malone[25]) for ethnopharmacological research based on a sequential answering of eight critical questions. An asterisk (*) indicates the research professional responsible for answering the critical question posed for each stage of research. Arrows indicate the routes of interaction between researchers. (Used with permission.)

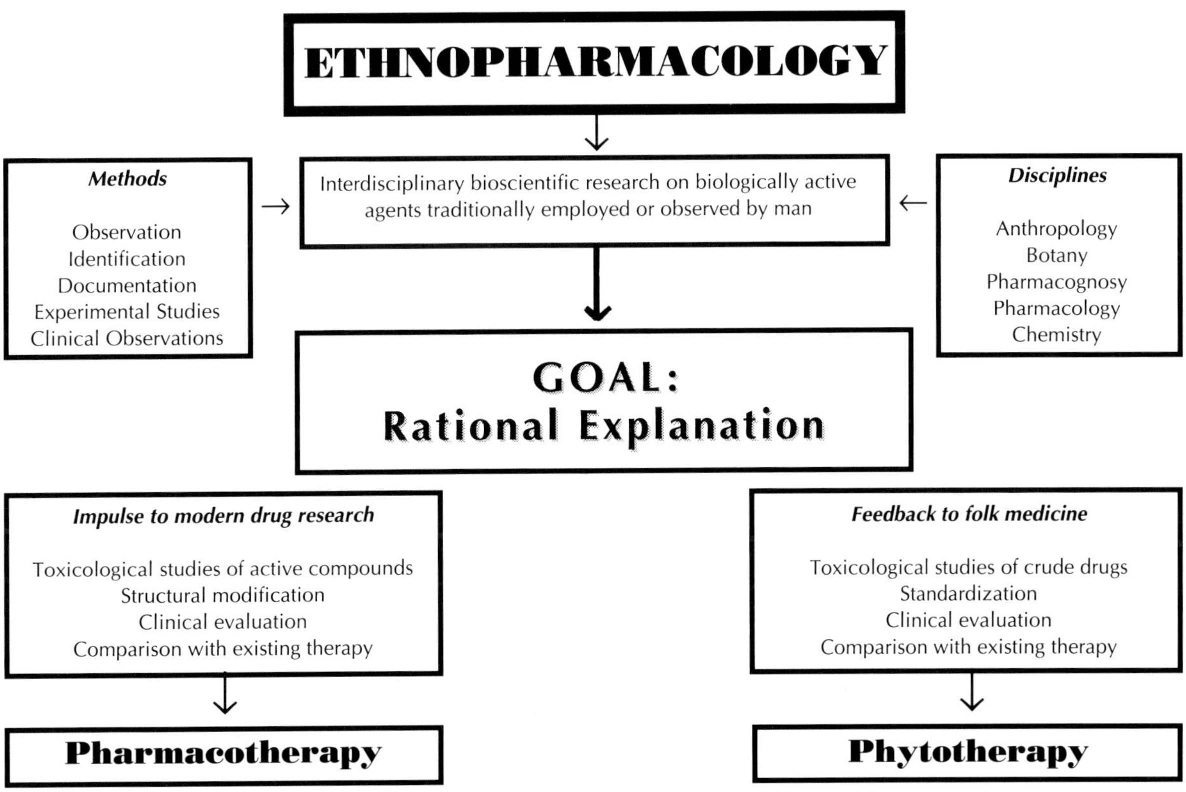

Figure 4.

ample is the clinical pharmacological evaluation of spray-dried juice of garlic.

In Sweden the medical advisers of SAREC (Swedish Agency for Research Cooperation with Developing Countries) have also stressed the importance of using clinical pharmacology at an early stage for medicinal plant projects sponsored by them in Somalia and Nicaragua. In the Central American project the antitussive effect of a cough syrup containing ipecacuanha extract and glaucin is tested, as well as the antidiarrhoic action of *Dorstenia* root.[30]

If the clinical pharmacology of a plant has given promising results, then it is meaningful to start an extensive phytochemical and pharmacological investigation.

The phytochemistry of a plant per se has, of course, its academic interest and value, but from a drug development point of view it is a waste of valuable resources and time to carry out extensive chemical studies, if the plant is not sufficiently active and has no therapeutic potential.

V. PILOT PLANT EXTRACTION

The quantities of drug material needed for clinical pharmacological research can almost never be prepared on a laboratory scale. This is a dilemma for the phytochemist of today. Whereas he can elucidate the structure on 50 to 100 mg of pure substance, the experimental and clinical pharmacological studies require at least 1 to 5 g. Since the interest of many organic chemists does not reach beyond the structure elucidation and synthesis of the isolated compounds, there is an enormous discrepancy between the number of medicinal natural products isolated and the number for which clinical pharmacological data are available. Quite simply, the chemist does not bother to prepare an extract or a pure substance in what he thinks is an enormous quantity. This attitude has been quite evident at several recent natural product symposia. However, whatever the attitude of the organic chemist might be, for the development of a plant-derived drug, gram quantities are needed.

For an extraction in the 10 to 25 L scale the well-known Büchi-apparatus is useful (Figure 5). A wider range of volumes and extraction methods is given by the multipurpose pilot-plant extraction unit CR611 (fabricated by Tournaire Fréres SA, Grasse, France) with the two standard volumes 80 and 250 L (Figure 6).

This unit permits water steam distillation of plant material as well as solvent extraction (to the left of Figure 6), and evaporation of solvent takes place in the kettle (to the right of Figure 6). In order not to lose too much solvent the temperature of the cooling water must not be too high (not over 15°C). This unit, in its two sizes, now works excellently in different parts of the world including Bangkok, Kathmandu, Eskesehiz, Leon, etc.

It is the firm opinion of the authors that all countries dealing with medicinal plants research should install at least one CR611 or similar unit for extraction of volatile as well as nonvolatile components of plant material on a pilot scale.

Figure 5. Büchi extractor.

VI. PROCESSING OF PLANT MATERIALS AND FORMULATIONS

The marketing of a medicinal plant of established value can be carried out at principally three different levels of "processing": dried plant material as a tea, an extract, and pure compound(s). These three possibilities will now be looked into in some detail. Which one of the three alternatives that might be appropriate for use depends mainly on the infrastructure of the country in question.

A. TEA

Tea is a simple galenical formulation, but the problem is to collect or cultivate the plant of correct identity, harvest it properly and put it on the market in a hygienic form. In at least two more primitive developing countries in Africa this mechanism will be adopted by packing the correctly identified plant material in a well-dried state in sealed plastic sacks. On each plastic sack would be printed in the local language the use of the drug and the dosage.

This might be considered very simple, but because many drugs with questionable identity of source and poor quality (moldy and adulterated) are sold in the markets of Africa the proposed procedure represents a definite advantage. Such a procedure is proposed for countries where no extraction facili-

Figure 6. Multipurpose pilot-plant extraction unit CR 611 (Tournaire Fréres SA, Grasse, France).

ties exist, but it goes without saying that there will be the need for people who can correctly identify the collected plants. In Lomé (Togo) Sandberg has encountered knowledgeable "guerriseurs" capable of identifying medicinal plants, including knowledge of their Latin names.

For such a project rather modest facilities are needed: a vehicle for collection, a clean space for drying and dispensing into plastic bags, a balance, a machine for printing the text on the plastic bags and a machine for sealing the bags. Collaboration with an agronomist will enable the "guerriseur" to cultivate suitable medicinal plants besides the collection from wild sources.

B. EXTRACTS

These represent a more advanced step in the processing of medicinal plants, which requires skilled and trained manpower and suitable equipment.

The simplest thing is the preparation of a water-ethanol extract of plant material. The solvent is evaporated and the remaining extract is used, i.e., a nonstandardized extract. An example of this type of extract is the Calaguala-extract (from dried leaves of a *Polypodium* species), manufactured in Tegucigalpa, Honduras, for the treatment of psoriasis.

A standardized extract is one containing a fixed content of active ingredients. This means that an assay method (generally a chemical method) must be used to determine the amount of active substance(s) in the extract. The Indian company Cipla manufactures in their extraction unit in Bangalore a senna extract (from leaves of *Cassia angustifolia*, from cultivations in the classical Tinnevelly region) with a definite content of sennosides A + B.

A more complex standardized extract is manufactured in a factory south of Seoul, Korea, where the Swiss company Pharmaton S.A. manufactures a standardized ginseng extract with a total content of 4% (3.9 to 4.1) ginsenosides and approximately the same proportions between the main ginsenosides. This constant proportion is obtained by mixing, before extraction, calculated amounts of different batches of ginseng roots, the ginsenosides of which have been previously determined by HPLC technique, applied to samples of ginseng roots of different origin.

C. THE PURE ACTIVE COMPOUND

This is the most advanced step in the extraction procedure, which sometimes requires advanced technology. When an active substance is isolated from a plant material on a laboratory scale, there are three possibilities for obtaining the desired therapeutic substances on an industrial scale.

1. The plant is cultivated on a large scale and the active substance is extracted: *Catharanthus roseus* is cultivated in south India and Madagascar, and the dimeric alkaloids vinblastine and vincristine are extracted; *Cephaelis ipecacuanha* is cultivated in southern Nicaragua and northern India for extraction of emetine (which might be converted to 2-hydroemetine); *Stevia rebaudiana* is cultivated in northern Thailand for extraction of stevioside, etc.
2. A plant containing a precursor to the desired therapeutic agent is cultivated on a large scale. The precursor is extracted and then converted to the desired therapeutic agent by chemical steps. Thus, *Voacanga africana* and *V. thouarsii* are cultivated in Cameroon. The seeds contain the indole alkaloid tabersonine, which is extracted and chemically converted to the therapeutically useful vincamine[46] (Figure 7).

   Almost classical is the extraction of diosgenin from the tubers of *Dioscorea* species, a procedure that was started by Syntex in Mexico, where the conversion to intermediary steroid hormones has been carried out. Cipla cultivates *Dioscorea floribunda* as a 2-year crop in Bangalore. The tubers are extracted in the same place and the diosgenin is further worked up to 16-DPA and to various steroid hormones. It is economically most feasible for a company to have its own starting material for hormone production. Unless the diosgenin is processed to 16-DPA the extraction of diosgenin is not economically worthwhile today.

   A new source for diosgenin is *Costus speciosus*, already cultivated in India. In northern Thailand another, possibly better, species of *Costus* is now under cultivation. In spite of the enormous amount of work carried out in Hungary, the U.S.S.R., and India, *Solanum laciniatum* and *Solanum khasianum* have failed on an

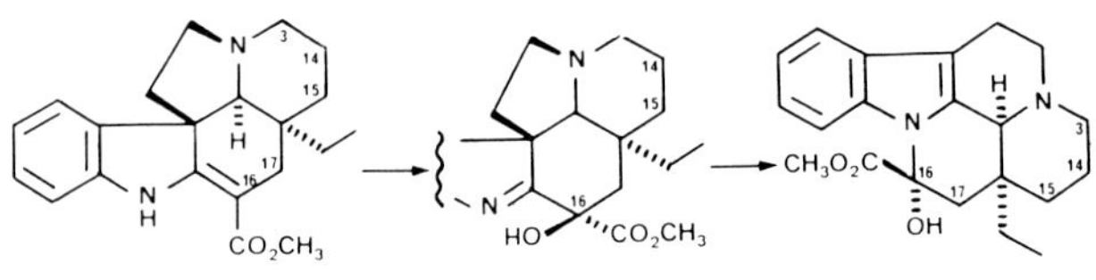

Figure 7. Conversion of tabersonine (I) to vincamine (II).

industrial scale for the extraction of solasodine, which chemically is a better precursor than diosgenin for the production of 16-DPA.

3. Total *synthesis* is the third alternative for industrial production of an active substance that was originally isolated from plant material. The synthetic approach is used for natural products with a simple structure such as papaverine and ephedrine, or when the plant material contains a very complicated mixture of active and inactive compounds and therefore is too expensive to separate and purify.

VII. CONCLUDING REMARKS

1. The value of *the interdisciplinary approach* cannot be overemphasized. Farnsworth et al.[41] have shown that many criteria can be used for the selection of potential sedative plants. Their approach (see Appendix 3) gives points to plants fulfilling the various criteria and results in a ranking of species. Above all, however, this example shows in a nutshell that a profound knowledge of the literature and of the different disciplines involved is necessary in this type of research.
2. The correct identification of species and preparation of voucher specimens is equally important in basic research and drug production.
3. The biological variation within the species calls for standardization of dosage form preparations whenever possible. This is necessary to achieve reliability and for the integration of traditional medicine with modern medicine.
4. The clinical evaluation will involve difficulties but it is necessary to assure acceptable safety, efficacy, and quality, and to determine effective and lethal doses. More medicinal plants would be evaluated if the costs and risks of early clinical research could be reduced.

Appendix 1

FORM FOR ETHNOPHARMACOGNOSTIC INVENTORY OF PLANTS USED IN TRADITIONAL MEDICINE

Botanical Data

Collection number: ________ Collector: ________
Place of collection: ________ Local contact person: ________
Vernacular name(s) in local language/dialect: ________

Latin name: ________ Plant Family: ________
Identification performed by: ________

Type of plant: ________
Part(s) of plant used: ________
Conditions of the collection: ________
Type of drug used: ________

Usages

Modes:
1. Crushed: juice of plant, cataplasm, dry powder.
2. Drink: macerate, infusion, decoction, type of solvent.
3. Fumigation (inhalation of smoke).
4. Ash.

Utilization: for drink, in external washing, for rubbing, for inhalation, for clyster.

Properties:

Observations:

Production

Wild plant, cultivated plant: ______________________
Size of population: ______________________
Size of cultivation: ______________________
Ecological conditions: ______________________
Soil: ______________ Altitude (meters above sea level): ______________
Light: ______________ Rainfall: ______________
Type of vegetation: ______________________

Importance of the present production: ______________________
Organization for marketing: ______________________
Price and availability: ______________________
Possibility of expansion of the production: ______________________
By collection: ______________________
By cultivation: ______________________

Observations:

Adapted from Delaveau and Bailleul[43] and Penso.[8]

Appendix 2
MEDICINAL PLANT STUDY CHECKLIST[a]

Plant Drug

Plant	Scientific name (binomial) with authority(ies) Family Common name(s) with translation and meaning Name of collector(s), collection number, code designation of herbarium where specimen(s) have been deposited (Ex. *Smith 879*, TEX) Geographic distribution Local abundance Habitat
Collection	Part of plant collected Preferred collection time and stage of development Special procedures Storage of plant
Extraction and processing	Plant(s) used Plant part(s) (type and amount) Solvent(s) (type and amount) Heat (type and amount) Length of processing Equipment Dosage form (powder, decoction) Storage
Drug therapy	Dosage (amount and timing) Route of administration and method of use (e.g., external, poultice, internal, infusion, ingested)

Appendix 2 (Continued)
MEDICINAL PLANT STUDY CHECKLIST[a]

	Disorder(s) treated Therapeutic activity — information from informant(s) and personal observations, medical records of patients, and the scientific literature Status of use (past or current use) Adjunct therapy (e.g., diet, sweat, baths, prayers)
General community and health characteristics	
Medical beliefs and setting	Practitioners, range of specialties, selection, training, competence (as judged by the community and researcher(s)) Diagnostic tools, techniques, and criteria for illness Criteria for cure of an illness Cause(s) of illness (natural, supernatural) Therapeutic setting — location, objects, and people present
Community characteristics	Major causes of death and disease Nutrition level Sanitation level Standard of living (e.g., income, housing) Culture and language Natural environment (major plant and animal communities, geographical characteristics)

[a]According to Croom,[14] with permission.

Appendix 3
CRITERIA FOR SELECTION OF POTENTIAL SEDATIVE-CONTAINING PLANTS[a]

Points assigned	Criteria
10	Plant reported in the literature as having some degree of CNS depressant activity as a result of biological evaluation in laboratory animals.
6	Plant for which credible folkloric references were found in the litera ture, in which some type of sedative activity was mentioned for the plant.
5	Plant taxonomically related to a group of plants that was well known to yield effective sedative principles (at either the family or genus level).
4	Plant for which literature reports indicated that alkaloids were present in the plant or had been isolated from the plant.
3	Plant classified in a group that had received little or no chemical or biological investigation, as evidenced from literature files or general knowledge.
2	Plant for which two or more species of the genus in question had been assigned both 10 and 6 points as indicated above.
-5	Plant for which literature sources directly indicated any type of toxicity ascribed to the plant, its extracts, or isolated compounds, either experimentally or on the basis of folklore.

[a]According to Farnsworth et al.,[41] with permission.

REFERENCES

1. **Bruhn, J.G., and Mellstrand, T.,** Natural products as drugs, *Sven. Farm. Tidskr.*, 81, 335, 1977 (In Swedish).
2. **Sticher, O.,** Die Natur als Quelle von Arzneistoffen (Nature as a source of drugs), *Pharm. Acta Helv.*, 49, 228, 1974.
3. **Chikov, P.,** Herbal Medicines in the Soviet Union, *World Health* (WHO), September 1973, 18.
4. **Bézanger-Beauquesne, L., Pinkas, M., and Torck, M.,** *Les Plantes dans la Thérapeutique Moderne*, Maloine, Paris, 1975.
5. **Farnsworth, N.R.,** The role of medicinal plants in drug development, in *Natural Products and Drug Development*, Krogsgaard-Larsen, P., Christensen, S.B., and Kofod, H., Eds., Munksgaard, Copenhagen, 1984, 17.
6. **Farnsworth, N.R., and Bingel, A.S.,** Problems and prospects of discovering new drugs from higher plants by pharmacological screening, in *New Natural Products and Plant Drugs with Pharmacological, Biological or Therapeutical Activity*, Wagner, H., and Wolff, P., Eds., Springer-Verlag, Berlin, 1977, 1.
7. **Bannerman, R.H.,** Organization and cooperation for the development of studies on plants used in traditional medicine: some guidelines for research and studies, *J. Ethnopharmacol.*, 2, 189, 1980.
8. **Penso, G.,** The role of WHO in the selection and characterization of medicinal plants (vegetable drugs), *J. Ethnopharmacol.*, 2, 183, 1980.
9. World Health Organization, *The Promotion and Development of Traditional Medicine*, WHO Technical Report Series 622, WHO, Geneva, 1978.
10. **Anand, N., and Nityanand, S.,** Integrated approach to development of new drugs from plants and indigenous remedies, in *Natural Products and Drug Development*, Krogsgaard-Larsen, P., Christensen, S.B., and Kofod, H., Eds., Munksgaard, Copenhagen, 1984, 78.
11. **Liang Huang,** Drug research based upon the leads of the Chinese traditional medicine, in *Natural Products and Drug Development*, Krogsgaard-Larsen, P., Christensen, S.B., and Kofod, H., Eds., Munksgaard, Copenhagen, 1984, 94.
12. **Ogura, M., Cordell, G.A., Quinn, M.L., Leon, C., Benoit, P.S., Soejarto, D.D., and Farnsworth, N.R.,** Ethnopharmacologic studies. I. Rapid solution to a problem — oral use of *Heliopsis longipes* — by means of a multidisciplinary approach, *J. Ethnopharmacol.*, 5, 215, 1982.
13. **Farnsworth, N.R.,** The development of pharmacological and chemical research for application to traditional medicine in developing countries, *J. Ethnopharmacol.*, 2, 173, 1980.
14. **Croom, E.M., Jr.,** Documenting and evaluating herbal remedies, *Econ. Bot.*, 37, 13, 1983.
15. **Delaveau, P.,** Evaluation of traditional pharmacopoeias, in *Natural Products as Medicinal Agents*, Beal, J.L., and Reinhard, E., Eds., Hippokrates Verlag, Stuttgart, 1981, 395.
16. **Bruhn, J.G., and Holmstedt, B.,** Ethnopharmacology; Objectives, Principles and Perspectives, in *Natural Products as Medicinal Agents*, Beal, J.L., and Reinhard, E., Eds., Hippokrates Verlag, Stuttgart, 1981, 405.
17. **Xiao, Pei-Gen,** Traditional Experience of Chinese Herb Medicine, its Application in Drug Research and New Drug Searching, in *Natural Products as Medicinal Agents*, Beal, J.L., and Reinhard, E., Eds., Hippokrates Verlag, Stuttgart, 1981, 351.
18. **Lozoya, X., and Enríquez, R.,** *El Zapote Blanco — Investigación sobre una Planta Medicinal Mexicana*, Consejo Nacional de Ciencia y Tecnología, México, D.F., México, 1981.
19. **Sandberg, F., and Bruhn, J.G.,** Pharmacognostic Screening of Plant Materials, *Bot. Notiser*, 125, 370, 1972.
20. **Samuelsson, G., Kyerematen, G., and Farah, M.H.,** Preliminary chemical characterization of pharmacologically active compounds in aqueous plant extracts, *J. Ethnopharmacol.*, 14, 193, 1985.
21. **Strömbom, J., and Bruhn, J.G.,** Alkaloids of *Pachycereus pecten-aboriginum*, a Mexican cactus of ethnopharmacologic interest, *Acta Pharm. Suec.*, 15, 127, 1978.
22. **Fairbairn, J.W.,** Perspectives in research on the active principles of traditional herbal medicine. A botanical approach: identification and supply of herbs. *J. Ethnopharmacol.*, 2, 99, 1980.
23. **Tempesta, E.,** Evaluation of local resources in traditional medicine, *J. Ethnopharmacol.*, 2, 163, 1980.
24. **Sandberg, F., and Cronlund, A.,** An ethnopharmacological inventory of medicinal and toxic plants from equatorial Africa, *J. Ethnopharmacol.*, 5, 187, 1982.
25. **Malone, M.H.,** The pharmacological evaluation of natural products — general and specific approaches to screening ethnopharmaceuticals, *J. Ethnopharmacol.*, 8, 127, 1983.
26. **Trotter, R.T., Logan, M.H., Rocha, J.M., and Boneta, J.L.,** Ethnography and bioassay: combined methods for a preliminary screen of home remedies for potential pharmacological activity, *J. Ethnopharmacol.*, 8, 113, 1983.
27. **Malone, M.H.,** Pharmacological approaches to natural product screening and evaluation, in *New Natural Products and Plant Drugs with Pharmacological, Biological or Therapeutical Activity*, Wagner, H., and Wolff, P., Eds., Springer-Verlag, Berlin, 1977, 23.
28. **Penso, G.,** *Index Plantarum Medicinalium Totius Mundi Eorumque Synonymorum* (INPLAMED), Org. Ed. Med. Farm., Milano, 1983, 1026 pp.
29. **Esposito-Avella, M., Brown, P., Tejeira, I., Buitrago, R., Barrios, L., Sanchez, C., Gupta, M.P., and Cedeño, J.,** Pharmacological Screening of Panamanian Medicinal Plants. Part 1, *Int. J. Crude Drug Res.*, 23, 17, 1985.
30. **Kyerematen, G., Medina, E., Saavedra, D., and Sandberg, F.,** Antidiarrhoeal properties of *Dorstenia drakena* L., *Acta Pharm. Suec.*, 22, 31, 1985.
31. **Dhawan, B.N., Patnaik, G.K., Rastogi, R.P., Singh, K.K., and Tandon, J.S.,** Screening of Indian plants for biological activity. VI., *Indian J. Exp. Biol.*, 15, 208, 1977.
32. **Verpoorte, R., Tjin a Tsoi, A., van Doorne, H., and Baerheim Svendsen, A.,** Medicinal plants of Suriname. I. Antimicrobial activity of some medicinal plants, *J. Ethnopharmacol.*, 5, 221, 1982.
33. **Akubue, P.I., Mittal, G.C., and Aguwa, C.N.,** Preliminary pharmacological study of some Nigerian medicinal plants. I., *J. Ethnopharmacol.*, 8, 53, 1983.
34. **Tcheknavorian-Asenbauer, A., and Wijesekera, R.O.B.,** Medicinal and Aromatic Plants for Industrial Development, UNIDO Document IO. 505, 3 June 1982 (Mimeographed), Vienna, 106 pages.
35. **Kong, Y.-C., Ng, K.-H., But, P.P.-I., Li, Q., Yu, S.-X., Zhang, H.-T., Huang, C.-C., Cheng, K.-F., Soejarto, D.D. Kan, W.-S., and Waterman, P.G.,** Sources of the antiimplantation alkaloid yuehchukene in the genus *Murraya*, *J. Ethnopharmacol.*, 15, 195, 1986.
36. **Spjut, R.W., and Perdue, R.E., Jr.,** Plant folklore: a tool for predicting sources of antitumor activity?, *Cancer Treat. Rep.*, 60, 979, 1976.
37. **Gran, L.,** The Uteroactive Principle of "Kalata-Kalata" (*Oldenlandia affinis* DC), Dissertation, University of Bergen, 1973.
38. **Farnsworth, N.R., and Kaas, C.J.,** An approach utilizing information from traditional medicine to identify tumorinhibiting plants, *J. Ethnopharmacol.*, 3, 85, 1981.
39. **Sofowora, A.,** The present status of knowledge of the plants used in traditional medicine in Western Africa: a medical approach and a chemical evaluation, *J. Ethnopharmacol.*, 2, 109, 1980.
40. **Suwanlert, S.,** A study of kratom eaters in Thailand, *Bull. Narcotics*, 27, 21, 1975.
41. **Farnsworth, N.R., Bederka, J.P., Jr., and Moses, M.,** Modern approaches for selecting biologically active plants. I. CNS depressants, *J. Pharm Sci.*, 63, 457, 1974.
42. **Soejarto, D.D., Bingel, A.S., Slaytor, M., and Farnsworth, N.R.,** Fertility regulating agents from plants, *Bull. WHO*, 56, 343, 1978.
43. **Delaveau, P., and Bailleul, F.,** Fiche d'enquête pour l'exploration des pharmacopées traditionnelles, *Plant. Med. Phytother.*, 13, 75, 1979.
44. **Hedberg, I., Hedberg, O., Madati, P.J., Mshigeni, K.E., Mshiu, E.N., and Samuelsson, G.,** Inventory of plants used in traditional medicine in Tanzania. III. Plants of the families Papilionaceae-Vitaceae, *J. Ethnopharmacol.*, 9, 237, 1983.

45. **Landgren, B.M., Aedo, A.R., Hagenfeldt, K., and Diczfalusy, E.,** Clinical effects of orally administered extracts of *Montanoa tomentosa* in early human pregnancy, *Am. J. Obst. Gynecol.*, 135, 480, 1979.
46. **Bisset, N.G.,** Phytochemistry and pharmacology of *Voacanga* species, in *Series of Revisions of Apocynaceae XV*, Leeuwenberg, A.J.M., Ed., Agricultural University Wageningen Papers 85-3, Wageningen, 1985, 99.
47. **Arseculeratne, S.N., Gunatilka, A.A.L., and Panabokke, R.G.,** Studies on medicinal plants of Sri Lanka: occurrence of pyrrolizidine alkaloids and hepatotoxic properties in some traditional medicinal herbs, *J. Ethnopharmacol.*, 4, 159, 1981.

Chapter 16

INFORMATION ON MEDICINAL PLANTS

Y. Chadha and Gian Singh

I. INTRODUCTION

Mankind has a long history in using herbal medicine. Rigveda and Ayurveda (4500-1600 B.C.) reveal that ancient Indians had rich knowledge of the medicinal value of plants. The famous Indian physician, Charaka (1000-600 B.C.) describes the uses and mode of application of some 350 medicinal plants in his well-known Sanskrit work, *Charaka Samihta*. This medical treatise is considered to be a compilation of the deliberations of a congress of sages from different parts of India for the exchange of medical knowledge. In fact, herbal medicine has played a dominant role in medical care in all cultures since ancient times. The Shen Nu Materia Medica of China was written in the first and second centuries.

In modern times, the use of natural herbal medicines declined with the development of the synthetic drug industry. This was especially so in the developed countries, but in the developing countries, traditional herbal medicine continued to play an important role, running parallel to modern medical practices. A study conducted in Indonesia in 1983 has shown that about 70% of the population at Ujung Pandang, the capital of South Sulawesi, regularly use traditional prescriptions.[1] Recently, however, there has been a marked increase in the use of preparations based on medicinal plants in the developed countries, more so as these are generally regarded as being free from the side effects associated with synthetic drugs. Health care products, natural cosmetics, and health foods are becoming increasingly popular. The observed efficacy of many traditional medicines is another important reason for the revival of interest in medicinal plants. Drugs such as atropine, reserpine, quinine, and morphine were discovered through the scientific study of traditional remedies. (China and to some extent the U.S.S.R. have now adopted an integrated system of allopathic, traditional, and folk systems of medicine.)

With the discovery of vincristine and vinblastine from *Catharanthus roseus* as clinically useful antitumor agents[2] in 1960, a major effort was started by the National Cancer Institute (U.S.) for the screening of plant extracts for anticancer activity. Under the National Cancer Chemotherapy Program, 120,000 plant extracts have been tested from 35,000 different species obtained from worldwide sources.[3] Some were selected only on folklore evidence of their medicinal value. Likewise, the search for antimicrobial, antistress, and antifertility principles from plants has made significant contributions to the literature on medicinal plants. The scientific search for new drugs from existing indigenous healing preparations continues. Marvin H. Malone[4] believes that natural product research has the best chance for discovering new prototype drugs that will be useful in the clinics.

Several international agencies are now taking an interest in the promotion of the use of and research on medicinal plants in the developing countries. In the declaration of World Health Organization (WHO) on primary health care for all by the year 2000, medicinal plants have been assigned an important role. Dr. Holfdan Mohler, Director General of WHO, has pointed out that "there is no doubt that the judicious use of such herbs, flowers and other plants for palliative purposes in primary health care can make a major contribution towards reducing the developing countries' drug bill". These countries are being encouraged to intensify research on medicinal plants so that traditional and new herbal medicines could be put to better use. To assist in this programme, WHO has compiled a comprehensive inventory of 21,000 medicinal plant species (including synonyms) used in the world, based on literature from 91 countries.[5]

UNESCO sponsored two regional networks for promoting cooperation among research and training institutions concerned with medicinal and aromatic plants: the South and Central Asia Medicinal and Aromatic Plants (SCAMAP) Network established in 1979; and the Regional Network for Chemistry of Natural Products in Southeast Asia (CNPSA) established in 1975. UNESCO is now seeking to establish an Asian and Pacific Information Network on Medicinal and Aromatic Plants (APINMAP). UNIDO has also been actively promoting the industrial utilization of medicinal plants in developing countries, and sponsored by the UNDP and donor countries has built up an impressive technical assistance programme.

In recent years agricultural research on medicinal plants has been intensified in many parts of the world. Newer strains have been developed which selectively produce higher contents of active constituents. The work on tissue culture of medicinal plants has also acquired significant dimensions. As a result of greater interest in the use of plants in medicine, cosmetics, health foods, and as a source of a newer class of organic compounds all over the world, the research efforts and the resultant literature in this field has increased substantially during the last 3 decades. In this context, systematic collection, processing, storage, retrieval, and dissemination of information on medicinal plants, particularly to developing countries, assumes considerable importance.

In the following pages are discussed the information sources on medicinal plants. The sources of information are classified here into Primary Publications (research periodicals, patents, reports, etc.), Secondary Publications (indexing and abstracting periodicals), Books, Data Bases, and Data on Trade and Statistics.

II. PRIMARY PUBLICATIONS

Primary publications such as research periodicals, research reports, conference proceedings, theses, and patents report new knowledge. Among these sources, periodicals are of the most importance. They are most easily accessible and have made the maximum contribution to the growth of literature in the field. There are about 600 periodicals published in the world which cover, *inter alia,* medicinal plants and their products. Many of these emanate from China and India and are devoted to broad disciplines such as phytochemistry, pharmacology, medicine, botany, and agriculture. There are only a few journals which carry research communications exclusively on medicinal plants and drugs derived therefrom. Among these are *Planta Medica, Plantes Medicinale et Phytotherapie, Fitoterapia, Farmatsiya, Zeitschrift fur Phytotherapie, Journal of Ethnopharmacology,* and *Journal of Scientific Research on Plants Medicine.* The significant journals in this field are listed in Table 1. Many of these journals are of rather recent origin.

Another class of primary sources are reports such as those prepared by national or international agencies on the distribution, ethnomedical uses, and prospects of cultivation and utilization of medicinal plants, e.g., reports of the Central Council of Research on Ayurveda and Siddha, New Delhi (available as booklets). Such reports are available from agencies such as Chemical Abstracts, Biological Abstracts, and Medicinal and Aromatic Plants Abstracts. Newsletters and annual reports of the institutes working on medicinal plants and journals devoted to farm education are also categorized as primary sources.

Current literature on patents appears in the form of abstracts, mainly in two places: abstract journals, and the official journals of the national patent offices. Of the commercial services, the most exhaustive service in worldwide patent literature is offered by Derwent Publications Ltd., London. Its products comprise the World Patent Index (WPI), the various Patents Abstracts publications, the special chemical patents documentation service, and the Central Patent Index. World Patent Index is described later under "Data Bases". International Patent Documentation Centre (INPADOC), Vienna, offers (in collaboration with World Intellectual Property Organization, Geneva) rapid access to the patent documents of over 47 countries. The International Patent Institute at The Hague also undertakes searches into patent specifications. Chemical Abstracts is another important source of patent documents relating to chemical aspects of medicinal plants. Some of the primary research periodicals report patent documents. *Referativnyi Zhurnal* and *Bulletin Signaletique* also cover patent literature very extensively.

III. SECONDARY AND TERTIARY PUBLICATIONS

This group of information sources includes abstracting and indexing periodicals, citation indexes, subject bibliographies, review journals, encyclopedias, dictionaries, pharmacopoeias, formularies, and other reference works.

Secondary publications, particularly abstracting periodicals, provide tools for hunting specific information on a topic and help in overcoming the language barrier to a certain extent. *Chemical Abstracts,* for example, covers literature in more than 50 languages, while the abstracts are published in English. Similarly, *Medicinal and Aromatic Plants Abstracts* (*MAPA*) covers literature in more than 20 languages. The important abstracting and indexing periodicals reporting literature on medicinal plants are listed in Table 2. Of these, MAPA and CROMAP emanating from India are the only two periodicals solely devoted to medicinal and aromatic plants. All others are devoted to broader subject areas such as agriculture, medicine, chemistry and chemical technology, biology, etc. Data bases of some of the abstracting periodicals are available in computer-readable forms and microforms. *Ulrich's Directory*[6] and *Encyclopedia of Information Systems and Services*[7] give a detailed list of abstracting and indexing services, and the data bases.

Current Awareness Services have been developed to overcome the time lags in publication of an article in a journal and its abstract in an abstracting journal. Among the better known current awareness services relevant to medicinal plants are *Chemical Titles, Current Abstracts of Chemistry and Index Chemicus (CAC & IC), Current Contents [Editions: Agriculture, Biology, and Environmental Sciences; Life Sciences; Physical, Chemical, and Earth Sciences;* and *Clinical Practice],* and *Index Medicus. CAC & IC,* published weekly by the Institute for Scientific Information (ISI), covers 100 core journals in organic chemistry, and gives abstracts of those articles reporting new compounds, reactions or syntheses. *Chemical Titles,* published biweekly by Chemical Abstracts Service, covers papers published in about 700 journals in the field of chemistry, and is the world's first computer-produced periodical. The weekly *Current Contents,* published by ISI, reproduce in their original format the tables of contents of important journals. Indexing journals covering literature on medicinal plants are included in Table 2.

IV. BOOKS

Books constitute an important source of information on medicinal plants. There are some ancient books of several countries in several languages which give valuable information on the traditional users of herbal medicines. Many modern treatises on medicinal plants and their efficacies have been brought out in India, China, and many other countries. In the category of books are included pharmacopoeias, national formularies, handbooks, manuals, catalogues, reports, dictionaries, encyclopedias, legislative books, etc. Their number is very large and it is not possible to list them here. However, there are sources which can lead one to important books on medicinal plants.

The *Union Catalogue of Chinese Medicine* has estimated that there are more than 700 books on Chinese medicine and about 10,000 pamphlets on Chinese prescriptions. More than 25 pharmacognostic books covering different districts in China have been published by central and provincial organizations in the last 30 years.[8] *Chinese Herbal Medicine* by Richard Hyatt[9] has appended a list of useful books on Chinese and Japanese medicine. *Checklist of Sanskrit Medical Manuscripts of India*[10] (in English and Hindi) provides information on about 1100 Sanskrit medical manuscripts available in 24 oriental and manuscript libraries in India. *Union Catalogue of Arabic and Persian Medical Manuscripts in the Libraries of Hyderabad* describes some 188 Arabic and 416 Persian medical manuscripts available in private and public libraries in Hyderabad.[11]

TABLE 1
Significant Primary Journals Reporting Work on Medicinal Plants

1. *Acta Chimica Sinica* (Beijing, China)
2. *Acta Pharmaceutical Hungarica* (Budapest, Hungary)
3. *Acta Pharmaceutical Jugoslavica* (Zagreb, Yugoslavia)
4. *Acta Pharmaceutical Sinica* (Beijing, China)
5. *Acta Pharmaceutical Suecica* (Stockholm, Sweden)
6. *Acta Pharmaceutical Toxicologia* (Copenhagen, Denmark)
7. *Acta Poloniae Pharmaceutica* (Warsaw, Poland)
8. *African Journal of Pharmacy and Pharmaceutical Sciences* (Apapa, Nigeria)
9. *Agents and Actions* (Basel, Switzerland)
10. *Agricultural and Biological Chemistry* (Tokyo, Japan)
11. *American Journal of Clinical Nutrition* (Bethesda, MD, USA)
12. *Anais De Farmacia E Quimica De Sao Paulo (Records of Pharmaceuticals and Chemicals of Sao Paulo)* (Sao Paulo, Brazil)
13. *Anales De Quimica, Serie C Bioquimica* (Madrid, Spain)
14. *Anales de la Real Academia de Farmacia* (Madrid, Spain)
15. *Ancient Science of Life* (Coimbatore, India)
16. *Annales Pharmaceutiques Francaises* (Paris, France)
17. *Antiseptic* (Madras, India)
18. *Archiv der Pharmazie (Weinheim)* (Weinheim, Germany)
19. *Archives of Dermatology* (Chicago, USA)
20. *Archives of Internationales de Pharmacodynamie et de Therapie* (Ghent, Belgium)
21. *Archives of Pharmacal Research (Seoul)* (Seoul, Korea)
22. *Arzneimittel Forschung* (Aulendorf, Germany)
23. *Atherosclerosis* (Amsterdam, Netherlands)
24. *Bangladesh Medical Journal* (Dacca, Bangladesh)
25. *Bangladesh Pharmaceutical Journal* (Dacca, Bangladesh)
26. *Biochemical and Biophysical Research Communications* (New York, USA)
27. *Biochemical Pharmacology* (Oxford, England)
28. *Boletim da Faculdade de Farmacia de Coimbra* (Coimbra, Portugal)
29. *Bollettino Chimico Farmaceutico* (Milan, Italy)
30. *Bolletino della Societa Italiana di Farmacia Ospedaliera* (Turin, Italy)
31. *British Medical Journal* (London, England)
32. *Bulletin of the Department of Medical Sciences* (Bangkok, Thailand)
33. *Bulletin of Faculty of Pharmacy Cairo University* (Cairo, Egypt)
34. *Bulletin of Medico-Ethnobotanical Research* (New Delhi, India)
35. *Bulletin on Narcotics* (Geneva, Switzerland)
36. *Carbohydrate Research* (Amsterdam, Netherlands)
37. *Chemical and Pharmaceutical Bulletin* (Tokyo, Japan)
38. *Chemical and Druggist* (London, England)
39. *Chemistry and Industry (London)* (London, England)
40. *Chemistry Letters* (Tokyo, Japan)
41. *Chinese Journal of Integrated Traditional and Western Medicine* (Beijing, China)
42. *Chinese Medical Journal* (Beijing, China)
43. *Clinical Pharmacology and Therapeutics* (St. Louis, MO, USA)
44. *Clinical Toxicology* (New York, USA)
45. *Clujul Medical* (Cluj-Napoca, Romania)
46. *Comptes Rendus de l'Academie Bulgare Des Sciences* (Sofia, Bulgaria)
47. *Comptes Rendus de l'Academie des Sciences, Paris* (Paris, France)
48. *Contraception* (California, USA)
49. *Current Medical Practice* (Bombay, India)
50. *Current Science* (Bangalore, India)
51. *Deutsche Apotheker Zeitung* (Stuttgart, Germany)
52. *Drug and Cosmetic Industry* (New York, USA)
53. *Drug Information Bulletin* (Maple Glen, USA)
54. *East African Pharmaceutical Journal* (Nairobi, Kenya)
55. *Eastern Pharmacist* (New Delhi, India)
56. *Economic Botany* (New York, USA)
57. *Egyptian Journal of Chemistry* (Cairo, Egypt)
58. *Egyptian Journal of Pharmaceutical Sciences* (Cairo, Egypt)
59. *Egyptian Journal of Phytopathology* (Cairo, Egypt)
60. *Ethnomedizin* (Hamburg, Germany)
61. *Experientia* (Basel, Switzerland)
62. *Farmacevtski Vestnik (Ljubljana)* (Ljubljana, Yugoslavia)
63. *Farmacia (Bucharest)* (Bucharest, Romania)
64. *Farmaco (Pavia)* (Pavia, Italy)
65. *Farmaco, Edizione Practica* (Milan, Italy)
66. *Farmaco Edizione Scientifica* (Milan, Italy)
67. *Farmatsevtichnyi Zhurnal* (Kiev, USSR)
68. *Farmatsiya (Moscow)* (Moscow, USSR)
69. *FEBS Letters* (Amsterdam, Netherlands)
70. *Fitoterapia* (Milan, Italy)
71. *Food and Chemical Toxicology* (Oxford, England)
72. *Gazzetta Chimica Italiana* (Rome, Italy)
73. *General Pharmacology* (Oxford, England)
74. *Geobios (Jodhpur)* (Jodhpur, India)
75. *Helvetica Chimica Acta* (Basel, Switzerland)
76. *Herba Hungarica* (Budapest, Hungary)
77. *Herba Polonica* (Warsaw, Poland)
78. *Heterocycles* (Sendai, Japan)
79. *Indian Drugs* (Bombay, India)
80. *Indian Drugs and Pharmaceuticals Industry* (Bombay, India)
81. *Indian Forester* (Dehradun, India)
82. *Indian Journal of Chemistry PtB* (New Delhi, India)
83. *Indian Journal of Experimental Biology* (New Delhi, India)
84. *Indian Journal of Forestry* (Dehradun, India)
85. *Indian Journal of Medical Research* (New Delhi, India)
86. *Indian Journal of Nematology* (New Delhi, India)
87. *Indian Journal of Pharmaceutical Sciences* (Bombay, India)
88. *Indian Journal of Pharmacology* (Varanasi, India)
89. *Indian Medical Gazzette* (Calcutta, India)
90. *Indian Perfumer* (Jammu, India)
91. *Indian Practitioner* (Bombay, India)
92. *International Journal of Crude Drug Research* (Lisse, Netherlands)
93. *Japanese Journal of Pharmacology* (Kyoto, Japan)
94. *Journal of African Medicinal Plants* (Cairo, Egypt)
95. *Journal of Agricultural and Food Chemistry* (Washington, D.C., USA)
96. *Journal of the Association of Official Analytical Chemists* (Washington, D.C., USA)
97. *Journal of Chemical Research (S)* (London, England)
98. *Journal of Chemical Society, Chemical Communications* (London, England)
99. *Journal of Chemical Society of Pakistan* (Karachi, Pakistan)
100. *Journal of Chemical Society, Perkin Transactions I* (London, England)
101. *Journal of Chromatography* (Amsterdam, Netherlands)
102. *Journal of Clinical Pharmacology* (Connecticut, USA)
103. *Journal of Drug Research (Cairo)* (Cairo, Egypt)
104. *Journal of Economic and Taxonomic Botany* (Jodhpur, India)
105. *Journal of Ethnopharmacology* (Lausanne, Switzerland)
106. *Journal of Food Science* (Illinois, USA)
107. *Journal of Food Science and Technology* (Mysore, India)
108. *Journal of Indian Chemical Society* (Calcutta, India)
109. *Journal of Korean Pharmaceutical Sciences* (Seoul, South Korea)
110. *Journal of the National Integrated Medical Association* (Mysore, India)
111. *Journal of the National Research Council of Thailand* (Bangkok, Thailand)
112. *Journal of Natural Products (Lloydia)* (Ohio, USA)
113. *Journal of Organic Chemistry* (Washington, D.C., USA)
114. *Journal of Pharmaceutical Sciences* (Washington, D.C., USA)
115. *Journal of Pharmaceutical Society of Korea* (Seoul, Korea)
116. *Journal De Pharmacie De Belgique* (Brussels, Belgium)

TABLE 1 (CONTINUED)
Significant Primary Journals Reporting Work on Medicinal Plants

117. *Journal of Pharmacy and Pharmacology* (London, England)
118. *Journal of Pharmacobio-Dynamics* (Tokyo, Japan)
119. *Journal of Pharmacology Experimental Therapeutics* (Baltimore, USA)
120. *Journal of Research in Ayurveda and Siddha* (New Delhi, India)
121. *Journal of Research Institute of Medical Sciences of Korea* (Seoul, South Korea)
122. *Journal of Science Food and Agriculture* (London, England)
123. *Journal of Scientific Research in Plants and Medicine* (Hardwar, India)
124. *Journal of Science Society of Thailand* (Bangkok, Thailand)
125. *Journal of Tropical Forestry (Jabalpur)* (Jabalpur, India)
126. *Khimiko-Farmatsevticheskii Zhurnal* (Moscow, USSR)
127. *Khimiya Prirodnykh Soedinenii* (Tashkent, USSR)
128. *Korean Journal of Internal Medicine* (Seoul, South Korea)
129. *Korean Journal of Pharmacognosy* (Seoul, South Korea)
130. *Korean Journal of Pharmacology* (Seoul, South Korea)
131. *Lancet* (London, England)
132. *Lawrence Review of Natural Products* (Collegeville, PA, USA)
133. *Lebensmittel-Wissenschaftlische und Technologie* (Zurich, Switzerland)
134. *Liebigs Annales Chemie* (Weinheim, Germany)
135. *Mahidol University Journal of Pharmaceutical Sciences* (Bangkok, Thailand)
136. *Medicina Traditional* (Mexico)
137. *Lipids* (Champaign, IL, USA)
138. *Nagarjun* (Calcutta, India)
139. *National Academy of Science Letters* (New Delhi, India)
140. *Naturwissenschaften* (Heidelberg, Germany)
141. *Naunyn-Schmiedeberg's Archives of Pharmacology* (Heidelberg, Germany)
142. *Nippon Nogeikagaku Kaishi* (Tokyo, Japan)
143. *Official Journal of Research Institute of Medical Sciences of Korea* (Seoul, South Korea)
144. *Pafai Journal* (Bombay, India)
145. *Pakistan Journal of Forestry* (Peshawar, India)
146. *Parfumerie und Kosmetik* (Heidelberg, Germany)
147. *Perfumer and Flavorist* (Illinois, USA)
148. *Parfums, Cosmetiques, Aromes* (Paris, France)
149. *Parfumes, Cosmetiques, Savons de France (Incorporating La France et ses Parfums)* (Paris, France)
150. *Pharmaceutica Acta Helvetica* (Zurich, Switzerland)
151. *Pharmaceutical Research Institute: Bulletin* (Osaka, Japan)
152. *Pharmaceutisch Weekblad* (Amsterdam, Netherlands)
153. *Pharmacological Research Communications* (London, England)
154. *Pharmacology* (Basel, Switzerland)
155. *Pharmazeutische Industrie* (Aulendorf, Germany)
156. *Pharmazeutische Rundschau* (Heusenstamm, Germany)
157. *Pharmazie in Unserer Zeit* (Weinheim, Germany)
158. *Pharmazeutische Zeitung* (Frankfurt, Germany)
159. *Pharmazie* (Dresden, Germany)
160. *Pharmindex* (Portland, OR, USA)
161. *Physicians' Drug Manual* (New York, USA)
162. *Phytochemistry* (Oxford, England)
163. *Plant Disease Reporter* (Washington, D.C., USA)
164. *Plant Science Letters* (Limerick, Ireland)
165. *Plant Medica* (Stuttgart, Germany)
166. *Plantes Medicinales et Phytotherapie* (Angers, France)
167. *Polish Journal of Pharmacology and Pharmacy* (Warsaw, Poland)
168. *Rastitel'nye Resursy* (Moscow, USSR)
169. *Research Communications in Chemical Pathology and Pharmacology* (New York, USA)
170. *Revista de Ciencias Farmaceuticas* (Sao Paulo, Brazil)
171. *Revista Colombiana de Ciencias Quimico Farmaceuticas* (Bogota, Columbia)
172. *Revista Cubana de Farmacia* (Havana, Cuba)
173. *Revista Farmaceutica* (Buenos Aires, Argentina
174. *Revista Latinoamericana de Quimica* (Monterrey, Mexico)
175. *Rheumatism* (New Delhi, India)
176. *Rivista Italiana Essenze Profumi, Piante Officinali, Aromi, Saponi, Cosmetici, Aerosol* (Review of the Italian perfume essences, herbs, aromas, soaps, cosmetics and aerosol industries) (Milan, Italy)
177. *Sachitra Ayurved* (Patna, India)
178. *Science and Culture* (Calcutta, India)
179. *Scientia Pharmaceutica* (Vienna, Austria)
180. *Scientia Sinica* (Beijing, China)
181. *Seoul University Journal* (Seoul, South Korea)
182. *Siriraj Hospital Gazette* (Bangkok, Thailand)
183. *Shoyakugaku Zasshi (The Japanese Journal of Pharmacognosy)* (Tokyo, Japan)
184. *Songklanaharian Journal of Science and Technology* (Bangkok, Thailand)
185. *Svensk Farmaceutisk Tidskrift* (Stockholm, Sweden)
186. *Taxon* (Utrecht, Netherlands)
187. *Tetrahedron* (Oxford, England)
188. *Tetrahedron Letters* (Oxford, England)
189. *Timisoara Medicala* (Timisoara, Romania)
190. *Toxicology* (Amsterdam, Netherlands)
191. *Toxicology and Applied Pharmacology* (New York, USA)
192. *Toxicology Letters* (Amsterdam, Netherlands)
193. *Toxicon* (London, England)
194. *Tropical Science* (Surrey, England)
195. *Vagbhata* (Trivandrum, India)
196. *Vestik Sel'Skokhozyaistvennoi Nauki* (Moscow, USSR)
197. *Yakhak Hoeji* (Seoul, South Korea)
198. *Yakugaku Zasshi (Pharmaceutical Society of Japan Journal* (Tokyo, Japan)
199. *Zeitschrift Naturforschungs Wissenschaftliche* (Tubingen, Germany)
200. *Zeitschrift fur Phytotherapie* (Munich, Germany)

TABLE 2
Important Abstracting and Indexing Journals for Medicinal Plants

1. *Agrindex* (Rome, Italy), Monthly.
2. *Biological Abstracts* (Philadelphia, USA), Semimonthly. Appropriate Sections: Agronomy, Botany, Economic Botany, Pharmacognosy and Pharmaceutical Botany, Pharmacology, Plant Physiology, Biochemistry & Biophysics, and Toxicology.
3. *Bulletin Signaletique* (Paris, France), Monthly. Appropriate Sections: Part 171: Chimie; Part 310: Genie Biomedicale Informatique Biomedicale; Part 330: Sciences Pharmacologique Toxicologie.
4. *Chemical Abstracts* (Columbus, OH, USA), Weekly. Appropriate Sections: 30: Terpenes & Terpenoids; 31: Alkaloids; 32: Steroids; 33: Carbohydrates; 34: Amino Acid Peptides & Proteins; 45: Industrial Organic Chemicals, Ethers, Fats and Waxes; 63: Pharmaceuticals; 64: Pharmaceutical Analysis, Pharmacology Plant Biochemistry, Biochemical Methods.
5. *Current Advances in Plant Science* (Oxford, England), Monthly. Relevant Headings are Secondary Products: Terpenoids, Alkaloids, Flavonoids, Phenolics and Others.

TABLE 2 (CONTINUED)
Important Abstracting and Indexing Journals for Medicinal Plants

6. *Current Contents* (Philadelphia, USA), Weekly. Editions: a. Agriculture, Biology and Environmental Sciences; b. Life Sciences; c. Physical, Chemical and Earth Sciences; and d. Clinical Practice.
7. *Current Research on Medicinal and Aromatic Plants* (Lucknow, India), Quarterly Abstracts are divided into 10 subheadings.
8. *Excerpta Medica* (Amsterdam, Netherlands), Monthly Section: Pharmacology and Toxicology.
9. *Horticultural Abstracts* (Slough, England), Monthly. Medicinal Plants mostly occur under Minor Temperate & Tropical Industrial Crops which are further subdivided under Condiments & Spice Plants; Essential Oil Plants; Insecticidal Pesticidal Plants, Medicinal Plants.
10. *Index Medicus* (Bethesda, MD, USA), Monthly. Information on medicinal plants is mainly found under: Plant extracts; Plant Poisoning; Plants, Medicinal; and Plants, Toxic.
11. *International Pharmaceutical Abstracts* (Washington, D.C., USA), Bimonthly
12. *Medicinal and Aromatic Plants Abstracts* (New Delhi, India), Bimonthly. World literature on medicinal plants is covered under 16 subject headings.
13. *Pharmaceutical Abstracts* (Austin, TX, USA), Quarterly.
14. Pharmazeutisches Jahrbuch: Referatesammlung des Internationalen Pharmazeutischen Schrifttums Im Jahre (Frankfurt/Main, Germany), Annual.
15. *Produits et Problemes Pharmaceutiques* (Paris, France), Monthly.
16. *Referativnyl Zhurnal: Farmakologiia, Khimioterapevticheskie Sredstva Toksikologiia;Refrativnyi Zhurnal: Biologiya* (Moscow, USSR), Monthly
17. *Referatovy Vyber Z Lekarenstvi: Pharmacy* (Prague, Czechoslovakia), Twice a month.
18. *Ringdoc Abstract Journal* (London, England), Weekly.

Glossary of Indian Medicinal Plants (1956) and its supplement (1969) by Chopra, Nayar, and Chopra, together cite 111 books, pharmacopoeias, manuals, catalogues, and floras.[12] *Handbook of Domestic Medicine and Common Ayurvedic Remedies,*[13] published by Central Council of Research in Indian Medicine and Homoeopathy (1978), refers to 75 books (including those in Sanskrit) on Indian medicinal plants. In a paper entitled "Information Sources and Services on Medicinal Plants" by Y. R. Chadha and K. Kashyapa,[14] 83 important books and 178 core periodicals reporting work on medicinal plants have been listed. A dictionary of herbs edited by Stuart[15] gives a list of 63 books in addition to other information on herbs.

Medical Botany[16] lists selected references primarily in English under the following geographical regions: General — Plants from more than one area — North America; Europe; Latin America (including Central and South America, West Indies); Middle East and North Africa; Sub-Saharan Africa; Southern Asia; Eastern Asia; and Australia and Oceania.

Pharmacopoeias provide authentic information on medicinal plants and their products officially recognized for use in health care preparations. Quality specifications for medicinal plants and their derivatives are well defined in the pharmacopoeias and national formularies. The first volume of the Indian Pharmaceutical Codex[17] contains 444 compound formulations; these include 21 categories and 351 single drugs of plant origin.

WHO Inventory of Medicinal Plants used in different countries lists 187 significant sources of information such as pharmacopoeias, formularies, and relevant legislative texts from 91 countries.[5]

New books published in the field are being noted and reviewed in a number of primary journals and, indexing and abstracting journals such as *Economic Botany, Journal of Ethnopharmacology, International Journal of Crude Drug Research, Deutsche Apotheker Zeitung, Plantes Medicinal et Phytotherapie, Chemical Abstracts, Biological Abstracts,* and *Medicinal and Aromatic Plants Abstracts.*

One encyclopaedic publication, *The Wealth of India — Raw Materials Series*[18] comprising 11 volumes deserves special mention. It contains monographic articles on some 5000 Indian plant species of economic importance which include all those used medicinally. Correct botanical names of the plants with their English and local names, areas of their occurrence, cultivation methods, medicinal properties, active principles and their pharmacological action, production and trade data, where available, are provided along with illustrations and relevant references. *The Wealth of India* is a comprehensive and authentic source of information on medicinal plants and covers the worldwide literature. This reference work is now under revision and Volume 1 of the revised series was published.

V. COMPUTERIZED DATA

Computer-readable files are being generated by several organizations, mainly in the U.S.. Medicinal & Aromatic Plants Information Service (MAPIS), provided by Publications & Information Directorate (CSIR), New Delhi, India, is one of the few data bases which is entirely on medicinal and aromatic plants. The NAPRALERT SYSTEM has a computerized data base which provides information on chemistry and pharmacology of natural products including those obtained from medicinal plants. Recently, a Computerized Data Base on Chinese Traditional Medicine has been established at the Chinese University of Hong Kong. Other data bases such as BIOSIS, CA SEARCH, CAB, MEDLINE, AGRINDEX, CHEMLINE are of more generalized nature. A brief description of the more important systems follows.

A. COMPUTERIZED INFORMATION ON CHINESE MEDICINAL MATERIAL

This is a computerized Chinese medicinal data base established at the Chinese University of Hong Kong. The system provides on-line information on all common and most widely used Chinese medicinal materials (1860 species), including botanical, chemical, pharmacological, and clinical reports. The data from selected recent treatises on Chinese medicine has been translated into English and stored into a computer using IBM STAIRS software. Medical Subject Headings (or MeSH) vocabularies used by Index Medicus and the Chemical Abstracts nomenclature have been used for storing the information. For Chinese herbal names, the Pin-yin transliteration system was used instead of the older Wade-Gile system, in addition to the Latin scientific names. Pin-yin was also used for traditional Chinese medical terms for which no suitable English expression exists.

The data (title, author, sources of publication, keywords, and abstracts) are being updated from 64 important Chinese periodicals reporting literature on medicinal plants. The data base is not yet available on-line outside the University or through any data service vendor.

B. MAPIS

This information service, established at the Publications & Information Directorate, CSIR, New Delhi, India has a data base on all aspects of medicinal and aromatic plants. A bimonthly abstracting journal, *Medicinal and Aromatic Plants Abstracts (MAPA)* has been issued from this data base since 1979. It covers about 600 journals from about 50 countries and 22 different languages. Each issue of *MAPA* contains 750 abstracts, a selective bibliography on one medicinal or aromatic plant, and abstracts of 15 patents. Bibliographical data for 1982 and 1983 are on magnetic tape, while from 1984 onward, full abstracts are being computerized. Mechanization of the data from 1979 to 1981 is in progress. Bibliographies prior to 1979 are searched manually. Services include supply of bibliographies, detailed information, as well as xerox copying of full documents.

C. NAPRALERT SYSTEM

A computerized data base on the chemistry and pharmacology of natural produces of plant, microbial, and animal (marine) origin is maintained at the College of Pharmacy, University of Illinois, Chicago. It has been given the acronym NAPRALERT (*Na*tural *Pr*oducts *Alert*). A systematic surveillance of the world literature on the chemistry and pharmacology of natural products has been in progress at the University of Illinois since 1975. A substantial amount of retrospective information has also been acquired and computerized for selected genera of plants and on the pharmacology activities of natural products. The system is maintained on a Computer Corporation of America model 204 data management system. Data from the years 1975 to 1984 culled from some 70,000 scientific articles covering information on 4600 genera of plants have been computerized. Reports on folkloric, ethnomedical, or traditional uses of plants have also been included.

D. CA SEARCH — COLUMBUS, OHIO

Chemical Abstracts Service, Biweekly. The file covers bibliographic references for items published in *Chemical Abstracts (CA)*. CA SEARCH draws upon articles that appear in 12,000 scientific and technical journals published in 140 nations and 50 languages, patents issued by 28 patent offices around the world, and material from conference proceedings, technical reports, dissertations, and new books. The data base is growing at the rate of 451,000 items annually. Titles, an uncontrolled list of keywords, source references, authors, patent information, and CA section numbers corresponding to major subject divisions in the published service are all searchable. No abstracts are contained on tapes — only references to volume abstract numbers in the published CA are included. On-line searching of the file since 1967 is available through the DIALOG, ORBIT, and BRS programmes.

E. CHEMLINE — BETHESDA, MD, NATIONAL LIBRARY OF MEDICINE (U.S.)

This is an on-line dictionary of more than 1 million chemical substances cited in TOXLINE, RTECS, MEDLINE, and TDB data bases and having a Chemical Abstracts Service registry number. The file is searchable by registry number, molecular formula and fragments, synonyms, Wiswesser Line Notations, ring analysis, and various other structural fragments.

F. MEDLINE (MEDLARS ON-LINE) BETHESDA, MD: NATIONAL LIBRARY OF MEDICINE (NIM), U.S.. MONTHLY

This interactive data base of the Medical Literature Analysis and Retrieval System (MEDLARS) contains all citations prepared for publications in Index Medicus plus other related files of information processed by or in conjunction with the National Library of Medicine. MEDLINE supplies references to articles from 3000 journals from 71 countries. More than 250,000 references are added annually to the fields. Items are extensively indexed using MeSH vocabulary. Authors, titles, abstracts, MeSH terms, language, and other bibliographic elements can be searched. The System is accessible to authorised centres through the National Library of Medicine computer facilities in Bethesda, MD, or the State University of New York at Albany. Libraries not authorised by the National Library of Medicine can access the file through the Bibliographic Retrieval System.

G. BIOSIS PREVIEWS, PHILADELPHIA, BIOSCIENCES INFORMATION SERVICE (U.S.)

Issued 36 times per year, this data base covers reports of original research in the biological and medical sciences. Approximately 9000 serials from more than 100 countries are being used for input. In 1984, 360,000 items were added to the BIOSIS Data Base. The data base contains a bibliographic citation, indexing terms (including Biosystematic and CROSS (subject) codes for each record in the data base. Abstracts are not available on the tape. The file is searchable by indexing terms, biosystematic and CROSS (subject) terms, author, source publication name, and journal CODEN. Both retrospective search and SDI services are available from the producer and from various subscribers to the data base tape. On-line searching of the files from 1969 to the present is available from DIALOG, Bibliographic Retrieval Services, System Development corporation (SDC), and the Canada Institute for Scientific and Technical Information (CISTI).

H. CAB ABSTRACTS. FARNHAM ROYAL, ENGLAND, COMMONWEALTH AGRICULTURAL BUREAU, MONTHLY

This data base offers comprehensive coverage of information in agriculture and areas of applied biology. Over 10,000 journals in 37 languages, books, reports, and other substantial papers are being included in the data base which is growing at the rate of about 180,000 references per year. Nearly 50 abstracting journals are issued from the data base. Total file or subfiles corresponding to an individual abstracting journals can be searched using authors, organizations, titles, abstracts, keywords, document language, publisher, journal name, data of publication, chapter headings, and subject codes corresponding to other headings in CAB journals.

I. DATA BASE ON THAILAND MEDICINAL PLANTS

For chemical research on Thailand medicinal flora, a bibliography of 42 volumes has been compiled having the title *Austrian-Thai Scientific Cooperation: The Chemistry of Medicinal Herbs in Thailand.* It includes computer print-outs obtained from the CA search in the DIALOG system and covers chemical research work from 1967 to 1982 on 1750 Thai plants.[19]

J. EXCERPTA MEDICA (EM), AMSTERDAM

Excerpta Medica provides access to basic research and clinical literature in biomedical fields through a variety of information activities. These include abstracting and indexing of worldwide relevant literature, maintenance of the computer-readable Excerpta Medica data base, and a subset file on drugs called Drugdoc. The EM data base is available online through Lockheed/DIALOG, the German Institute for Medical Documentation and Information (DIMDI), and Hoechst AG Drug Information Center. In addition, Excerpta Medica produces about 44 special services or abstracts bulletins and bibliographies. Input includes more than 35000 biomedical periodicals, books, congress proceedings, symposia, and other relevant literature. The data base holds approximately 3 million items up to 1984 with an annual addition of 250,000 items. Published two to six times per year.

K. CURRENT AGRICULTURAL RESEARCH INFORMATION SYSTEM (CARIS), FAO, ROME

CARIS provides collection, processing, storage, and dissemination of information on current agricultural research institutions, workers, and projects in developing countries. A decentralized system, CARIS consists of regional and national centres located throughout the world that collect information from their areas for merging and redistribution by the CARIS Coordinating Centre in Rome. Primary products of the System include print directories of ongoing research and a machine-readable data base of developing information. It covers research carried out in or on behalf of developing countries in the subject areas of agriculture, animal production, forestry, inland fisheries, and food.

Computerized files are maintained in ISIS/ISO format. Batch mode searches from CARIS files are conducted by the Coordinating Centre on request. Services are available to FAO member countries.

L. INTERNATIONAL INFORMATION SYSTEM FOR THE AGRICULTURAL SCIENCES AND TECHNOLOGY (AGRIS), FAO, ROME

AGRIS is designed to carry out the collection, storage, and retrieval of information related to currently produced agricultural documents. AGRIS is a decentralized system of nearly 130 national and regional input centres throughout the world, with the AGRIS Coordinating Centre in Rome responsible for receiving, integrating, and processing bibliographic data. Chief products of the system are a monthly hardcopy index and a corresponding computer-readable data base. The scope includes agriculture, food, human nutrition environment, economics, animal science, veterinary science, fisheries, forestry, industry, geography, geology, and human needs. Online access and SDI services are offered from the AGRIS data base at those centres with computer services. The AGRIS Output Tape, an accumulation of document indexes, is published monthly and can be obtained from the AGRIS Coordinating Centre. Services are primarily for use by the network centres. The data base is also available online through ESA/IRS, DIMDI, and the International Atomic Energy Agency.

M. WPI (WORLD PATENT INDEX) LONDON, DERWENT PUBLICATIONS, U.K. WEEKLY

A massive file covering about 6 million patent references with an annual addition of 300,000 patents. Records are searchable by patent number, assignee, subject classifications (both Derwent codes and International Patent Classifications), priorities, and titles. The file is available from DIALOG, upon approval from the data base producer, System Development Corporation of Japan and Telesystems Questel.

VI. TRADE AND STATISTICAL DATA

It is not easy to access the world trade in medicinal plants in terms of volume or value because trade statistics do not identify all plants individually, and statistics do not reveal medicinal and other usage separately among the listed plants. However, data compiled by the United Nations Statistical Office and International Trade Centre, UNCTAD/GATT, Geneva indicated that the value of total imports of selected medicinal plants increased from \$355 million in 1960 to \$551 million in 1980. India, South Korea, and China are the major exporters of medicinal plants. Six developed countries, viz., the U.S., the U.K., France, Germany, Switzerland, and Japan continue to be the major importers of medicinal plants.

Although over 400 species of medicinal plants are traded in Western Europe alone (Hamburg being the trade centre) in quantities ranging from a few kilograms to many tonnes, only a limited number of these plants enter the trade in significant quantity and their primary use even then may not be strictly medicinal.[20] It has been estimated that some 6000 to 7000 tons of herbs are extracted annually for the ingredients of medicines used in the U.K. The Food Additives Committee of the U.K. includes 572 herbs on its list, while a similar list prepared under the auspices of the European Economic Community includes over 430 herbs.[21] The current status of the medicinal plant industry in India and the world trade in medicinal plants was reviewed in some recent publications.[20,22-23]

Trade catalogues issued by various drug producing firms, directories of manufacturers of phytochemicals and finished products, suppliers of raw materials, cultivars, etc., provide useful information.

Martindale's Extra Pharmacopoeia[24] and *American Drug Index*[25] are good sources of information of the actual users (buyers) of medicinal plants. Another useful directory is the one prepared by Kapoor and Mitra,[26] which provides names of the Indian companies manufacturing composite drugs along with the components (herbal) of these drugs. Advertisements of the products by the companies published in commercial or research journals are still other sources to become familiar with plants used in commercial preparations and the firms using them.

World trade in medicinal plants was reviewed by the International Trade Centre, UNCTAD/GATT, Geneva, in their 1982 publication, "Markets for Selected Medicinal Plants and their Derivatives". This publication, while discussing the trade in general for specific plants, also lists the sources of information.[20]

The commercial information on those medicinal plants which are also used as sources of essential oils or as spices is more easily available.[27,28] Some of the trade information sources have also been described in other publications.[29,30] Important agencies and publications having trade information are listed in Tables 3A to C.

TABLE 3A
Important Periodicals Reporting Trade and Statistics on Medicinal Plants

1. *Agricultural Situation in India* (New Delhi, India), Monthly. Commodity reviews on arecanut & spices. Statistical information regarding production and area under cultivation of these crops are given.
2. *Chemical Marketing Reporter* (New York, USA), Weekly. International coverage with emphasis on the United States. Covers chemical industry including medicinal plants and essential oils, spices and perfumery.
3. *Commodity Week* (Herts, UK), Weekly, the Saturday edition of *The Public Ledger*.
4. *Current Research on Medicinal & Aromatic Plants (CROMAP)* (Lucknow, India), Bimonthly. Reports market trends, production, price, exports, imports, supliers, growers and consumers of medicinal plants.
5. *Dragoco Report* (Totowa, USA), Monthly.
6. *Drug and Cosmetic Industry* (New York, USA), Monthly.
7. *Drug Information Bulletin* (Cincinnati, USA), Quarterly.
8. *Drug Merchandising* (Toronto, Canada), Monthly.
9. *Drug News* (Bombay, India), Monthly.
10. *Drug Trade News* (New York, USA), Twice a month.
11. *Farmacia Nueva* (Madrid, Spain), Monthly.
12. *Farmacista Sociale* (Aversa Caserta, Italy), Monthly.
13. *Farmaco (Pavia)* (Pavia, Italy), Monthly.
14. *Immex* (Paris, France), Monthly.
15. *IMS Pharmaceutical Market Letter* (London, England), Weekly.
16. *Informacion Farmaceutica* (Mexico City, Mexico), Monthly.
17. *The International Customs Journal* (Brussels, Belgium).
18. *International Pharmaceutical Federation: Information Bulletin* (Hague, Netherlands), Quarterly.
19. *Japanese Weekly on Pharmacy and Chemistry/Yakugyo Shimbun* (Osaka, Japan), Weekly.
20. *Journal of the Chemists' and Druggists' Association* (Bombay, India), Monthly.
21. *Marches Tropicaux et Mediterraneens* (Paris, France), Weekly.
22. *Monthly Statistics of the Foreign Trade of India* (Calcutta, India), Monthly.
23. *Nard Journal* (Chicago, USA), Bimonthly.
24. *Pharmaceutical Marketing and Media* (Stanford, CT, USA), Twice a month.
25. *Pharmaceutische Weekblad* (Hague, Netherlands), Weekly.
26. *The Public Ledger* (Daily) (Herts, UK), Daily.
27. *Pharmacy News* (Ludhiana, India), Monthly.
28. *Quimp Quarterly* (Quick index of medicinal preparations) (Karachi, Pakistan), Quarterly.
29. *Scrip World Pharmaceutical News* (London, England), Weekly.
30. *Search: Drugs Division* (New Jersey, USA), Monthly.
31. *Unlisted Drugs* (New Jersey, USA), Monthly.
32. *World Trade Annual* (New York, USA), Annual.

TABLE 3B
Important Institutions (Bodies) Having Trade and Statistical Information on Medicinal Plants

International Organizations

1. Commission Internationale Des Industries Agricoles Et Alimentaires/International Commission For Agricultural and Food Industries; 24, rue de Teheran F-75008 Paris, France.
2. International Trade Center (UNCTAD/GATT) Documentation Service; Villa le Bocage, Polais des Nations 1211 Geneva 10, Switzerland.
3. Food and Agriculture Organization of the United Nations (FAO) Committee on Commodities; via delle Terme di Caracalla CH-1211 Geneva 22, Switzerland.

TABLE 3B (CONTINUED)
Important Institutions (Bodies) Having Trade and Statistical Information on Medicinal Plants

4. Joint FAO/WHO Codex Alimentarius Commission Committee on Food Additives; via delle Terme di Caracalla I-001/00 Rome, Italy.
5. United Nations, Department of Economic and Social Affairs, New York, USA.
6. Untied Nations Economic and Social Council (ECOSOC) Population Commission, Commission on Narcotic Drugs, International Narcotics Control Board New York, New York 10017, USA.
7. United Nations Industrial Development Organization (UNIDO), Department of Industrial Operations, P.O. Box 300, A1400, Vienna International Center, Vienna Austria.
8. World Health Organization (WHO), Division of Prophylactic and Therapeutic Substances Expert Committee on Specifications for Pharmaceutical Preparations; Villa le Bocage, Polais des Nations 1211 Geneva 10, Switzerland.

National Organizations

9. Agroexport: State Trade Organization For the Export of Agricultural and Food Products; 4 Shkurti 6 Tirana, Albania.
10. Groupement des Fabricants D'Essences, Huiles Essentielles, Extraits, Produits Chimiques Aromatiques et Colorants "Aroma"; 49, Square Marie-Louise B-1040 Brussels, Belgium.
11. International Customs Tariff Bureau; Brussels, Belgium.
12. Ministry of Trade, Industry and Forests, Department of Trade and Industry; Taschichhodzong, Thimbu, Bhutan.
13. Centre for Scientific, Technical and Economic Information in Agriculture and Forestry, Agricultural Academy George Dimitrov; Boulevard Dragan Tsankov 6, Sofia, Bulgaria.
14. Pharmachim: State Foreign Trade Corporation; Ilienspo Chaussee 16, Sofia, Bulgaria.
15. Pharmaceutical Indusries Corporation; 204 u. Wisara Road, Rangoon, Burma.
16. Statistics Canada, External Trade Division, Imports by Commodities; Ottawa, Canada.
17. China National Native Produce and Animal By-Products Imports and Exports Corporation; 82 Tung-an-men Street, Beijing, China.
18. Brancheforningen For Soebe-, Parfumeri-, Toilet-OG Kemisktekniske Artikler (SPT) (Trade association for soap, perfumes, toilet goods and chemotechnical products):Strodamsvej 52, DK-2100 Copenhagen, Denmark.
19. Centro De Informacion de Drogas, Universidad Nacional; Pedro Henriquez Urena, Santo Domingo, Dominican Republic.
20. Association des Delegues des Organisations Professionnelles des Producteurs et Ramasseurs de Plantes Medicinales et Aromatiques de la Cee/ Association of Delegates of the Professional Organizations of Producers and Collectors of Medicinal and Aromatic Plants of the EEC; 6, Boulevard Marechal-Joffre F-91490 Milly-La-Foret, France.
21. Centre de 'Etude des Plants Medicinales Faculte Mixte de Medicine et de Pharmacie; Boulevard Daviers F-49000 Angers, France.
22. Chambre Syndicale de la Phytopharmacie et de la Protection des Plantes; 2, rue Denfert Rocherau 92100 Boulogne Billancourt, France.
23. Syndicate des Importateurs-Exportateurs de Matieres Premieres Aromatiques; 2, rue de Penthieure F-75008 Paris, France.
24. Basic Chemicals, Pharmaceuticals and Soaps Export Promotion Council; Plot No. 56, Ashok Chambers, Fifth Floor, Broach Street, Bombay 9, India.
25. Department of Commercial Intelligence and Statistics. Govt. of India, Calcutta, India.
26. India Institute of Foreign Trade, New Delhi, India.
27. Japan Information Centre of Science and Technology; 5-2, 2-Chome, Nagata-Cho Chiyoda-ku Tokyo 100, Japan.
28. Ministry of Finance, Customs Bureau, Japan Exports and Imports (Tokyo), Japan.
29. Agricultural Marketing Centre; P.O. Box 565, Amman, Jordan.
30. Korea Scientific and Technological Information Centre (KORSTIC); C.P.O. Box 1229 Seoul, South Korea.
31. Drug Research Production Unit; Faculty Pharmacy, University Ife, Ile-Ife Nigeria.
32. Institute of Traditional Medicine; University of Ife, Faculty of Pharmacy Ife, Nigeria.
33. Bureau of Plant Industry, Department of Agricultural and Natural Resources; Manila, Philippines.
34. Trustul Plafar (Medicinal plants trust); Bd. Hristo Bolev No. 1, Bucharest, Romania.
35. Ceylon Institute of Scientific and Industrial Research, Natural Products Section; P.O. Box 787, Colombo 7, Sri Lanka.
36. Department of Customs, *Foreign Trade Statistics of Thailand,* Bangkok, Thailand.
37. Direction generale des douanes, *Statistique Annuelle du Commerce Exterieur de la Suisse*; Berne, Switzerland.
38. Department of Trade, *Overseas Trade Statistics of the United Kingdom*; London, U.K.
39. General Produce Brokers' Association of London; Plantation House, Mincing Lane, London EC3M 3LR, U.K.
40. Tropical Research and Development, Industrial Technology Department; Culham, Abingdon, Berkshire, U.K.
41. Drug Information Association College of Pharmacy, University of Cincinnati; Cincinnati, Ohio 45221 (USA).
42. Flavour and Extract Manufacturer's Association of The United States; 900 17th Street, N.W. Washington, D.C. 20006 USA.
43. United States Department of Commerce, Bureau of the Census, *US Imports for Consumption and General Imports: Tsusa Commodity by Country of Origin* (FT 846) Washington, D.C., USA.
44. European Communities Statistical Office, *Analytical Tables of Foreign CST*, Vol. 1. Congress/products, Luxembourg.

TABLE 3C
Directories and Guides, Etc., for Trade Information on Medicinal Plants

1. *ABC Europ Production,* ABC-Verlagshaus, Postfach 4034, Berliner Allee 8, D-6100 Darmstadt, Germany, Annual 2 vols. In English, French, German, Italian, Portuguese and Spanish. Universal register of European exports.
2. *AIMO Directory of Indian Manufacturers,* The All Indian Manufacturers Organization, 4th Ed. 1976. Jeevan Sahakar, Sir Pherozshah Mehta Road, Fort, Bombay. Chemical Industries (Alkaloids, Herbal drugs, individual spices & plant products, e.g., *Glycyrrhiza* extract).
3. *American Druggist Blue Book,* American Druggist Blue Book, 224 West 57th Street, New York, New York 10019, USA, Twice a year. Information on pharmaceuticals in United States.
4. *Annuaire De La Droguerie Francaise,* Louis Johanet, 51, rue Boursault, 75017 Paris, France. Annual. Information on pharmaceuticals in United States.

TABLE 3C (CONTINUED)
Directories and Guides, Etc., for Trade Information on Medicinal Plants

5. *Annual Bulletin of Trade in Chemical Products,* United Nations Sales Section, New York, New York 10017, USA. Prepared by the Economic Commission for Europe. In English, French, and German.
6. *Annual Drug Data Report*, Medicamento de Actualidad Apartado de Correos 1641, Barcelona, Spain. Twice a year. In English. Information on pharmaceuticals in Spain.
7. *Annual Report for Pharmaceutical Preparations in Japan,* reproduced in your pharmaceuticals market in Japan, published by JETRO, Tokyo, Japan.
8. *Annual Survey of Industries,* No.60 Chemicals and Chemical Products, Annual. In English. Gives statistics on pharmaceuticals. Available from Manager of Publications, Civil Lines, Delhi 110006, India.
9. *Apotheker Jahrbuch (Druggists Yearbook),* Wissenschaftaliche Verlagsges-ellschaft mbH Birkenwaldstrasse 44, 7000 Stuttgart 1, Germany. Annual. In German. Information on pharmaceuticals in Germany.
10. *Cadastro Brasileiro De Materias Primas Farmaceuticas, Por Produto, Por Fabricante* (Brazilian directory of raw materials for pharmaceutical industry, listed by product and by manufacturing). Associacao Brasileira da Industria Farmaceutica, Avenida Beira Mar 262 Rio De Janeiro, Brazil. Annual. In Portuguese. Information on products and manufacturers of pharmaceuticals in Brazil.
11. *Canadian Chemical Pharmaceutical and Product Directory,* Lloyed Publications of Canada, P.O. Box 262, West Hill, Ontario M1E 4R4, Canada. Annual. Lists Canadian manufacturers, distributors and wholesalers of pharmaceuticals, chemicals raw materials, and products.
12. *Canadian Trade Index,* The Canadian Manufacturers' Association, 67 Yonge Street, Toronto, Ontario M5E 1J9, Canada. Annual. Text in English; indexes in French and Spanish. Guide to buying and selling in Canada. Lists manufacturers and products.
13. *Chemical Economics Handbook,* Chemical Information Services, Stanford Research Institute, Menlo Park, California 94025, USA. Collection of reports and data sheets updated monthly. Present and future status of chemical raw materials, primary and intermediate chemicals and product groups.
14. *Chemical and Medical Formulary of India* (1976-77). Ed. R.K. Mehta. Basic and Business Publications Bombay, India. 176 pp.+ 44 pp (index). Description, composition, indications, contraindications, precautions, dosage, packing & storage conditions, other products of bulk drugs. Indexes: Companywise, productwise, & therapeutic (disease-wise).
15. *Chemical Week Buyers' Guide,* McGraw-Hill, New York, Annual. 1984. Lists buyers of plant drugs and finished products under Botanicals & Extracts, individual plant/product names such as *Psyllium, Rauwolfia,* essential oils, gums and chemical constituents such as menthol, reserpine, etc.
16. *Chemist and Druggist Directory,* Annual. 1983. Buyers' Guide: Botanical drugs; Botanical extracts; Botanicals; individual chemical/plant name; Crude drugs, Herbal Extracts, Herbal Preparations, Medicinal; Herbal relief for high BP; Herbal Remedies; Herbal Skin Care; Herbal Tablets; Herbs, medicinal & culinary.
17. *Commodity Trade Statistics,* United Nations, Sales Section, New York, New York 10017, USA. Prepared by the Statistical Office of the United Nations. Each issue provides important export statistics on commodities for different groups of countries. Commodities are arranged by Standard International Trade Classification (SITC).
18. *Compendium of Pharmaceuticals and Specialities,* Canadian Pharmaceutical Association, 175 College Street, Toronto 23, Canada. Annual. Pharmaceuticals and specialities available in Canada; list of manufacturers and products.
19. *Directory of Crude Drugs and Aromatic Plants Dealers.* Eds. Shah, N.C., and Virmani, O.P., Central Institute of Medicinal and Aromatic Plants, New Delhi.
20. *Directory of Indian Exporters,* Suppl. to Eighth Edition 1974. Govt. of India, Department of Commercial Intelligence and Statistics, Calcutta-1 India. Commodity-wise list of Exporters, Industrial plant names, Crude drugs, Crude Drug & Herbs, Herbs and Roots.
21. *Drug and Cosmetic Catalog,* Drug Markets, Inc. 101 West 31st Street, New York, New York 10001, USA. Every two years. Alphabetical list of suppliers of raw materials, containers and machinery for manufacturers of drug and cosmetic supplies.
22. *Drugs in Current Use and New Drugs,* Springer Publishing Co., Inc. 200 Park Avenue, New York, New York 10003, USA, Annual. Worldwide information on pharmaceuticals.
23. *Drug Development Communications* (New York, USA), Irregular. Information on pharmaceuticals manufactured in United States.
24. *Export Prospects of Selected Medicinal Herbs in West Germany, USA and Japan.* Indian Institute of Foreign Trade, New Delhi, India, English, 1980. Demand patterns and future market prospects of cultivated medicinal herbs of Indian origin. Processed herbs in the form of alkaloids, glycosides and formulated drugs are also covered.
25. *Export Statistics. Basic Chemicals, Pharmaceuticals & Cosmetics,* Export Potential Council, Jhansi Castle, 4th Floor, 7 Cooperage Road, Bombay 400039, India. English, 2 Volumes.
    Vol- 1 — 1972-1975; Vol. II — 1974-1977. Vol. 1 — Itemwise exports with a breakdown of the countries along with panelwise and region-wise exports.
    Vol. II — Regionwise and countrywise exports with commoditywise breakdown. Covered under Essential Oils and Crude Drugs.
26. *Exporters' Directories, Basic Chemicals, Pharmaceuticals and Cosmetics* Export Promotion Council, "Jhansi Castle" 4th Floor, 7, Cooperage Road, Bombay- 400 039, India, 1978, 178 pp. Classified list of Export products, Alphabetical list of Exporters & product index Phytochemicals (individual names), Herbal preparations, Crude Drugs/Medicinal Herbs.
27. *Ferguson's Ceylon Directory 1973-1974.* The Associated Newspapers of Ceylon Ltd. Lake House, D.R. Wijenardene, Mawatha, Colombo 10, Sri Lanka. Irregular 1974. Approx. 2000 pp. In English. Lists of manufacturers by products.
28. *Herbal Drugs in Indian Pharmaceutical Industry,* (Eds. S.L. Kapoor and R. Mitra) National Botanical Research Institute, Lucknow, India. 1979, 85 pp. 1650 herbal medicines manufactured by 160 pharmaceutical concerns use nearly 540 botanical species in those formulations. Seventeen drugs used most commonly in pharmaceutical preparations have been listed along with the composition in which they are used.
29. *HKIMS (Hong Kong Index of Medical Specialities),* Scientific Publications, Cuscaden House Hotel, Room 204, 40 Cuscaden Road, Singapore 10, Singapore.
30. *How to Find Out in Pharmacy: A Guide to Sources of Pharmaceutical Information* (Ed A.L. Brun), Pergamon Press Ltd, Headington Hill Hall, Oxford OX3 OBW, England, 1969, 139 pp. Six chapters giving information on journals, institutions, bibliographies, law etc. Emphasis on British and American sources.
31. *IIMS (Indonesian Index of Medical Specialities),* Scientific Publications, Cuscaden House Hotel, Room 204, 40 Cuscaden Board, Singapore 10, Singapore. Every three years. In English. Information on pharmaceuticals available in Indonesia.
32. *Importers' Directory, Basic Chemicals, Pharmaceuticals and Cosmetics,* Export Promotion Council, 7, Cooperage Road, Bombay-400 039, India, 1975, 271 pp. List of products and importers thereof. See Ayruvedic medicines, crude drugs & industrial product names.

TABLE 3C (CONTINUED)
Directories and Guides, Etc., for Trade Information on Medicinal Plants

33. *Index of New Products*, Pharmaceutical Society of Great Britain, London, U.K., Annual. A card service providing information on new pharmaceutical products; gives manufacturers and comprehensive technical data.
34. *Indian Industries 15th Ed. 1980*. Indian Industries Publications, A/3 Everest Tardeo Road, Bombay 400 034. Chemicals (alkaloids, essential oils, papain, etc).
35. *Indian Pharmaceutical Guide*, Pamposh Publications, E-38 Hauz Khas, New Delhi 110 016, India. Annual. In English. Information on pharmaceuticals available in India.
36. Indian Pharmaceutical Industry, Development Council, Drugs and Pharmaceuticals Directorate General of Technical Department, Govt. of India, New Delhi, India. Irregular. 1969. 189 pp. Lists pharmaceutical manufacturers alphabetically, gives data on products, capacity, trade names, etc.
37. *L'Informatore Farmaceutico (Pharmaceutical Information)*, Organizzazione Editoriale, Medico-Farmaceutica, Via Edola 42, 20125 Milan, Italy. Annual. In Italian. Lists drugs and manufacturers in Italy; suppliers and producers of raw materials.
38. *Kelly's Manufacturers and Merchants Directory, 1985*, Kelly's Directories, Windsor Court, East Grindstead House, East Grindstead, West Sussex RH 19 1XB, London, UK. Annual. see: Drug (Crude/Botanical/Medicinal) Merchants, Grinders or Millers; Medicinal Non-Ethical Preparation Manufacturers; Herb Merchants or Producers.
39. *Kline Guide to the Chemical Industry*, P. Noble. 3rd Ed. 1977, Charles H. Kline and Co., Inc., 369 Passaic Avenue, Fairfield, New Jersey 07006, USA. Gives marketing, economic, and company data on the entire chemical industry including fine chemicals.
40. *Kompass Directories*, Kompass International AG Neuhausstrasse 4, 8044 Zurich, Switzerland. Multilingual. Indexed according to product classification. Alphabetical list of firms, trademarks, suppliers, manufacturers and importers.
41. *Manufacturers of Pharmaceuticals and Medicines*, Statistics Canada, Publications Distribution Unit Ottawa, Ontario, K1A, Canada. Annual.
42. *Manufacturing Industries, No. 3: Chemicals, Drugs and Medicines*, Commonwealth Bureau of Census and Statistics, P.O. Box 17, Canberra, A.C.T. 2600 Australia. Annual.
43. *Markets for Selected Essential Oils and Oleoresins (Les Marches De Dierses Huiles Essentielles Et Oleoresines)*, International Trade Centre (UNCTAD/GATT) 14, route des Morillons CH-1211 Geneva 22, Switzerland, 1974. 323 pp. Studies the market for certain essential oils in Canada, France, Germany, Japan, United Kingdom, United States.
44. *Markets for Selected Medicinal Plants and their Derivatives*, International Trade Centre UNCTAD/GATT, Geneva, Switzerland, 1982, 206 pp.
45. *MIMS (Monthly Index of Medicinal Specialities):* Annual Compendium, Haymarket Publishing Ltd, Gillow House, 5 Winsley Street, London W1A 2HG, England. Lists manufacturers alphabetically; index of trade names.
46. *Modern Drug Encyclopedia and Therapeutic Index*, Dun. Donnelley Publishing Corporation, 666 Fifth Avenue, New York, New York 10019, USA. Every three years. Information on pharmaceuticals world-wide.
47. *National Formulary*, American Pharmaceutical Association, 2215 Constitution Avenue, NW Washington, DC 20037 USA. Every five years. An official compendium of standards for selected drugs on the basis of therapeutic value, recognised under federal and state food and drug laws. Covers United States.
48. *Nihon Yakugyo Boeki Kyokai: Japta List: J (Japanese drug directory)*, Japan Pharmaceutical Traders' Association, Kanda Izumicho, Chiyoda-ku, Tokyo, Japan. Irregular. In Japanese.
49. *North American Reference Encyclopedia of Drugs and Drug Abuse*, North American Publishing Co. 134 N. 13th Street, Philadelphia, Pennsylvania 19167, USA. Irregular.
50. *OPD Chemical Buyers Directory*, Special Issue of *Chemical Marketing Reporter*, Schnell Publishing Company, Inc., 100 Church Street, New York, New York 10007, USA. Annual Lists alphabetically suppliers and products.
51. *Pharmaceutical Directory of India*, Star Publications, 103 Modi Street, Bombay 400001, India. Irregular.
52. *Principal Drugs: An Alphabetical Guide*, Comp. by SJ Hopkins, Faber and Faber Ltd 3 Queen Square, London WC1, England. Irregular. 1973. 192 pp.
53. *Privredni Adresar SFRJ 1975 (Trade directory of Yugoslavia)*, Privredni Pregled, Marsala Birjuzova 3-5, Belgrade, Yugoslavia, 1975. 1969 pp. Text in Serbo-Croatian; indexes in English, German, Russian and Serbo-Croatian. Lists manufacturers and institutions according to activities; list of products. See: Medicaments and medicinal herbs, p. 801.
54. *Report on the Market Orientation Tour for Phytochemicals, Bulk Drugs and Drug Intermediates*. Basic Chemicals, Pharmaceuticals and Cosmetics Export Promotion Council, Jhansi Castle, 7, Cooperage Road, Bombay 400039, India. 1981. Report of a market survey of the developed countries for medicinal plants and their derivatives including basic drugs and intermediates. Countries covered are Japan, USA, UK, Germany and Switzerland.
55. *Russian Drug Index*, National Library of Medicine, 8600 Rockville Pike, Bethesda, Maryland 20014, USA. Irregular. Names of drugs developed in USSR; Drug and pharmacological indexes.
56. *Short Market Surveys Medical/Aromatic Plants and Plant Extracts in the Fed. Rep. Germany: France — 1981;* International Trade Centre, (UNCTAD/GATT), Palais des Nations, 1211 Geneva 10, Switzerland.
57. *Slam: Trade Yearbook of Africa*, Office Slam, P.O. Box 14126, Madrid 27, Spain. Irregular. In English, French and Spanish. For each African country, manufacturers, exporters and importers are listed by product.
58. *Societes et Fournisseurs D'Afrique Noire et de Madagascar*, Ediafric - La Documentation Africaine, 57, Avenue d'Iena, 75016 Paris, France. Trade directory of West Africa and Madagascar. Enterprises listed by country, with index.
59. *Soviet Trade Directory*, Flegon Press, 43 Queensborough Terrace, London W2, England. Irregular. Text in English; indexes in English and Russian. Covers industry in Union of Socialist Republics. Arranged in broad subject areas.
60. *Statistics of Foreign Trade/Statistiques Du Commerce Exterieur*, OECD, Chateau de la Muette 2, rue Andre Pascal, Paris, France, Monthly. In English and French (Series A of the OECD bulletins of foreign trade statistics). See: Pharmaceuticals, Index No. 54.
61. *Trade Directories of the World*, Croner Publications, Inc. 211-03 Jamaica Avenue Queens Village, New York 11428, USA, Irregular. Approx. 125 pp. In English. Loose-leaf plant with monthly supplements. Lists over 1700 business and trade directories. Worldwide coverage, arranged by geographical area, countries, trade, and professions. Includes index directories.
62. *Trado Asian & African Directory of Exporters-Importers & Manufacturers*, Trado Publications Pvt Ltd, C-6 Safdarjung Development Area, Community Centre, New Delhi-110016 (India). Deals with Spices, Herbs and Crude drugs, Drugs & Herbs.
63. *Tropical Products Institute Market Reports*, Tropical Development and Research Institute, 56-62 Gray's Inn Road, London WC1X 8LU, England. Reports giving data on production of specific essential oils, herbs, menthol, etc.

TABLE 3C (CONTINUED)
Directories and Guides, Etc., for Trade Information on Medicinal Plants

64. *United Nations Permanent Central Narcotics Board: Estimated World Requirements of Narcotic Drugs,* United Nations, New York, New York 10017, USA. Annual. In English, French, Spanish.
65. *Wer Liefert Was? (Who Supplies What?),* Leipziger Messeamt Messehaus am Market 1,701 Leipzig, Germany. 1975. 38th ed. 960 pp. Text in German; indexes and headings in English, French, German and Russian. Product section arranged by broad categories with names and addresses of firms. See: Pharmaceutical products and preparations.
66. *World Pharmaceutical Firms,* Noyes Data Corporation, Park Ridge, New Jersey 07656, USA, Every three years.
67. *World-Wide Chemical Directory,* IPC Business Press, 40 Bowling Green Lane London EC1R One, England. 1977. 424 pp. Information on more than 10,000 chemical companies and organizations in 95 countries. Lists trade and professional associations related to the chemical and pharmaceutical industries. Alphabetical index of companies.
68. *Yearbook of International Trade Statistics,* United Nations, New York, New York, 10019, USA. Prepared by the Statistical Office of the United Nations, 2 Vols. Vol. 1: Trade, by country, Vol. 2: Trade, by commodity.

REFERENCES

1. **Effendi, M.I., Sudriman, I., and Masbi, M.,** *Proc. 3rd Symp. on Medicinal Plants Res.*, Gadjah Moda Univ., Indonesia, 1983, 240. Cited by Achmad, S.F., Retrospect and Prospect of Medicinal Plants Works in Indonesia *in Proc. Fifth Asian Symp. Medicinal Plants and Spices*, Seoul, Republic of Korea, 1984, 145.
2. **Johnson, I.S.,** in *Cancer Medicine*, 2nd ed., Holand, J.F., and Frei, E., Eds. Lea and Febiger, New York, 1982, 910, Cited by Frei, E. National Cancer Chemotherapy Programme, *Science,* 217(4560), 600, 1982.
3. **Frei, E.** National Cancer Chemotherapy Programme, *Science*, 217(4560), 600, 1982.
4. **Malone, M.H.,** Pharmacological Evaluation of Natural Products — General and Specific Approaches to screening Ethnopharmaceuticals, J. *Ethnopharmacol.*, 8, 127, 1983.
5. **Penso, G.,** *WHO Inventory of Medicinal Plants Used in Different Countries*, World Heath Organization, Geneva, 1980.
6. *Ulrich's International Periodicals Directory*, 22nd ed., R.R. Bowker Company, New York, 1983.
7. *Encyclopedia of Information Systems and Services*, 6th ed, Scmittroth, J., Jr., Lucas, A.F., Novallo, A., and Marcaccio, K.Y., Gele Research Co., Michigan, 1985-86.
8. **Chang, H.S.,** Computerized Information on Chinese Medicinal Material, in *Proc. Fifth Asian Symp. Medicinal Plants and Spices*, Seoul, Republic of Korea, 1984, 291.
9. **Hyatt, R.,** *Chinese Herbal Medicine*, Thorson Publishers, New York, 1978, 159.
10. *Checklist of Sanskrit Medical Manuscripts of India*, Central Council for Research in Ayurveda & Siddha, New Delhi.
11. *Union Catalogue of Arabic and Persian Medical Manuscripts in the Libraries of Hyderabad*, Central Council for Research in Ayurveda & Siddha, New Delhi.
12. **Chopra, R.N., Nayar, S.L., and Chopra, I.C.,** *Glossary of Indian Medicinal Plants*, CSIR, New Delhi, 1956, Suppl., 1969.
13. *Handbook of Domestic Medicine and Common Ayurvedic Remedies*, Central Council of Research in Indian Medicine and Homoeopathy, New Delhi, 1978.
14. **Chadha, Y.R., and Kashyapa, K.,** Some Important Books and Serials on Medicinal Plants, presented at Symposium on Exploration and Improvement of Medicinal Plants, Jadavpur Univ., Calcutta, 1980.
15. **Stuart, M.,** Ed., *The Colour Dictionary of Herbs & Herbalism*, Orbis Publishing, London, 1982.
16. **Lewis, W.H. and Elvin-Lewis, M.P.F.,** *Medical Botany, Plants Affecting Man's Health*, Wiley Interscience, New York, 1977.
17. **Mukerji, B.,** *The Indian Pharmaceutical Codex, Vol. 1 - Indigenous Drugs*, Council of Scientific & Industrial Research, New Delhi, 1953.
18. *The Wealth of India — A Dictionary of Indian Raw Materials & Industrial Products*, Publications & Information Directorate, New Delhi. Raw Materials Series — 11 Vols and 2 Supplements, 1948-1976.
19. **Manunapichu, K.,** IOCD and Chemistry of Medicinal Plants, in, *Proc. Fifth Asian Symp. Medicinal Plants and Spices*, Seoul, Republic of Korea, 1984, 105.
20. *Markets for Selected Medicinal Plants and their Derivatives*, UNCTAD/GATT, Geneva, 1982.
21. **Phillipson, J.D.,** The Pros and Cons of Herbal Remedies, *Pharmaceutical J.*, Oct. 3, 1981.
22. **Virmani, O.P.,** Current Status of Medicinal Plant Industry in India, *Indian Drugs*, 17, 318, 1980.
23. **Sundaresh, I.,** Export Potential of Medicinal Plants and their Derivatives from India, in *Cultivation & Utilization of Medicinal Plants.* Atal, C.K., and Kapur, B.M., Eds., Regional Research Laboratory, Jammu-Tawi, 1982, 800-823.
24. *Martindale, The Extra Pharmacopoeia*, 27th ed., Wade, E., and Reynolds, F.E.F., Eds., Pharmaceutical Press, London, 1979.
25. *America Drug Index*, 27 ed., Billups, N.F., and Billups, S.M., Eds., J.B. Lippincott Company, Philadelphia, 1983.
26. **Kapoor, S.L., and Mitra, R.** *Herbal Drugs in Indian Pharmaceutical Industry*, National Botanical Research Institute, Lucknow, 1979.
27. *UNIDO Guides to Information Sources No. 38. Information Sources on Essential Oils,* United Nations, New York, 1981.
28. *UNIDO Guides to Information Sources No. 28. Coffee, Cocoa, Tea and Spices Industry*, United Nations, New York.
29. *UNIDO Guides to Information Sources No. 20. Information Sources on the Pharmaceutical Industry*, United Nations, New York, 1976.
30. Pandita (Kaul), Naina, Guide to Information Sources on Medicinal and Aromatic Plants, Project Report submitted in partial fulfilment of the Training Course in Information Science at INSDOC, New Delhi, 1984.

INDEX

A

B

D

E

H

I

M

S

T

U

V